International Scientific Conference Energy Efficiency in Transport (EET 2020)

IOP Conference Series: Materials Science and Engineering
Volume 1021

Kharkiv, Ukraine
18 – 20 November 2020

ISBN: 978-1-7138-4551-5
ISSN: 1757-8981

Printed from e-media with permission by:

Curran Associates, Inc.
57 Morehouse Lane
Red Hook, NY 12571

Some format issues inherent in the e-media version may also appear in this print version.

Additional copies of this publication are available from:

Curran Associates, Inc.
57 Morehouse Lane
Red Hook, NY 12571 USA
Phone: 845-758-0400
Fax: 845-758-2633
Email: curran@proceedings.com
Web: www.proceedings.com

TABLE OF CONTENTS

IOP Conf. Series: Materials Science and Engineering 1021 (2021) 011001 doi:10.1088/1757-899X/1021/1/011001

PREFACE

The International Scientific Conference "**ENERGY EFFICIENCY IN TRANSPORT**" (EET-2020) will be held on November 18-20, 2020 in Ukrainian State University of Railway Transport at the address: Feuerbach square 7, Kharkiv, Ukraine.

Ukrainian State University of Railway Transport, "UKRZALIZNYTSIA", Transport Academy of Ukraine, Building Academy of Ukraine, Academy of Engineering sciences of Ukraine, Instytut Kolejnictwa (Warszawa), University of Zilina in Zilina, Technical University of Sofia, Belarusian State University of Transport, Vilnius Gediminas Technical University, Azerbaijan Technical University, Dmytro Motornyi Tavria State Agrotechnological University, International scientific magazine "Friction and Wear", "Association of Industrial Hydraulics Specialists", State Enterprise plant "ELECTROTYAZHMASH", Communal Enterprise KHARKIV METROPOLITAN, LLC "Corporation KRT", Fuchs Mastyla Ukraina LLC, PJSC ELACS, Agrinol company, Kyiv Central Design Bureau of Valves, LLC "Ukrainian-British joint venture AZMOL-British Petrochemicals", LLC "MAPEI Ukraine", LLC "Construction Group International", LLC "BOMIS Ltd", LLC "Nolimit Germany", LLC "BASF", LLC "LIRA-SAPR" will be the organizers of the conference.

The plenary session and three sections will be work at the conference: section 1 - "Energy efficiency of rolling stock and transportation"; section 2 - "Energy management of rolling stock and structures of transport infrastructure"; section 3 - "Resource and energy savings in transport". 270 participants will present 90 reports at the conference.

The detailed information about the conference is available on the website http://eet-conf.kart.edu.ua/en/

EET-2020 COMMITTEE

CONFERENCE CHAIR

Chair **Sergiy Panchenko,**
D.Sc. (Tech), Prof., Rector, Ukrainian State University of Railway Transport (Kharkiv).

Co-chair: **Volodymyr Zhmak**, Acting Chairman of the Management Board JSC "Ukrzaliznytsia" (Kyiv).
Frantisek Buresh, Member of the Management Board JSC "Ukrzaliznytsia" (Kyiv).
Andrzej Żurkowski,
Prof, Dr.,Directorof Track Institute, Warszawa, (Poland).
Asta Radzevičienė,
Prof, Dr. Vice-Rector for International Relations Vilnius Gediminas Technical University (Lithuania)

Jozef Jandačka,
Prof. Ing., PhD., Rector University of Zilina (Slovak Republic).

Vilayat Mammad oglu Valiyev, Prof, Dr. Rector of Azerbaijan Technical University (Azerbaijan).

Deputy chair: **Glib Vatulia,**
D.Sc. (Tech),Prof., Vice-Rector, Ukrainian State University of Railway Transport (Kharkiv).

IOP Conf. Series: Materials Science and Engineering 1021 (2021) 011001 doi:10.1088/1757-899X/1021/1/011001

SCIENTIFIC COMMITTEE

Section «Energy efficiency of rolling stock and transportation»

Mykhailo Babaiev	D.Sc. (Tech), Prof., Head of Electrical Power Engineering, Electrical Engineering and Electromechanics Department, Ukrainian State University of Railway Transport (Kharkiv)
Vasyl Blyndiuk	D.Sc. (Tech), Prof., Professor of Electrical Power Engineering, Electrical Engineering and Electromechanics Department, Ukrainian State University of Railway Transport (Kharkiv)
Serhiy Buryakovskiy	D.Sc. (Tech), Prof., Director of Research and Design Institute "Lightning", National Technical University «Kharkiv Polytechnic Institute» (Kharkiv)
Hennadii Hetman	D.Sc. (Tech), Prof., Head of Railways Electric Rolling Stock Department, Dnipro National University of Railway Transport (Dnipro)
Oksana Hoholyuk	D.Sc. (Tech), Prof., Head of Theoretical and General Electrical Engineering Department, Institute of Energy and Control Systems of National University "Lviv Polytechnic" (Lviv)
Borys Liubarskyi	D.Sc. (Tech), Prof., Head of Electric Transport and Diesel Engineering Department, National Technical University «Kharkiv Polytechnic Institute» (Kharkiv)
Ihor Orlovskyi	D.Sc. (Tech), Prof., Associate Professor of Telecommunications, Computer Science and Electrical Engineering Faculty, UTP University of Science and Technology (Poland)
Mykola Ostroverkhov	D.Sc. (Tech), Prof., Head of Theoretical Electrical Engineering Department, Ukraine National Technical University "Kyiv Politechnic Institute" (Kyiv)
Andrii Prokhorchenko	D.Sc. (Tech), Prof., Associate Professor of Operational Management Department, Ukrainian State University of Railway Transport (Kharkiv)
Galina Sivyakova	D.Sc. (Tech), Prof., Head of Power Industry and Automation Technical Systems Department, Karaganda State Industrial University (Kazakhstan)
Petro Stakhiv	D.Sc. (Tech), Prof., Professor of Theoretical and General Electrical Engineering Department, Institute of Energy and Control Systems, National University "Lviv Polytechnic" (Lviv)
Tomasz Staśkiewicz	Mgr inż. (Tech), Wydział Inżynierii Lądowej i Transportu, Politechnika Poznańska (Polska)
Mykola Khvorost	D.Sc. (Tech), Prof., Head of Occupational Safety and Health Department, National University of Urban Economy in Kharkiv (Kharkiv)
Oleksii Chornyi	D.Sc. (Tech), Prof., Professor, Director of Electromechanics, Energy Saving and Control Systems Institute, Kremenchug National University M. Ostrogradsky (Kremenchuk)
Volodymyr Shysh	PhD (Tech), Associate Prof., Deputy Director of Development and Technical Policy Department, JSC "Ukrzaliznytsia" (Kyiv)
Serhiy Yatsko	PhD (Tech), Associate Prof., Associate Professor of Electrical Power

IOP Conf. Series: Materials Science and Engineering 1021 (2021) 011001 doi:10.1088/1757-899X/1021/1/011001

Engineering, Electrical Engineering and Electromechanics Department, Ukrainian State University of Railway Transport (Kharkiv)

Section «Energy management of rolling stock and structures of transport infrastructure»

Anatoly Falendish	D.Sc. (Tech), Prof., Head of Heat Engines and Energy Management Department, Ukrainian State University of Railway Transport (Kharkiv)
Alexander Aleksakhin	PhD (Tech), Associate Prof., Associate Professor of Thermophysics, Molecular Physics and Energy Efficiency Department, V. N. Karazin Kharkiv National University (Kharkiv).
Oleg Vasilenko	PhD (Tech), Senior Lecturer of Heat Engines and Energy Management Department, Ukrainian State University of Railway Transport (Kharkiv).
Alexander Vrublevsky	dr hab. inż., Profesor Katedra Budowy, Eksploatacji Pojazdów i Maszyn, Uniwersytet Warmińsko-Mazurski w Olsztynie (Polska)
Valery Deshko	D.Sc. (Tech), Prof., Head of Heat Engineering and Energy Saving Department of the Energy Saving and Energy Management Institute, National Technical University of Ukraine "Igor Sikorsky Kiev Polytechnic Institute" (Kyiv)
Yevhen Dudka	Deputy head of Energy Management Department, JSC "Ukrzaliznytsya" (Kyiv).
Artur Kagramanyan	PhD (Tech), Associate Prof., Associate Professor of Heat Engines and Energy Management Department, Ukrainian State University of Railway Transport (Kharkiv).
Alexey Lobyak	PhD (Tech), Associate Prof., Head of Structural Mechanics and Hydraulics Department, Ukrainian State University of Railway Transport (Kharkiv).
Mykhailo Meleshko	Deputy head of Technical Service, Regional Branch "Odessa railway" of JSC "Ukrzaliznytsya" (Odesa).
Alexander Ogar	D.Sc. (Tech), Prof., Head of the Department of Railway Stations and Nodes of the Ukrainian State University of Railway Transport (Kharkiv).
Anatoly Pavlenko	dr nauk tech., Politechnika Świętokrzyska Wydział Inżynierii Środowiska, Geomatyki i Energetyki Katedra Fizyki Budowli i Energii Odnawialnej (Polska)
Andrii Pidopryhora	Head of the Department of Control using Energy Resources of the Locomotive State Department of South Railway
Mykola Sotnik	D.Sc. (Tech), Associate Prof., Associate Professor of Applied Hydroaeromechanics Department, Sumy state University (Sumy)

Section «Resource and Energy Savings in Transport»

Serhii Voronin	D.Sc. (Tech), Prof., Head of Construction, Track and Handling Machinery Department, Ukrainian State University of Railway Transport (Kharkiv).
Tetiana Butko	D.Sc. (Tech), Prof., Head of Railway Operation Management Department, Ukrainian State University of Railway Transport (Kharkiv).
Oleksandr Dykha	D.Sc. (Tech), Prof., Head of Wear Resistance of Machines Department, Khmelnytskyi National University (Khmelnytskyi).

IOP Conf. Series: Materials Science and Engineering 1021 (2021) 011001 doi:10.1088/1757-899X/1021/1/011001

CONFERENCE SECRETARIAT

CONTACS

All correspondence should be addressed to:

RDTSB_CONF Secretariat, Feuerbach square, 7, 61050, Kharkiv, Ukraine
Tel: +38 (057) 730-1937, 730-1066, 730-1078, 730-1074
Tel/fax +38 (057) 771-4683
e-mail: eet-conf@kart.edu.ua
http: http://eet-conf.kart.edu.ua/en/

IOP Conf. Series: Materials Science and Engineering 1021 (2021) 011002 doi:10.1088/1757-899X/1021/1/011002

Peer review declaration

All papers published in this volume of *IOP Conference Series: Materials Science and Engineering* have been peer reviewed through processes administered by the Editors. Reviews were conducted by expert referees to the professional and scientific standards expected of a proceedings journal published by IOP Publishing.

- **Type of peer review:** Double-blind
- **Describe criteria used by Reviewers when accepting/declining papers. Was there the opportunity to resubmit articles after revisions?**

  See attached reviewing form. Yes
- **Conference submission management system:**

  Registration and abstract submission via conference web site http://eet-conf.kart.edu.ua/en/registration-en, papers submission and further communication via e-mail eet-conf@kart.edu.ua
- **Number of submissions received:** 96
- **Number of submissions sent for review:** 89
- **Number of submissions accepted:** 56
- **Acceptance Rate (Number of Submissions Accepted / Number of Submissions Received X 100):** 58,3%
- **Average number of reviews per paper:** 2
- **Total number of reviewers involved:** 36
- **Any additional info on review process (ie plagiarism check system):** UNICHECK
- **Contact person for queries:**

  Glib Vatulia
  glebvatulya@gmail.com,

  Olena Kaliuzhna
  kaliuzhna.el@gmail.com

The composite functional statistical criterion for evaluating the effectiveness of the automatic control process and system

A I Akimov[1], J A Akimova[2], V V Panchenko[1] and M M Odiehov[1,3]

[1] Department of Electric Power Engineering, Electrical Engineering and Electromechanics Ukrainian State University of Railway Transport, Feyerbakh Square 7, 61050, Kharkiv, Ukraine
[2] Department of Higher Mathematics Ukrainian State University of Railway Transport, Feyerbakh Square 7, 61050, Kharkiv, Ukraine

[3] Email: odegov@kart.edu.ua

Abstract. The process of managing and controlling equipment is a broader concept than data transmission. It includes collecting the information about the controlled parameters, transferring the information, its processing, the development of control actions and their transfer to the object in order to bring the controlled parameters back to normal. These tasks are solved by the control and monitoring system. Modern ACMS have not only a larger number of elements, but also a more complex internal structure. Even if all elements of ACMS function are in a reliable way in the process of its operation, the performance of the task assigned to the system cannot be treated as a reliable event. One of the essential features of ACMS is that they are meant to improve the quality of the controlled objects. To quantify the quality of complex systems, effectiveness indicators are used. Indicators of the quality of the system performance of the assigned functions are called the effectiveness of system operation or simply effectiveness. The main idea of evaluating the effectiveness of ACMS is to get the opportunity to compare systems that are identical in purpose, even if they differ from each other in the principles of design and operation. There are certain connections between the individual elements of ACMS, the system as a whole and the controlled object through which they interact with each other. From the point of view of evaluating ACMS effectiveness, the information content of these links is crucial, since the processes of automatic control, for example, in the field of electrical energy production and distribution, are carried out on the basis of information received about the controlled object.

1. Introduction

Analysis of achievements in the field of automatic control and monitoring systems shows that modern manufacturers pay attention to the strength of the products offered and characterize them according to the criteria:
- calculated multiplication of the system accuracy indicator by the rate of its operation;
- by signal volume;
- by the probability of erroneous message reception.

And the average consumer has to use these parameters when choosing complex and expensive products.

Currently, there is no rigorous mathematical substantiation of a generalized criterion for assessing the effectiveness of a process and an automatic control and monitoring system. Which should characterize the information capacity of the process and the automatic control and monitoring system.

Thus, the article raises the topical issue of the mathematical substantiation of evaluating the effectiveness of the control and management process carried out by the ACS through the introduction of the entropy of the controlled object, taking into account the entropy of the ACS, caused by errors in the implementation of the monitoring and control process of each subsystem of the object.

2. The rationalization technique for a composite cross-section

2.1 Problem statement

Modern automatic checkout and control systems are gaining wider and wider popularity in various areas of the national economy in general and in transport sphere in particular. At the same time, we require definite criteria to generally evaluate the effectiveness of the process and automatic control and monitoring systems (ACMS) themselves.

2.2 Analysis of the recent researches and publications

The problem has been analyzed in detail in [1-13]. The scientists have not suggested a profound solution to the task under study. There have been developed some criteria to evaluate the effectiveness of the automatic control and monitoring process. However, there are several of them, which causes certain difficulties in evaluating the effectiveness of the process and ACMS.

2.3 The aim of the article

The article is aimed at selecting and justifying a composite criterion to evaluate the effectiveness of the process and automatic control and monitoring system.

2.4 The main body

A composite criterion for the evaluation of the effectiveness has to characterize the information capacity of the process and the system of automatic control and monitoring.

Modern automatic control and monitoring systems use discrete messages based on the fact that arbitrarily complex messages can be transmitted using a sequence of binary digits 0 and 1.

A binary sequence can be used to transmit an event message selected from $N=2^m$ possible events, where $m = \log_2 N$ is the length of the sequence (the number of bits in a binary number). Thus, the length of the sequence required to transmit a certain message is proportional to the logarithm of the number of possible messages. It is obvious that the maximum amount of information H_{max}, contained in a message is proportional to its length, i.e.

$$H_{max} = \log_2 N. \tag{1}$$

Value H_{max} indicates an upper bound on the amount of information that can be contained in a message.

The actual amount of information depends not only on the number of possible messages, but also on their probabilities, since messages are chosen from a set of possible non-uniform messages. On the grounds of the mentioned above, the concept of a quantitative measure of information is introduced.

$$H(X) = -\sum_{i=1}^{N} P(X_i)\log_2 P(X_i), \tag{2}$$

where $P(X_i)$ is a probability of the i-message from the set N.

In the case when the probabilities of all messages, except one, turn to zero, i.e. if $P(X_i) = 0$ and if $P(X_k) = 1\ (k \neq i)$, the amount of information is zero. If it is known in advance that all possible messages in the set *N* are equally probable, the amount of information is maximum.

Thus, in all cases, except limiting ones ($N_{min} = 0$ and $H_{max} = \log_2 N$), the inequality holds true

$$0 < H < \log_2 N .(3) \tag{3}$$

Value H is called the entropy of the random value X. It is a measure of the initial uncertainty of the result when selecting messages X from the set of possible messages. Message selection completely removes this uncertainty and delivers an amount of information equal to H.

The amount of information received by such a system in the process of monitoring and controlling an object over a time interval (t,τ) is equal to the change in the amount of entropy:

$$I(X,t,\tau) = H_0(X,t,\tau) - H_k(X,t,\tau), \tag{4}$$

where $H_0(X,t,\tau)$ is the entropy, which describes the uncertainty of the object under control and ACMS before monitoring and controlling (X is the set of object states at the time point t, corresponding to the moment when monitoring is stopped, τ is the time point corresponding to the moment when control system stops operating; $H_k(X,t,\tau)$ is the conditional entropy of the controlled object and ACMS in the implementation of the monitoring and control process.

Each state of the controlled object at any time t is characterized by the probability of completing the task in the time interval (t,τ):

$$P(t,\tau) = P_{BO}(t,\tau) \cdot P_{\Pi O}(t,\tau), \tag{5}$$

where $P_{BO}(t,\tau)$, $P_{\Pi O}(t,\tau)$ are respectively the probabilities of the absence of sudden and gradual failures of the controlled object and the control and monitoring system.

If sudden and gradual failures are independent, the entropy of the control and monitoring process is the sum of the entropies caused by these failures:

$$H_0(X,t,\tau) = H_{BO}(X,t,\tau) + H_{\Pi O}(X,t,\tau), \tag{6}$$

where $H_{BO}(X,t,\tau) = -\{P_{BO}(X,t,\tau)\log_2 P_{BO}(X,t,\tau) + (1 - P_{BO}(X,t,\tau))\log_2(1 - P_{BO}(X,t,\tau))\}$ in accordance with the expression (2);

$H_{\Pi O}(X,t,\tau) = -\int_{-\infty}^{\infty} \cdots \int f_{\Pi O}(X,t,\tau)\log_2 f_{\Pi O}(X,t,\tau)dX$ is the entropy caused by the influence of the gradual failures on the output parameters of the control and monitoring process (here $f_{\Pi O}(X,t,\tau)$ is the distribution density of the output parameter X, subjected to the influence of gradual failures).

The optimal level of the controlled object uncertainty is selected from the condition ensuring a given probability of the task completion by the object:

$$\begin{aligned} H_{O\Pi T}(X,t,\tau) = -\{P(X,t_{kk},\tau_{kPO})\log_2 P(X,t_{kk},\tau_{kPO}) + \\ + [1 - P(X,t_{kk},\tau_{kPO})]\log_2[1 - P(X,t_{kk},\tau_{kPO})]\}, \end{aligned} \tag{7}$$

where $P(X,t_{kk},\tau_{kPO})$ is the probability that the object as well as control and monitoring system will complete the task completion in the time interval (t_{kk},τ_{kPO}), which starts at the time moment when monitoring stops t_{kk} (monitoring end), and stops at the time moment when the controlled object stops operating τ_{kPO} (end of the object operation).

The control time τ_k should be kept to a minimum so that the maximum amount of information with the greatest reliability can be obtained per unit of time. At the same time, this minimum should not exceed the permissible time for taking the object out of the readiness state, as this will lead to an unjustified complication of the control and monitoring system.

The composite criterion of effectiveness should characterize the information capacity of the process and the automatic control and monitoring system.

The amount of information received by the ACMS in the process of monitoring and controlling an object over a time interval is determined in accordance with the dependence:

$$I(t,\tau) = H_0(t,\tau) - H(t,\tau), \quad (8)$$

where $H_0(t,\tau)$ is the entropy that describes the uncertainty of the controlled object and ACMS before control and monitoring starts. It is determined by the formula (6);
$H(t,\tau)$ is the remaining entropy of the controlled object and ACMS after the process of object controlling and monitoring is completed. It is determined by the formula (7).

The equality (8) describes the real information capacity of ACMS. The potential possibility of the system is the value

$$I_{\Pi}(t,\tau) = H(t,\tau). \quad (9)$$

The effectiveness of ACMS from the informational point of view can be evaluated by the criterion

$$Э_I(t,\tau) = \frac{I_P(t,\tau)}{I_{\Pi}(t,\tau)}, \quad (10)$$

where $I_P(t,\tau) = I(t,\tau)$ is the real amount of information calculated according to (8).

Taking into account (8) and (9), we get expression (11) to evaluate the effectiveness of ACMS

$$Э(t,\tau) = \frac{H_0(t,\tau) - H(t,\tau)}{H_0(t,\tau)} \quad (11)$$

According to the expression (11) it is easy to notice that the ideal ACMS will have the effectiveness value $Э_I(t,\tau) = 1$, and the real one – the value $Э_I(t,\tau) < 1$. However, the criterion (11) does not take into account the dynamics of the control and monitoring process, complexity and cost of ACMS. These disadvantages can be eliminated by introducing a different ratio for ACMS effectiveness

$$Э(t,\tau) = \frac{K_I(t,\tau)}{K_{IO}(t,\tau)}, \quad (12)$$

where $K_I(t,\tau) = \frac{I_{\max}(t,\tau)}{C_{\Sigma}(t,\tau)}$ is the maximum amount of information obtained by ACMS per unit of its total cost;

$I_{\max}(t,\tau) = \sum_{i=1}^{m} I_{i\max}(t,\tau)$ is the maximum amount of information obtained when monitoring the process by m parameters;

$K_{IO}(t,\tau) = \frac{I_{\max\max}(t,\tau)}{C_{\min}(t,\tau)}$ is the maximum average amount of information received by ACMS when controlling the process by m parameters per unit of the minimum total cost of the idealized ACMS;

$I_{\max\max}(t,\tau) = \sum_{i=1}^{m} I_{i\max\max}(t,\tau)$ is the maximum average amount of information obtained when the object is controlled by m parameters.

Taking into account the abovementioned relations (8), (9), (12) and the equality $I_{\max\max}(t,\tau) = m$ we get the final formula to evaluate the effectiveness ACMS:

$$Э(t,\tau) = \frac{\sum_{i=1}^{m}[H_{0i}(t,\tau) - H_i(t,\tau)]C_{\min}(t,\tau)}{m\sum_{i=1}^{m} C_i(t,\tau)}. \quad (13)$$

3. Conclusion

Thus, to evaluate the effectiveness of the monitoring and control process carried out by ACMS, it is necessary:

- to determine the entropy of each subsystem of the object and ACMS before the control process starts;
- to determine the entropy of the controlled object, taking into account the entropy ACMS, caused by the errors taking place during the process of monitoring and control of each subsystem of the object;
- to calculate the initial cost $C_{\min}(t,\tau)$ and the final cost C of ACMS;
- to make calculations using the formula (13).

It is worth mentioning that the variation range of the composite statistical criterion corresponds to condition $0 \le Э(t,\tau) \le 1$.

Advanced ACMS with high efficiency have a composite index of effectiveness close to one.

References

[1] O.Fomin, A. Akimov, Ju Akimova, A. Mastepan. The criteria choice of evaluating the effectiveness of the process and automatic control systems.//Procedia Computer Science, 2019, Vol. 149, p.246-251.

[2] Avduevskii V. Reliability and effectiveness in engineering. Volume 1. Methodology. Organization. Terminology / V. Avduevskii, I. Appolonov, E. Barzilovich et. al. Ed. A. Rembeza. – Moscow: Mashinostroenie, 1986, 224 p.

[3] Papkov B. Reliability and effectiveness of power supply / B. Papkov, D. Pashali // Ufa: Ufa Sate Aviation Technical University, 2005, 380 p.

[4] Muginshtein L.A. Power optimal traction calculation for operation of trans of increased mass and length [Text] / L.A. Muginshtein, L.A. Yabko // IHHA, June 22-55, Shanghai, China, 2009.

[5] Fokin J. Reliabilty and effectiveness of electrical networks. – Moscow: Vysshaia shkola, 1989, 149 p.

[6] SEL-421 Reference Manual, p. R.1.109. Available at http://www.seline.com/instructionmanual.htm.

[7] M.Gugerty,R.Jenkins, and D. Dulezilek, "Case Study Comparison of Serial and Ethernet Digital Communications Tehnologies for Transfer of Relay Quantities", proceeding of the Distribu TECH Conference and Exhibition, San Diego, CA, February 2007.

[8] Liamets Y., Podchivaline A., Chevelev A., Nudelman G., Zakonjsek J. Informational task of relay protection// CIGRE SC BS Colloquium, Sydney, Australia, 2003. Paper 213.

[9] J.Lewis Blackburn, Tomas J. Domin/ Protective Relaying Principles and Applications / 2007.

[10] Prutik A., Shmojlov A. Algorithm for evaluating technical effectiveness of relay protection and automation. Scientific problems of transport in Siberia and the Far East, №1, 2009, p. 192-195.

[11] Tadjibaev A. Methods and tools for evaluating the state of power equipment / A. Tadjibaev, F. Khalilov, G. Smirnov. Issue 8. – Saint Petersburgh, 1999,174 p.

[12] Matveev V. Evaluation of the efficiency of the power supply system in Topka / V. Matveev, M.

Eremeev, M. Vasenin // Bulletin of the Kuzbass State Technical University, 2010, №6, p. 74-78.

[13] Akimov O. Evaluating the effectiveness of power supply systems on electrified railways / O. Akimov, D. Kukushkin, A. Popadin, A. Pryschepa // Systems of information processing, coll. of scien. papers Kharkiv Air Force University, 2016, Issue 7(114), p.12-14.

Determination of parameters for automatic decompression of heat and power facility cylinders

O O Anatskyi[1,4], D O Aulin[1], S V Bobrytskyi[2], Y V Zhovtiy[1] and P V Chernenko[3]

[1]Department of Maintenance and repair of rolling stock, Ukrainian State University of Railway Transport, 61050 Feierbakh Sq. 7, Kharkiv, Ukraine
[2]Department of Mechanics and Machine Design, Ukrainian State University of Railway Transport, 61050 Feierbakh Sq. 7, Kharkiv, Ukraine
[3]Department of armored vehicles, National Academy of National Guard of Ukraine, 61001 Square of the Defenders of Ukraine 3, Kharkov, Ukraine

[4]Email: anatskiyaleksandr@gmail.com

Abstract. This paper analyzes various types of startup systems used for the heat and power facility start-up. Factors impacting startup characteristics of the heat and power facility were considered. Methods for the modernization of the startup system and possible revamping of auxiliary starting aid devices of locomotive diesel engines without significant design changes by means of cylinder test valve opening were identified. Kinematic diagrams of device gearing for automatic decompression of the heat and power facility cylinders were developed and analyzed. The device requirements were determined, namely, response time, cylinder test valve open and closing torque overriding T_{min}=11 Nm. Based on the necessity of torque backup at least 30 % the gear ratio required for the selected engine and its dimensions was calculated.

1. Introduction

In order to start-up the heat and power facility (Internal Combustion Engine) the electric energy (the electric startup) and the pneumatic power (the pneumatic startup) are mainly used. In rare cases the explosive energy (the pyrotechnical startup), the rotational energy (the inertial startup) or the spring strain energy, etc. All listed startup systems face the high torque developed at the diesel engine start-up for overriding the crankshaft rotation antitorque moment resulting in the increased wear of shaft drive friction couples [1].

Locomotives use the electric startup system of diesel engines being the most important component of the system providing the diesel engine start-up. Articles concerned with the heat and power facility start-up repeatedly stated that the startup period is a transient process in the heat and power facility operation and the necessity of deeper studying of factors impacting the carburation and the startup itself, determining of technical solutions for improvement of the system startup characteristics [2].

Technical solutions for improvement of the system startup characteristics impacting the ignition and combustion of fuel mixture at the startup can be divided into controllable in the course of operation solutions and structural solutions that are impossible to control. For instance, the quality of fuel system operation depends on the fuel type, its cetane number and temperature. During the operation such parameters can be controlled by adding highly inflammable liquids [3], selection of fuel with corresponding cetane number [4], its heating before feeding to the engine, including the use of microwave radiation [5]. In addition, there is the common improvement of startup system using the boosted air preparation: heating by heating plugs; particularly promising are systems of boosted air heating upstream the vortex tubes' startup device [6]. In our opinion, a quire promising is the startup process improvement due to the cylinder decompression without significant design changes of the diesel engine design by means of the cylinder test valve opening.

2. The basic part

The one of possible design startup improvement solutions is the modernization of the heat and power facility startup system.

In the course of development of the heat and power facility startup system modernization the decomposition diagram (Figure1) was developed that allowed to separate hierarchical elements with certain number of decomposition elements for further studies. On the first study stage it was discovered that for resource saving at the locomotive diesel engine startup in order to eliminate negative impact of startup peak currents and accumulator battery life extension is possible to use the starting aid device. The proposed device design consists of step-geared engine that is interlocked with the cylinder test valve. The gear drive allows reducing the rotation speed and increasing the engine torque [7]. Moreover, the startup device design shall include the manual cylinder test valve drive in order to increase the reliability and possibility to conduct process operations during rheostatic testing and inspection [1].

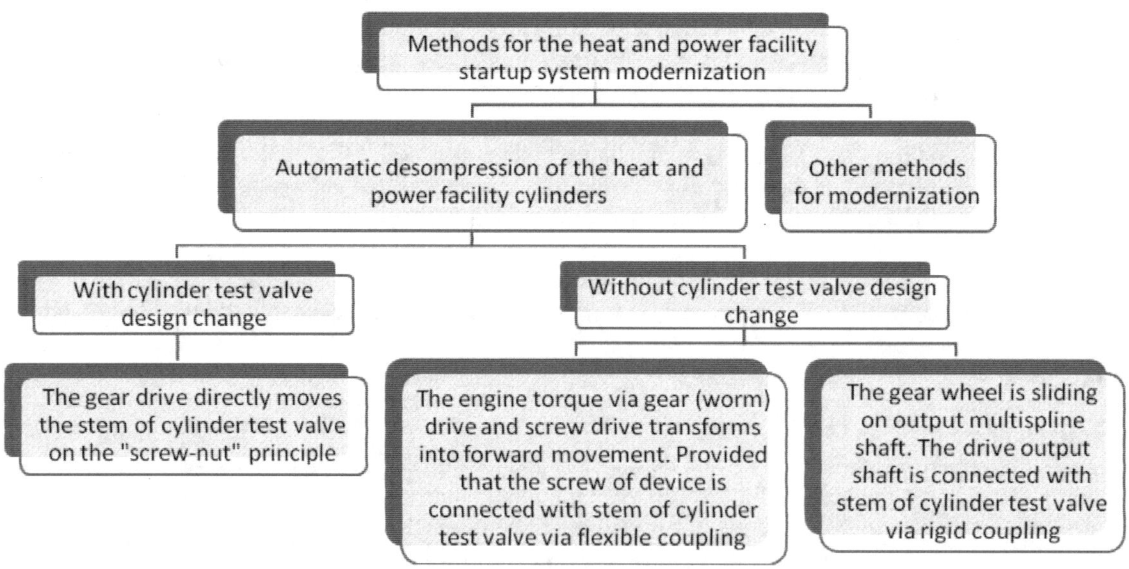

Figure 1. Methods for the heat and power facility startup system modernization.

It should be noted that the use of such devise is possible in two options:

- The first option foresees the necessity of stock cylinder test valve design change;
- The second option does not fore see the design change.

At the start of works on the drive arrangement the mentioned criteria were determined considering which such works will be performed. Among such criteria are:

- The cylinder test valve closing and opening speed, V;
- The torque required for cylinder test valve closing and opening, T;
- The drive dimensions, LxBxH.

The drive gear with cylinder test valve design change was selected, i.e. the gear directly moves the stem of cylinder test valve. Since the device should be fit in dimensions between the engine body and the locomotive hood doors, simplify the drive design, eliminate excess connections and improve operating conditions [8, 9].

On the next study stage, the step engine was selected considering its characteristics kinematic diagrams were further developed (Figure 2, 3, 4) [10, 11].

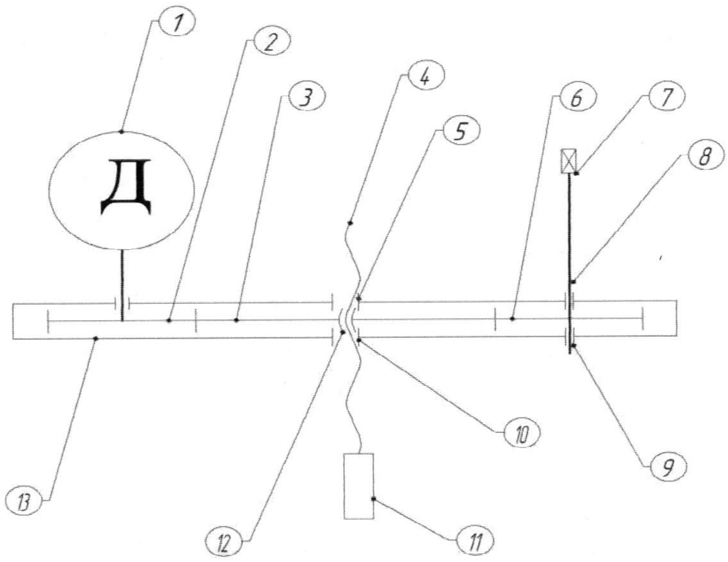

Figure 2. Kinematic diagram of spur gear drive:
1 – Electric motor; 2 – Spur gear wheel; 3 – Spur gear; 4 – Driving screw; 5, 8, 9, 10 – Bearings; 6 – Hand pinion; 7 – Shaft end for free handle; 11 – Cylinder test valve stem; 12 – Driving nut, 13 – Body.

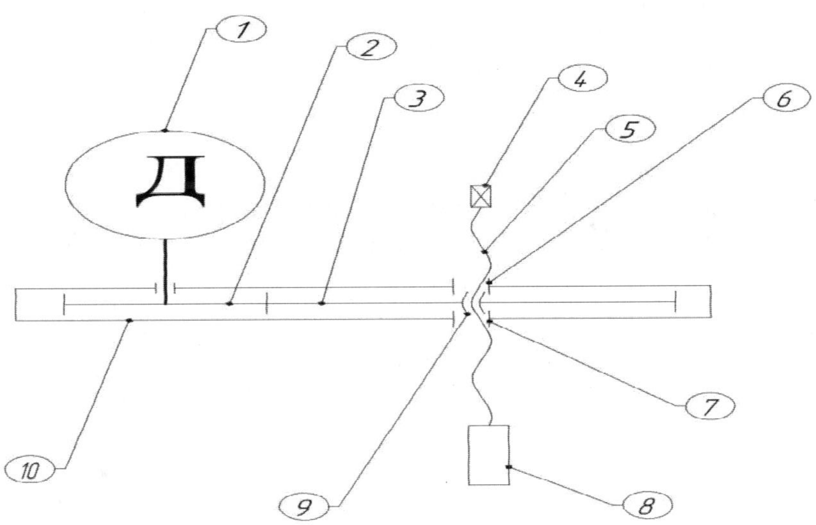

Figure 3. Kinematic diagram of spur gear drive without hand pinion:
1 – Electric motor; 2 – Spur gear wheel; 3– Spur gear; 4 – Driving screw end for free handle; 5 – Driving screw; 6, 7 – Bearings; 8 – Cylinder test valve stem; 9 – Driving nut, 10 - Body.

The kinematic diagram of spur gear drive (Figure 2) is characterized by large dimensions due to additional hand pinion.

IOP Conf. Series: Materials Science and Engineering 1021 (2021) 012002 doi:10.1088/1757-899X/1021/1/012002

In order to decrease the device dimensions, it was decided to pivot from hand pinion (Figure. 3) and move hand drive free handle to gear shaft. However, the application of such arrangement of spur gear still has not allowed to fit in dimensions.

The further study of different gear drives has shown the necessity to use the worm gear and made it possible to provide the transmission with high gear ratio resulting in the gear small size.

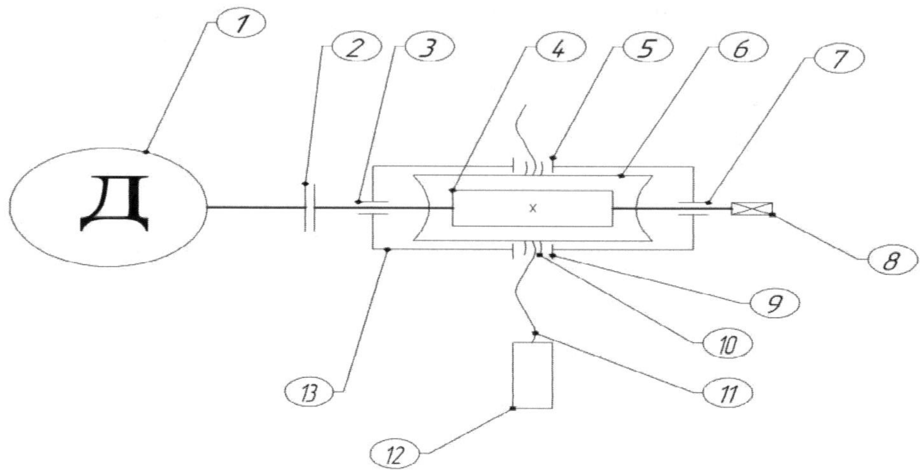

Figure 4. Kinematic diagram of worm gear drive:
1 – Electric motor; 2 – Coupling; 3, 5, 7, 9 – Bearings; 4 – Worm gear; 6 – Worm gear wheel; 8 – Shaft end for free handle; 10 – Driving nut; 11 – Driving screw; 12 – Cylinder test valve stem, 13 – Body.

After the gear drive selection, the works were focused on the determination of drive device parameters itself, among which is the gear ratio u, that shows how the rotating speed decreases and the torque on the cylinder test valve stem increases in comparison with the engine. Provided that the following shall be considered:

$$u \rightarrow 1 \Rightarrow V \rightarrow \max,\ T \rightarrow \min,\ LxBxH \rightarrow \min;$$

$$u \rightarrow \max \Rightarrow V \rightarrow \min,\ T \rightarrow \max,\ LxBxH \rightarrow \max;$$

The measuring of required torque has shown the following value T_{min}=11 Nm. Due to the necessity of torque backup at least 30 % and maximum increase the closing speed of the cylinder test valve, decrease the dimensions the gear ratio required for the selected engine was calculated (Figure 5).

As the graph shows, gear ratios u=2 and u=4 don't correspond the set parameters and the possible reliable operation starts at u=5; for this reason the gear ratio u=8 was chosen for the device implementation because it completely complies with the backup criterion, its response time and fits in the required dimensions.

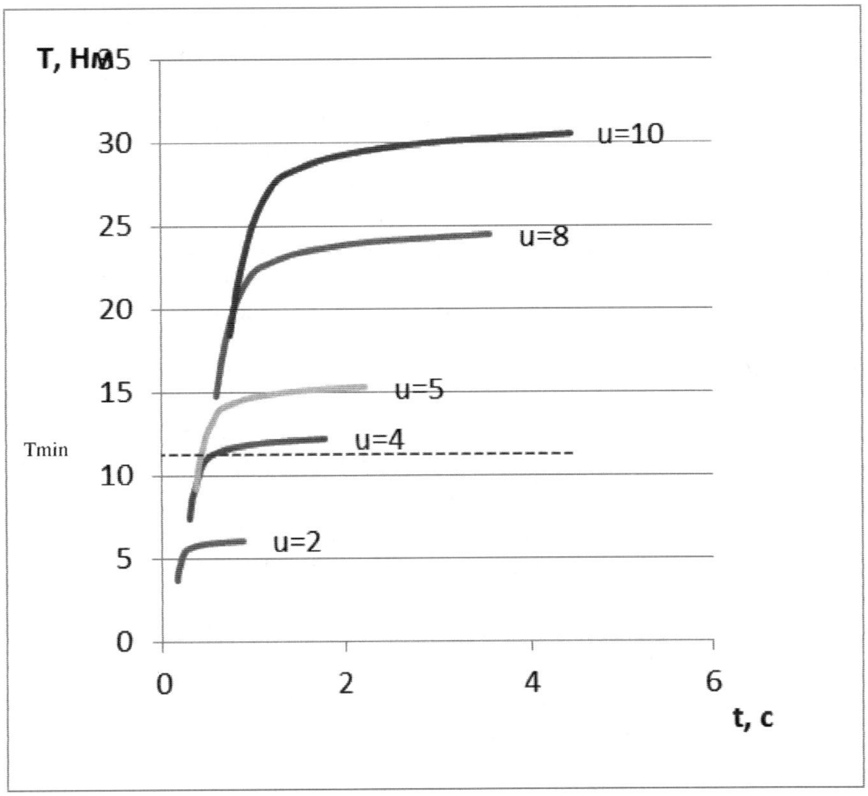

Figure 5. Cylinder test valve closing time depending on gear ratio.

3. Conclusions

The examined locomotive diesel engine startup systems have numerous disadvantages that adversely effect the whole heat and power facility technical conditions. Among such adverse factors are:

- The direct connection to accumulator battery at the traction generator startup;
- The sharp increase of startup current of accumulator battery resulting in the lifetime decrease;
- The additional wear of diesel engine due to increased starting torque.

In order to mitigate adverse effects, it was suggested to improve the startup process using the automatic decompressor; determined technical requirements to the device and its optimal response time; calculated the gear ratio of the cylinder test valve opening mechanism required for the chosen engine and its dimensions.

On the basis of these data it can insisted on the necessity to continue the work on the heat and power facility startup system optimization. I.e. it is required to change the startup system control parameters and further modernization of hardware design solutions for the heat and power facility startup easement.

References

[1] Aulin D, Zinkivskyi A, Anatskyi O and Kovalenko D. 2018 Measures for Resource Saving for Diesel Locomotives International Journal of Engineering & Technology 7 152-156

[2] Gurusamy, S.K. , Rajagopalan, C. , Sateesh Kumar, R. , Ashok, B. Reducing starting current for existing commercial vehicle engines 4th Symposium on International Automotive Technology, SIAT 2015; ARAI Campus Pune; India; 23 January 2015.

[3] Md Nurun Nabi, Mohammad G.Rasul. Notable reductions in blow-by and particle emissions during cold and hot start operations from a turbocharged diesel engine using oxygenated fuels. Fuel Processing Technology Volume 203, 15 June 2020, 106394.
[4] Jiaqiang E, Guanlin Liu, Zhiqing Zhanga, Dandan Hana, Jingwei Chena, Kexiang Weib, Jinke Gonga, Zibin Yinc Effect analysis on cold starting performance enhancement of a diesel engine fueled with biodiesel fuel based on an improved thermodynamic model. Applied Energy Volume 243, 1 June 2019, Pages 321-335.
[5] Smolin, A.A. Analiz parametrov sistemyi toplivopodachi s ustanovkoy SVCh nagreva topliva /A.A. Smolin// Materialyi 8-y Mezhdunarodnoy tehnicheskoy nauchno -prakticheskoy konferentsii «Nauchnaya industriya evropeyskogo kontinenta», T. 25. - Praga, 2012 - S.27 - 31.
[6] Adem Celika, Mehmet Yilmaz, Omer Faruk Yildiz. Improvement of diesel engine startability under low temperatures by vortex tubes. Energy Reports Volume 6, November 2020, Pages 17-27.
[7] Zablonskyi K I 1999 *Detali Mashyn* (Odesa: AstroPrynt) 402
[8] S V Bobrytskyi, O A Logvinenko, O O Anatskyi, I M Yehorova. Assessment of modern rolling stock pulling drive gear reliability.TRANSBUD-2019 IOP Conf. Series: Materials Science and Engineering 708 (2019) 012002. doi: 10.1088/1757-899X/708/1/012002
[9] Kinytskyi Ya.T. 2002 *Teoriia mekhanizmiv i mashyn* (Kyiv: Naukova Dumka) 650
[10] DSTU EN ISO 3952-2:2018 *Kinematic diagrams – Graphical symbols – Part 2* (Kyiv:Derzhspozhyvstandart Ukrainy) 16
[11] DSTU EN ISO 3952-2:2018 *Kinematic diagrams – Graphical symbols – Part 3* (Kyiv:Derzhspozhyvstandart Ukrainy) 12

Comparative analysis of the reliability of electric locomotives based on semi-Markov models of restorable systems

O A Dyshin[1], K Sh Pashayeva[2,4] and C G Aslanov[3]

[1]Azerbaijan State University of Oil and Industry, Azerbaijan Engineering Academy, 10 D. Aliyeva, Baku, Azerbaijan
[2]Process Automation Engineering Department, Baku Higher Oil School, 30 Xocali avenue, Baku, Azerbaijan
[3]CJSC "Azerbaijan Railways", 230 D. Aliyeva, Baku, Azerbaijan

[4]E-mail: is_kamalya@yahoo.com

Abstract. Based on the general theory of semi-Markov recovery processes and their application to Markov recovery processes, the applicability of stationary reliability indicators to assessing the reliability of electric locomotives with an asynchronous motor, the functioning of which is considered as a superposition of independent alternating semi-Markov processes, is shown. By calculating the main indicators of system reliability in a steady state, a comparative analysis of the operational reliability of electric locomotives with an asynchronous DC and AC motors, used respectively on the Azerbaijan and Kazakhstan railways, was carried out.

1. Introduction

The foundations of the theory of semi-Markov processes (SMP) and their applied possibilities are described in detail in the monograph [1], in which classes of semi-Markov processes were considered that simulate the behavior of recoverable systems with fully accessible (unrestricted) recovery and are implemented as a superposition of independent semi-Markov processes with a finite number of states, in particular, a superposition of alternating renewal processes with constraints such as the exponentiality of the original distributions.

In [2], super-positions of both independent and dependent SMPs with a common phase space are introduced as SMPs that simulate the functioning of restored systems with various types of redundancy, a limited and unlimited number of recovery devices (RD). In principle, in this class one can study any of the systems given, for example, in the reference book [3] without additional restrictions such as the exponentiality of the original distributions. As in the well-known work [4], it is noted that the applied capabilities of the SMP are limited for systems with a finite number of states over a long time interval.

An effective way of specifying the SMP is Markov renewal processes (MRPs) - two-dimensional Markov chains, the first component of which (in turn, being a Markov chain) describes the state of the system, and the second fixes the sojourn times of this system in states. The semi-Markov property of the system states is that the residence time in the state depends only on the given (and possibly the next) state and does not depend on the previous evolution of the system. Generally speaking, the physical state of a real system does not possess the semi-Markov property. However, if in a certain way "expand" the physical state $\{l_1, \ldots, l_n\}$ (n – the number of possible states of the system at each moment of time), adding to it some vector $x = \{x_1, \ldots, x_n\}$, $x_1 > 0$, then in many cases it is possible to

achieve that the new extended state $\{l_1, ..., l_n, x_1, ..., x_n\}$ will already have a semi-Markov property. This allows us to describe the functioning of a number of restored systems using PMB [5].
The problem of modeling complex technical recoverable systems using semi-Markov processes remains relevant at the present time, being the subject of research by scientists from the union of independent states (UIS) countries [6-8] and far abroad [9-13].

In this work, to analyze the reliability of operation of electric locomotives with an asynchronous motor, an approach based on semi-Markov processes and Markov recovery processes is used, the mathematical apparatus for studying which was developed in [1] and was further developed in [2,14,15]. By calculating the main reliability characteristics of restored systems in a steady state (stationary availability factor, mean time between failures and mean time to recovery), a comparative analysis of the operational reliability of electric locomotives with DC and AC induction motors, currently used at the Azerbaijan and Kazakhstan Railways, was carried out.

2. Processes of Markov recovery

A jump-like homogeneous Markov process can be constructively given by a stochastic kernel that determines the transition probabilities of the Markov chain, which specifies changes in the states of the process, and a non-negative function on states that specifies the parameters of exponentially distributed random variables that determine the residence times in the states of the process between neighboring jumps.

The concept of semi-Markov processes is a generalization of this construction of jump-like processes. The change in the states of these processes is also controlled by a discrete Markov chain called an embedded Markov chain (EMC), and the sojourn times in states between adjacent jumps have arbitrary distribution functions that depend on the state of the process. Thus, for the constructive specification of semi-Markov processes, the initial material is the sequence $\{\xi_n, \Theta_n, n \geq 0\}$, the first component of which $\{\xi_n, n \geq 0\}$ forms a homogeneous Markov chain that fixes the state of the process at the nth step (after the nth jump), and the second component $\Theta_n \geq 0$ records the time the system is in the state ξ_n.
The two-component sequence $\{\xi_n, \Theta_n, n \geq 0\}$ is a homogeneous Markov chain, the transition probabilities of which do not depend on the values of the second component (the residence time in the previous state does not affect the residence time in this state and the change in this state). Such a Markov chain $\{\xi_n, \Theta_n, n \geq 0\}$ is called the Markov renewal process. The times Θ_n turn out to be conditionally independent random variables for a fixed trajectory sojourn of the embedded Markov chain $\{\xi_n, n \geq 0\}$ [11].

According to Kolmogorov's extension theorem (see, for example, [16]), a random process with a fixed phase space is uniquely (in a probabilistic sense) determined by specifying all its finite-dimensional distributions.

In the case of Markov chains with discrete time and phase space $(\mathrm{Y}, \mathcal{Y})$, where $(\mathrm{Y}, \mathcal{Y})$ is an arbitrary measurable space, all finite-dimensional distributions are calculated using the initial distribution $p_0(B)$, $B\epsilon\mathcal{Y}$, and transition probabilities $P(y, B)$, $y\epsilon Y$, $B\epsilon\mathcal{Y}$. So it is enough to know only $p_0(B)$, $B\epsilon\mathcal{Y}$, and $P(y, B)$, $y\epsilon Y$, $B\epsilon\mathcal{Y}$ in order to completely define a Markov chain with discrete time.
Let's $Y = Z \times [0, \infty)$,$\mathcal{Y} = \mathcal{Z} \times \sigma$, where Z-σ-algebra of subsets from Z, and σ - σ-algebra of Borel sets on $[0, \infty)$; $\times$ - is the sign of the Cartesian product of spaces.

Consider a two-component Markov chain $\{\xi_n, \Theta_n, n \geq 0\}$, $\xi_n\epsilon\mathcal{Z}$, $\Theta_n\epsilon[0, \infty)$, for which $P(\xi_0\epsilon B) = p_0(B), B\epsilon Z$; $P(\xi_{n+1}\epsilon B, \Theta_{n+1} \leq t | \xi_0, \Theta_0, \xi_1, \Theta_1, ..., \xi_n = z, \Theta_n) = P\{\xi_{n+1}\epsilon B, \Theta_{n+1} \leq t | \xi_n = z\}$ moreover $P\{\xi_{n+1}\epsilon Z, \Theta_{n+1} < \infty | \xi_n = z\} = 1$, where $p_0(\cdot)$ –some distribution on Z. A characteristic feature of the Markov chain $\{\xi_n, \Theta_n, n \geq 0\}$ expressed in the fact that the dependence of the probability of an event $\{\xi_{n+1}\epsilon B, \Theta_{n+1} \leq t\}$ from the past is realized only depending on the first component ξ_n at the previous moment.

IOP Conf. Series: Materials Science and Engineering 1021 (2021) 012003 doi:10.1088/1757-899X/1021/1/012003

3. Semi-Markov processes

When modeling the evolution of refurbished technical systems (RTS), it is of interest to calculate the characteristics not only at the moments of state change, but also at any current moments of time t. With every MRP $\{\xi_n, \Theta_n, n \geq 0\}$, which is given by the right-hand sides (1) and (2), is connected with (at least one) random process with continuous time $\xi(t)$ with the same as for $\{\xi_n, n \geq 0\}$ phase space. In order to associate a single random process with continuous time with the indicated PMW, we define on the trajectory $\xi_0, \xi_1, \xi_2, \ldots$ nested in the given MRP Markov chain $\{\xi_n, n \geq 0\}$ family of conditionally independent random variables (RV) $\zeta_{\xi_n, \xi_{n+1}}$ with conditional distribution functions (DF) [1]:
$P\{\Theta_{n+1} \leq t | \xi_n, \xi_{n+1}\} = F_{\zeta_{\xi_n, \xi_{n+1}}}(t)$. Through RV $\zeta_{\xi_n, \xi_{n+1}}$ the so-called counting process is introduced: $\nu(t) = max\{n: \sum_{k=0}^{n} \zeta_{k, \xi_{k+1}} \leq t\}$; Random process $\xi(t) \stackrel{\text{def}}{=} \xi_{\nu(t)}$, where $\{\xi_n, n \geq 0\}$– Markov chain embedded in PMW $\{\xi_n, \Theta_n, n \geq 0\}$, called semi-Markov process (SMP) [1].

3.1. Superposition of independent semi-Markov processes

Consider a system of n independently operating elements (subsystems), the functioning of each of which is described by the SMP $\xi^i(t)$ with arbitrary phase space $(\mathrm{Z}^{(i)}, \mathcal{Z}^{(i)})$, $i = \overline{1,n}$ (processes $\xi^i(t)$) collectively independent, $\mathcal{Z}^{(i)}$ - σ- algebra of subsets $\mathrm{Z}^{(i)}$. Foe every $\mathrm{Z}^{(i)}$ there is a partition $\mathrm{Z}^{(i)} = Z_1^{(i)} \cup Z_0^{(i)}$, $Z_1^{(i)} \cap Z_0^{(i)} = \emptyset$, $i = \overline{1,n}$ where $Z_1^{(i)} \epsilon \mathcal{Z}^{(i)}$; is interpreted as a set of working states of the i-th element; $Z_0^{(i)} \epsilon \mathcal{Z}^{(i)}$ – set of failure states of the i-th element.
Let $\zeta_i(t) = 1$ if $\xi^i(t) \epsilon Z_1^{(i)}(t)$; $\zeta_i(t) = 0$ if $\xi^i(t) \epsilon Z_0^{(i)}(t)$; $\zeta(t) = \{\zeta_1(t), \zeta_2(t), \ldots, \zeta_n(t)\}$

Vector process $\zeta(t)$ characterizes the state of the system at the moment of time according to the states (operability or inoperability) of its elements. The set of possible values of the process ζ (t) is the set of different binary vectors $d = (d_1, d_2, \ldots, d_n)$ from $(0,0, \ldots 0)$ to $(1,1, \ldots 1)$. We denote it by $D = \{d\}$. Consider the concept of a system failure: on the set of binary vectors D, the function $g(d)$ such that

$$g(d) = \begin{cases} 1, if\ the\ system\ is\ operational\ for\ a\ given\ combination\ of\ states\ its\ elements \\ 0, otherwise \end{cases}$$

The set of values of the vector d for which the system is operational is denoted by D_1, and the set of values at which the system is inoperable is denoted by D_0, i.e. $D_1 = \{d: g(d) = 1\}$, $D_0 = \{d: g(d) = 0\}$. By assumption $D = D_1 \cup D_0, D_1 \cap D_0 = \emptyset$. Let us investigate the reliability characteristics of such a system, in particular, the average stationary (steady-state) residence times of the process $\zeta(t)$ in sets D_1 и D_0 (mean time between failures and mean time to recover the system, respectively) and the limit value at $t \to \infty$ the likelihood of catching the process $\zeta(t)$ in the set D_1 (stationary availability factor).
By introducing additional components $u^{(i)}(t)$ the problem can be reduced to the study of the characteristics of the CMP. Let be $u^{(i)}(t)$ – lack of progress $\xi^{(i)}(t)$ at the moment t, i.e. $u^{(i)}(t) = t - sup\{u: u \leq t, \xi^{(i)}(u) \neq \xi^{(i)}(t)\}$. Then the process is undershot $\xi^{(i)}(t)$ at time t of the closest "in the past" to time t jump of one of the processes $\xi^{(i)}(t), k = \overline{1,n}$, equal to $\upsilon^{(k)}(t) = u^{(i)}(t) - u(t)$, $u(t) = \min_{i=1,2,\ldots,n} u^{(i)}(t)$.

Superposition of (semi-Markov) processes $\xi^{(i)}(t), i = \overline{1,n}$, called [2] $2n$- component process $\xi(t) = \left(\xi^{(1)}(t), \ldots, \xi^{(n)}(t); \upsilon^{(1)}(t), \ldots, \upsilon^{(n)}(t)\right)$

The availability factor, mean time between failures and mean recovery time in the stationary mode of the PMP ξ (t) can be determined, respectively, by the following formulas [2]:

$$K_\Gamma = \sum_{d \epsilon D_1} \prod_{i=1}^{n} T_{d_i}^{(i)} \left\{ \prod_{i=0}^{n} T_1^{(i)} + T_0^{(i)} \right\}^{-1} \tag{1}$$

$$T_1 = \sum_{d \epsilon D_1} \prod_{i=1}^{n} T_{d_i}^{(i)} \left\{ \sum_{d \epsilon D_0'} \prod_{i=1}^{n} T_{d_i}^{(i)} \sum_{j \epsilon I(d)} \frac{1}{T_0^{(j)}} \right\}^{-1} \tag{2}$$

$$T_0 = \sum_{d \epsilon D_0} \prod_{i=1}^{n} T_{d_i}^{(i)} \left\{ \sum_{d \epsilon D_0'} \prod_{i=1}^{n} T_{d_i}^{(i)} \sum_{j \epsilon I(d)} \frac{1}{T_0^{(j)}} \right\}^{-1}, \tag{3}$$

where $T_1^{(i)}$ is mean time between failures of the i-th element; $T_0^{(i)}$ is average recovery time of the i-th element; $T_{d_i}^{(i)} = T_1^{(i)}$ for $d_i = 1$ and $T_{d_i}^{(i)} = T_0^{(i)}$ for $d_i = 0$; D_0' is the set of vectors dϵD$_0$ such that a change in the value of some one component from zero to one transforms the vector d into the set D_1; $I(d)$ is the set of numbers of vector components $d \epsilon D_0'$, changing the value of each of which from zero to one transforms the vector d into the set D_1.

Formulas (1) - (13) give exact relations for calculating the stationary reliability indicators of systems with independently functioning elements, and $K_\Gamma = T_1/(T_1 + T_0)$ [3] .

4. System "p of n" with shutdown after failure

Consider a system of n independently functioning elements. Suppose that the functioning of each i-th element is described by an alternating recovery process with uptime (RPU) $\alpha_1^{(i)} \left(P\left\{ \alpha_1^{(i)} \le t \right\} = F_1^{(i)}(t) \right)$ and recovery time (RT) $\alpha_0^{(i)} \left(P\left\{ \alpha_0^{(i)} \le t \right\} = F_0^{(i)}(t) \right)$. We will consider the system to be operational when it works at least $p(1 \le p \le n)$ elements (in the example of the functioning of a railway locomotive, which we consider below, $p = n$). For definiteness, we assume that at the initial moment of time $t = 0$ all elements begin to work. Let us determine the main reliability characteristics of the system: stationary availability factor K_Γ, mean time between failures T_1 and mean recovery time T_0.

Suppose the following conditions are met:

A$_1$. Random variables (RV) $\alpha_1^{(i)}$ – RPU elements, $\alpha_0^{(i)}$ – RT elements, $i = \overline{1,n}$ (n– number of elements in the system) are independent in the aggregate and have limited average $0 \le M\alpha_1^{(i)} = T_1^{(i)} < \infty, 0 \le M\alpha_0^{(i)} = T_0^{(i)} < \infty$.

A$_2$. Distribution functions (DF) RV $\alpha_1^{(i)}$ и$\alpha_0^{(i)}, i = \overline{1,n}$, $F_1^{(i)}(t)$ и $F_0^{(i)}(t)$ respectively, are absolutely continuous with respect to the Lebesgue measure.

When conditions are met A_1, A_2 right [2] following expressions for K_Γ, T_1 and T_0 respectively, equivalent to expressions (1)-(3):

$$K_\Gamma = \sum_{|d| \ge p} \prod_{i=1}^{n} T_{d_i}^{(i)} \left\{ \prod_{i=0}^{n} T_1^{(i)} + T_0^{(i)} \right\}^{-1} \tag{4}$$

$$T_1 = \left(\sum_{|d| \ge p} \prod_{i=1}^{n} T_{d_i}^{(i)} \right) \left\{ \sum_{|d| = p} \prod_{i=0}^{n} T_{d_i}^{(i)} \sum_{\{j : d_j = 1\}} 1/T_1^{(i)} \right\}^{-1} \tag{5}$$

$$T_0 = \left(\sum_{|d| \le p} \prod_{i=1}^{n} T_{d_i}^{(i)} \right) \left\{ \sum_{|d| = p-1} \prod_{i=0}^{n} T_{d_i}^{(i)} \sum_{\{j : d_j = 0\}} 1/T_0^{(i)} \right\}^{-1} \tag{6}$$

Let us show the application of formulas (4) - (6) for calculating the stationary reliability indicators of electric locomotives with an asynchronous traction motor in order to compare the reliability of electric locomotives with DC and AC motors used, respectively, on the Azerbaijan and Kazakhstan railways. The calculations will be carried out on the basis of the available data for 2016 in the first case for malfunctions of a small toothed gear($i = 1$), motor axial bearing($i = 2$), anchor bearing ($i = 3$), traction motor ($i = 4$) and in the second case is due to malfunctions of a small toothed gear($i = 1$), compressor ($i = 2$), pantograph($i = 3$), traction motor($i = 4$). Let us denote by $\alpha_{0,j}^{(i)}$ и $\alpha_{1,j}^{(i)}$– the recovery times (RT) and the failure-free operation times (FOT) of the i -th element of the j -th electric locomotive.

Let there be the following data on the downtime (in hours) of DC electric locomotives operated in 2016 on the Azerbaijan Railway (AR) for j - electric locomotives ($j = 1, \ldots, 11$): 1) VL11-274 :$\alpha_{0,1}^{(3)} = 78{,}4$; 2) VL11-341 : $\alpha_{0,2}^{(2)} = 98{,}5$; $\alpha_{0,2}^{(4)} = 98; 60{,}25$; 3) VL11-351 :$\alpha_{0,3}^{(2)} = 8; 85{,}4$; $\alpha_{0,3}^{(3)} = 62$; $\alpha_{0,3}^{(3)} = 28$; 4) VL11-353 : $\alpha_{0,4}^{(1)} = 118,$; 5) VL11-364 :$\alpha_{0,5}^{(3)} = 74$; 6) VL11-402: $\alpha_{0,6}^{(1)} = 110{,}25; 101{,}5$; $\alpha_{0,6}^{(2)} = 21{,}4$; 7) VL11-460: $\alpha_{0,7}^{(2)} = 89{,}7$; 8) VL11-461: $\alpha_{0,8}^{(1)} = 342$; 9) VL11-462: $\alpha_{0,9}^{(2)} = 58; 58{,}6$; 10) VL11-463 : $\alpha_{0,10}^{(2)} = 250$; $\alpha_{0,10}^{(4)} = 56{,}5$; 11) VL11-464 : $\alpha_{0,11}^{(2)} = 39; 59{,}4$.

Relevant data on the downtime of AC electric locomotives in operation in 2016. on the Kazakhstan Railways (KR):

1) KZ8:$\alpha_{0,1}^{(3)} = 8$; $\alpha_{0,1}^{(4)} = 4$; 2) KZ 9:$\alpha_{0,2}^{(2)} = 4$; $\alpha_{0,2}^{(4)} = 3$; 3) KZ 12:$\alpha_{0,3}^{(1)} = 40; 46$; $\alpha_{0,3}^{(2)} = 14$;$\alpha_{0,3}^{(3)} = 5$,$\alpha_{0,3}^{(4)} = 64$; 4) KZ 15:$\alpha_{0,4}^{(1)} = 48$; $\alpha_{0,4}^{(2)} = 22$; 5) KZ 17:$\alpha_{0,5}^{(1)} = 200$; 6) KZ 18:$\alpha_{0,6}^{(1)} = 72$; 7) KZ 20:$\alpha_{0,7}^{(1)} = 150$; $\alpha_{0,7}^{(2)} = 19$;$\alpha_{0,7}^{(3)} = 13$; 8) KZ21:$\alpha_{0,8}^{(3)} = 8$; $\alpha_{0,8}^{(4)} = 4$; 9) KZ 27:$\alpha_{0,9}^{(2)} = 45$;$\alpha_{0,9}^{(3)} = 10$; $\alpha_{0,9}^{(4)} = 15$; 10) KZ29:$\alpha_{0,10}^{(2)} = 25$;$\alpha_{0,10}^{(4)} = 7$; 11) KZ34:$\alpha_{0,11}^{(2)} = 8$; $\alpha_{0,11}^{(3)} = 11$.

According to the data on the Azerbaijan Railways (shortly AR) we have (taking into account 365x12x24 = 8760 working hours per year), denoting through $T_{0,j}^{(i)}$ and $T_{1,j}^{(i)}$– respectively, the average VR and the average FOT of the i-th element of the j-th electric locomotive.

$T_{1,1}^{(i)} = 4340{,}8\ (i = \overline{1,4})$; $T_{0,1}^{(3)} = 78{,}4$, $T_{0,1}^{(i)} = 0 (i \neq 3)$; 2) $T_{1,2}^{(i)} = 2125{,}8\ (i = \overline{1,4})$; $T_{0,2}^{(1)} = 0$,$T_{0,2}^{(2)} = 98{,}5$; $T_{0,2}^{(3)} = 0$; $T_{0,2}^{(3)} = 0$; $T_{0,2}^{(4)} = (38 + 60{,}25)/2 = 49{,}125$; 3)$T_{1,3}^{(i)} = 1715{,}3\ (i = \overline{1,4})$; $T_{0,3}^{(1)} = 0$;$T_{0,3}^{(2)} = 46{,}7$; $T_{0,3}^{(3)} = 62$; $T_{0,3}^{(4)} = 28$; 4) $T_{1,4}^{(i)} = 4321\ (i = \overline{1,4})$, $T_{0,4}^{(1)} = 118$; $T_{0,4}^{(i)} = 0\ (i \neq 1)$; 5) $T_{1,5}^{(i)} = 4343\ (i = \overline{1,4})$; $T_{0,5}^{(i)} = 74$; $T_{0,5}^{(i)} = 0\ (i \neq 3)$; 6) $T_{1,6}^{(i)} = 2131{,}7\ (i = \overline{1,4})$, $T_{0,6}^{(1)} = 105{,}875$;$T_{0,6}^{(2)} = 21{,}4$; $T_{0,6}^{(3)} = T_{0,6}^{(4)} = 0$; 7) $T_{1,7}^{(i)} = 4335{,}15\ (i = \overline{1,4})$; $T_{0,7}^{(2)} = 89{,}7$;$T_{0,7}^{(i)} = 0\ (i \neq 2)$; 8) $T_{1,8}^{(i)} = 4209\ (i = \overline{1,4})$; $T_{0,8}^{(1)} = 342$; $T_{0,8}^{(i)} = 0\ (i \neq 1)$; 9) $T_{1,9}^{(i)} = 2881{,}13\ (i = \overline{1,4})$; $T_{0,9}^{(1)} = 58{,}3$; $T_{0,9}^{(i)} = 0\ (i \neq 2)$; 10) $T_{1,10}^{(i)} = 2817{,}83\ (i = \overline{1,4})$; $T_{0,10}^{(1)} = 0$; $T_{0,10}^{(3)} = 0$;$T_{1,10}^{(4)} = 56{,}5$; 11) $T_{1,11}^{(i)} = 2145{,}15 (i = \overline{1,4})$; $T_{0,11}^{(1)} = 81$; $T_{0,11}^{(2)} = 49{,}2$;$T_{0,11}^{(3)} = T_{0,11}^{(4)} = 0$.

Since for the considered system with disconnection of elements $T_{1,j}^{(1)} = T_{1,j}^{(2)} = T_{1,j}^{(n)} = const = T_{1,j}^{syst}$, где $T_{1,j}^{syst}$– mean (for failures) mean time between failures of the system, determined by the expression $T_{1,j}^{syst} = \left(8760 - T_{0,j}^{syst}\right)/\left(r_j + 1\right)$, $T_{0,j}^{syst} = \sum_{i=1}^{n} T_{0,j}^{(i)}$, $r_j = \sum_{i=1}^{n} r_j^{(i)}$,

where $r_j^{(i)}$– number of refusals i-th element, then the values $T_{1,j}^{syst}/n$ и$T_{0,j}^{syst}/n$ can be interpreted as the mean (by elements) MTBF and system recovery time and denote $T_{1,j}^{elem} = T_{1,j}^{syst}/n$ и$T_{0,j}^{elem} = T_{0,j}^{syst}/n$, wherein $K_{\Gamma,j}^{elem} = T_{1,j}^{elem}/\left(T_{1,j}^{elem} + T_{0,j}^{elem}\right)$. Thus, denoting the parameters $K_{\Gamma,j}$, $T_{1,j}$ and $T_{0,j}$ in (11)-(13) as system parametrs $K_{\Gamma,j}^{syst}$, $T_{1,j}^{syst}$ и $T_{0,j}^{syst}$, we come to the conclusion that for the system j with disconnection the relations

$$K_{\Gamma,j}^{syst} = K_{\Gamma,j}^{elem},\ T_{1,j}^{syst} = T_{1,j}^{elem},\quad T_{0,j}^{syst} = T_{0,j}^{elem}$$

and formulas (13) - (15) are equivalent to the following simple formulas

$$K_{\Gamma,j} = T_{1,j}^{syst}/\left(T_{1,j}^{syst} + T_{0,j}^{syst}\right) = T_{1,j}^{elem}/\left(T_{1,j}^{elem} + T_{0,j}^{elem}\right), \quad (7)$$

$$T_{1,j} = T_{0,j}^{syst}/n = T_{1,j}^{elem}, \quad (8)$$

$$T_{0,j} = T_{0,j}^{syst}/n = T_{0,j}^{elem} \quad (9)$$

This means that for the considered system with disconnection of elements, the average stationary reliability characteristics of the system are equal to the corresponding characteristics of the elements.

Applying formulas (7) - (9) to the data on the AR, we get:

1) $K_{\Gamma,1} = 0{,}982$; $T_{1,1} = 1085{,}2$; $T_{0,1} = 19{,}6$; 2) $K_{\Gamma,2} = 0{,}935$; $T_{1,2} = 531{,}45$;$T_{0,2} = 36{,}9$; 3) $K_{\Gamma,3} = 0{,}926$; $T_{1,3} = 428{,}825$; $T_{0,3} = 34{,}175$; 4) $K_{\Gamma,4} = 0{,}973$;$T_{1,4} = 1080{,}25$; $T_{0,4} = 29{,}5$; 5) $K_{\Gamma,5} = 0{,}983$; $T_{1,5} = 1085{,}75$; $T_{0,5} = 18{,}5$; 6) $K_{\Gamma,6} = 0{,}9$; $T_{1,6} = 532{,}295$; $T_{0,6} = 58{,}287$; 7) $K_{\Gamma,7} = 0{,}98$; $T_{1,7} = 1083{,}787$; $T_{0,7} = 22{,}425$; 8) $K_{\Gamma,8} = 0{,}925$; $T_{1,8} = 1052{,}25$; $T_{0,8} = 85{,}5$; 9) $K_{\Gamma,9} = 0{,}96$; $T_{1,9} = 720{,}282$;$T_{0,9} = 19{,}43$; 10) $K_{\Gamma,10} = 0{,}9$; $T_{1,10} = 704{,}457$; $T_{0,10} = 76{,}625$; 11) $K_{\Gamma,11} = 0{,}923$; $T_{1,11} = 536{,}287$; $T_{0,11} = 44{,}85$.

Now calculating the average (over all $n = 11$ electric locomotives) values of the coefficients K_Γ, T_1 and T_0, i.e.

$$\bar{K}_\Gamma = \frac{1}{n}\sum_{j=1}^{n} K_{\Gamma,j}\,, \qquad T_1 = \frac{1}{n}\sum_{j=1}^{n} T_{1,j} \quad \text{и} \quad T_0 = \frac{1}{n}\sum_{j=1}^{n} T_{0,j}\,,$$

will get

$$\bar{K}_\Gamma = 0{,}944,\ \bar{T}_1 = 803{,}71,\quad \bar{T}_0 = 40{,}48 \quad (10)$$

According to the data on the Kazakhstan Railway (KR), as a result of the calculation, we obtain:

$$\bar{K}_\Gamma = 0{,}965,\ \bar{T}_1 = 665{,}56 \quad \bar{T}_0 = 17{,}52. \quad (11)$$

Comparing the average indicators (10) and (11), we come to the following conclusion: although the mean time between failures $\bar{T}_1$ for DC electric locomotives of AR is more than for AC electric locomotives of KR, by about 140 hours, their average availability factor $\bar{K}_\Gamma$ is less 2% and the average recovery time is 13 hours longer, which indicates the greater reliability of AC electric locomotives in comparison with DC electric locomotives.

5. Conclusion

The operation of electric locomotives with an induction motor based on data on unscheduled repairs associated with violations of individual elements of an electric locomotive can be represented as a recoverable system with disconnection after failure of one element (or several) of all remaining elements. Under the conditions A_1 and A_2 for the obtained semi-Markov process of restoration, the formulas for calculating the reliability indicators in the steady state (stationary) mode are applicable. To calculate these indicators, simplified formulas are obtained in the work, which can be used for a comparative analysis of the reliability levels of various types of complex systems with shutdown in case of failures, having non exponential distribution functions of the residence times in states.

References

[1] Korolyuk V S Turbin A F 1976 *Semi-Markov processes and their applications*(Kyiev: Naukova Physics) p 183.

[2] Korlat A N Kuznetsov V N Novikov M M and Turbin A F 1991 *Semi-Markov models of recoverable systems and queuing systems* (Chisinau: Shtinitsa) p 276.

[3] Kozlov B A and Ushakov I A 1975 *Handbook for calculating the reliability of radio electronics and automation equipment* (Moscow: Sov.radio).

[4] Beichelt F and Franken P 1988 *Reliability and maintenance Mathematical approach* translated from it. (Moscow: Radio and communication).

[5] Silvestrov D S 1980 *Semi-Markov processes with a discrete set of states* (Moscow: Sov.radio) p 272.

[6] Kopp V Ya Obzherin Yu E and Peschansky A I 2000 *Stochastic models of automated production systems with time redundancy* (Sevastopol) p 285.

[7] Zeleny O V Nosovskiy A V and Stadnik O A 2007 Semi-Markov models in the problems of assessing the reliability and risk from NPP operation *Problems of safety of nuclear power plants I of Chernobyl* **7** pp 30-40.

[8] Borisevich A V and Dyakin N V 2015 Semi-Markov model for assessing the reliability indicators of an uninterruptible power supply of a data center *Modern scientific research and innovations* **8 Part I** pp 1-8.

[9] Papadoponlou A and Vassilou R-C G 1999 Cantinuous Time Non-Homogeneous Semi-Markov Systems Semi-Markov Models and Applications Kluwer Acad. Publ. pp 241-251.

[10] Limnios N and Oprisan 2001 *Semi-Markov Processes and Reliability* (Boston, Birkhauser).

[11] Grabski F 2003 The Reliability of an Object with Semi-Markov Failure Rate *Appl.math.Comput*. **135** pp 1-16.

[12] Grabski F 2007 Applications of Semi-Markov processes in reliability *RTA* **3-4 December-Special Issue** pp 60-75.

[13] Grabski F 2011 Semi-Markov failure rates processes *Applied Mathematics and Computation* **217** pp 9956-9965.

[14] Korolyuk V S Turbin A F 1978 *Phase enlargement of complex systems (*Kyiev: Vishcha school) p 110.

[15] Korolyuk V S and Turbin A F 1982 *Markov recovery processes in problems of system reliability*. (Kyiev: Naukova Physics) p 235.

[16] Gikhman I I and Skorokhod A V 1965 *Introduction to the theory of stochastic processes* (Fizmatgiz, Moscow)

IOP Conf. Series: Materials Science and Engineering 1021 (2021) 012004 doi:10.1088/1757-899X/1021/1/012004

Determination of the regularities of the covered wagon supporting structure dynamic loading when conducting fire from it in the horizontal plane

O Fomin[1], **A Lovska**[2,6], **J Gerlici**[3], **Yu Fomina**[3], **D Fedosov-Nikonov**[4], **P Prokopenko**[1]

[1]Department of Cars and Carriage Facilities, State University of Infrastructure and Technologies, Kyrylivska str., 9, Kyiv, Ukraine, 04071
[2]Department of wagons, Ukrainian State University of Railway Transport, Feuerbach sq., 7, Kharkiv, Ukraine, 61050
[3]Department «Transport and handling equipment», University of Zilina in Zilina, Univerzitna 8215/1, 010 26 Zilina, Slovakia, 01026
[4]State Enterprise "Ukrainian Research Institute of Wagon Building", Prikhodko str, 33, Kremenchuk, Ukraine, 39621

[6]alyonalovskaya.vagons@gmail.com

Abstract. A study of the supporting structure dynamic loading of a covered wagon for the military equipment transportation and for firing has been carried out. A mathematical model has been drawn for this purpose, taking into account the vertical and angular displacements relative to the longitudinal axis (side rolling) of the wagon supporting structure when firing. Accelerations as the components of the dynamic load acting on the wagon supporting structure have been determined. It was found that the accelerations acting on the wagon supporting structure in the vertical plane were about 6.0 m/s^2 (0.6 g). In the horizontal plane, the accelerations were 2.6 m/s^2 (0.26 g). That is, the wagon movement can be assessed as "satisfactory". To reduce the wagon dynamic loading during the conduct of fire, it is proposed to install elastic vibration dampers under the support sectors. In this case, the horizontal acceleration was about 2 m/s^2 (0.2 g). Taking into account the proposed technical solutions, the wagon dynamic load is reduced by about 25% in comparison with the case of the rigid interaction of the sector with the frame. The strength analysis of the covered wagon bearing structure has been carried out. It was found that the maximum equivalent stresses are concentrated in the area of the front stops and are about 315 MPa. The conducted research will contribute to the creation of recommendations for the design of covered wagons for the military equipment transport and for firing, as well as to increase the efficiency of the railway transport operation..

1. Introduction

The railway industry development at the present stage of the transport system functioning requires the commissioning of rolling stock innovative structures, in particular, wagons. In this case, special attention should be paid to their supporting structures. It is not only about improving technical and economic indicators and introducing new progressive materials, but also about creating multifunctional structures.

Multifunctionality of the supporting structure means the ability to perform more than one function. Therefore, it is proposed to put into operation wagons that would provide the possibility not only of

IOP Conf. Series: Materials Science and Engineering 1021 (2021) 012004 doi:10.1088/1757-899X/1021/1/012004

transporting goods, but also of using, for example, for the country military-strategic purposes. That is a very urgent and important task today.

It is known that flat wagons are most widely used for the military equipment transportation. However, for the transportation of military equipment that needs atmospheric precipitation protection, the use of these wagon types is not advisable. Therefore, it is proposed the commissioning of covered wagons for the military equipment transportation, as well as the possibility of conducting fire effects while moving. It is important to note that the issue of designing such wagons has not yet received due attention. In addition, the regulatory framework in accordance with which the modern wagon structures are designed does not take into account the loads that can act on them during the conduct of fire. This makes necessary research in this direction and the creation of appropriate recommendations.

Features of bench tests for fatigue resistance of covered wagons dummy units are described in [1]. The static and dynamic loads of the prototypes were carried out on the hydroimpulse test bench for testing large-sized structures. The study of the reliability of the obtained results are presented.

The results of a numerical and experimental study of the wagon supporting structure strength are given in papers [2, 3]. A covered wagon of type 11-9962 was chosen as a prototype. The method of electrical strain gauging was used as an experimental research method. The carried out studies confirmed the feasibility of the technical solutions adopted while designing.

It is important to note that when designing and calculating wagons, the standard load values were taken into account. That is, the overtime load regimes of the wagon bearing structure, including during the military equipment transportation and the conduct of fire, were not taken into account.

Prospects for the use of extra-fine steels in rail car building are illustrated in [4]. The analysis of the characteristics of steels with a high strength class, which are manufactured in Russia and abroad, is carried out. An assessment of the reduction in the structure weight when using high-strength steels is given. However, the authors did not consider the prospects for the introduction of these steels in specialized wagon designs, including for the transportation of military equipment and fire exposure.

The features of the strength analysis of a modern covered wagon structure are highlighted in [5]. In this case, the means of computer simulation of strength by the finite element method were used. However, the loads that may act on a wagon during the transportation of military equipment are not taken into account when calculating the wagon load-bearing structure. It is also not specified in the paper, if it is possible to conduct the fire from this wagon type while moving.

Research on optimization of the wagon supporting structure is carried out in a paper [6]. To reduce the material consumption of the wagon, it is proposed to manufacture its supporting structure from aluminium panels. The component of these panels is a material with elastic characteristics, which helps to reduce the dynamic loading of the supporting structure in operation. At the same time, the authors limited themselves to taking into account the standard values of the loads acting on the wagon in operation.

The determination of the dynamic loading of the wagon supporting structures is carried out in [7, 8]. Mathematical models that allow to obtain accelerations as components of the dynamic load acting on the wagon supporting structures are presented. The obtained results have been confirmed by computer simulation using the finite element method. However, the case of wagon overtime loading when transporting military equipment and conducting fire was not taken into account.

The purpose of the article is to determine the regularities of the dynamic loading and strength of the supporting structure of a covered wagon for military equipment transportation and for conducting fire in a horizontal plane while moving. To achieve this, the following tasks have been identified:

- to create a computer model of the supporting structure of a covered wagon for military equipment transportation and for conducting fire;
- to carry out the dynamic loading mathematical simulation of the supporting structure of a covered wagon for military equipment transportation and for conducting fire;
- to determine the main strength indicators of the supporting structure of a covered wagon for military equipment transportation and for conducting fire, and to determine the natural oscillation frequencies of the covered wagon supporting structure.

2. Computer model creation of the supporting structure of a covered wagon for military equipment transportation and for conducting fire

It is proposed to create a specialized covered wagon (Figure 1) to transport military equipment and to conduct fire while moving. A covered wagon of the type 11-217 was used as a prototype. Graphic work on the model designing was carried out in the SolidWorks software. The peculiarity of the wagon is that its supporting structure has a sliding roof. This makes possible to conduct fire in the vertical plane.

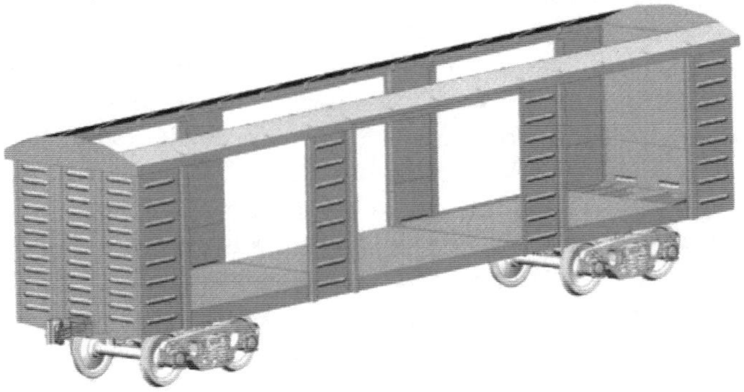

Figure 1. Covered wagon for military equipment transportation and for conducting fire.

Also, the wagon is equipped with a movable double-folding door for loading military equipment, as well as for conducting fire in a horizontal plane. In this case, military equipment is located on special support areas on the wagon frame. It provides for possible to equip the support areas with wheel stops, as well as with other devices for fastening military equipment. The wagon frame is identical in design to the prototype wagon frame. The wagon load-bearing structure is supported by typical bogies model 18-100. The possibility of using bogies of more promising designs under the wagon body is envisaged.

2. Dynamic loading mathematical simulation of the supporting structure of a covered wagon for transporting military equipment and for conducting fire

In order to determine the wagon dynamic loading during the conduct of fire, mathematical simulation has been carried out. The studies were carried out in a plane coordinate system. The design scheme is shown in Figure 2. It has been taken into account that shots are fired from three military installations simultaneously. The muzzle energy of one piece of military equipment is about 90 kJ. When formulating differential equations of motion, it was assumed that the wagon supporting structure leans on two typical bogies model 18-100 having the corresponding technical characteristics.

In this case, the differential equations of motion describing the oscillations of the wagon supporting structure during fire shots in the horizontal plane have the form:

$$\begin{cases} M_{CW} \cdot \ddot{q}_1 + c \cdot q_1 = P_{FR}\left(sign\left(\dot{\Delta}_1 - \dot{\Delta}_2\right) + sign\left(\dot{\Delta}_1 + \dot{\Delta}_2\right)\right) + P_f, \\ I_{CW} \cdot \ddot{q}_2 + c \cdot b \cdot q_2 = P_{FR} \cdot b\left(sign\left(\dot{\Delta}_1 - \dot{\Delta}_2\right) + sign\left(\dot{\Delta}_1 + \dot{\Delta}_2\right)\right) + M_f, \end{cases} \tag{1}$$

$$\Delta_1 = q_1 - b \cdot q_2, \; \Delta_2 = q_1 + b \cdot q_2, \tag{2}$$

where M_{CW} is the wagon supporting structure weight; I_{CW} is the inertia moment of the wagon supporting structure relative to the longitudinal axis; c is the stiffness of the spring set coils; P_{FR} is the dry friction force in the spring set; b is the half-width of the wagon supporting structure; P_f is the force that is transmitted to the wagon supporting structure when conducting fire; M_f is the force moment acting on the wagon supporting structure when conducting fire.

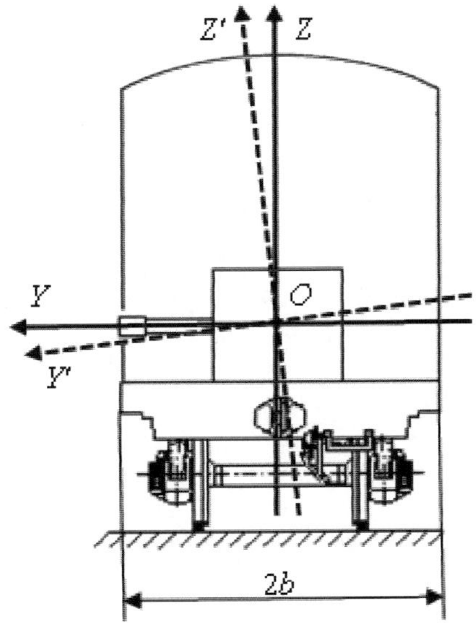

Figure 2. Design scheme of a covered wagon.

The calculation has been carried out by the Runge-Kutta method in the MathCad software [9 – 11]. The calculation results are shown in Figures 3, 4. It is found that the accelerations acting on the wagon supporting structure in the vertical plane are about 6.0 m/s^2 (0.6 g). The accelerations are 2.6 m/s^2 (0.26 g) in the horizontal plane. That is, the wagon motion can be assessed as "satisfactory".

In order to reduce the wagon dynamic loading during the conduct of fire in motion, the case of installing elastic vibration absorbers under the support areas is also considered. In this case, the mathematical model mentioned above is reduced to the form:

$$\begin{cases} M_{CW} \cdot \ddot{q}_1 + c \cdot q_1 = P_{FR}\left(sign\left(\dot{\Delta}_1 - \dot{\Delta}_2\right) + sign\left(\dot{\Delta}_1 + \dot{\Delta}_2\right)\right) + P_f, \\ I_{CW} \cdot \ddot{q}_2 + c \cdot b \cdot q_2 = P_{FR} \cdot b\left(sign\left(\dot{\Delta}_1 - \dot{\Delta}_2\right) + sign\left(\dot{\Delta}_1 + \dot{\Delta}_2\right)\right) + M_f - M_a, \end{cases} \quad (3)$$

where M_a is the moment acting on the support area when conducting fire.

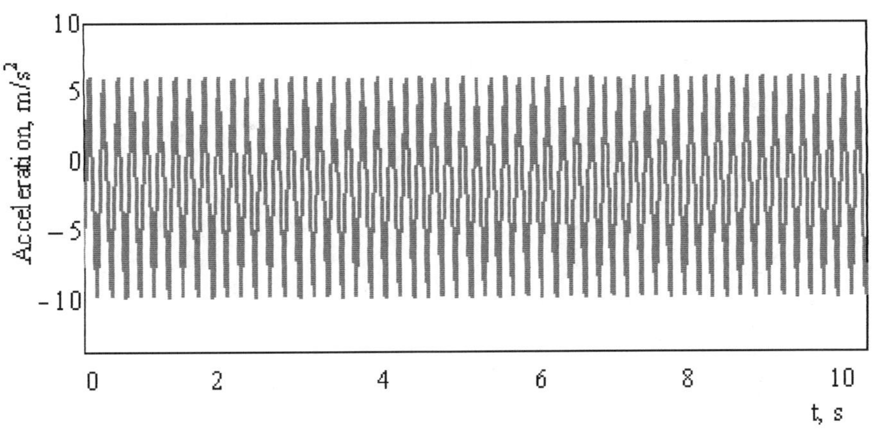

Figure 3. Accelerations acting on the covered wagon supporting structure in the vertical plane.

IOP Conf. Series: Materials Science and Engineering 1021 (2021) 012004 doi:10.1088/1757-899X/1021/1/012004

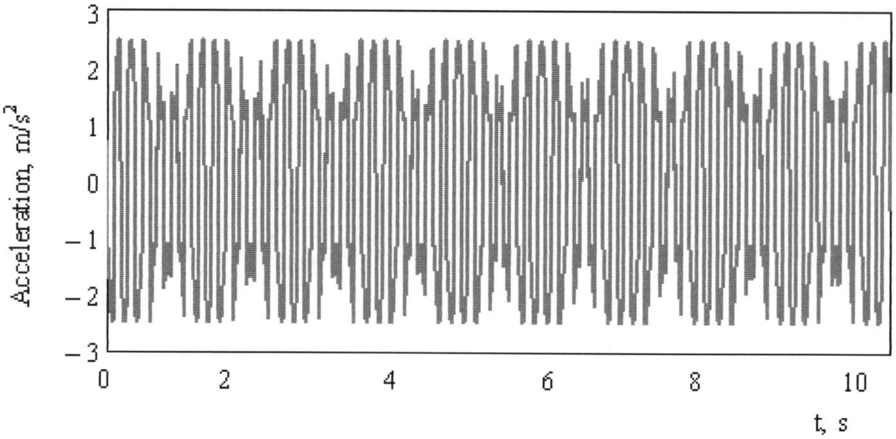

Figure 4. Accelerations acting on the covered wagon supporting structure in the horizontal plane.

The computational scheme of the support area is shown in Figure 5. When carrying out the calculations, it has been taken into account that the impact absorbers work during angular movements of the wagon. That is, when formulating differential equations of motion, the angular stiffness of the impact absorbers has been taken into account.

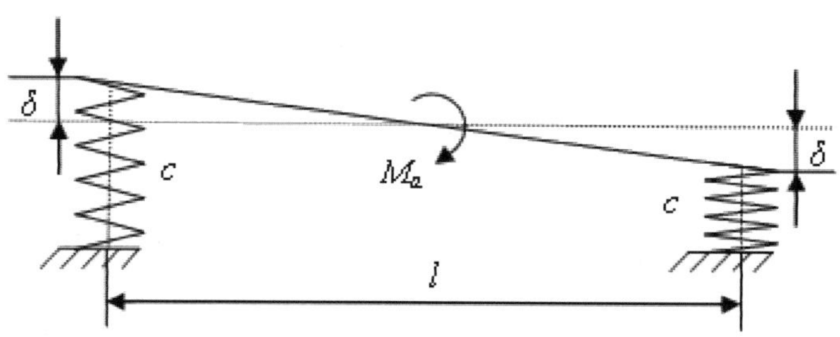

Figure 5. Design scheme of the support area.

The calculation results made it possible to establish that the impact absorbers stiffness should be in the range from 3 kN/m to 10 kN/m. The calculation results for the impact absorber stiffness of 3 kN/m are shown in Figures 6, 7.

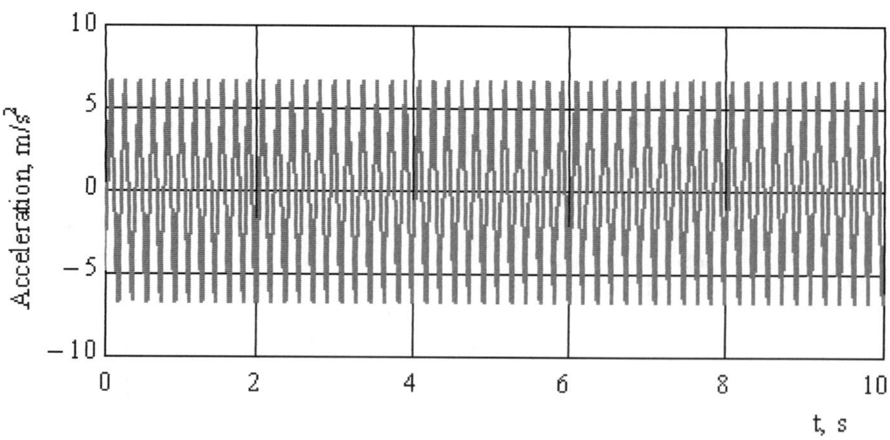

Figure 6. Accelerations acting on the covered wagon supporting structure in the vertical plane.

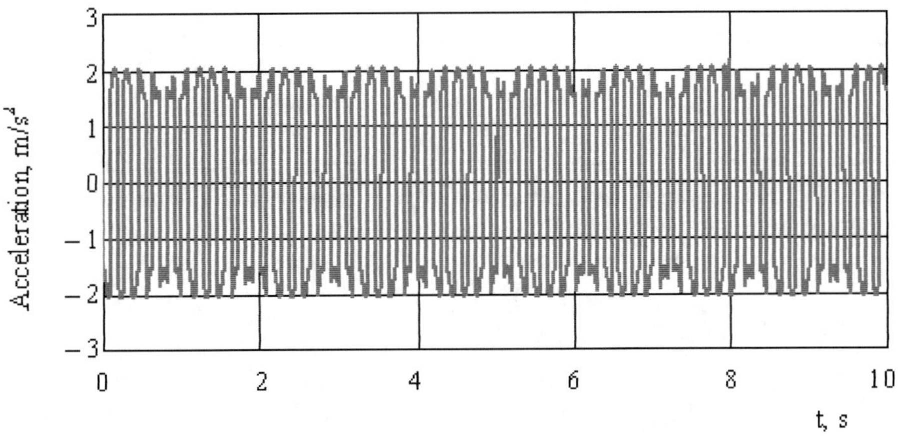

Figure 7. Accelerations acting on the covered wagon supporting structure in the horizontal plane.

In this case, the horizontal accelerations were about 2 m/s^2 (0.2 g). So, with the use of impact absorbers in the supporting sectors, it becomes possible to reduce the wagon dynamic loading in comparison with the case of the rigid interaction of the sector with the frame by almost 25%.

The study of accelerations acting on the wagon supporting structure at different angles of firing from the military equipment was also carried out (Figure 8). The calculation results are shown in Figure 9.

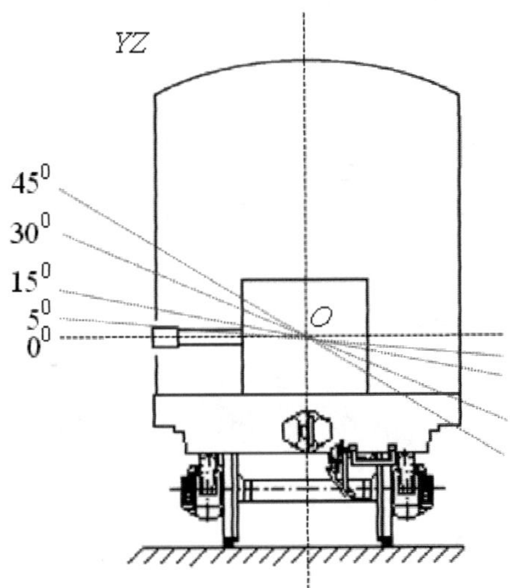

Figure 8. Scheme of conducting fire from the military equipment.

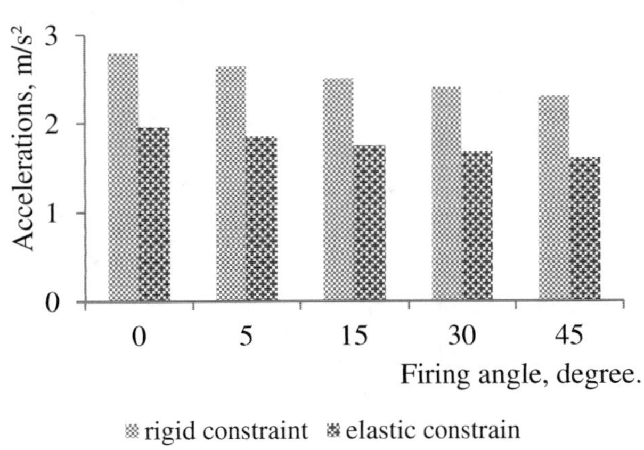

Figure 9. Accelerations acting on the wagon supporting structure in the horizontal plane.

It can be seen from Figure 10 that the magnitude of the horizontal accelerations decreases when the firing angle increases. In this case, the value of vertical accelerations increases slightly.

3. The main strength indicators determination of supporting structure of a covered wagon for the military equipment transportation and for conducting fire

To determine the strength of the covered wagon supporting structure, a finite element analysis has been performed using the CosmosWorks software [12 – 15]. When creating a finite element model, spatial isoparametric tetrahedrons have been used. The optimal number of elements was determined by the graphical analytical method [16 – 19].

The finite element model of the covered wagon supporting structure is shown in Figure 10. The number of nodes in the model is 241723, and the number of elements is 693226. The maximum element size is 80 mm, and the minimum one is 16 mm. The percentage of elements with an aspect ratio of less than three is 18.8 and the percentage of elements with an aspect ratio of more than ten is 33.6. The minimum number of elements in a circle is 22; and the ratio of increasing the elements size is 1.8. The material of the wagon supporting structure is steel 09G2S. The model was fixed in the zones of supporting on the bogies.

Figure 10. Finite-element model of the covered wagon supporting structure.

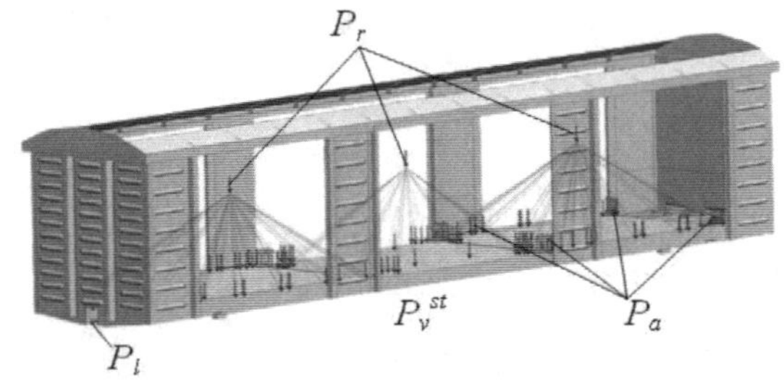

Figure 11. Design scheme of the covered wagon supporting structure.

When drawing up the design diagram of the covered wagon supporting structure, it has been taken into account that a vertical static load P_v^{st} and the load acting on the support areas when firing, which was considered as a remote one P_y, act on it (Figure 11). It is taken into account that the longitudinal load P_l, which is assumed to be 2.5 MN, acts on the front stops of the automatic coupling. The forces acting on the support areas P_a in the places where military equipment is attached to them are also taken into account. This force was decomposed into components, taking into account the angles of the fixing means in space.

The calculation results made it possible to establish that the maximum equivalent stresses are concentrated in the area of the front stops and are about 315 MPa (Figure 12) and do not exceed the permissible values [20 – 22]. The maximum displacements are about 3 mm (Figure 13). The maximum deformations are $5.46 \cdot 10^{-3}$.

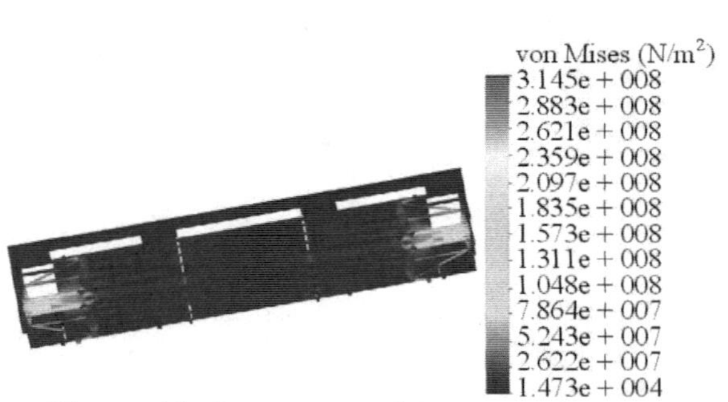

Figure 12. Stress state of the covered wagon supporting structure (bottom view).

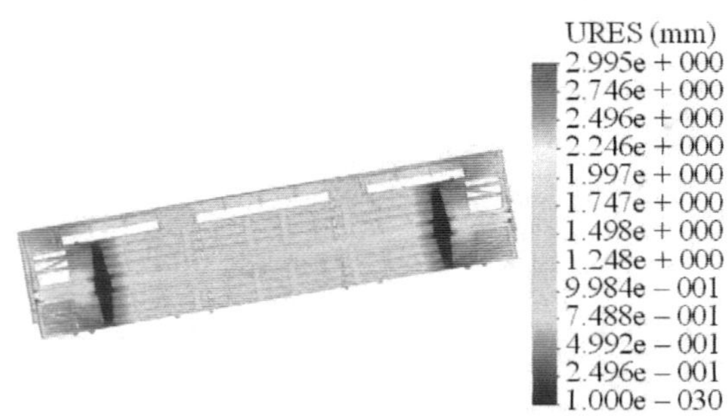

Figure 13. Displacements in the nodes of the covered wagon supporting structure (bottom view).

Also, within the framework of the study, a modal analysis of the supporting structure of a covered wagon has been carried out. The values of natural oscillation frequencies are within the permissible limits, since the first natural frequency has a value of more than 8 Hz [18, 19].

Conclusions

1. A computer model of the supporting structure of a covered wagon for military equipment transportation and for conducting fire has been created. A special feature of the wagon is that its

supporting structure has a sliding roof, which allows conducting fire in a vertical plane. The wagon supporting structure is equipped with a movable double-folding door for loading military equipment, as well as for conducting fire in a horizontal plane. Military equipment is placed on special support areas on the wagon frame.

2. Dynamic loading mathematical simulation of the supporting structure of a covered wagon for transporting military equipment and for conducting fire has been carried out. The studies have been carried out in a plane coordinate system. It has been taken into account that the muzzle energy of one piece of military equipment is about 90 kJ. It was found that the accelerations acting on the wagon supporting structure in the vertical plane are about 6.0 m/s^2 (0.6 g). The accelerations are 2.6 m/s^2 (0.26 g) in the horizontal plane. That is, the wagon motion can be assessed as "satisfactory".

In order to reduce the wagon dynamic loading during conduct of the fire while moving, the case of installing elastic vibration absorbers under the support areas has been also considered. The results of the calculation made it possible to establish that the stiffness of the impact absorbers should be in the range from 3 kN/m to 10 kN/m. When using impact absorbers in the support areas, the wagon dynamic load is reduced by about 25%. in comparison with the case of the rigid interaction of the sector with the frame.

3. Determination of the main strength indicators of the load-bearing structure of a covered wagon for military equipment transportation and for conducting fire has been carried out. The calculation has been carried out using the finite element method in the CosmosWorks software. The calculation results made it possible to establish that the maximum equivalent stresses are concentrated in the area of the front stops and are about 315 MPa. The maximum displacements are about 3 mm. The maximum deformations are $5.46 \cdot 10^{-3}$.

The numerical values of the natural oscillation frequencies of the covered wagon supporting structure during conduct of the fire have been determined. It was found that the values of natural oscillation frequencies are within the permissible limits.

The conducted research will contribute to the creation of recommendations for designing of covered wagons for military equipment transportation and for conducting fire, as well as to increasing the efficiency of the railway transport operation.

Acknowledgments

This study was conducted in the framework of a scientific theme "Development of conceptual frameworks for restoring the efficient operation of obsolete freight cars" (Project registration number: 2020.02/0122).

References

[1] Bityuckij A A, Afanasev, Hilov I A and Guskov V I 2015 Stendovye ispytaniya na soprotivlenie ustalosti maketnyh uzlov krytogo vagona dlya perevozki avtomobilej Transport Rossijskoj Federacii **3(58)** pp 81–85

[2] Chepurnoj A D, Litvinenko A V, Baranov A N, Shejchenko R I and Bondarenko M A 2014 Eksperimentalnye issledovaniya gruzovogo vagona Visnik NTU «HPI» **22 (1065)** pp 44–61

[3] Chepurnoj A D, Shejchenko R I, Graborov R V, Tkachuk N A, Bondarenko N A., Grabovskij A V and Lunev E A 2017 Chislennye issledovaniya gruzovogo vagona Visnik NTU "HPI" **23 (1245)** pp 47–55

[4] Boronenko Yu P and Filippova I O 2015 Ispolzovanie vysokoprochnyh stalej v vagonostroenii Transport Rossijskoj Federacii **3(58)** pp 16–19

[5] Raffaele Sepe and Angela Pozzi 2015 Static and modal numerical analyses for the roof structure of a railway freight refrigerated car *R. Sepe et alii, Frattura ed Integrità Strutturale* **33** pp 451–462. doi: 10.3221/IGF-ESIS.33.50

[6] Hyun-Ah Lee, Seong-Beom Jung, Hwan-Hak Jang, Dae-Hwan Shin, Jang Uk Lee, Kwang Woo Kim and Gyung-Jin Park 2016 Structural-optimization-based design process for the body of a railway vehicle made from extruded aluminum panels *Journal of Rail and rapid transit* **11**

[7] Fomin O, Lovska A, Píštěk V and Kučera P 2020 Research of stability of containers in the

combined trains during transportation by railroad ferry *MM SCIENCE JOURNAL* **1** pp 3728 – 3733
[8] Fomin O, Lovska A 2020 Improvements in passenger car body for higher stability of train ferry *Engineering Science and Technology an International Journal* https://doi.org/10.1016/j.jestch.2020.08.010
[9] Kliuiev S 2018 Experimental study of the method of locomotive wheel-rail angle of attack control using acoustic emission *Eastern-European Journal of Progressive Technologies* **2/9 (82)** pp 69–75 doi: 10.15587/1729-4061.2018.122131
[10] Kondratiev A V, Gaidachuk V E, Kharchenko M E 2019 Relationships between the ultimate strengths of polymer composites in static bending, compression, and tension *Mechanics of Composite Materials* **52 2** pp 259–266
[11] Pistek V, Klimes L, Mauder T and Kucera P 2017 Optimal design of structure in rheological models: An automotive application to dampers with high viscosity silicone fluids Journal of Vibroengineering 19 6 pp 4459–4470
[12] Alyamovskiy A A 2007 SolidWorks/COSMOSWorks 2006 – 2007 Inzhenernyiy analiz metodom konechnyih elementov (Cham: Moscow DMK) p 784
[13] Alyamovskiy A A 2010 COSMOSWorks Osnovyi rascheta konstruktsiy na prochnost v srede SolidWorks (Cham: Moscow DMK Press) p 785
[14] Dižo J Steišunas S Blatnický M 2016 Simulation analysis of the effects of a rail vehicle running with wheel flat *Manufacturing Technology* **16 5** pp 889-896
[15] Dižo J Harušinec J Blatnický M 2015 Multibody system of a rail vehicle bogie with a flexible body *Manufacturing Technology* **15 5** pp 781-788
[16] Fomin O, Lovska A, Píštěk V and Kučera P 2019 Dynamic load computational modelling of containers placed on a flat wagon at railroad ferry transportation VIBROENGINEERING PROCEDIA **29** pp 118–123
[17] Vatulia G L, Lobiak O V, Deryzemlia S V , Verevicheva M A and Orel Ye F 2019 Rationalization of cross-sections of the composite reinforced concrete span structure of bridges with a monolithic reinforced concrete roadway slab *IOP Conference Series: Materials Science and Engineering* **664** 012014 doi:10.1088/1757-899X/664/1/012014
[18] Fomin O, Lovska A, Radkevych V, Horban A, Skliarenko I and Gurenkova O 2019 The dynamic loading analysis of containers placed on a flat wagon during shunting collisions *ARPN Journal of Engineering and Applied Sciences* **14 21** pp 3747–3752
[19] Vatulia G Komagorova S Pavliuchenkov M 2018 Optimization of the truss beam. Verification of the calculation results *MATEC Web of Conferences* **230** 02037 doi: 10.1051/matecconf/201823002037
[20] DSTU 7598:2014 2014 Vagoni vantazhni. Zagalni vimogi do rozrahunkiv ta proektuvannya novih i modernizovanih vagoniv koliyi 1520 mm (nesamohidnih) p 162
[21] GOST 33211-2014 2014 Vagonyi gruzovyie. Trebovaniya k prochnosti i dinamicheskim kachestvam p 54
[22] DIN EN 12663-2-2010 2010 Railway applications – Structural requirements of railway vehicle bodies – Part 2: Freight wagons p 54

IOP Conf. Series: Materials Science and Engineering 1021 (2021) 012005 doi:10.1088/1757-899X/1021/1/012005

Determination of residual resource of flat wagons load-bearing structures with a 25-year service life

O Fomin[1], G Vatulia[2], M Horbunov[3], A Lovska[4,6], V. Píštěk[5], P. Kučera[5]

[1]Department of Cars and Carriage Facilities, State University of Infrastructure and Technologies, Kyrylivska str., 9, Kyiv, Ukraine, 04071
[2]Department of Structural Mechanics and Hydraulics, Ukrainian State University of Railway Transport, Feuerbach sq., 7, Kharkiv, Ukraine, 61050
[3]Department of railway and road transport, lift and care systems, Volodymyr Dahl East Ukrainian National University, Central pr., 59a, Severodonetsk, Ukraine, 93400
[4]Department of wagons, Ukrainian State University of Railway Transport, Feuerbach sq., 7, Kharkiv, Ukraine, 61050
[5]Faculty of Mechanical Engineering, Institute of Automotive Engineering, Brno University of Technology, Technická, 2896/2, Brno, Czech Republic, 61669

[6]alyonalovskaya.vagons@gmail.com

Abstract. The article determines the residual resource of flat wagons load-bearing structures with a 25-year service life. For this purpose, field studies of the technical condition of the load-bearing structures of flat wagons based on the wagon depot of the Foundation of the Southern Railway branch of JSC "Ukrzaliznytsia" were carried out. The flat wagon of model 13-401 was chosen as the subject. It is established that the design service life of the load-bearing structures of the studied flat wagons, taking into account the extension of their operation is not less than 18 years. The strength of the load-bearing structure of the flat wagon in the environment of SolidWorks Simulation (CosmosWorks) software by the finite element method is calculated. It is established that the maximum equivalent stresses occur in the zone of interaction of the spine beam with the pivot be equal to 337.5 MPa. Therefore, the strength of the load-bearing structure of the flat wagons is provided. The estimation of the course of the flat wagons is determined. The research was carried out in the vertical plane during the movement of the butt rail track. The calculation was performed in the MathCad software environment by the Runge-Kutta method. The assessment of the wagon is "excellent". The conducted researches will promote increase of efficiency of combined transportations in the international communication and strengthening of foreign economic positions of Ukraine in the market of transport services.

1. Introduction

Ukraine has long been a crossroads of strategically important international transport corridors. Today, the link provides transport logistics between Europe and Asia.

In order to increase the volume of traffic through the territory of Ukraine, combined transport systems have been developed. One of the most promising among such systems are container and piggyback.

It is known that one of the most popular types of wagons in the international combined service are flat wagons. To increase the efficiency of the use of flat wagons in international traffic, their modernization is carried out, which consists in the installation of stationary or folding fitting stops on the load-bearing structure. This allows adapting the load-bearing structures of universal flat wagons to combined transport, in particular container.

At the same time, the level of replenishment of the wagon fleet of Ukraine in recent years is quite insignificant. In this regard, it is possible to determine the residual life of the operated flat wagons, consider the possibility of extending their service life in order to be able to be involved in international traffic. This will ensure the efficiency of combined transportation in international traffic and strengthen Ukraine's foreign economic position in the transport services market.

The substantiation of the extension of the useful life of wagons for the transportation of pellets is covered in [1]. To assess the possibility of further operation of wagons, an experimental determination of the level of load and stress state of the load-bearing structures of wagon bodies based on static, impact and strength tests.

Prediction of the residual life of the hopper dispenser wagon after long operation, taking into account the actual physical and mechanical characteristics of the material of the load-bearing structures is carried out in [2]. A procedure for forecasting the residual life of the load-bearing structure of the wagon has been developed. The results of virtual and experimental studies of the strength of the load-bearing structure of the wagon are presented. However, these studies were performed on hopper wagons. The authors did not study the issue of extending the service life of flat wagons.

The need to adjust the Regulations on extending the service life of freight wagons plying in international traffic is considered in [3]. The expediency of introducing requirements for rolling stock, the service life of which is extended, is substantiated.

Substantiation of options for extending the service life of specialized flat wagons is covered in [4]. Methods of comparative assessment of options for extending the service life of flat wagons, which take into account the economic component of these types of work. Research has been conducted on universal flat wagons, as well as double-decker flat wagons for wagon transportation. Studies on the extension of the service life of flat wagons for the transportation of containers have not been considered in the work.

The study of the possibility of extending the service life of the bodies of universal gondola wagons, which have exhausted their regulatory life, is carried out in [5]. The specified dynamic loadings acting on a bearing design of a gondola wagon taking into account real wear of its elements are defined. However, the authors did not pay attention to the study of the possibility of extending the service life of flat wagons.

Substantiation of measures to extend the service life of freight wagons was carried out in [6]. The scheme of technical diagnostics of freight wagons is developed. It is concluded that it is possible to extend the service life of freight wagons only through the implementation of high-quality technical diagnostics. However, there is no applied use of the proposed measures.

The purpose of the article is to determine the residual life of load-bearing structures of flat wagons with a 25-year service life to justify further use in the international combined service. To achieve this goal, the following tasks are set:

- investigate the technical condition of the main load-bearing elements of the frames of flat wagons with a 25-year service life;
- to calculate the design service life of the load-bearing structure of the flat wagon;
- calculate the strength of the load-bearing structure of the flat wagon;
- assess the progress of the flat wagon.

2. Presentation of the main material of the article

In order to verify the possibility of extending the service life of the load-bearing structure of the flat wagon, the method described in [7] was used. This technique allows determining the residual life of the elements of the load-bearing structures of wagon bodies, taking into account the probability of their failure in operation and the maximum values of equivalent stresses. As input parameters to this technique, statistical data of geometric dimensions of the main load-bearing elements of the frames of flat wagons (spine beam and main longitudinal) with a service life of about 25 years were used. The flat wagon of model 13-401 of the Dneprodzerzhinsk wagon-building plant was chosen as the researched one (Figure 1).

As is known, the backbone and the main longitudinal beams of the flat wagon 13-401 consist of two I-beams №60, which has a variable length. In cantilever parts of a frame height of an I-beam makes 360 mm.

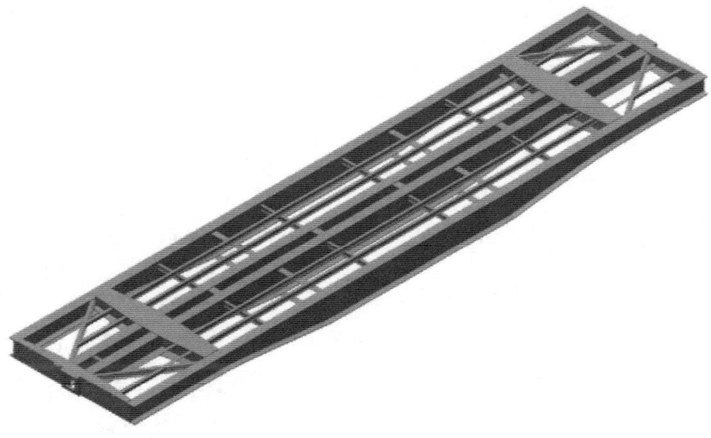

Figure 1. The load-bearing structure of the flat wagon model 13-401.

To justify the extension of the service life of the flat wagon, field studies of the technical condition of the main load-bearing elements of the frame were carried out based on the wagon depot. The value measured during field studies is the thickness of the I-beam rack (Figure 2).

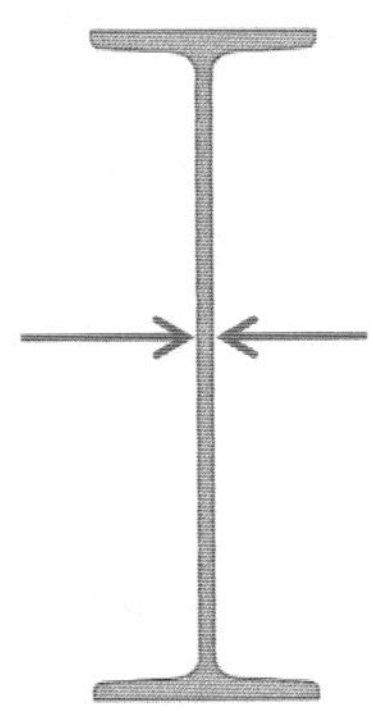

Figure 2. The area of measurement of the thickness of the rack of the I-beam.

The number of measurements is 20. The measurement results are given in tables 1, 2.

Table 1. The results of measurements of wear of the I-beam of the spine beam of the frame of the flat wagon 13-401.

Sequence number of the wagon	1	2	3	4	5	6	7	8	9	10
Measurement value, mm	11.74	11.74	11.74	11.74	11.72	11.72	11.72	11.7	11.7	11.72
Sequence number of the wagon	11	12	13	14	15	16	17	18	19	20
Measurement value, mm	11.72	11.72	11.76	11.74	11.74	11.72	11.7	11.7	11.72	11.72

Table 2. The results of measurements of wear of the I-beam of the longitudinal beam of the frame of the flat wagon 13-401.

Sequence number of the wagon	1	2	3	4	5	6	7	8	9	10
Measurement value, mm	11.7	11.69	11.7	11.71	11.71	11.7	11.72	11.7	11.7	11.72
Sequence number of the wagon	11	12	13	14	15	16	17	18	19	20
Measurement value, mm	11.72	11.72	11.72	11.7	11.71	11.7	11.7	11.7	11.72	11.72

An important step in field research is to determine the number of statistics needed to obtain an adequate result. The required number of static data was determined by the formula [8]

$$n = \frac{t^2 \cdot \sigma^2}{\delta^2}, \tag{1}$$

where t is determined from the ratio F(t) = γ/2; F(t) is Laplace function, tabular value; σ is standard deviation of the random variable under study, which must be known a priori, even before the experimental measurements; δ^2 is absolute error of measurement result.

In technical measurements, the value of measurement reliability is usually taken equal to $p = 0.95$. When the number of measurements is n = 20, we take the Student's distribution parameter t_γ equal to 2.1

Based on the calculations it is established that for the samples given in tables 1 and 2 the number of measurements is sufficient to obtain an adequate assessment of the result.

The used technique in comparison with an expert assessment essentially increases reliability of the chosen decision on appointment of new service life of the wagon. In addition, this technique allows you to determine the residual service life with the same reliability for any number of wagons.

In accordance with the method, the probability of failure P_i elements of the load-bearing structure of the flat wagon in operation is determined by the formula

$$P_i = \frac{\sum_{1}^{k} \frac{R_H}{R}}{k}, \tag{2}$$

where k is the number of studied wagons; R_H is the number of defective elements of one type in the wagon; R is the total number of elements of one type.

Estimation of design service life was determined by the formula [7]

$$T_k = \frac{\left(\frac{\sigma_{a_iN}}{[n]}\right)^m \cdot N_0}{N_{cI} \cdot \sum_j \left(\sigma_{a_j}^I\right)^m \cdot P_j^I + N_{cII} \cdot \sum_k \left(\sigma_{a_k}^{II}\right)^m \cdot P_k^{II}}, \tag{3}$$

where σ_{aiN} is the average value of the endurance limit; $[n]$ is allowable margin of safety; N_0 is test base; N_{cI}, N_{cII} are the number of eponymous faulty structural elements of the wagon under study; m is an indicator of the degree of fatigue curve; σ_{aj}^{I}, σ_{ak}^{II} are the ultimate strength of the material under study; P_j^I, P_k^{II} are the probability of failure of the structural element.

The initial data have the following values:

- the average value of the endurance limit σ_{aiN} =245 MPa;
- allowable safety factor [n]=2;
- test base N_0=10^7;
- the number of faulty structural elements of the wagon of the same name under study is equal to 1 (spine beam) and 2 (longitudinal beam);
- indicator of the degree of fatigue curve m = 2;
- the strength limit of the load-bearing structure is equal to 490 MPa for steel 09G2D.

The obtained results allowed to conclude that the design service life of the ridge and longitudinal beams of the studied flat wagons, taking into account the extension of their operation is not less than 18 years.

At the next stage, the main indicators of the strength of the load-bearing structure of the flat wagon with the highest amount of wear recorded during field research (Tables 1, 2) were determined. It is established that the mass of the load-bearing structure of the flat wagon, taking into account the wear of its elements, is 3.5% less than the load-bearing structure of the flat wagon, which has nominal dimensions. The strength calculation was performed by the finite element method, which is implemented in the SolidWorks Simulation (CosmosWorks) software environment [9 – 13]. The finite element model of the load-bearing structure of the flat wagon is shown in Figure 3. When determining the number of grid elements, the graph analytical method was used [14 – 17]. Ten-node isoperimetric tetrahedral were used as finite elements. The number of grid elements was 838084, nodes – 274584. The maximum size of the element was 100 mm, the minimum – 20 mm, the maximum aspect ratio – 4702.9, the percentage of elements with a side ratio less than 3 – 39.3, more than 10 – 10.2 . The model was fixed in the areas of support of the load-bearing structure on the bogies.

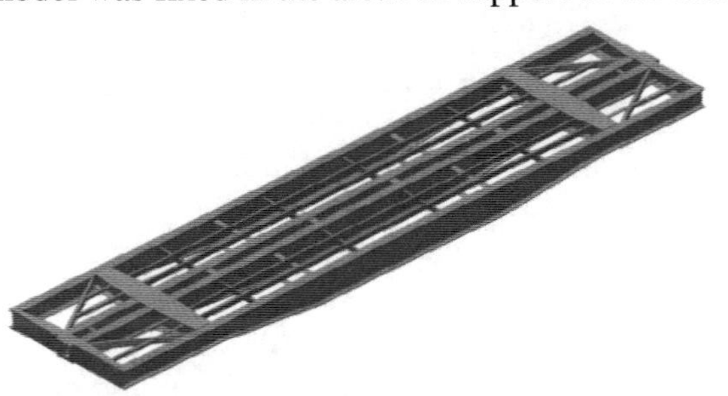

Figure 3. Finite-element model of the load-bearing structure of the flat wagon.

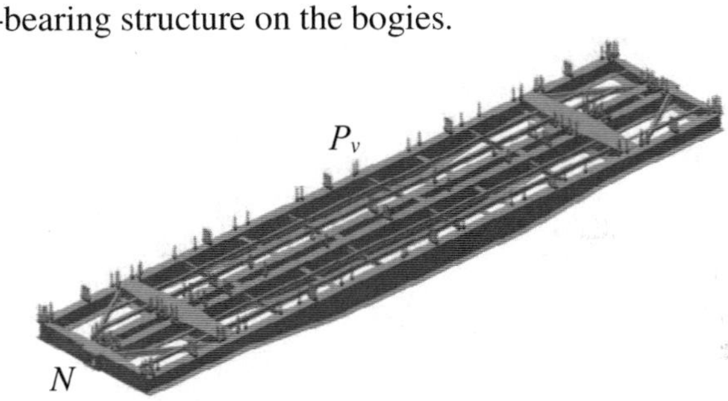

Figure 4. The design scheme of the load-bearing structure of the flat wagon in the I design mode.

When compiling the design scheme of the load-bearing structure of the flat wagon (Figure 4), the following loads were taken into account: vertical P_v and longitudinal N. It is taken into account that the full load capacity of the flat wagon loaded with conditional cargo is used. The calculation was performed for I and III calculation modes.

The results of the calculation of the strength of the load-bearing structure of the flat wagon in the I design mode "impact" are shown in Figures 5, 6.

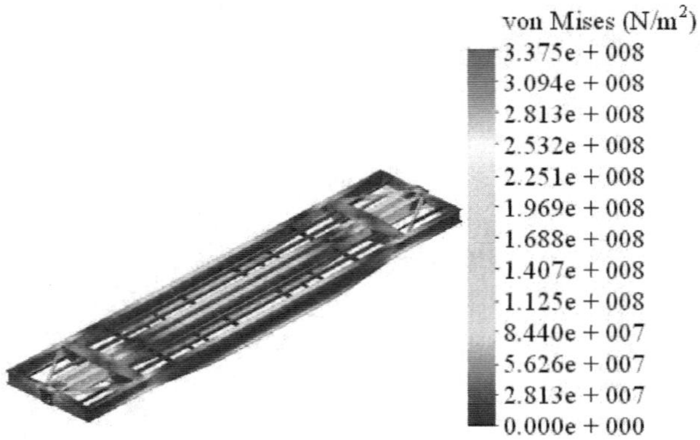

Figure 5. The stress state of the load-bearing structure of the flat wagon.

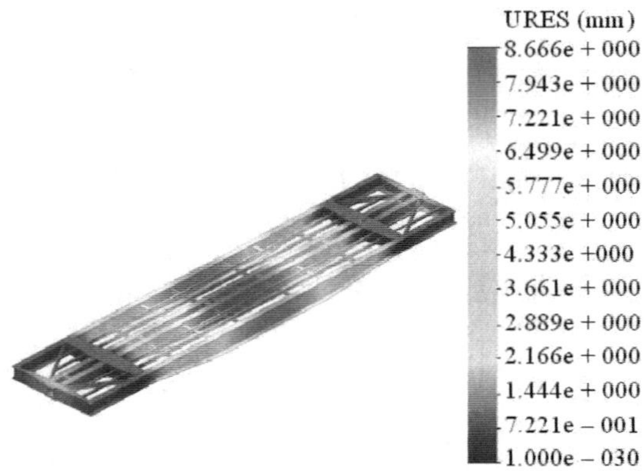

Figure 6. Movement in the nodes of the load-bearing structure of the flat wagon.

The results of the calculation of the strength of the load-bearing structure of the flat wagon at I and III calculation modes are given in Table 3.

Table 3. Indicators of strength of the load-bearing structure of the flat wagon under operational load conditions.

Strength indicators	Load mode				
	I			III	
	shock	compression	jerk-stretching	blow-compression	jerk-stretching
Stress, MPa	337.5	313.6	296.8	287.98	280.8
Movement in knots, mm	8.6	8.4	8.3	8.1	7.8

Based on the calculations, we can conclude that the maximum equivalent stresses occur in the area of interaction of the spine beam with the pivot is not equal to 337.5 MPa. The maximum displacements in the nodes of the load-bearing structure of the flat wagon are recorded in the middle part of the frame – 8.6 mm. Therefore, the strength of the load-bearing structure of the flat wagon is provided [18 – 20]. In order to use the flat wagon for container transportation, it is possible to install fitting stops on its load-bearing structure.

To determine the assessment of the flat wagon, mathematical modeling of the dynamic load of its load-bearing structure was performed. For this purpose, the mathematical model given in [21] was used. The model takes into account the movement of the wagon in the vertical plane when moving the butt rail track. The calculation is performed for the case of empty wagon movement in the MathCad software environment. The Runge-Kutta method was used as a calculation. Based on the calculations, the accelerations that act on the load-bearing structure of the wagon in the center of mass (Figure 7), in the areas of support on the running gear (Figure 8).

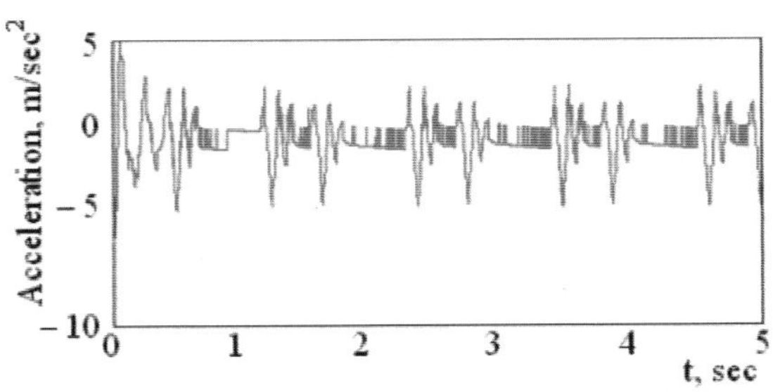

Figure 7. Acceleration of the load-bearing structure of the flat wagon in the center of mass.

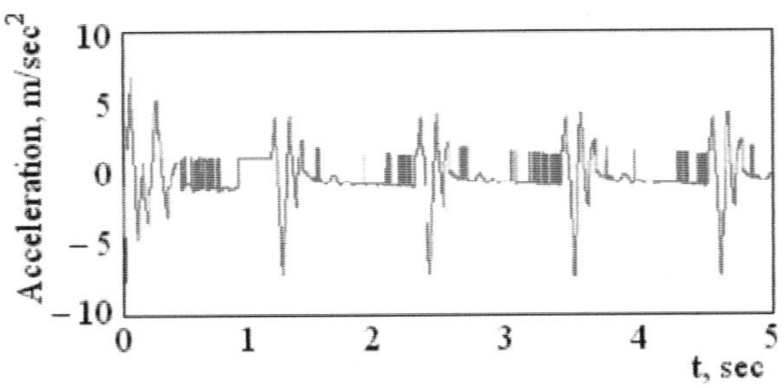

Figure 8. Acceleration of the load-bearing structure of the flat wagon in the areas of support on the bogies.

From Figure 7 shows that the acceleration of the load-bearing structure of the flat wagon in the center of mass is 5 m/s^2 (0.5g). The acceleration of the load-bearing structure of the flat wagon in the areas of support on the carts is about 7.5 m/s^2 (0.75g), (Figure 8). The acceleration of the carts is about 10 m/s^2 (1.0 g). The obtained results allow us to conclude that the assessment of the wagon is "excellent".

Conclusions

1. The technical condition of the main bearing elements of frames of flat wagons with 25-year service life is investigated. At the same time, field research was conducted based on the wagon depot of the Basis of the Southern Railway branch of JSC Ukrzaliznytsia. The thickness of the leg of the dorsal ridge and the main longitudinal beams of the frame is taken into account as the studied value.

2. The design service life of the load-bearing structure of the flat wagon is calculated. The obtained results allowed to conclude that the design service life of the ridge and longitudinal beams of the studied flat wagons, taking into account the continuation of their operation is not less than 18 years.

3. The calculation of the strength of the load-bearing structure of the flat wagon. The finite element method, which is implemented in the CosmosWorks software environment, is used as a calculation. It is established that the maximum equivalent stresses occur in the zone of interaction of the spine beam with the pivot be equal to 337.5 MPa. The maximum displacements in the nodes of the load-bearing structure of the flat wagon are recorded in the middle part of the frame – 8.6 mm. That is, the strength of the load-bearing structure of the flat wagon is provided.

4. An assessment of the progress of the flat wagon. The research was carried out in the vertical plane during the movement of the butt rail track. The acceleration of the load-bearing structure of the flat wagon in the center of mass was 5 m/s^2 (0.5g). The acceleration of the load-bearing structure of the flat wagon in the areas of support on the carts is about 7.5 m/s^2 (0.75g). The acceleration of the carts is about 10 m/s^2 (1.0 g). Therefore, the assessment of the wagon is "excellent".

The conducted research will help to increase the efficiency of the use of flat wagon, and hence the transportation process in international traffic.

Acknowledgments: The authors gratefully acknowledge funding from the Specific research on "Innovative principles for creating resource-saving structures of railroad cars based on the refined dynamic loads and functionally adaptive flash-concepts", which is funded from the State budget of Ukraine in 2020 and "Development of conceptual frameworks for restoring the efficient operation of obsolete freight cars" (Project registration number: 2020.02/0122).

References

[1] Anofriev V G, Rejdemejster A G, Kalashnik V A and Kuleshov V P 2016 K voprosu prodleniya poleznogo sroka sluzhby vagonov dlya perevozki okatyshej *Nauka ta progres transportu. Visnik Dnipropetrovskogo nacionalnogo universitetu zaliznichnogo transportu* **3 (63)** pp 148–160 doi 10.15802/stp2016/74749

[2] Putyato A V, Konovalov E N and Afanaskov P M 2016 Prognozirovanie ostatochnogo resursa vagona hopper-dozatora posle dlitelnoj ekspluatacii s uchetom fakticheskih fiziko-mehanicheskih harakteristik materiala nesushej konstrukcii *Mehanika mashin, mehanizmov i materialov* **1(34)** pp 26–35

[3] Boronenko Yu P, Tretyakov A V, Zharova E A 2012 O korrektirovke Polozheniya o prodlenii sroka sluzhby gruzovyh vagonov, kursiruyushih v mezhdunarodnom soobshenii *Evraziya vesti Moskva* **10** pp 13–14

[4] Zharova E A 2008 Obosnovanie variantov prodleniya srokov sluzhby specializirovannyh vagonov-platform (Cham: Sankt Peterburg PGUPS) p 16

[5] Okorokov A M, Fomin O V, Lovska A O, Vernigora R V, Zhuravel I L and Fomin V V 2018 Research into a possibility to prolong the time of operation of universal semi-wagon bodies that have exhausted their standard resource *Eastern-European journal of enterprise technologies* **3/7(93)** pp 20–26 doi: 10.15587/1729-4061.2018.131309

[6] Sapronova S Yu, Bulich D I and Tkachenko V P 2017 Prodovzhennya terminu ekspluataciyi vantazhnih vagoniv *Visnik Shidnoukrayinskogo Nacionalnogo universitetu imeni Volodimira Dalya* **3 (233)** pp 179–182

[7] Afanasev A V 2008 Sovershenstvovanie metodiki ocenki tehnicheskogo sostoyaniya i raschetnogo obosnovaniya prodleniya sroka sluzhby poluvagonov *Transport Urala* **1** pp 49–52

[8] Kosmin V V 2007 Osnovy nauchnyh issledovanij (Moscow: GOU "Uchebno-metodicheskij centr po obrazovaniyu na zheleznodorozhnom transporte") p 271

[9] Fomin O, Lovska A 2020 Improvements in passenger car body for higher stability of train ferry Engineering Science and Technology an International Journal https://doi.org/10.1016/j.jestch.2020.08.010

[10] Fomin O, Lovska A, Píštěk V and Kučera P 2020 Research of stability of containers in the combined trains during transportation by railroad ferry *MM SCIENCE JOURNAL* **1** pp 3728 – 3733

[11] Fomin O, Lovska A, Píštěk V and Kučera P 2019 Dynamic load computational modelling of containers placed on a flat wagon at railroad ferry transportation *VIBROENGINEERING PROCEDIA* **29** pp 118–123

[12] Dižo J Steišunas S Blatnický M 2016 Simulation analysis of the effects of a rail vehicle running with wheel flat *Manufacturing Technology* **16 5** pp. 889-896.

[13] Dižo J Harušinec J Blatnický M 2015 Multibody system of a rail vehicle bogie with a flexible body *Manufacturing Technology* **15 5** pp. 781-788.

[14] Fomin O, Lovska A, Radkevych V, Horban A, Skliarenko I and Gurenkova O 2019 The dynamic loading analysis of containers placed on a flat wagon during shunting collisions *ARPN Journal of Engineering and Applied Sciences* **14 21** pp 3747–3752

[15] Vatulia G L, Lobiak O V, Deryzemlia S V, Verevicheva M A and Orel Ye F 2019 Rationalization of cross-sections of the composite reinforced concrete span structure of bridges with a monolithic reinforced concrete roadway slab *IOP Conference Series: Materials Science and Engineering* **664** 012014 doi:10.1088/1757-899X/664/1/012014

[16] Vatulia G Komagorova S Pavliuchenkov M 2018 Optimization of the truss beam. Verification of the calculation results *MATEC Web of Conferences* **230** 02037 doi: 10.1051/matecconf/201823002037

[17] Kondratiev A V, Gaidachuk V E, Kharchenko M E 2019 Relationships between the ultimate strengths of polymer composites in static bending, compression, and tension *Mechanics of Composite Materials* **52 2** pp 259–266

[18] DSTU 7598:2014 2014 Vagoni vantazhni. Zagalni vimogi do rozrahunkiv ta proektuvannya

novih i modernizovanih vagoniv koliyi 1520 mm (nesamohidnih) p 162
[19] GOST 33211-2014 2014 Vagonyi gruzovyie. Trebovaniya k prochnosti i dinamicheskim kachestvam p 54
[20] DIN EN 12663-2-2010 2010 Railway applications – Structural requirements of railway vehicle bodies – Part 2: Freight wagons p 54
[21] Domin Yu V and Chernyak G Yu 2003 Osnovi dinamiki vagoniv (Cham: Kiyiv KUETT) p 269

Improvement of the grain cargo handling technology on the basis of resource-saving

Anton Kovalov[1], Svitlana Prodashchuk[1], Anna Kravets[1,3], Dmytro Mkrtychian[1] and Mikola Prodashchuk[2]

[1] Department of Freight and Commercial Railway Operation Management, Ukrainian State University of Railway Transport, 7 Feuerbach Sq., Kharkiv, Ukraine
[2] Theoretical and Applied Computer Science, V.N. Karazin Kharkiv National University, 4 Svobody Sq. Kharkiv, Ukraine

[3] E-mail: docent.kravets.uvkr@kart.edu.ua

Abstract. The article considers the state of grain cargo transportation and related technical and technological problems. As a solution, the possibility of using multiple grain shipping containers and the introduction of a new modern grain cargo handling technology at the station based on inventory management and proper selection of vehicles was considered. The considered measures will contribute to simplifying, improving and reducing the number of cargo operations performed during grain handling, making their implementation more practical and energy-efficient.

1. Introduction
Grain cargo transportation takes a significant volume in the range of cargoes carried by transport and is one of the strategic products offered by the Ukrainian economy today. Grain has a significant impact on the development of the agricultural sector and is the main export product of the country. Grain cargo is mainly exported through seaports, while grain is delivered to ports mainly by rail (61%), road (36%) and river transport (3%). Therefore, the purpose of the article is to study the methods of loading grain cargo into a vehicle and organize optimal conditions to obtain the most effective quality indicators.

2. General part
The volume of grain cargo transportation is constantly growing [1]. The paper analyzes statistical data on the volume of grain and milling products transportation from 2009 to 2020 (Figure 1) [2]. The analysis shows that the volume of transportation has still been growing until now. Therefore, the transportation technology always has some shortcomings and, accordingly, the task arises to develop new resource-saving technologies.

The main problems in the grain transportation are: seasonal shortage of grain cars [3] and imperfect technology of grain loading complexes when such goods are transported in containers. The issue of seasonality of this cargo is considered in the model of cargo rolling stock management, which takes into account the parameter of uneven transportation [1].

The use of containers for loading grain has very good prospects, especially in the organization of export transportation, as labor-intensive and time-consuming cargo transshipment in ports are thus avoided [4]. Also, fewer grain cars which are not always available are required and more platforms are used to transport containers, that are much cheaper than grain trucks and more versatile at any season. In addition, the likelihood of using containers increases that usually return empty after unloading in areas where grain cargo may be loaded.

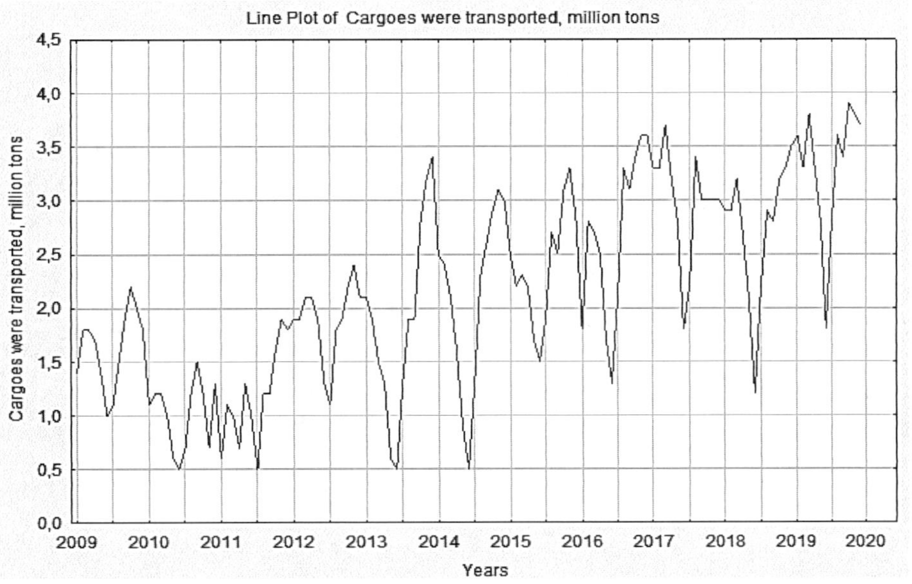

Figure 1. Transportation volumes of grain and milling products from 2009 to 2020.

One of the most important components in the entire cargo delivery chain is the organization of container loading at the points of departure. The loading technology for grain transportation containers can have various forms. Containers can be loaded directly on the site of the cargo owner and delivered to the railway station by vehicles designed specifically for container transportation, or grain cargo can be delivered by the owner directly to the grain warehouse of the station, where containers will be loaded and then placed on railway platforms. In [5, 6], the optimal placement of infrastructure facilities for primary grain handling is considered, while the feasibility of using various types of transport and forming an effective chain of grain transportation is discussed in [7, 8].

Therefore, the task is to develop a resource-saving technology for loading grain cargo, which will take into account as many factors as possible that influence the total cost of transportation, such as the number of loading and unloading machines (LUM), the number of deliveries to the warehouse complex and its operating time a day [9, 10].

To formalize the extreme task of improving the grain handling technology, a model of the functioning of the cargo front line (grain warehouse) has been developed.

To build a model, operating cost during performing work was selected as the optimization criterion, as it is one of the most important resource-saving components (table 1). The controlled parameter of the system is the handling capacity of the cargo front line Q_p, determined by the number of LUM M, deliveries of cars to the cargo front line Y, and the operating time during the day T. The problem of choosing a rational version of the operating technology for grain cargo handling is formulated as follows: for the specified characteristics of the input flow of requirements and the type of service system, parameters should be determined that correspond to the minimum value of the optimization criterion R_i.

Additional complication is the fact that there may be more than one option of the warehouse operation, depending on the capacity of the handled containers per day, the handling capacity of LUM per day, and the warehouse capacity. Therefore, the following conditions are taken into account when the optimal operating technology is determined:

- the capacity of containers processed per day is less than the handling capacity of LUM per day, the warehouse capacity is sufficient for storing cargo, i.e. $\sum_{i=1}^{f} N_{ki} Qн_i^{конт} \leq \sum_{i=1}^{f} M_i P_{mi} T_i, m_i P_{cmi} \leq Q_{sk}^{o}$, cars are delivered to the warehouse by routes.

Table 1. Operating costs of grain handling at points of departure

Total cost	Indicator	Cost incurred during grain cargo handling	Indicator
Cost related to performing technological operations on the i-th cargo front line, UAH	$R_i^{техн}$	Cost related to the delivery, arrangement, collection and pick-up of cars from the i-th freight front line, UAH	$W_i^{ман}$
		Cost related to the downtime of cars under cargo operations on the I-th cargo front line, UAH	$W_i^{н(в)}$
Cost related to cars waiting for service on the I-th freight front line and accumulation, UAH	$R_{очі}^{ваг}$	Cost related to idle time of cars, because the i-th freight front line is in operation, UAH	$W_i^{вф}$
		Cost related to cars waiting for performing cargo operations when they arrive at the station during periods when the cargo front line is not in operation; UAH	$W_i^{во}$
		Cost related to cars waiting for maintenance, UAH	$W_i^{обсл}$
Maintenance cost of technical equipment of the I-th cargo front line, UAH	$R_{утi}^{тех}$	Depreciation cost of LUM on the I-th cargo front line, taking into account the coefficient of efficiency of capital investments, UAH	W_i^{HPM}
		Labour cost of employees who operate LUM on the I-th cargo front line, UAH	$W_i^{пр}$
		Maintenance cost of the I-th cargo front line and the labor cost of employees who operate it, UAH	$W_i^{вф}$
Cost associated with idle time of technical equipment on the I-th cargo front line, UAH	$R_{очі}^{тех}$	Cost associated with idle time of motor vehicles when loading or unloading cars on the I-th freight front line, UAH	$W_{ai}^{н(в)}$
		Cost associated with idle time of vehicles because LUM is in operation, UAH	W_{ai}^{HPM}
		Cost associated with LUM waiting for performing cargo operations, due to the random coming of cars to the I-th cargo front line, UAH	$W_{HPMi}^{н(в)}$
Cost associated with the delay in processing transportation documents for cargo arriving or departing from the i-th cargo front line, and cargo en route, UAH	$R_i^{затр}$	Cost associated with the delay in processing transportation documents for cargo arriving or departing from the I-th cargo front line, UAH	$W_i^{док}$
		Cost associated with the delay of cargo, UAH	$W_i^{вант}$

Combined probability of a simultaneous event

$$p_1 = p(\sum_{i=1}^{f} N_{ki} Q\textit{н}_i^{\textit{конт}} \le \sum_{i=1}^{f} M_i P_{mi} T_i, m_i P_{cmi} \le Q_{sk}^{o}) =$$
$$= p(\sum_{i=1}^{f} N_{ki} Q\textit{н}_i^{\textit{конт}} \le \sum_{i=1}^{f} M_i P_{mi} T_i) \cdot p(m_i P_{cmi} \le Q_{sk}^{o} / \sum_{i=1}^{f} N_{ki} Q\textit{н}_i^{\textit{конт}} \le \sum_{i=1}^{f} M_i P_{mi} T_i), \quad (1)$$

where $p(\sum_{i=1}^{f} N_{ki} Q\textit{н}_i^{\textit{конт}} \le \sum_{i=1}^{f} M_i P_{mi} T_i)$ is the probability that the capacity of containers handled per day is less than the handling capacity of LUM per day;

$p(m_i P_{cmi} \le Q_{sk}^{o} / \sum_{i=1}^{f} N_{ki} Q\textit{н}_i^{\textit{конт}} \le \sum_{i=1}^{f} M_i P_{mi} T_i)$ is the conditional probability of the event in which the amount of cargo in a single serve is less than the storage capacity provided that the capacity of containers handled per day is not more than handling capacity for LUM per day, i.e. $\sum_{i=1}^{f} N_{ki} Q\textit{н}_i^{\textit{конт}} \le \sum_{i=1}^{f} M_i P_{mi} T_i$;

N_{ki} is number of containers handled per day, cont/day;

$Q\textit{н}_i^{\textit{конт}}$ is container load capacity, kg;

P_{mi} is LUM productivity, t/h;

$$p_1 = p(\sum_{i=1}^{f} N_{ki} Q\textit{н}_i^{\textit{конт}} \le \sum_{i=1}^{f} M_i P_{mi} T_i) \cdot p(m_i P_{cmi} \le Q_{sk}^{o}) \cdot \varphi(\sum_{i=1}^{f} N_{ki} Q\textit{н}_i^{\textit{конт}} \le \sum_{i=1}^{f} M_i P_{mi} T_i, m_i P_{cmi} \le Q_{sk}^{o}), \quad (2)$$

where $p(m_i P_{cmi} \le Q_{sk}^{o})$ is the probability of the volume of cargo in the delivery not exceeding the warehouse capacity;

$\varphi(\sum_{i=1}^{f} N_{ki} Q\textit{н}_i^{\textit{конт}} \le \sum_{i=1}^{f} M_i P_{mi} T_i, m_i P_{cmi} \le Q_{sk}^{o})$ is a function that expresses the relationship between the probabilities $p(\sum_{i=1}^{f} N_{ki} Q\textit{н}_i^{\textit{конт}} \le \sum_{i=1}^{f} M_i P_{mi} T_i)$ and $p(m_i P_{cmi} \le Q_{sk}^{o})$;

- the capacity of containers handled per day is less than the handling capacity of LUM per day, the warehouse capacity is insufficient for storing $\sum_{i=1}^{f} N_{ki} Q\textit{н}_i^{\textit{конт}} \le \sum_{i=1}^{f} M_i P_{mi} T_i, m_i P_{cmi} \le Q_{sk}^{o}$

$$p_1 = p(\sum_{i=1}^{f} N_{ki} Q\textit{н}_i^{\textit{конт}} \le \sum_{i=1}^{f} M_i P_{mi} T_i), m_i P_{cmi} \le Q_{sk}^{o}) =$$
$$= p(\sum_{i=1}^{f} N_{ki} Q\textit{н}_i^{\textit{конт}} \le \sum_{i=1}^{f} M_i P_{mi} T_i) \cdot p(m_i P_{cmi} \le Q_{sk}^{o} / \sum_{i=1}^{f} N_{ki} Q\textit{н}_i^{\textit{конт}} \le \sum_{i=1}^{f} M_i P_{mi} T_i), \quad (3)$$

$$p_1 = p(\sum_{i=1}^{f} N_{ki} Q\textit{н}_i^{\textit{конт}} \le \sum_{i=1}^{f} M_i P_{mi} T_i) \cdot p(m_i P_{cmi} \le Q_{sk}^{o}) \cdot \varphi(\sum_{i=1}^{f} N_{ki} Q\textit{н}_i^{\textit{конт}} \le \sum_{i=1}^{f} M_i P_{mi} T_i), m_i P_{cmi} > Q_{sk}^{o}); \quad (4)$$

- the capacity of containers handled per day is more than the handling capacity of LUM per day, the warehouse capacity is sufficient for storing cargo $\sum_{i=1}^{f} N_{ki} Q н_i^{конт} > \sum_{i=1}^{f} M_i P_{mi} T_i, m_i P_{cmi} > Q_{sk}^0$

$$p_1 = p(\sum_{i=1}^{f} N_{ki} Q н_i^{конт} > \sum_{i=1}^{f} M_i P_{mi} T_i), m_i P_{cmi} \le Q_{sk}^o) = \\ = p(\sum_{i=1}^{f} N_{ki} Q н_i^{конт} > \sum_{i=1}^{f} M_i P_{mi} T_i) \cdot p(m_i P_{cmi} \le Q_{sk}^o / \sum_{i=1}^{f} N_{ki} Q н_i^{конт} > \sum_{i=1}^{f} M_i P_{mi} T_i), \quad (5)$$

$$p_1 = p(\sum_{i=1}^{f} N_{ki} Q н_i^{конт} > \sum_{i=1}^{f} M_i P_{mi} T_i) \cdot p(m_i P_{cmi} \le Q_{sk}^o) \cdot \varphi(\sum_{i=1}^{f} N_{ki} Q н_i^{конт} > \sum_{i=1}^{f} M_i P_{mi} T_i), m_i P_{cmi} \le Q_{sk}^o); \quad (6)$$

- the capacity of containers handled per day is greater than the handling capacity of LUM per day, the warehouse capacity is insufficient for storing cargo $\sum_{i=1}^{f} N_{ki} Q н_i^{конт} > \sum_{i=1}^{f} M_i P_{mi} T_i), m_i P_{cmi} \le Q_{sk}^o$

$$p_1 = p(\sum_{i=1}^{f} N_{ki} Q н_i^{конт} > \sum_{i=1}^{f} M_i P_{mi} T_i, m_i P_{cmi} > Q_{sk}^o) = \\ = p(\sum_{i=1}^{f} N_{ki} Q н_i^{конт} > \sum_{i=1}^{f} M_i P_{mi} T_i) \cdot p(m_i P_{cmi} > Q_{sk}^o / \sum_{i=1}^{f} N_{ki} Q н_i^{конт} > \sum_{i=1}^{f} M_i P_{mi} T_i), \quad (7)$$

$$p_1 = p(\sum_{i=1}^{f} N_{ki} Q н_i^{конт} > \sum_{i=1}^{f} M_i P_{mi} T_i) \cdot p(m_i P_{cmi} > Q_{sk}^o) \cdot \varphi(\sum_{i=1}^{f} N_{ki} Q н_i^{конт} > \sum_{i=1}^{f} M_i P_{mi} T_i), m_i P_{cmi} > Q_{sk}^o). \quad (8)$$

Rate setting condition: $\sum_{i=1}^{4} p_i = 1 \cdot$

Operation of the stochastic nature m_{and} i Q_{sk}^o are practically independent of each other, functions $\varphi(\sum_{i=1}^{f} N_{ki} Q н_i^{конт} \le \sum_{i=1}^{f} M_i P_{mi} T_i, m_i P_{cmi} \le Q_{sk}^o)$, $\varphi(\sum_{i=1}^{f} N_{ki} Q н_i^{конт} \le \sum_{i=1}^{f} M_i P_{mi} T_i), m_i P_{cmi} > Q_{sk}^o)$, $\varphi(\sum_{i=1}^{f} N_{ki} Q н_i^{конт} > \sum_{i=1}^{f} M_i P_{mi} T_i), m_i P_{cmi} \le Q_{sk}^o)$, $\varphi(\sum_{i=1}^{f} N_{ki} Q н_i^{конт} > \sum_{i=1}^{f} M_i P_{mi} T_i), m_i P_{cmi} > Q_{sk}^o)$ are assumed to be equal to one.

The continuity and reliability of the handling process depends on the level, rational use of the available equipment and the adopted technology. The conditions for the smooth operation of subsystems are determined by the load level of the interacting channels, which are calculated using load intensities. When cargo is handled:

$$\frac{24}{Y_i} \ge M[t_{onp}] + M[t_{np}] + M[t_{он(в)}] + m_i k_{sdv} M[t_{н(в)}] + M[t_{onep}] + m_i M[t_{nep}] + M[t_{ocy}] + M[t_{cy}], \quad (9)$$

$$M_i P_m \geq \frac{m_i P_{st} + N_a P_a}{M[t_{н(в)}] + M[t_a]}, \tag{10}$$

where Y_i is the number of deliveries to the warehouse per day;

$M[t_{onp}], M[t_{он(в)}], M[t_{oper}], M[t_{ocy}]$ are, respectively, the mathematical expectation of the waiting time for the performance of technological operations: spotting-arrangement, cargo handling, rearrangeent, collection-picking up;

m_i is the number of cars delivered to the cargo front line, cars;

k_{sdv} is coefficient of double operations;

$M[t_{np}], M[t_{н(в)}], M[t_{nep}], M[t_{cy}]$ is, respectively, the mathematical expectation of the time of performance of technological operations: spotting-arrangement, cargo handling, rearrangeent, collection-picking up;

M_i is the number of LUM on the I-m cargo front line, pcs;

P_m is productivity of LUM operating on the I-m cargo front line, t/h;

P_{st} is static load of the car, t/car;

N_a is number of vehicles handled on the I-th cargo front, pcs;

P_a is vehicle performance, t/h;

$M[t_a]$ is mathematical expectation of performing cargo operations with the vehicle.

Condition (10) indicates that the performance of LUM should match the rate of delivery of cars and vehicles to the cargo front line.

If the conditions for the smooth operation of subsystems is complied with, the efficiency of the station is increased because of the reduced negative impact of cause-and-effect relationships and rational use and loading of handling devices.

The minimum values $M_i^{\min}$, $Y_i^{\min}$ are determined by the requirement to perform a given workload, and maximum values are determined by the availability of technical means.

$$M_i^{\min} = \frac{k_i Q_i}{P_i T_i}, \tag{11}$$

$$Y_i^{\min} = \frac{m_i l_{ваг}}{l_{вф}}, \tag{12}$$

where k_i is a coefficient that takes into account additional cargo handling operations;

Q_i is volume of cargo arrival or departure, t;

P_i is operational capacity of the LUM, t/h;

T_i is time of operation of the cargo front during the day, h; respectively, the number of LUM, deliveries to the I-th cargo front line and cars arriving at the I-th cargo front line per day, pcs;

$l_{ваг}, l_{вф}$ is respectively the length of the car and the I-th cargo front line, m.

When more than one freight front lines are handled with a single shunting locomotive, the operation parameters must be estimated as a single complex for all front lines at the same time, because the optimization criterion takes into account the costs associated with cars waiting for delivery and depends on the number of deliveries to all front lines.

The objective function of the model has the form

$$R = \sum_{i=1}^{f} R_i(M_i, Y_i, T_i) \rightarrow \min , \qquad (13)$$

under the limitations
$$\begin{cases} \dfrac{k_i Q_i}{P_i T_i} \le M_i \le M_i^{\max}; \\ \dfrac{m_i l_{ваг}}{l_{вф}} \le Y_{and} \le Y_i^{\max}; \\ T_i^{\min} \le T_i \le 24; \end{cases} \qquad (14)$$
$$m_i \ge 0;$$

Since each component of the function is a complex functional, the article shows the model implicitly. The function will further be considered explicitly.

As an example, the optimal functioning parameters of the grain warehouse based on a specific railway station are estimated. The obtained results are shown in the Figure 2.

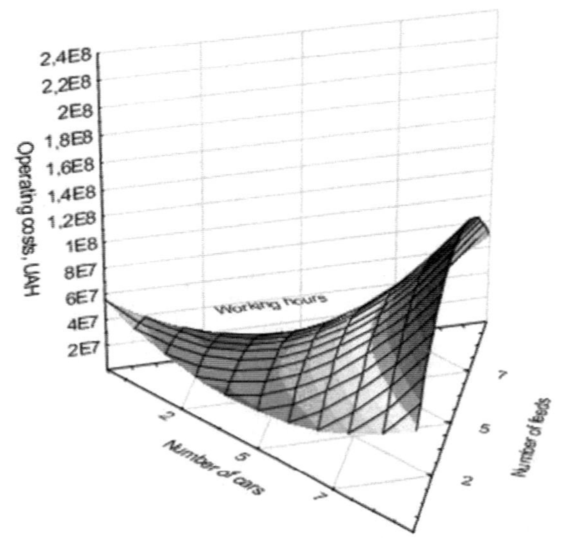

Figure 2. Result of estimation of optimal functioning parameters.

The results of estimations proved the presence of the extreme minimum, which allows us to determine the optimal number of loading and unloading machines, the number of deliveries and the operating time of the warehouse complex during the day, both in the stand-by and operational mode.

The proposed approaches can be used for the operational adjustment of the current situation, taking into account the predicted arrival of grain cargo for loading. It can be integrated into the decision support system in the corresponding automatic workstations of operational employees of the station: the station attendant, shunting dispatcher, acceptance/delivery inspector, process engineer, and others.

3. Conclusion

The volumes of grain cargo transportation and their seasonality have been analyzed. It is found that during peak loading periods there is an acute shortage of grain cars. Therefore, one of the ways to eliminate this shortage is to use grain shipping containers.

Possible variants of grain warehouse operation are determined, that take into account the ratio of the main parameters that affect the warehouse operation, including container capacity, warehouse capacity and handling capacity of LUM.

The operation technology of the departure point for grain cargo handling is formalized on the basis of a stochastic programming model, which allows to minimize the total cost of loading and unloading operations and determine the optimal number of LUM, the number of deliveries and the working time of the warehouse complex during the day.

4. References

[1] Butko T, Prodaschuk S, Bogomazova G, Prodaschuk M and Purii R 2017 Improvement of technology for management of freight rolling stock on railway transport *Eastern-European Journal of Enterprise Technologies* **3(87)**, 4-11

[2] 2020 Freight shipped by rail and type by freight State Statistics *Service of Ukraine* January July

[3] Myamlin S V, Kozachenko D N and Vernigora R V 2013 Problems and Prospects of Transportation of Grain Freights by Rail in Ukraine *Railway Transport of Ukraine* **2(99)**, 32-34

[4] Shuldiner Yu. V., Oliynyk Yu. O. Improvement of grain cargo transportation by rail *Car Fleet, Railway publishing house «Rolling stock»* **7**, 16-19

[5] Seyed Mohammad Nourbakhsh, Yun Bai, Guilherme D N Maia, Yanfeng Ouyang and Luis Rodriguez 2016 Grain supply chain network design and logistics planning for reducing post-harvest loss *Biosystems Engineering* **151**, 105-115

[6] Kozachenko D, Vernigora R and Rustamov R 2017 Creation of export-oriented network of grain elevators in Ukraine *Science and Transport Progress. Bulletin of Dnipropetrovsk National University of Railway Transport* **2 (68)**, 56-68

[7] Michael F Hyland, Hani S Mahmassani, Lama Bou Mjahed 2016 Analytical models of rail transportation service in the grain supply chain: Deconstructing the operational and economic advantages of shuttle train service Transportation Research Part E: Logistics and Transportation Review **93**, 294-315

[8] Elvis Ndembe and John Bitzan 2018 Grain freight elevator consolidation, transportation demand, and the growth of shuttle facilities *Research in Transportation EconomicsNovember* **71**, 54-60

[9] Bagdonienė D and Mazūra S 2004 The optimisation of loading facilities at the terminal *Transport Vol XIX* **6**, 239-251

[10] Pavlenko O and Volkova T 2020 Improvement of the grain delivery quality management by the ukrainian territory *Municipal economy of cities: Engineering science and architecture* **154**, 216-222

IOP Conf. Series: Materials Science and Engineering 1021 (2021) 012007 doi:10.1088/1757-899X/1021/1/012007

Main criteria of complex evaluation of subway train power facility technological productive potential

I I Kulbovskyi[1,5], O V Aharkov[2], V S Kharuta [3], M M Halushko [4]

[1] Department of Automation and Computer-Integrated Technologies of Transport, State University of Infrastructure and Technologies, Kirilivska str., 9/3, 04080 Kyiv, Ukraine
[2] Department of Theoretical and Applied Mechanics, State University of Infrastructure and Technologies, Kirilivska str., 9/3, 04080 Kyiv, Ukraine
[3]Department of Transport Law and Logistics, National Transport University, Omelianovycha-Pavlenka str., 1, 04080 Kyiv, Ukraine
[2] Department of Telecommunication Technologies and Automation, State University of Infrastructure and Technologies, Kirilivska str., 9/3, 04080 Kyiv, Ukraine

[5]E-mail: kulbovskiy@ukr.net

Abstract. Maintaining the train safety and passenger transportation safety is regulated by the technical condition of rolling stock, which meets the requirements of regulatory documentation. Modern conditions of operation of traction rolling stock units (UTRS) are characterized by insufficient use of production and technological potential.
One of the main reasons for this situation is the lack of necessary scientific and methodological developments that describe and improve system methods, models and criteria for the effectiveness of the production processes of UTRS.

1. Introduction

The present stage of activity of the enterprises is characterized by competition growth that demands continuous process of improvement and optimization of productions [1-4]. Along with such measures for optimization of functioning of the enterprises as resource-saving and energy efficiency [5-14], level of automation and mechanization, more productive is use of complex indicators of efficiency of activity of the enterprises and their divisions [15-16].

The greatest efficiency of use of complex indicators can be reached at the enterprises with complex administrative structure and a large number of structural divisions. It is possible to carry transport enterprises with branched structure to such enterprises, in particular the subways.

For example, in the Kyiv subway, except divisions that directly provide transportations of passengers, expenses for operation, repair and complex service of artificial constructions, track elements [17-18] and tunnels, and rolling stock in which service three train depots – Darnytsa, Obolon and Kharkivska – are engaged in constantly growing.

Wide range of the equipment that is served by the locations of structural divisions becomes the reason of low efficiency of functioning of the specified divisions.

Criteria of efficiency of complex assessment of production and technological potential use of UTRS in real to the functioning environment in work were formulated and offered. The mathematical model of calculation and level of use of production and technological potential and quality of management of productions of UTRS was developed.

2. Methods and course of the study

Production and technological capacity of divisions of the traction rolling stock represents a possibility of performance of useful work with maximum efficiency of use of the fixed business assets and also human, material and financial resources.

For definition of PTP UTRS it is used following expression:

$$W_i = W_{1i},\ i=1,2,\ldots,N \tag{1}$$

where W_i is the size that characterizes the production and technological potential of PTRS on a period i; W_{1i} is useful work which is performed by repair shop on i period.

Size W_i and its components W_{1i} can be expressed in natural and monetary value. It is expedient to express the size W_i in terms of money as it allows to present various physical quantities by means of uniform indicators (expenses on work implementation; Income, the job got for performance, etc. Irrational use of the equipment testifies to insufficient the level of use of fixed assets of UTRS.

Lack of scientific and methodical capacity of the enterprise; Lack of scientifically based lists of works which could be executed in UTRS (except the established types of activity); Lack of effective mechanisms of management of productions of UTRS in the conditions of the market relations, cause irrational use of UTRS.

For the purpose of improvement of efficiency of functioning of processes of repair and the current maintenance of the traction rolling stock it is necessary to carry out the analysis of the parameters influencing the size W1i which decides on the help of the following expression:

$$\begin{aligned} W_{1i} &= \left(\sum_{j=1}^{M}\sum_{i=1}^{N} h_{ji}L_{ji}\right)(1+r_i) + \left(\sum_{k=1}^{K} a_k c_k q_k\right)(1+r_i) + \left(\sum_{j=1}^{M} b_j \tilde{c}_j \tilde{q}_j\right)(1+r_i) + \left(\sum_{i=1}^{N}\sum_{p=1}^{P} c_{1ip}Q_{ip}\right)(1+r_i) = \\ &= (1+r_i)\left(\sum_{j=1}^{M}\sum_{i=1}^{N} h_{ji}L_{ji} + \sum_{k=1}^{K} a_k c_k q_k + \sum_{j=1}^{M} b_j \tilde{c}_j \tilde{q}_j + \sum_{i=1}^{N}\sum_{p=1}^{P} c_{1ip}Q_{ip}\right); \end{aligned} \tag{2}$$

$i= 1,2,\ldots, M;\ i= 1,2,\ldots, N;\ k= 1,2,\ldots, K;\ p= 1,2,\ldots, P,$

where r_i is the set level of profitability of functioning of productions to repair and service of the traction rolling stock on i period $(0< r_i < 1)$;

c_{1ip} is the cost of the rendered service p on i period;

Q_{ip} is the number of the rendered services and the performed works in the environment of functioning of productions on repair flowing to deduction traction rolling stock on i period.

h_{ji} may include such standard costs: the salary with charge of the technological personnel occupied with repair and the current maintenance of the traction rolling stock on i period; spare parts; operational materials; electric energy; thermal energy; expenses on compressed air. The norm of expenses h_{ji} can be determined from expression

$$h_{ji}=h_{3ij}k_{3i}+h_{Mj}k_{Mi}+h_{3rj}k_{3ri}+h_{eij}k_{ei}+h_{Tij}k_{Ti}+h_{CTij}k_{CTi}, \tag{3}$$

where h_{3ij}, h_{Mj}, h_{3rj} is the standard of expenses on the salary, materials and spare parts for name mechanisms j in the basic period, respectively;

k_{3i}, k_{Mi}, k_{3ri} is the index of growth of expenses on the salary, materials, spare parts in i period, respectively;

h_{eij}, h_{Tij}, h_{CTij} are standards of expenses on electric, thermal energy and compressed air in the basic period, respectively;

k_{ei}, k_{Ti}, k_{CTi} are respectively indexes of growth of standard costs, on electric, thermal energy and compressed air in i period.

Size $\sum_{i=1}^{K} a_k c_k q_k (1+\alpha_k)$ depends on the cost of the equipment and coefficient of its intensity of use α_k.

Size $\sum_{i=1}^{N}\sum_{p=1}^{P} c_{1in}Q_{in}$ depends on the number of the rendered services and the performed works Q_{ip}.

So, UTRS making sizes W_{1i} of production and technological potential depends on efficiency of use of human, material and energy resources and extension of the nomenclature of the performed services and works due to increase in coefficient of intensity of use of the equipment.

For realization of reserves of use of production and technological potential there is a need of introduction of standard target methods of management of productions of UTRS which provide formation of scientifically based standards of expenses of the resources necessary for implementation of production economic activity, and development of the corresponding mechanisms purposeful on achievement of the high end results. One of the main criteria of efficiency of use of production and technological potential of UTRS at realization of standard target methods of management is the size

$$E_{\Phi i} = \left| W_{1i} - \overline{W}_{1i} \right|, \partial \text{ля}\, W_{1i} \rangle \overline{W}_{1i} \tag{4}$$

The value of $E_{\Phi i}$ for some interval $[O,T]$ can be expressed using the expression

$$E_{\Phi i} = \int_O^T \varphi_1\left(i, W_1, \overline{W}_1\right)\left|W_1(i) - \overline{W}_1(i)\right| di \tag{5}$$

where $\varphi_1\left(i, W_1, \overline{W}_1\right)$ weight function, which shows that the same deviations at different points in time are not the same $\varphi_1\left(i, W_1, \overline{W}_1\right) \leq 1$.

It can be noted that the work of the management system of production processes of UTRS is reduced to minimizing the functionality

$$E_{\Phi i} = \min_{U_i}\left[\int_O^T \varphi_1\left(i, W_1, \overline{W}_1\right)\left|W_1(i) - \overline{W}_1(i)\right| di\right] \tag{6}$$

The effectiveness of UTRS production process management can be assessed using the following parameter:

$$e_{\Phi} = \frac{\int_O^T \varphi_1\left(i, W_1, \overline{W}_1\right)\left|W_1(i) - \overline{W}_1(i)\right| di}{\int_O^T \overline{W}_1(i) di} \tag{7}$$

Considering the stochastic nature of size e_{Φ} it can be written:

$$\bar{e}_{\Phi} = M\left\{e_{\Phi}\right\} = M\left\{\frac{\int_O^T \varphi_1(i, W_1, \overline{W}_1)\left|W_1(i) - \overline{W}_1(i)\right| di}{\int_O^T \overline{W}_1(i) di}\right\} \tag{8}$$

The obtained mathematical expressions for assessment of production and technological potential and effective management of processes of repair shops, allow solving problems of optimization.

Indicators of level of use of PTP have to be operated whether that is allowed to establish the necessary value and to support its state to change it in the necessary direction. Most fully the level of use of PTP can be estimated by the system of the indicators classified by the principle of characteristics of the made products, objects and means of labor and also action methods regarding work. Concerning repair production the indicators characterizing an object of the labor can enter such system; means of labor; action methods regarding work.

Increase in efficiency and quality of use of production and technological potential of UTRS provides, first of all, development of means of labor and establishment of rational ratios between

IOP Conf. Series: Materials Science and Engineering 1021 (2021) 012007 doi:10.1088/1757-899X/1021/1/012007

human and automated by work. For a research of technological level of production at all stages of life cycle, it is necessary to have the set (standard) model of a condition of PTP. In the presence of such model it is possible to define as far as to it there corresponds the current situation of production.

Figure 1 show the list of major factors and indicators under the influence of which the production and technological potential of UTRS, the level of its development and extent of use is formed.

Significantly the quantitative and qualitative structure of processing equipment influences on the level of use of technological potential production.

The most dynamic factor determining the level of use of production and technological potential is the technological level of processes of production. The package of measures for optimization of technological processes leads to decrease in labor input, increase in labor productivity and quality of the performed works.

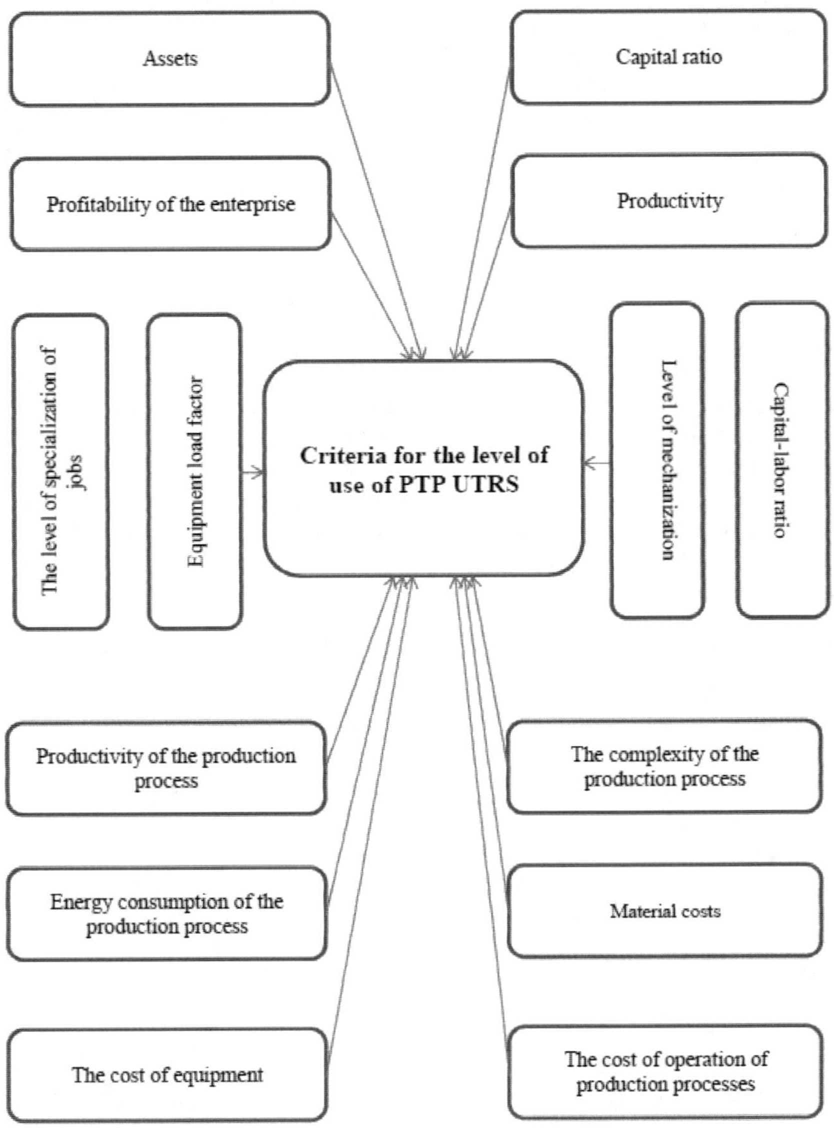

Figure 1. Major factors and indicators under the influence of which the production and technological potential of UTRS, the level of its development and extent of use is formed.

IOP Conf. Series: Materials Science and Engineering 1021 (2021) 012007 doi:10.1088/1757-899X/1021/1/012007

Each of the presented factors is characterized by a set of indicators which quantity depends on a real condition of the concrete enterprise. Definitions of influence of all factors and indicators on the level of use of production and technological potential is relevant to the scientifically applied to the task which can be solved on the basis of economic-mathematical modeling.

On the basis of the carried-out analysis it is possible to draw a conclusion that systematization and classification of the main criteria of efficiency and quality of use of production and technological capacity of divisions of the traction rolling stock allows passing to realization of system model of functioning of UTRS in modern operating conditions of the enterprise.

3. Conclusions

Summing up, the level of utilization of production and technological potential of the units for maintenance of the subway traction rolling stock influences on the quantitative and qualitative composition of technological equipment, and the complex of measures for optimization of technological processes leads to reduction of complexity, increase of labor productivity and quality of the performed works. The mathematical model for estimation of production technological potential and efficiency of management of production processes of PTS is offered which gives an opportunity to solve problems of optimization with the use of mathematical model of enterprise functioning.

References

[1] Carpintero S 2011 The Competitive Advantages of The Spanish Companies in the International Toll Road Industry *Journal of Civil Engineering and Management* **17(4)** pp 483-493.

[2] Wielgross T et al 2010 Consolidation drives competition at the local level *Railway Gazette International* Oct pp 60-62.

[3] Nakano H 2014 A study on Internationalisation Strategies of Global Enterprises Overseas of Infrastructure Management (Japanese) *Journal of Transport Policy Studies* **17 (1)** pp 12-23.

[4] Nakano H 2014 Globally competitive strategies of Infrastructure management in the transport sector, (Japanese) *Nihon Hyoronsya*, December 2014

[5] Arena M Azzone G and Conte A 2013 A streamlined LCA framework to support early decision making in vehicle development *Journal of Cleaner Production* **41** pp 105–113

[6] Jayal A D Badurdeen F Dillon O W and Jawahir I S 2010 Sustainable manufacturing: Modeling and optimization challenges at the product, process and system levels *Journal of Manufacturing Science and Technology* **2(3)** pp 144–152.

[7] Hoseini Rahdar M Heidari M Ataei A and Choi J-K 2016 Modeling and optimization of R-717 and R-134a ice thermal energy storage air conditioning systems using NSGA-II and MOPSO algorithms *Applied Thermal Engineering* **96** pp 217–227.

[8] Kulbovskyi I Bakalinsky O Sorochynska O Kharuta V Holub H and Skok P 2019 Implementation of innovative technology for evaluating high-speed rail passenger transportation. *EUREKA: Physics And Engineering* **0(6)** pp 63-72.

[9] Gonzalez Hernandez A Paoli L and Cullen J M 2018 How resource-efficient is the global steel industry? *Resources, Conservation and Recycling* **133** pp 132–145.

[10] Corcelli F Ripa M and Ulgiati S 2018 Efficiency and sustainability indicators for papermaking from virgin pulp – An emergy-based case study. Resources *Conservation and Recycling* **131** pp 313–328.

[11] Fomin O Sulym A Kulbovskyi I Khozia P and Ishchenko V 2018 Determining rational parameters of the capacitive energy storage system for the underground railway rolling stock *Eastern-European Journal Of Enterprise Technologies* **2 (1 (92))**, pp 63-71.

[12] Jonek-Kowalska I 2019 Efficiency of Enterprise Risk Management (ERM) systems. Comparative analysis in the fuel sector and energy sector on the basis of Central-European companies listed on the Warsaw Stock Exchange *Resources Policy* **62** pp 405–415.

[13] Wang J Yang F Zhang X and Zhou Q 2018 Barriers and drivers for enterprise energy efficiency: An exploratory study for industrial transfer in the Beijing-Tianjin-Hebei region *Journal of Cleaner Production* **200** pp 866–879.
[14] Choi J-K Thangamani D and Kissock K 2019 A systematic methodology for improving resource efficiency in small and medium-sized enterprises *Resources, Conservation and Recycling* **147** pp 19–27.
[15] Centobelli P Cerchione R and Esposito E 2018 Aligning enterprise knowledge and knowledge management systems to improve efficiency and effectiveness performance: A three-dimensional Fuzzy-based decision support system *Expert Systems with Applications* **91** pp 107–126.
[16] Kulbovskyi I Sapronova S Holub H Tkachenko V Afanasieva I фтв Safronov O 2019 Development of a model for managing the quality of repair and maintenance of rolling stock in transport infrastructure projects *Transport Means – Proceedings of the International Conference* pp 201-205.
[17] Nakano H 2017 A Study on the Features of the Evolution Processes and Business Models of Global Enterprises in the Transport Sector *Transportation Research Procedia* **25** pp 3769–3788.
[18] Boiko V Molchanov V Tverdomed V and Oliinyk O 2018 Analysis of vertical irregularities and dynamic forces on the switch frogs of the underground railway *MATEC Web of Conferences* **230** 01001.
[19] Aharkov O V Tverdomed V M Boiko V D Kovalchuk V V and Strelko O H 2019 Influence of the structural design of rail fastenings on ensuring the stability of track gauge in operating conditions *IOP Conference Series: Materials Science and Engineering* **708 (1)** 012001.

IOP Conf. Series: Materials Science and Engineering 1021 (2021) 012008 doi:10.1088/1757-899X/1021/1/012008

Prospects of introducing alternative sources of energy on Ukrainian railways

V V Panchenko [1,2], A S Masliy [1] and R O Kharin[1]

[1] Department of Electric Power Engineering, Electrical Engineering and Electromechanics Ukrainian State University of Railway Transport, Feyerbakh Square 7, 61050, Kharkiv, Ukraine

[2] Email: vlad_panchenko@ukr.net

Abstract: The potential of solar and wind energy was investigated, the existing premises of using the electric power obtained from non-conventional power sources of non-traction consumers for power supply of an electric traction system were analyzed. The influence of solar generation sources connected with the traction substation tires on the stability of non-traction consumers was determined. Possible options of connecting renewable energy sources to the railway power systems were examined. Calculations for determining the number of solar panels for compensating the auxiliary power consumption of the traction substation were made. The modeling of the possibility of implementing alternative energy sources in a traction substation in the PVsyst program was carried out. In Ukraine, the change of conditions for the provision of electricity supply services was caused by the rapid growth and development of renewable energy sources, in connection with which the number of suppliers increased. The question about the possibility of using alternative electricity sources to compensate the auxiliary power consumption of traction substations arises. Therefore, it is relevant to solve the problem of joint work of auxiliary power consumers together with non-traditional energy sources. The auxiliary power consumption of the traction substation include: heating drives and tanks of transformers; cooling down trip coils, relays, control panels, alarm systems, alarm and centralization and blocking devices; heating substation premises, store rooms and shower room. In order to implement alternative energy on the railway, it is necessary to analyze the auxiliary power consumption of the traction substation; to analyze the possibility of placing alternative power sources on the traction substation premises; to investigate how the powering system of the auxiliary power consumption of the traction substation can operate (both directly and in parallel mode).

1. Introduction.

In Ukraine, the amount of electricity obtained from non-traditional sources is growing. At the same time, alternative sources of electricity are developing rapidly, the number of suppliers is increasing and the conditions for providing electricity supply services are changing [1]. This process is conditioned by the difficult environmental situation, environment protection procedures and the complexities of energy policy in the world. Over the past few years, Ukraine has made significant progress in the development of alternative energy sources, in particular solar and wind power plants. However, the introduction of innovative energy-saving technologies using alternative renewable energy sources is still at the initial stage in our country. Today, the Ukrainian railways almost exclusively use traditional power supplies in their power supply systems. Yet, the rapid development of alternative energy is pushing for the modernization of railway power supply systems through the introduction of non-conventional power sources [2]. We believe that the introduction of solar and wind energy sources is the most acceptable solution to the issue. The main directions for introducing alternative energy on the railway are as follows:

- power supply of traction power supply systems from the external power system, in which, alternative sources of energy operate along with traditional ones;
- power supply to compensate power auxiliary consumption of a traction substation, objects of infrastructure of non-traction consumers both separately, and in a parallel mode of work [3].

It is especially important in the regions where the external electricity supply of railways is unstable, so the use of independent sources of electricity is crucial to ensure the traction and meet the needs of non-traction railway consumers [4]. Nowadays two documents regulate the requirements for connecting solar and wind power plants to industrial power grids in Ukraine:

- requirements for wind and solar photovoltaic power plants for connection to external electrical networks;
- connection of wind energy facilities to electrical networks. Procedure and requirements. The document [5] defines the basic requirements for wind and solar photovoltaic power plants (i.e. for power plants that use static electronic energy converters) with a capacity of 150 kW for connection to public grid.

Connection to electrical networks of other types of renewable sources with synchronous generators that are directly connected to electrical networks is regulated by other normative documents in force in Ukraine.

2. Analysis of the recent researches and publications.

Researching the possibility of introducing alternative sources of energy to compensate power auxiliary consumption of traction substations is topical and promising. Over the past years a number of experimental researches have been conducted on introducing alternative energy sources on Ukrainian railways. The analysis of the recent researches and publications, which initiated the solution of this problem, showed that such scientists as M. Pastushenko, O. Polyakh, V. Sychenko, O. Bondar, Y. Honcharov, E. Kosarev, M. Prykhoda, T. Reshetnyak and others paid their attention to various aspects of renewable energy sources. The following tasks have been solved:

- analysis of the solar generation impact on the operation of traction substations on electrified railways;
- prospects for the introduction of renewable energy sources on the railway transport of Ukraine;
- the possibility of using alternative sources of energy to compensate power auxiliary consumption of traction substations, etc.

On the territory of Ukraine the conditions for the provision of electricity supply services have been changed, which caused a rapid increase and development of renewable energy sources. Therefore, it is necessary to solve the problem of joint work of auxiliary consumers with non-traditional energy sources.

3. Defining the aim and objectives of the study.

The aim is to study the possibility of introducing alternative energy sources into the power supply network of electrified railways by equipping a DC traction substation No. 18 in Zmiiv, Kharkiv region with a solar power plant to compensate its power auxiliary consumption.

To study the possibility of introducing alternative energy sources on the railway the following tasks should be solved:

- to analyze power auxiliary consumption of the traction substation;
- to analyze the possibility of locating alternative power sources on the premises of the traction substation;
- to investigate the operation modes of the power supply system to compensate power auxiliary consumption of the traction substation (both directly and in a parallel mode).

IOP Conf. Series: Materials Science and Engineering 1021 (2021) 012008 doi:10.1088/1757-899X/1021/1/012008

4. Presenting the main material.

The object of the study was a DC traction substation No. 18, Zmiiv, Kharkiv region, its auxiliary power consumption is 27,513 kWh in the period from March to May. For the spring period, the average auxiliary power consumption is 9171 kWh per month, i.e. 299.05 kWh (12.46 kWh) per day.

Let's calculate the number of solar panels to provide power for auxiliary power consumption. It is necessary to determine the type of solar panel to perform the calculation. Jinko Solar JKM305M-60 Eagle 305 W solar panels cost not much, but are of high efficiency.

The total power of the solar battery consists of the output power of individual photovoltaic modules. The output current of the solar photovoltaic modules is determined by the number of elements connected in parallel, and the output voltage – by the number of elements connected in series. Having the nominal power of the photovoltaic station and the power of one photo module, we determine the required number of photo modules:

$$N^{сб} = \frac{P_{nom}}{P_1^{pm}} \tag{1}$$

where P_{nom} is nominal power of the photovoltaic station [W]

P_1^{pm} is nominal power of a photo module [W]

$$N^{pm} = \frac{15000}{305} = 49 \text{ pcs.}$$

Let us recalculate the total number of photo modules, taking into account the way they are connected to the inverter. The number of modules connected in series:

$$N_{ser}^{pm} = \frac{U_{inv}}{U_{\max}^{pm}}, \tag{2}$$

where U_{inv} is the output voltage of the inverter [V]

$U_{\max}^{pm}$ is the voltage of a photovoltaic module [V]

$$N_{ser}^{pm} = \frac{230}{22,8} \approx 10 \text{ pcs.}$$

the power of the batteries connected in series:

$$P_{ser}^{pm} = N_{ser}^{pm} \cdot P_1^{pm}, \tag{3}$$

$$P_{ser}^{pm} = 10 \cdot 305 = 3050 \text{ W}$$

the number of photovoltaic modules connected in parallel:

$$N_{par}^{pm} = \frac{P_{\max}^{sys}}{P_{ser}^{pm}}, \tag{4}$$

where $P_{\max}^{sys}$ is the power of the settlement system;

$$N_{par}^{pm} = \frac{12,46}{3,13} = 4 \text{ pcs.}$$

the total number of photovoltaic modules in the system:

IOP Conf. Series: Materials Science and Engineering 1021 (2021) 012008 doi:10.1088/1757-899X/1021/1/012008

$$N^{pv} = N^{pm}_{par} \cdot N^{pm}_{ser}, \quad (5)$$

$$N^{сб} = 10 \cdot 4 = 40 \text{ pcs.}$$

the total number of photovoltaic modules:

$$S_{pvm} = S_1 \cdot N^{pm}, \quad (6)$$

$$S_{pvm} = 1,63 \cdot 40 = 65,2 \text{ m}^2$$

The total area of the roof of the traction substation No. 18, Zmiiv is 378 m^2, i.e. the area required for solar-battery cells will occupy 17.24% of the total roof area [6].

Let's determine the number of solar panels, their location and model a solar power plant to compensate the auxiliary power consumption of the traction substation using the software package PVsyst.

Let's have a look at the traction substation No. 18, Zmiiv of the Regional Branch «Pivdenna Zaliznytsia» of Ukrzaliznytsia JSC (Figure 1).

Figure 1. Traction substation No. 18, Zmiiv.

The initial data of the selected solar panels, inverters, traction substation were entered into the PVsyst program (Figure 2). We selected solar panels of Jinko Solar JKM305M-60 Eagle 305 [7] W brand. They are modern single-crystal solar batteries that:

- cost not much and are of high efficiency;
- withstands harsh operating conditions, high salt spray performance and ammonia resistance;
- uses the latest technology to increase the efficiency of the module, which makes it ideal for roof installation;
- minimizes the possibility of rupture of conductive track network.

Solar panels work only when there is some sunlight. Dynamic current significantly decreases in the cloudy weather or in the darkness. To compensate this effect, the battery must accumulate the electricity produced by these modules during the day [8]. There are different types of batteries. Therefore, rechargeable batteries for photovoltaic systems must meet certain requirements:

- low level of self-discharge;
- ability to work in modes;
- deep discharge;

- operate with small charge currents;
- operate at below-freezing temperature (for year-round systems);
- minimum service requirements [9].

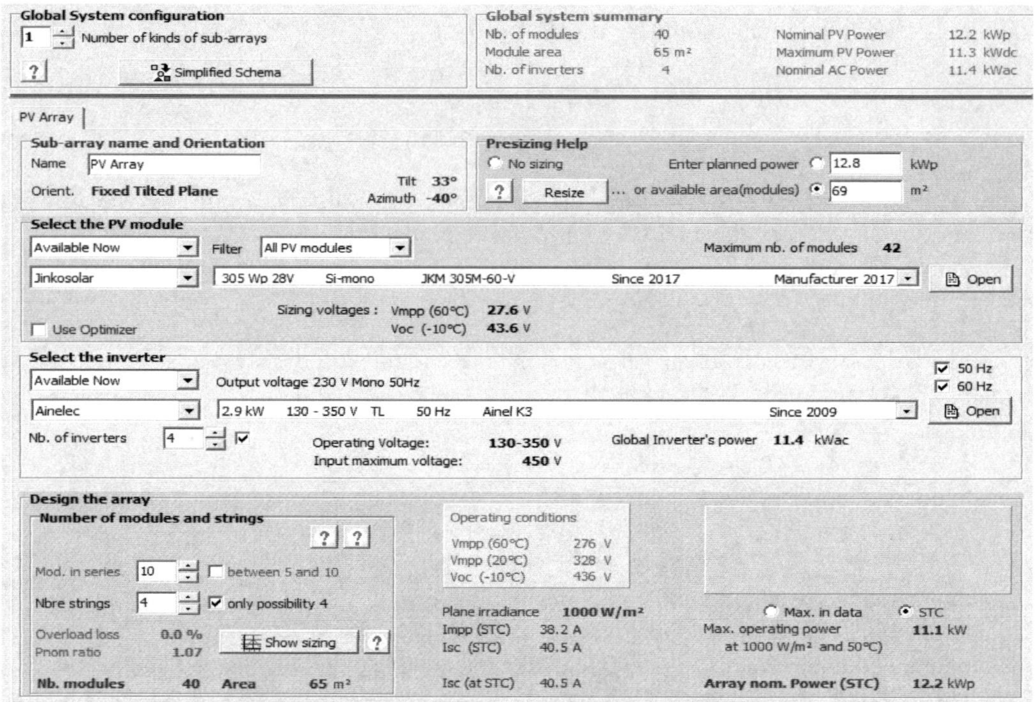

Figure 2. Initial data of a solar substation.

We calculated the number of solar panels and determined their location with the maximum efficiency in real operating conditions (Figure 3). Solar panels are located in 4 rows of 10 photo modules each with a total capacity of 12.20 kW and occupy a total area of 65 m^2.

We chose an Ainelec K3 130-350V/50Hz inverter with a maximum efficiency of 98%.

The solar battery tilt is 33°(the optimal tilt for the eastern region of Ukraine and the location of the substation) and the azimuth is 40°. Such parameters guarantee the highest "potential" indicator of energy production [10].

PVSYST V6.86	20/02/20	Page 1/5

Grid-Connected System: Simulation parameters

Project :	**Zmiyiv**				
Geographical Site		**Zmiyiv**		Country	**Ukraine**
Situation		Latitude	49.67° N	Longitude	36.35° E
Time defined as		Legal Time	Time zone UT+2	Altitude	85 m
		Albedo	0.20		
Meteo data:		**Zmiyiv**	NASA-SSE satellite data 1983-2005 - Synthetic		

Simulation variant :	**Traction substation №18 Zmiyiv**				
		Simulation date	20/02/20 14h56		

Simulation parameters	System type	**Building system**		
Collector Plane Orientation	Tilt	33°	Azimuth	-40°
Models used	Transposition	Perez	Diffuse	Perez, Meteonorm
Horizon	Free Horizon			
Near Shadings	Linear shadings			
User's needs :	Unlimited load (grid)			

PV Array Characteristics					
PV module	Si-mono	Model	**JKM 305M-60-V**		
Original PVsyst database		Manufacturer	Jinkosolar		
Number of PV modules		In series	10 modules	In parallel	4 strings
Total number of PV modules		Nb. modules	40	Unit Nom. Power	305 Wp
Array global power		Nominal (STC)	**12.20 kWp**	At operating cond.	11.06 kWp (50°C)
Array operating characteristics (50°C)		U mpp	289 V	I mpp	38 A
Total area		Module area	**65.5 m²**	Cell area	57.0 m²
Inverter		Model	**Ainel K3**		
Original PVsyst database		Manufacturer	Ainelec		
Characteristics		Operating Voltage	130-350 V	Unit Nom. Power	2.85 kWac
Inverter pack		Nb. of inverters	4 units	Total Power	11.4 kWac
				Pnom ratio	1.07

PV Array loss factors				
Thermal Loss factor	Uc (const)	20.0 W/m²K	Uv (wind)	0.0 W/m²K / m/s
Wiring Ohmic Loss	Global array res.	127 mOhm	Loss Fraction	1.5 % at STC
Module Quality Loss			Loss Fraction	-0.8 %
Module Mismatch Losses			Loss Fraction	1.0 % at MPP
Strings Mismatch loss			Loss Fraction	0.10 %
Incidence effect, ASHRAE parametrization	IAM =	1 - bo (1/cos i - 1)	bo Param.	0.05

Figure 3. Characteristics of the substation modeling.

The calculations show that this solar substation completely covers the value of the auxiliary power consumption of the design traction substation. The auxiliary power consumption of the traction substation is 11.08 kW, the solar substation will generate 12.20 kW, which completely covers the electricity consumption.

The location of solar panels on the roof of the traction substation is illustrated by the 3D model in PVsyst program (Figure 4). The model of the traction substation is as close as possible to reality. The solar panels are located on the roof of the traction substation with a tilt of 33° and are directed to the east to maximize energy production.

Grid-Connected System: Near shading definition

Project : **Zmiyiv**
Simulation variant : **Traction substation №18 Zmiyiv**

Main system parameters	System type	**Building system**		
Near Shadings	Linear shadings			
PV Field Orientation	tilt	33°	azimuth	-40°
PV modules	Model	JKM 305M-60-V	Pnom	305 Wp
PV Array	Nb. of modules	40	Pnom total	**12.20 kWp**
Inverter	Model	Ainel K3	Pnom	2850 W ac
Inverter pack	Nb. of units	4.0	Pnom total	**11.40 kW ac**
User's needs	Unlimited load (grid)			

Perspective of the PV-field and surrounding shading scene

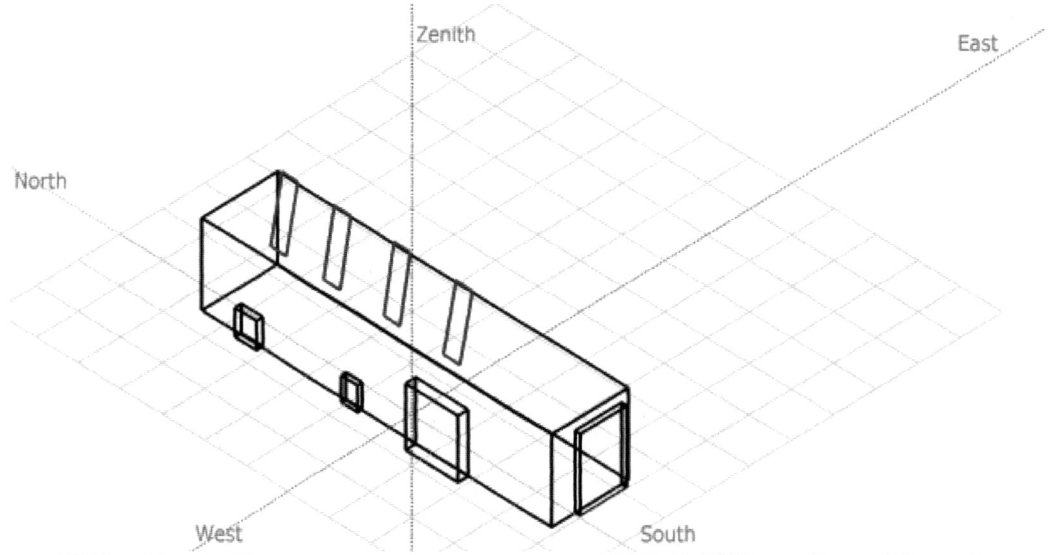

Figure 4. 3D model of the traction substation withy solar panels installed.

The efficiency of the solar substation for the calendar year is illustrated by figure 5, where the total energy obtained from alternative sources is 13.92 MW per year. The performance ratio of the solar substation is 84.57%. The average indicator of alternative energy obtained is 3.13 kW per day, the average loss of alternative energy is 0.57 kW per day.

PVSYST V6.86		20/02/20	Page 3/5

Grid-Connected System: Main results

Project : **Zmiyiv**
Simulation variant : **Traction substation №18 Zmiyiv**

Main system parameters	System type	**Building system**		
Near Shadings	Linear shadings			
PV Field Orientation	tilt	33°	azimuth	-40°
PV modules	Model	JKM 305M-60-V	Pnom	305 Wp
PV Array	Nb. of modules	40	Pnom total	**12.20 kWp**
Inverter	Model	Ainel K3	Pnom	2850 W ac
Inverter pack	Nb. of units	4.0	Pnom total	**11.40 kW ac**
User's needs	Unlimited load (grid)			
Main simulation results				
System Production	**Produced Energy**	**13.92 MWh/year**	Specific prod.	1141 kWh/kWp/year
	Performance Ratio PR	84.57 %		

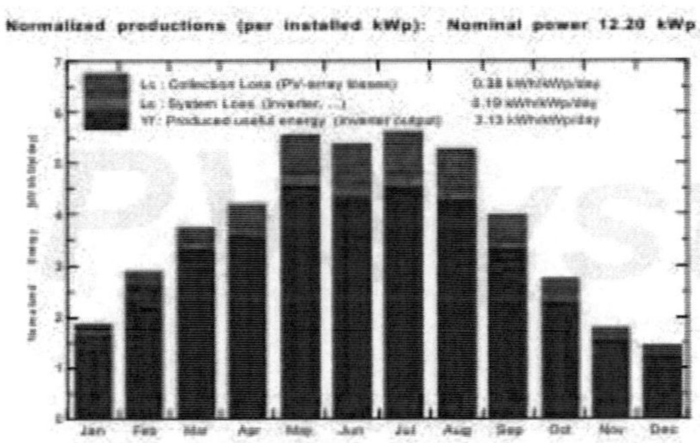

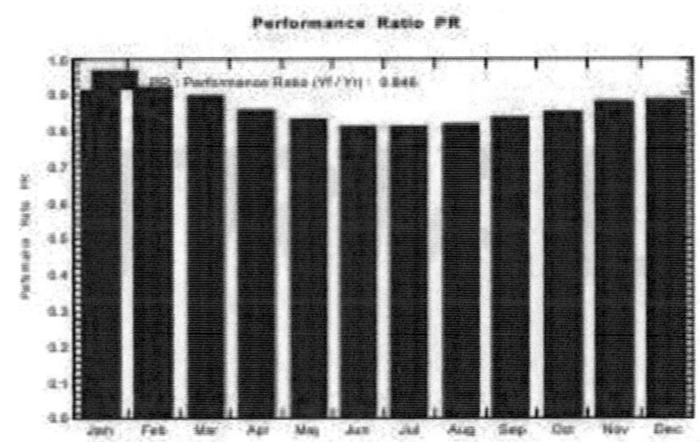

Traction substation №18 Zmiyiv
Balances and main results

	GlobHor kWh/m²	DiffHor kWh/m²	T_Amb °C	GlobInc kWh/m²	GlobEff kWh/m²	EArray MWh	E_Grid MWh	PR
January	36.9	22.32	-5.24	57.5	54.3	0.680	0.640	0.912
February	56.8	30.52	-4.86	80.4	76.6	0.953	0.902	0.920
March	94.6	50.22	0.72	115.1	110.4	1.333	1.262	0.899
April	117.6	63.30	9.83	125.9	120.9	1.392	1.312	0.854
May	166.8	77.50	16.58	171.2	164.4	1.830	1.731	0.829
June	163.8	80.70	20.21	160.4	153.7	1.679	1.584	0.810
July	172.4	79.98	22.71	174.1	167.1	1.818	1.719	0.809
August	151.3	69.13	22.13	162.9	156.7	1.709	1.618	0.814
September	104.7	51.60	16.21	119.7	114.9	1.291	1.220	0.836
October	65.1	35.03	9.06	84.5	80.6	0.936	0.879	0.852
November	35.7	22.30	0.55	53.2	50.3	0.613	0.573	0.883
December	27.9	17.98	-4.57	43.9	41.1	0.510	0.475	0.886
Year	1193.5	600.47	8.69	1348.8	1290.9	14.745	13.916	0.846

Legends:			
GlobHor	Horizontal global irradiation	GlobEff	Effective Global, corr. for IAM and shadings
DiffHor	Horizontal diffuse irradiation	EArray	Effective energy at the output of the array
T_Amb	T amb.	E_Grid	Energy injected into grid
GlobInc	Global incident in coll. plane	PR	Performance Ratio

PVsyst Evaluation mode

Figure 5. Simulation results for power generation by a solar power station.

The simulation results indicate that the most productive period for the production of alternative energy is from June to August.

IOP Conf. Series: Materials Science and Engineering 1021 (2021) 012008 doi:10.1088/1757-899X/1021/1/012008

The average energy obtained through this period is 4.20 kW per day, the average loss per day is 0.92 kW. In winter, efficiency decreases but does not disappear.

The least favorable period for alternative energy production is the period from November to January. The amount of energy received is 1.6 kW per day, the average loss per day is 0.15 kW. The traction substation was modeled taking into account all climatic conditions of Kharkiv region. The results of calculations of the shading factor at different times of the year are shown in Figure 6.

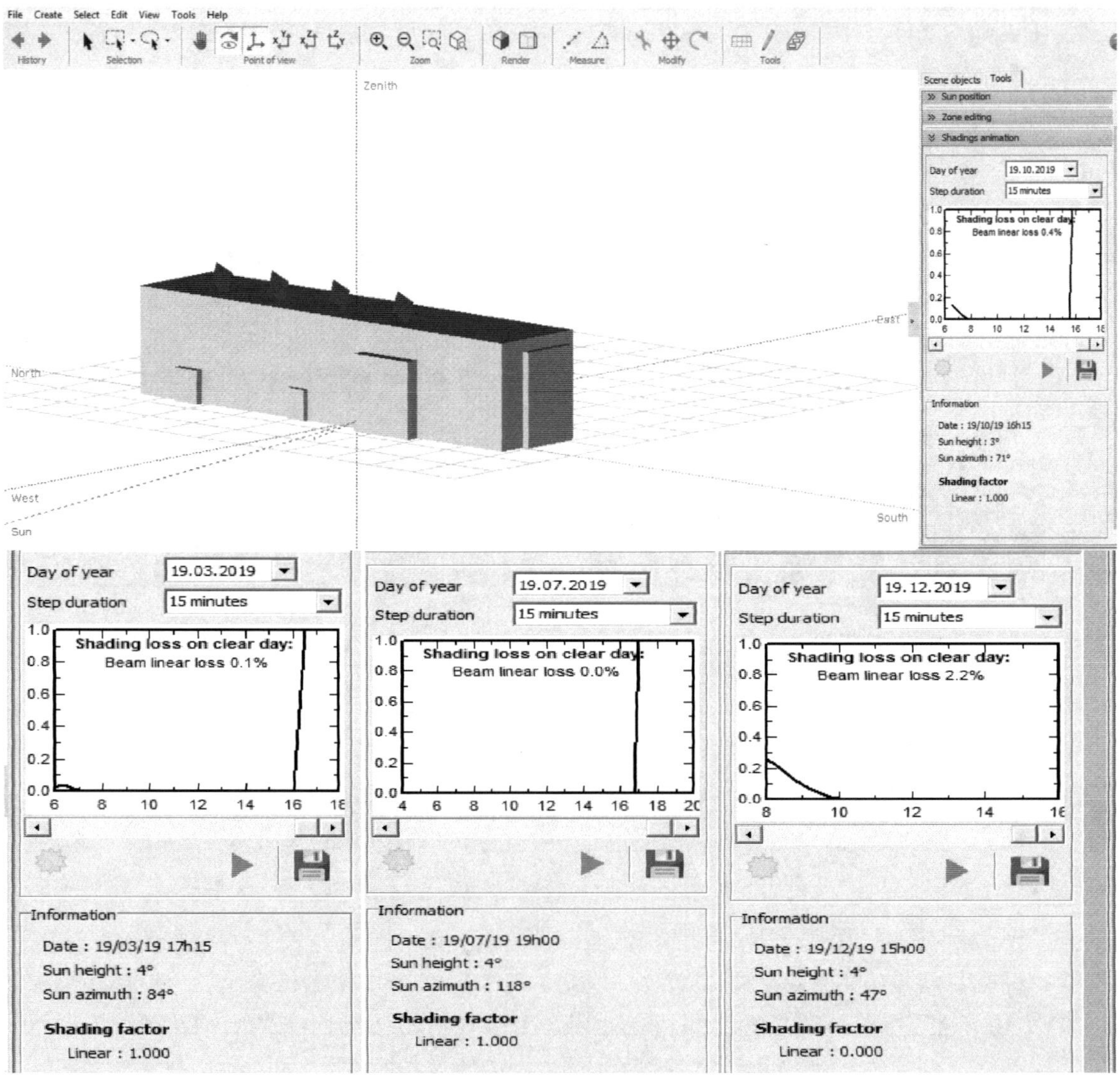

Figure 6. Seasonal shading factor

The greatest shading of solar panels occurs in the period from November to March, when it is about 40% of daylight, and the smallest shading of solar panels is observed from April to September – up to 15%.

5. Conclusions

The current preconditions for the use of electricity obtained from non-traditional sources for power supply to non-traction consumers have been analyzed.

Having summed up studying the joint work of the auxiliary consumption of the traction substation with non-traditional energy sources, we have:

- calculated the number of solar panels and their ability to supply auxiliary power of the traction substation;
- modeled alternative energy production and defined shading factor for each season;
- determined that the total energy obtained from alternative sources is 13.92 MW per year, which is sufficient to compensate power consumption of the traction substation.

References

[1] http://zakon4.rada.gov.ua/laws/show/601-17 [Elektronnyi resyrs], Zakon Ukrainy, shchodo vstanovlennya «Zelenogo» taryfu

[2] Analiz roboti gospodarstva elektrifikacії ta elektropostachannja v 2013 roci. – K. : TOV «VD «Manufaktura», 2014. - 256 s.

[3] Sychenko V, Bondar O and Prykhoda M 2015 Analysis of the solar generation influence on the operation of traction substations on electrified railways *Lighting Eng. and Power Eng.* **1** 10-7.

[4] Kovalev O.P. Vozobnovljaemye istochniki jenergii i jenergoobespechenie avtonomnyh potrebitelej // Trudy DVGTU. Vyp. 134. Teplojenergetika. – Vladivostok: Izd-vo DVGTU, 2003. – S. 16 – 20.

[5] Requirements for wind and solar photovoltaic power plants for connection to external electrical networks http://www.uself.com.ua/fileadmin/documents/Wind_and_Solar_PV_Tech_Req__Final__Version_Ukrainian.pdf

[6] Pastushenko M 2013 Prospects for implementing renewable energy sources on Ukrainian railways *En Sav. Power Eng. En. Audit* **12** 45-51

[7] Solar panels Jinko Solar Eagle PERC 60 JKM305M-60 https://7-vz.com/ua/product/polikristallicheskiy_modul_jinko_solar_eagle_perc_60_jkm305m_60/

[8] Alokhin, V.A. Sleepy energy fields (in Russian) / V.A. Sleepy energy fields. Alokhin // Zvistki Tul'skogo power university. Technical sciences. - 2013. - No. **12**. - C. 3 8.

[9] Poliakh O, Kuhaienko J and Reshetniak T Opportunities for using alternative sources of energy for auxiliary power consumption of a traction substation 2016 *Proc. IX Int. Sc.-Pr. Conf. TRANSELECTRO–2016 (Dnipro)* (Dnipro: DNURT) p. 26

[10] Goncharov J, Sychenko V, Bosyi D, Pastushenko M and Kosarev E 2014 Increasing efficiency of traction energy operation when using renewable sources of energy *Problemy Kolejnictwa* **162** 65-82

IOP Conf. Series: Materials Science and Engineering 1021 (2021) 012009 doi:10.1088/1757-899X/1021/1/012009

Development of a methodology for determining an energy efficient technology for the freight transportation on a single-track railway line

A. Prokhorchenko[1], O. Malakhova[1], D. Gurin[1], G. Sikonenko[1], and G. Prokhorchenko[1]

[1] Department of Management of Operational Work, Ukrainian State University of Railway Transport Feierbakh sq., 7, Kharkiv, 61050, Ukraine

E-mail: prokhorchenko@kart.edu.ua

Abstract. In the context of the search for strategies for the development of railway infrastructure, it is important to theoretically substantiate the choice of constructive and organizational measures that will allow the development of promising passenger and freight traffic. This study proposes an approach to determining the rational parameters of energy efficient technology for the transportation of freights on the site in the context of the implementation of one of the reconstruction measures - modern dispatch centralization of train traffic control. The main aim of research is to develop a methodology for determining an energy efficient technology for transporting freights on a single-track railway line. It is proposed to use analytical methods for calculating the average speed of trains with a combination of micromodeling of traction characteristics of freight trains of different mass. Theoretical calculations of the change in the section speed were carried out when the dispatch centralization was introduced, taking into account the choice of various options for the mass of freight trains. The calculations performed for a single-track line and the dependences of the cost change per one shipped carriage and the section speed on the train mass were obtained, which made it possible to determine the optimal parameters of the energy-efficient technology for the transportation of freights by the line. The proposed method for determining the energy efficient technology for the transportation of freights can be used in the technical and economic calculations of projects for the implementation of various systems of train dispatch control on single-track railway lines.

1. Introduction

In the context of the search for strategies for the development of railway infrastructure, it is important to theoretically substantiate the choice of constructive and organizational measures that will allow the development of promising passenger and freight traffic [1]. One of the elements of the railway infrastructure that often become limiting in the network and need to improve the organization of transportation or technical equipment is single-track lines. To increase the competitiveness of rail transportation, limit impact, reduce costs, it is important to determine the feasibility of implementing the proposed measures, taking into account the efficient use of energy, resources – energy efficiency [2, 3, 4, 5]. Considering that in the field of train operation there is a significant potential for reducing energy consumption without any significant capital costs, this study proposes to consider in detail the approach to determining the rational parameters of energy efficient technology for transporting freights on the site in the context of the introduction of modern dispatch centralization of train traffic control.

2. Literature review and problem statement

Taking into account the necessity to search for measures to improve the competitiveness of rail transport in the context of the importance of reducing costs and reducing environmental impact, the number of studies devoted to improving energy conservation and energy efficiency in the railway industry is greater [6, 7]. There are quite a lot of studies aimed at finding energy efficient technologies for the transportation of freights by optimizing train operation strategies [8, 9, 10]. This is driven by the high potential for reducing energy costs through improved driving and driven by the significant rate of cost reduction in road transport, which is one of the main competitors for the railway.

It is reviewed that the calculation of energy efficient train speed profiles is usually called the train trajectory optimization problem. According to [11], determining an energy efficient train path can lead to potential energy savings in the range of 5-20%. In the work [12] it is proposed to optimize the costs of train movement by minimizing its thrust energy based on solving a nonlinear optimization problem by methods of evolutionary search – Genetic Simulated Annealing, Firefly, and Big Bang-Big Crunch algorithms. Within the framework of the problem statement, speed limits, profile, norms of travel time and train weight were taken into account, and punctuality as a penalty factor. Research [13] in determining energy efficient trajectories of train movement focused on taking into account the effect of forecast changes in wind speed and direction, which is available before departure, on train movement. In [15], optimization modeling is applied to solve the issues of aerodynamics and energy efficiency of railway intermodal trains. The authors note that the potential to reduce fuel consumption per intermodal train is 15,000,000 gallons per year, with a corresponding savings of 28 million USD. This confirms the importance of introducing energy efficient technologies for the freights transportation.

However, the focus on improving driving by better adapting to track profile does not allow for the interdependence of train traffic in the flow, and therefore has limited potential for cost savings. Much higher energy costs arise from improperly planned stops of freight trains in lines under overtaking or crossing operations on single-track lines. Under such conditions, studies are aimed at finding balanced train schedules, reducing the number of stops and delays, and making it possible to determine energy efficient train trajectories, taking into account the traffic plan on the railway line, or the network polygon, are promising and relevant. It should be noted that the most difficult task is to build a train schedule for a single track line [14]. In [16], an optimization model is proposed, taking into account a possible speed limit to find the optimal speed for each train on a single-track railway line. An interesting study is [17], which notes the importance of the influence of train length on energy efficiency. It is proposed to expand the mathematical model of scheduling, which takes into account train length, energy efficiency and minimizes not only delays, but also energy consumption.

One of the measures to increase the capacity of a single-track railway line is the introduction of more modern train control systems [18]. The construction of an energy efficient technology for the transportation of freights on a railway line when introducing a new dispatching system requires a combination of the above approaches. It is important to determine the optimal train trajectories, taking into account their interdependence in the train schedule. In the work [19], relying on the simulation of train movements using the "OpenTrack" software, scenarios of the equipment of the railway line with ETCS Level 2 and ETCS Level 3 (L3) are investigated. The OpenTrack software application is developed by the Swiss Federal Institute of Technology's Institute for Transportation Planning and Systems (ETH IVT) [4]. This is a train movement model, refers to a micro-level simulation, designed for traction calculations, train scheduling and simulation based on train databases, infrastructure and timetables defined by the user.

However, these methods and software products can't be used for the 1520 mm railway track of the post-Soviet railway space, in particular, the Ukrainian railway and the like. Due to the various technical characteristics of the infrastructure and the peculiarities of the organization of transportation.

The Ukrainian railway system belongs to the railways of mixed movement of passenger and freight trains on the same infrastructure with partial observance of the freight train timetable. This makes it difficult to determine the average speed of train movement due to the lack of a timetable for the movement of freight trains on the site after the implementation of the proposed centralization dispatch system. Under such conditions, approaches with a combination of micro- and macromodeling of train traffic are more acceptable to determine an energy-efficient technology for transporting freights. One of the software applications that allows for traction calculations is the "Traction calculations" module of the "Railway" grapho-analytical system GAS [20, 21].

To solve the problem in this work, it is proposed to use analytical methods for calculating the average speed of trains with a combination of micromodeling of traction characteristics of freight trains of different weights. The work carried out realistic calculations for a single-track line and obtained the dependences of the cost change per one shipped car and the section speed on the train mass, which made it possible to determine the optimal parameters of the energy-efficient technology for the transportation of freights by the line. The rest of the articles are structured as follows. Line 4 presents the results. After that, a conclusion is made and instructions for further research are given.

3. Materials and methods

Considering that one of the least costly measures to increase the throughput and carrying capacity of a single-track line is the introduction of dispatch centralization, it is proposed to study in more detail the possibility of searching for rational parameters of energy-efficient technology for transporting freights on a single-track line. Using the example of the Poltava - Kremenchuk - Burty line of the regional branch "Southern Railway" of Ukrzaliznytsia JSC, it is proposed to theoretically substantiate the introduction of dispatch centralization with the search for an energy efficient technology for transporting freights on the line. The Poltava - Kremenchuk - Burty line is a double-track line with a limiting element, the Kremenchuk - Kriukov-on-Dnipro stretch, which is single-track. It is for this stretch that experimental calculations have been proposed. Line Kremenchug - Burty electrified with alternating current. The line Kremenchuh - Kriukov-on-Dnipro is equipped with automatic blocking, and all stations are equipped with electrical interlocking.

To search for an energy efficient technology for the transportation of freights with the introduction of dispatch centralization, the study proposed to find a balance between the energy costs arising when a line of freight trains of different mass and length moves along and their effect on the section speed, which characterizes the quality of the organization of the transportation process on the line. In this study, the section speed is understood as the average speed of trains on the line, taking into account the time spent on acceleration, deceleration and parking at separate points.

Taking into account the limited initial data for technical and economic calculations with sufficient accuracy for practical purposes, the site speed (V_s) can be found using the speed coefficient (β_x) by the formula

$$V_s = \beta_x \cdot V_m, \tag{1}$$

where β_x – speed coefficient; V_m – train movement by the line, excluding stops at intermediate stations and time for acceleration and deceleration, km/h.

The speed factor makes it possible to obtain the most qualitative characteristic of the impact of measures for the implementation of dispatch centralization on the section speed. The speed factor reflects the effect on the section speed of both the total duration of train stops at intermediate stations and the time spent on stops to complete crossings and overtakes.

To determine the estimated sectional speed V_s, it is proposed to use the analytical dependence of the speed coefficient proposed by Professor B. Maksimovich [22]

$$\beta_x = \frac{V_s}{V_t} = \frac{T'_{tt} + T''_{tt}}{T'_{tt} + T''_{tt} + T_{st}} = 1 - \frac{T_{st}}{T'_{tt} + T''_{tt} + T_{st}}, \tag{2}$$

where $T'_{tt} + T''_{tt}$ is the travel time of a pair of trains on the line without stopping at intermediate stations ("clean" travel time); T_{st} is the total stopping time of a pair of trains at intermediate stations, including the time for acceleration and deceleration; V_t is technical speed, km/h.

From formula (2) it can be seen that in order to establish the dependence of the local speed on the factors determining it, it is necessary to find the dependence of the value on the same factors. This time for freight trains for single-track lines can be expressed as the sum of the parking time at crossing and under overtaking

$$T_{st} = t_{cr} K_{cr} + t_{ov} K_{0v}, \tag{3}$$

where K_{cr} is the total number of train crossings on a pair of freight trains; K_{0v} is the number of overtaking freight trains by passenger per pair of freight trains; t_{cr}, t_{ov} is average duration of the train stop, respectively, for crossing and overtaking, min.

The approximate number of crossings on the line of freight trains with freight trains can be determined (Fig. 1) by the formula

$$k_{cr}^{fr} = \frac{T'_{tt} + T''_{tt} + T_{st}}{I_{gr}} - \delta k = \frac{(T'_{tt} + T''_{tt} + T_{st}) N_{fr}}{1440} - \delta k, \tag{4}$$

where I_{gr} is estimated graphical interval for the possible departure of freight trains in this direction, min, $I_{gr} = \frac{1440}{N_{fr}}$; δk is correction due to non-multiples time $T'_{tt} + T''_{tt} + T_{st}$ of the interval I_{gr} and varies widely $0 \div 1$.

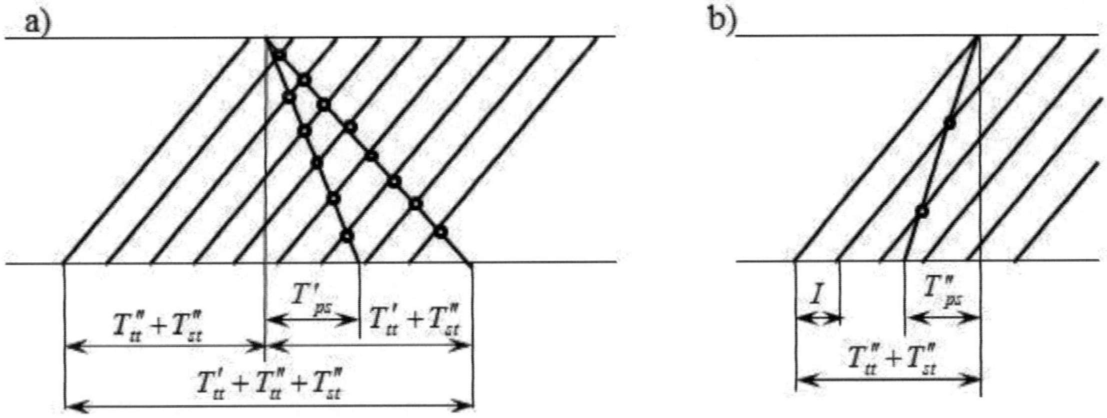

Figure 1. Scheme of the number of crossing and overtaking of trains on the line: a – scheme of the number of crossings; b – scheme of the number of overtaking

When freight trains cross each other, one of them, as a rule, is passed through a separate point continuously, so the number of crossings (4) will also determine the number of stops per pair of freight trains.

The number of crossings of freight trains with one passenger train will be

$$k'_{cr} = \frac{T'_{tt} + T''_{st} + T_{ps}}{1440} - \delta k\,, \tag{5}$$

where T_{ps} is travel time on a line of a passenger train of one direction, min; T''_{st} is stopping time for freight trains, per train, min.

Number of crossings of freight trains with a pair of passenger trains

$$k^{PS}_{cr} = k'_{cr} + k''_{cr} = \frac{(T'_{tt} + T''_{tt} + T_{st} + T'_{ps} + T''_{ps})N_{fr}}{1440} - 2\delta k\,. \tag{6}$$

Let us denote the ratio of the travel time of a pair of passenger and freight trains $\frac{T_{ps}' + T_{ps}''}{T_{tt}' + T_{tt}''}$ by Δ. Then formula (6) can be represented as:

$$k^{PS}_{cr} = \frac{\left[(T'_{tt} + T''_{tt}) + \Delta + T_{st})\right]N_{fr}}{1440} - 2\delta k\,. \tag{7}$$

Since freight trains pass them when crossing with passenger trains, formula (7) also expresses the number of stops of freight trains on a couple of passenger trains, that is $k^{ps}_{cr} = k^{ps}_{ov}$.

Substituting dependence (4, 7) into formula (2), we obtain a detailed expression for determining the speed coefficient β_x

$$\beta_x = 1 - \frac{N_{fr}t_{cr} + N_{PS}\left[(1+\Delta)t_{cr} + (1-\Delta)t_{ov}\right]}{1440 - N_{PS}(t_{ov} - t_{cr})}\,. \tag{8}$$

Formula (7) only very roughly reflects the dependence of the speed coefficient on the main factors that determine the conditions for the passage of trains on the line: the size of the movement of freight and passenger trains and the duration of train stops during crossing and overtaking. To obtain a calculation formula from it, it is necessary to take into account: the coincidence of crossings with overtaking, the time intervals between trains at the extreme race, lines, the ratio of the duration of parking under overtaking and when crossing, and some other conditions and provisions. The influence of all factors on the speed factor can be taken into account by the appropriate type of graph.

To clarify the calculations, taking into account the type of schedule, the section speed coefficient on a single-track line with a normal (non-batch) traffic schedule can be obtained on the basis of formula (8), taking into account the fact that more accurately the number of train stops for crossing will be determined from the expression:

$$k_{cr} = k^{fr}_{cr} + k^{PS}_{cr} - k_C\,, \tag{9}$$

where k_C – the number of crossings that are combined with overtaking.

As the analysis of train schedules shows, about 90% of overtaking is combined with crossed trains. Hence:

$$k_C = 0{,}9k_{ov} = \frac{0{,}9(1-\Delta)(T'_{fr} + T''_{fr})N_{PS}}{1440} - \delta k \cdot 1{,}8\frac{N_{PS}}{N_{fr}}\,, \tag{10}$$

When this correction is included in formula (8) and after transformations, let's obtain the formula:

$$\beta_X = 1 - \frac{(0{,}75 + 0{,}3\gamma_f^{GR})[(N_{fr} + 1{,}2\Delta N_{PS})t_{cr} + (1 - 0{,}7\Delta)N_{PS}t_{ov6}]}{1440}, \tag{11}$$

where γ_f^{GR} - the fill factor of the graph.

Usually, the average value of the graph filling can be accepted $\gamma_f^{GR} = 0{,}5$. Substituting these values in (11), let's obtain:

$$\beta_X = 1 - \frac{(0{,}75 + 0{,}3 \cdot 0{,}5)[(N_{fr} + 1{,}2\Delta N_{PS})t_{cr} + (1 - 0{,}7\Delta)N_{PS}t_{ov}]}{1440}, \tag{12}$$

$$\Downarrow$$

$$\beta_X = 1 - \frac{0{,}9 \cdot [(N_{fr} + 1{,}2\Delta N_{PS})t_{cr} + (1 - 0{,}7\Delta)N_{PS}t_{ov}]}{1440}, \tag{13}$$

$$\Downarrow$$

$$\beta_X = 1 - \frac{0{,}9/0{,}9 \cdot [(N_{fr} + 1{,}2\Delta N_{PS})t_{cr} + (1 - 0{,}7\Delta)N_{PS}t_{ov}]}{1440/0{,}9}, \tag{14}$$

$$\Downarrow$$

$$\beta_X = 1 - \frac{1 \cdot [(N_{fr} + 1{,}2\Delta N_{PS})t_{cr} + (1 - 0{,}7\Delta)N_{PS}t_{ov}]}{1600}. \tag{15}$$

where N_{fr} - the number of pairs of freight trains; N_{PS} - the number of pairs of passenger trains; t_{cr}, t_{ov} - respectively, the average stops of freight trains under crossing and overtaking, min.

Taking into account that, within the framework of the problem statement, it is necessary to calculate the section speed depending on the mass of the freight train, which determines the time of its movement T'_{tt} or T''_{tt}, in expression (15), by the variables, the duration of the crossing operation t_{cr} and the overtaking operation t_{ov}, and the size of freight trains N_{fr} on the line.

The most probable value of the average duration of train stops under the crossing can be taken equal to

$$t_{cr} = \frac{t_{cr}^{\min} + t_{cr}^{\max}}{2} = \frac{k_{cr}^{fr} t_{cr}^{fr} + k_{cr}^{PS} t_{cr}^{PS}}{k_{cx}^{вант} + k_{cx}^{пас}}. \tag{16}$$

Under such conditions, after the calculations of the search for the weighted average time of the duration of the crossing operation between trains of different categories for different variants of the mass and length of freight trains, it is possible to calculate the section speed according to expression (1).

4. Results

To determine the travel time and power consumption of a freight train on the line, depending on its weight, traction calculations were carried out for the Kremenchuk - Burty line with the maximum stretch Kriukov-on-Dnipro - Burty, the steering grade of calves i_{st} =8.3 ‰. The calculations were carried out using the GAS "Railway" software package [20, 23, 24]. The calculations were carried out for freight locomotives – VL80T throughout the entire line. The results of the performed thrust calculations are shown in table 1.

Based on the results of traction calculations and the found running speeds of freight trains of different weights in the study with the above proposed approach (see line 3), the precinct speed calculations are carried out. The dependences of the costs of train traction and section speed on the mass of a freight train on the line are shown in Figure 2.

Table 1. Results of the performed traction calculations for the maximum distance Kremenchuk-Burty

Parameters	Train mass, t						
	4600 t	5000 t	6000 t	6100 t	6200 t	6300 t	7000 t
	m=57 c.w.	m=57 c.w.	m=65c.w.	m=67c.w.	m=69c.w.	m=71c.w.	m=77c.w.
Distance, km	19,374						
steering grade	ict = 8,3 ‰						
multiplicity of thrust			Kriukov-on-Dnipro station – Burty station push locomotive				
Total electricity consumption, kWh	1865,33	1940,23	2568,15	2602,5	2635,45	2663,39	2913,23
Electricity consumption for additional traction, kW*h	-	-	967,91	983,15	999,19	1010,9	1121,66
Electricity cost for train traction, UAH	1427,17	1510,379	2795,668	2851,895	2902,72	2950,737	3340,875
Travel time, min	25,79	26,24	26,65	26,81	26,92	27,07	27,91
Technical speed V_{tech}, km/h	43,1	43,1	43,1	43,1	43,1	43,1	41,5

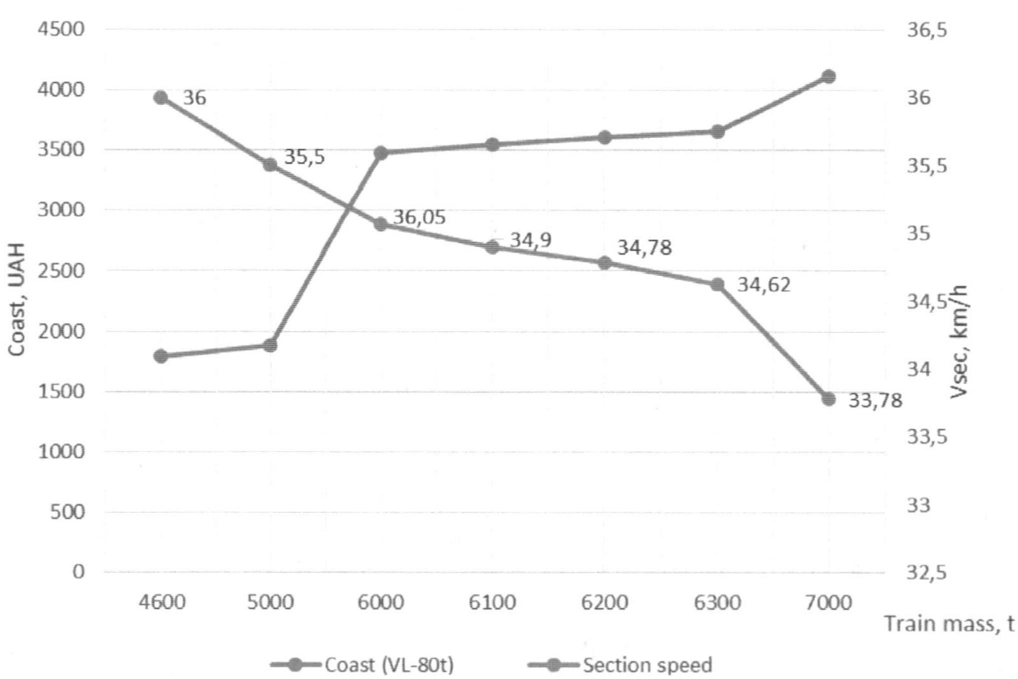

Figure 2. Dependence of electricity costs for train traction and section speed on train

IOP Conf. Series: Materials Science and Engineering 1021 (2021) 012009 doi:10.1088/1757-899X/1021/1/012009

weight

The dependence is shown in Figure 2 assesses the technology of freights transportation on the site under the conditions of the current dispatch control system for train traffic. However, in order to assess measures to increase the throughput and carrying capacity of a railway line, it is important to assess the change in the operating parameters of the line in the context of the introduction of dispatch centralization (DC). After the implementation of the DC, it is planned to improve the technical and technological capabilities of train dispatching due to the possibility of implementing zero-point crosses. This is possible in the conditions of automatic control of train movement on the line and improvement of the quality of planning of train traffic, including by eliminating negotiations between the train dispatcher and the station attendants. The main condition for ensuring a zero-point crossing of trains is the presence of two tracks on the Kriukov-on-Dnipro - Burty line. An analysis of the conditions for the implementation of a zero-point crossing of trains in accordance with the Instructions for Determining Station and Interval Intervals [25] confirms the possibility of performing this operation on the boundary element of the line. The specified conditions correspond to the track and technical support for the implementation of DCs on the tracks and separate points of the Kremenchug-Burty line. Based on the Instruction (SD 0034) and data on the permissible speeds and distances of the track development of the race and the receiving-departure tracks on the line, the calculation of station intervals and section speed after the DC implementation was carried out.

According to the calculations, the dependence of the cost per wagon dispatched and the section speed on the train weight on the single-track line Kremenchuh-Burty is found after the introduction of dispatch centralization, which is shown in Figure 3.

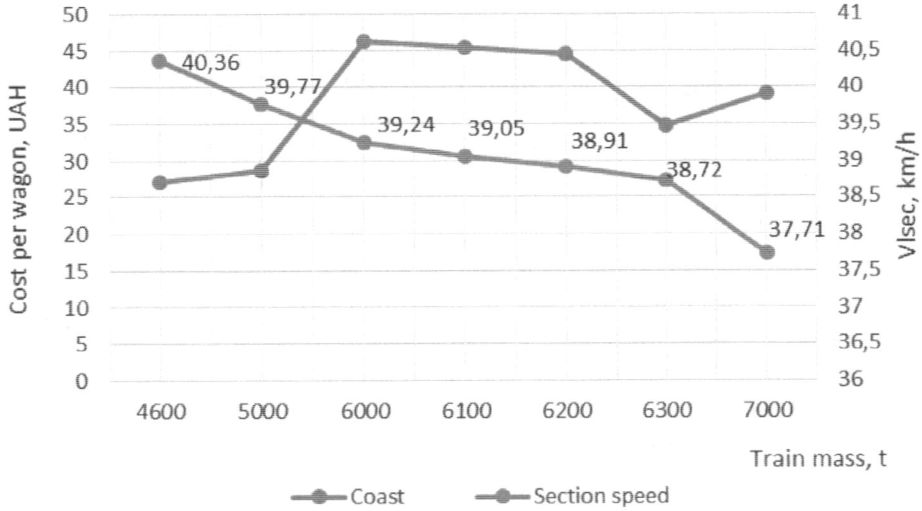

Figure 3. Dependence of the cost per wagon dispatched and the section speed on the train weight on the Kremenchuk-Burty single-track line after the DC implementation.

According to the calculations, the introduction of dispatch centralization and the organization of the passage of trains according to the zero-point crossing scheme will increase the section speed, reduce the time of line movement and reduce the cost of electricity for train traffic.

Figure 3 shows that, as with the DC implementation (Fig. 2), there is a decrease in the section speed with an increase in the train mass, however, the total electricity consumption for traction of trains and the cost per carriage decreases. The most rational mass from the point of view of increasing the energy efficiency of freight traffic on the line, the rational mass of a freight train is 6300 tons and is achieved by using a push locomotive (double traction).

5. Conclusion

Based on the results of this study, a method is proposed for determining an energy efficient technology for transporting freights on a single-track line. Using the example of the Kremenchuk - Burty line of the regional branch "Southern Railway" of JSC "Ukrzaliznytsia", it is proposed to theoretically substantiate the introduction of dispatch centralization with the search for an energy efficient technology for transporting freights on the line. Experimental calculations have shown that the introduction of modern systems of centralized dispatch and dispatch control at the Poltava -Pivdenna - Kremenchuk - Burty lines will reduce the time for making managerial decisions and their implementation, especially in non-standard and emergency situations, which will increase the section speed. The most rational mass of a freight train has been found from the point of view of increasing the energy efficiency of freight traffic on the line after the DC implementation, which is 6300 tons. When operating freight trains weighing 6300 tons, after the implementation of the DC, it will allow reaching a section speed of 38.72 km/h, which is 10.6% more than the section speed under the conditions of the current train traffic dispatch control system. Comparison of the proposed methodology for calculating the section speed based on establishing the running speeds of trains based on traction calculations and determining the weighted average duration of crossing and overtaking operations with the real data of the line operation confirms the adequacy of the calculation results. The discrepancy between the empirical and calculated data of the section speed for the given volumes and characteristics of passenger and freight trains do not exceed an error of 1%. The proposed method for determining the energy efficient technology for the freights transportation can be used in the technical and economic calculations of the proposed projects for the implementation of various train dispatch control systems. In further studies, it is planned to improve this approach based on taking into account the most probable train delays on the line when calculating the section speed.

References

[1] Lovska A, Fomin O, Píštěk, V and Kučera P 2020 Dynamic Load Modelling within Combined Transport Trains during Transportation on a Railway Ferry. *Appl. Sci* **10** p 5710.

[2] ISO 50001:2011. Energy management systems – Requirements with guidance for use.

[3] Matsiuk V, Myronenko V, Horoshko V, Prokhorchenko A, Hrushevska T, Shcherbyna R, Matsiuk N, Khokhlacheva J, Biziuk I and Tymchenko N 2019 Improvement of efficiency in the organization of transfer trains at developed railway nodes by implementing a "flexible model" *Eastern-European Journal of Enterprise Technologies* **2/3(98)** pp 32-39.

[4] Nash A and Huerlimann D 2004 Railroad Simulation Using OpenTrack. *WIT Transactions on The Built Environment* **74** 10.2495/CR040051.

[5] Korol S O, Moroz M M, Korol S S, Serhiienko S A 2017 Method and device for increasing weight charging of four-stroke engine cylinders. *Naukovyi Visnyk Natsionalnoho Hirnychoho Universytetu* **(5)** pp 56-61.

[6] Khayyam S et al. 2016 Evolution of business model in railway industry in the presence of Energy management system. *CIRED Workshop* (Helsinki) 14-15 June 2016.

[7] Technologies and potential developments for energy efficiency and CO2 reductions in rail systems. International Union of Railways (UIC) (Paris) December 2016. https://uic.org/IMG/pdf/_27_technologies_and_potential_developments_for_energy_efficiency_and_co2_reductions_in_rail_systems._uic_in_colaboration.pdf

[8] Bodnar B Y, Kapitsa M I, Afanasov A M and Kyslyi D N 2015 Definition of Energy Saving Acceleration Modes of Trains. *Science and Transport Progress* **5(59)** 40-52. doi: 10.15802/stp2015/55359

[9] Scheepmaker G, Goverde R, and Kroon L 2016 Review of energy-efficient train control and timetabling. *European Journal of Operational Research* **257(2)** 355-376. https://doi.org/10.1016/j.ejor.2016.09.044

[10] Chevrier R 2010 An Evolutionary Multi-objective Approach for Speed Tuning Optimization with Energy Saving in Railway Management *Conf. Record - IEEE Conference on Intelligent Transportation Systems (ITSC). 13th International IEEE Conf.* Doi:

10.1109/ITSC.2010.5625026
[11] Hansen I A and Pachl J 2014 Railway Timetabling & Operations. *Eurailpress* (Hamburg).
[12] Keskin K and Karamancioglu A 2017 Energy-Efficient Train Operation Using Nature-Inspired Algorithms *Journal of Advanced Transportation* Article ID 6173795 https://doi.org/10.1155/2017/6173795
[13] Trivella A, Wang P and Corman F 2020 The impact of wind on energy-efficient train control., *EURO Journal on Transportation and Logistics*. 100013 https://doi.org/10.1016/j.ejtl.2020.100013
[14] Alaghband M and Moghaddam B 2020 An Optimization Model for Scheduling FreightTrains on a Single Rail Track. *Scientia Iranica. International Journal of Science & Technology*. Corpus ID: 209140712 Doi:10.24200/sci.2020.53538.3304 https://arxiv.org/pdf/1912.04454.pdf
[15] Lai Y, Barkan C and Önal H 2008 Optimizing the aerodynamic efficiency of intermodal freight trains. *Transport. Res. E Logist. Transport. Rev*. **44 (5)** pp 820-834.
[16] Xu X, Li K, Yang L and Ye J 2014 Balanced train timetabling on a single-line railway with optimized velocity. *Applied Mathematical Modelling*.Volume **38** Issue 3 1 February 2014 pp 894-909.
[17] Toletti A, Martinis V and Weidmann U 2015 What about train length and energy efficiency of freight trains in rescheduling models? *18th Euro Working Group on Transportation, EWGT 2015, 14-16 July 2015, Delft, The Netherlands/ Transportation Research Procedia* **10** (2015) 584 – 594. doi: 10.1016/j.trpro.2015.09.012
[18] Miyoshi M, Takebal T and Miyatake M 2012 Development of an energy efficient train traffic control system for saving electricity *Computers in Railways XIII. WIT Transactions on The Built Environment*, WIT Press **127** pp 499-510.
[19] Wirth M and Schöbel A 2018 Mindestzugfolgezeiten bei ETCS Level 2 und Level 3 auf der Wiener S-Bahn-Stammstrecke *Signalling+Datacommunication (112)* **4/2020** pp 21-26.
[20] Prytula M G and Pasechnyk O A 2018 Development of the algorithms formation of energy-optimized trains traffic modes. *Science and Transport Progress. Bulletin of Dnipropetrovsk National University of Railway Transport* **6(78)** pp 82-100.
[21] Prytula M G and Pasechnyk O A 2019 Optimizational traction taskson the railway network. Science and Transport Progress. *Bulletin of Dnipropetrovsk National University of Railway Transport* **3(81)** pp 44-64.
[22] Maksimovich B M, Kochnev Ph P and Sotnikov I B 1960 District speed and factors affecting it. *Questions of the organization of train traffic* (Moscow: Transzheldorizdat).
[23] Instructions for using the GAS software product - "Railway" 2007 State Administration of Railway Transport of Ukraine (Lviv).
[24] Rules of traction calculations for train work 1985 (Moscow: Transport).
[25] Instructions for determining station and train intervals (SD-0034) 2001 (Kyiv: Transport of Ukraine) pp 28-65.

IOP Conf. Series: Materials Science and Engineering 1021 (2021) 012010 doi:10.1088/1757-899X/1021/1/012010

Investigation of the Influence of Dynamic Loads of Industrial Equipment on the Occurrence of Prolonged Yielding of their Foundation Soils

V A Aleksandrovych[1,2], O V Havryliuk[1]

[1] O.M. Beketov National University of Urban Economy in Kharkiv, 17, Marshal Bazhanov Street, Kharkiv, 61002, Ukraine

[2]Vadym.Aleksandrovych@kname.edu.ua

Abstract. The parameters of oscillations of the foundations and structures of the dynamic equipment of the new production workshop located in the Kharkiv region were experimentally investigated. For this purpose, appropriate measurements of the amplitude of oscillations in the vertical and orthogonal horizontal directions were performed. To compare the parameters of the amplitudes of forced oscillations and determine the damping properties of the applied rubber spacers placed between the frame and the slides, as well as the quenching effect of the foundation massif, several measurements of oscillation parameters were performed at the locations mentioned above. The amplitude of oscillations of the vertical component of the vibration load on different sides of the rubber dampers was reduced by 2-3 times, the horizontal component was completely attenuated. It was also established that there was no transmission of vibrations to the foundations of the production workshop. The safety of the measured parameters of oscillations from the point of view of long significant unstabilized subsidence of the foundations, both equipment and building, was determined. Recommendations for monitoring are given from the point of view of the possible deterioration of the properties of the used rubber dampers over time.

1. Introduction

The issue of foundations performance in terms of their perception of dynamic loads from existing industrial equipment remains one of the most serious and least studied issues in industrial construction. Despite a considerable amount of research on this topic, this situation still occurs due to the wide variety of possible machines and mechanisms and as a result the practical impossibility to predict what exactly the size, shape, and direction of dynamic loads will be generated by each machine. Therefore, when installing foundations that will transfer dynamic loads to the base, it is necessary to pay special attention to investigating the parameters of these loads.

The situation can be complicated by the possible water saturation of the base soils during operation due to seasonal or emergency groundwater level rise. When designing foundations for such equipment, in addition to the amount of static load, which must be distributed on the base soils, it is necessary to additionally determine the magnitudes, shapes and directions of forced oscillations that occur in both stationery and emergency modes of dynamic equipment. It is necessary to focus on both the technical specification data of the equipment manufacturer and the actual data obtained during the study of the operating parameters of similar equipment. Also, if possible, measures should be taken to reduce the amount of transmitted dynamic forces, such as balancing the moving and rotating parts, the installation of dampers, etc. After all, the dynamic loads transmitted to the ground of the foundation, in certain conditions, can lead to unstabilized subsidence lasting years and even decades of the mechanism operation, as well as affect the neighboring foundations of buildings and structures.

2. Definition of unsolved aspects of the problem

When designing foundations for equipment with dynamic loads, it is necessary to consider the effect of vibration on the soil base, especially if its base is composed of sandy soils, because this effect leads to the phenomena known in the literature [1-7] as vibration creep and soil thinning. These phenomena are characterized by a decrease in the strength of the base soils, increasing their compressibility and consequently cause cracks and damage to structural units when exceeding the limit of their strength from a combination of static and dynamic loads. Or they can lead to stoppage or damage of the technical equipment as a result of the offset alignment due to the difference in the yielding of the foundations of the equipment involved in one technical process. In addition to the frequency and amplitude of forced oscillations of the equipment, the vibration level is also affected by the type and size of the foundations, as well as the frequency of natural oscillations of structures and their elements and the presence of dampers. Thus, during designing, it is quite difficult to calculate the operational and non-standard parameters of oscillations of the foundations of dynamic equipment. Therefore, it is recommended to clarify these parameters, as well as post-commissioning monitoring.

3. Review of the latest research sources and publications

In modern research [1 - 2], there are mainly three phases of deformation of the base depending on the dynamics of the deformation of the soil base exposed to vibration loads. Phase 1, as a rule, occurs at relatively small static and dynamic loads, subsidence of the base in it occurs due to the reduction of soil porosity. Phase 2 is characterized by the development of significant plastic deformations in the soil mass which have a slight tendency to attenuate during the entire period of dynamic loads and almost instantly stabilize after their removal [3 - 7]. This subsidence mainly occurs in sandy (including after reaching their maximum density) water-saturated soils, but under certain conditions can occur in non-water-saturated sands and even in clay soils. Phase 3 is characterized as destructive without a tendency to stabilize. The loss of stability of the foundation occurs at a high rate, the soil in this case is considered by researchers as a viscous medium, and the phenomenon itself is called soil thinning [8 - 11].

4. Problem statement

Summing up the above initial data during the construction of a new production workshop of an industrial enterprise, the question arose of the possible adverse impact of technological equipment on the performance of the foundations of the equipment, as well as on the performance of foundations of the workshop under construction, given that there is the need for sand pillows under some foundations of the building as well as under the foundations of dynamic equipment. The purpose of the research is to measure the actual oscillations of the structures of dynamic equipment, their foundations and the foundations of the new workshop in order to determine the possible occurrence of additional to the design stabilized or unstabilized deformations of the soil base in the zone of dynamic influences.

5. Basic material and results

Studies of the parameters of dynamic oscillations were conducted on the production site of the existing yacht and boat manufacturing enterprise in the Kharkiv region. Since this type of production involves the use of polyester resins, other volatile substances and welding, the building was equipped with powerful ventilation systems that are driven by powerful industrial radial fans.

The building of the new production unit has a total size of 78.0 x 36.0 m. The objects of research are powerful radial fans placed along the rear facade of the building are installed on their own foundations (see Figure 1).

IOP Conf. Series: Materials Science and Engineering 1021 (2021) 012010 doi:10.1088/1757-899X/1021/1/012010

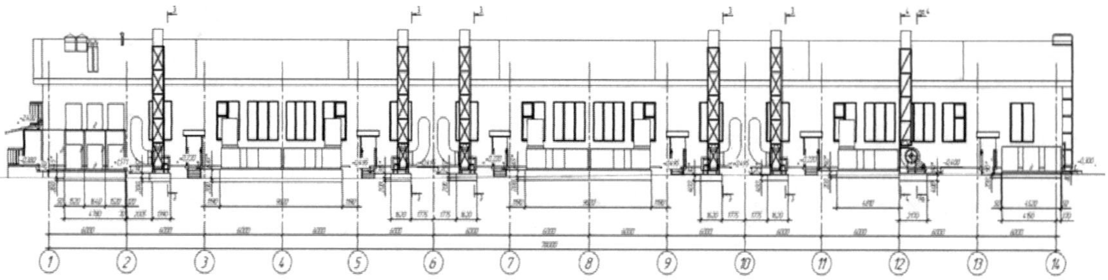

Figure 1. Layout of the ventilation systems along the front 1–14.

During the design of this building while performing engineering-geological surveys, the following engineering-geological elements were installed:

- engineering geological element EGE 1 - bulk soils are represented by loams with admixtures of construction and household waste from 10% to 20%, non-sedimentary. Layer thickness is 0.5-2.6 m;
- engineering geological element EGE 2 - soil-vegetation layer - loams with admixtures of plant remains. Layer thickness is 0.6-0.8 m;
- engineering geological element EGE 3 - semi-hard, soft-plastic, subsidence loams. Layer thickness is 3-3.2 m;
- engineering geological element EGE 4 - semi-hard, soft-plastic loams.

Groundwater at the search area as of August was not recorded to a depth of 7.0 m.

For the physico-mechanical characteristics of the base soils see Table 1.

EGE 3 was chosen as the bearing EGE for the foundations of the production building, and the foundations for radial fans were made according to the shallow structural scheme on a sand pillow (see Figure 2).

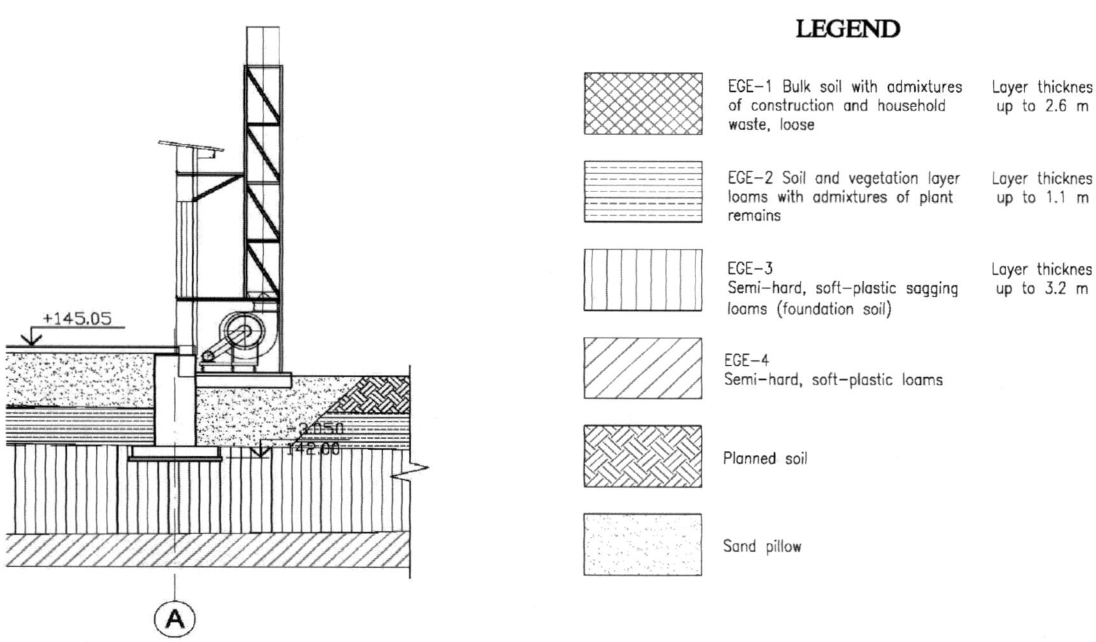

Figure 2. Installation of workshop foundations and radial fans on the engineering geological section of the construction site.

A set of equipment consisting of a VIP-2 vibrometer with a D21A displacement sensor and an Oscill digital oscilloscope connected to the vibrometer for spectral and visual analysis of oscillations, as well as for recording the oscillogram was used to determine the oscillation parameters. While performing the measurements, the parameters of oscillations along all 3 main axes of the orthogonal coordinate system were recorded. The Z axis was taken as vertical to the earth surface, and the X and Y axes form the conditional plane of the earth surface. Therefore, when indicating the direction of vibration measurement below, these assumptions should be kept in mind.

While performing the measurements, the following oscillation parameters were set: the amplitude of the forced oscillations of the support part of the fan (above the dampers) in the orthogonal directions X, Y and Z, the amplitude of the forced oscillations of the foundation in the orthogonal directions X, Y and Z, spectral analysis of oscillations with determination of their prevailing frequencies and determination of possible transmission of oscillations on the foundations and workshop structure.

During the development of the Heating and ventilation section of the project for the construction of a new production building, standard solutions proposed by the manufacturer of dynamic equipment were used, such as schemes of support of ventilation system elements on the foundation, installation and nomenclature of damping gaskets, placement of embedded parts, etc. Separately, metal frames were developed to support the exhaust pipes, which rest on the same slab foundations as the radial fans. All these structures are interconnected by pipelines and are in some interaction.

Table 1. Physico-mechanical characteristics of the base soils.

Engineering-geological elements (EGE) according to GOST 25100-95		Specific weight of soil, kN/m^3			Specific adhesion, kPa			The angle of internal friction, degree			on modulus,	Estimated resistance, kPa
		V_H	V_I	V_{II}	C_H	C_I	C_{II}	φ_n	φ_I	φ_{II}	E	R_0
EGE 1	Bulk soil with admixtures of construction and household waste, loose	15.2	12.1	-	-	-	-	-	-	-	-	100
EGE 2	Soil and vegetation layer - loams with admixtures of plant remains	18.3	16.2	-	-	-	-	-	-	-	-	100
EGE 3	Semi-hard, soft-plastic, sagging loams	26.7	18.9	14.9	0.02	-	-	17	16	14	10	110
EGE4	Semi-hard, soft-plastic loams	26.7	19.2	15.1	0.02	-	-	18	17	16	14	-

When determining the amplitudes of oscillations, the reference values were the readings of the scale of the device VIP-2. The calibration was performed for this device by the state enterprise "Kharkivstandartmetrologiya".

The operating frequency range of the device (with guaranteed constant error) is from 10 to 1000 Hz. The range of vibration velocity measurement is from 0.1 to 100 mm/s, vibration of oscillations – from 2 to 1000 μm.

The main element of the vibrometer is a vibration transducer which perceives mechanical oscillations and, by converting them into electrical signals, transfers them to the input of the measuring device.

The measuring device is made as a separate unit, on the front panel of which there are operating mode switches, the Input "Vibrator Inverter" for connecting a vibration transducer and the OUT

IOP Conf. Series: Materials Science and Engineering 1021 (2021) 012010 doi:10.1088/1757-899X/1021/1/012010

connector for connecting an oscilloscope. The measurement of oscillation parameters is performed using an analog scale with an arrow, the maximum deviation of which corresponds to the maximum value of the set measurement limit using the "Measurement Range" switch. Switching of vibration of oscillations and vibration acceleration modes is performed using the "Measurement Type" switch. When the oscilloscope is connected, it is possible to control the frequency for which the amplitude is measured, as well as to perform spectral analysis of oscillations to determine the predominant frequency and, if available, secondary harmonics. And if the scale of the oscilloscope is calibrated on the scale of the vibrometer, it is possible to perceive and analyze information on the magnitude of the amplitude of oscillations directly from its screen (see Figure 3).

Thus, it was determined that the maximum deviation of the arrow of the analog scale of the device in any mode of the switch "Measurement Range", when measuring the amplitude of oscillations, corresponds to the interval from -2 V to +2 V on the oscilloscope screen. Therefore, when adjusting the oscilloscope to work with the VIP-2 vibrometer, its sensitivity is set to a maximum of 2 V, which corresponds to the maximum of the selected measurement limit. Intermediate values, respectively, can be determined by interpolation.

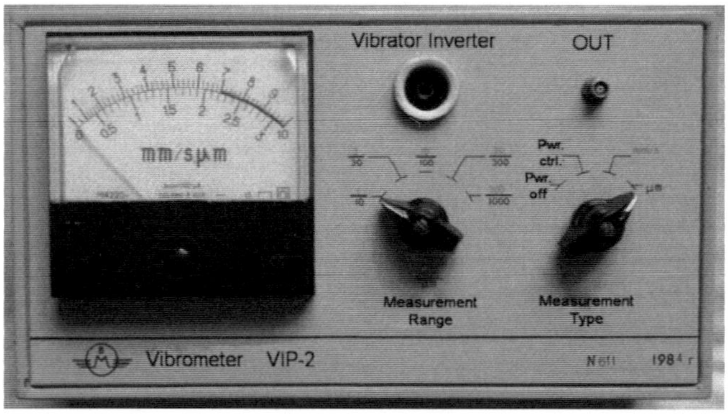

Figure 3. General view of the VIP-2 vibrometer.

When determining the oscillation parameters of the fan design, the vibrometer sensor was mounted on a frame rigidly attached to the fan, and when determining the oscillations of the foundations of this equipment – to the metal support slides welded to the embedded parts of the foundations (below complete rubber dampers) (see Figure 4). Therefore, we further consider the parameters of forced oscillations determined during the research.

Figure 4. General view of the research area.

IOP Conf. Series: Materials Science and Engineering 1021 (2021) 012010 doi:10.1088/1757-899X/1021/1/012010

During the research it was found that the maximum vibration of oscillations of oscillations registered by the VIP-2 device is 30 μm (corresponding to the amplitude of 15 μm), so the switch of the measurement limit was set to 30 μm. Accordingly, the indicators displayed on the oscilloscope screen were calibrated, namely: the set limit of vibration of oscillations measurements on the vibrometer 30 μm corresponds to the value of ± 2.0 V on the oscilloscope screen. The division value on the oscilloscope screen in voltage units is 0.5 V, so the division value in amplitude units is:

$$Div.value = \frac{30(\mu m)\cdot 0.5(V/div.)}{4(V)} = 3.75(\mu m\ /\ div.)\ , \quad (1)$$

First, the parameters of the forced oscillations of the support part of the radial fan, on which it is rigidly fixed and connected to the design of pipelines, were measured to determine the initial parameters of forced oscillations that occur during its operation in stationary mode (see Figures 5 - 7).

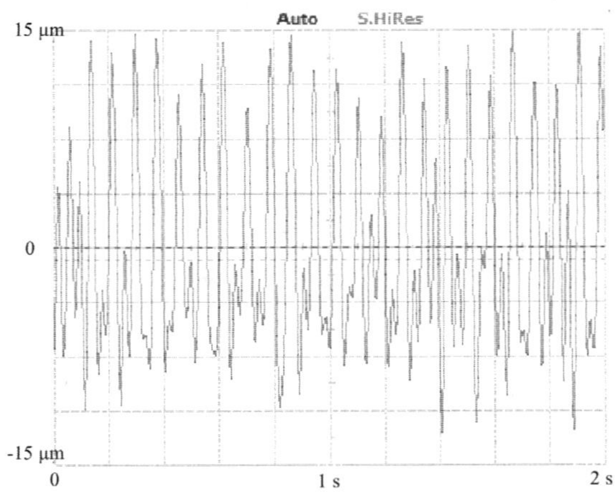

Figure 5. Oscillogram of oscillations of the support part along the Z axis. (the division value along the Y axis = 3.75 μm, along the X axis = 0.2 s).

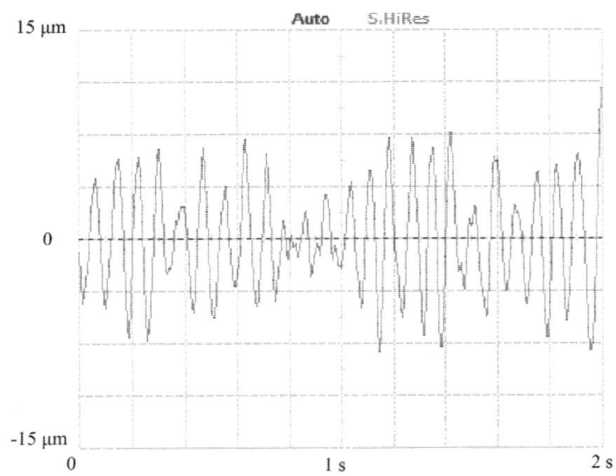

Figure 6. Oscillogram of oscillations of the support part along the X axis. (the division value along the Y axis = 3.75 μm, along the X axis = 0.2 s).

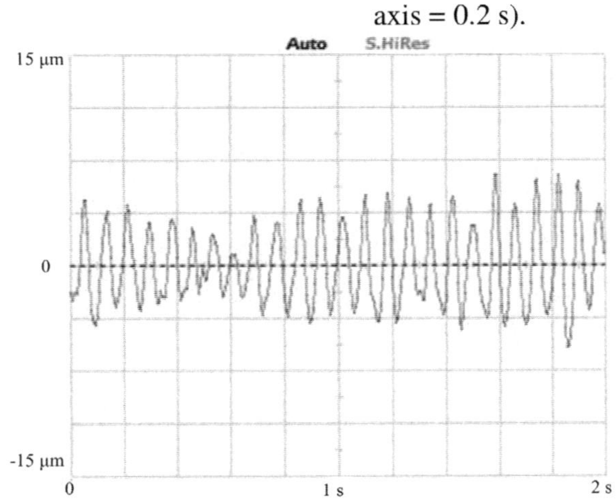

Figure 7. Oscillogram of oscillations of the support part along the Y axis. (the division value along the Y axis = 3.75 μm, along the X axis = 0.2 s).

Thus, the following was determined: the maximum amplitude of oscillations is 15 μm (corresponds to the vibration of oscillations equal to 30 μm) and is registered in the vertical direction to the earth's surface Z. In the horizontal directions X and Y, the measured amplitude was ~ 2 times smaller, about 7 μm.

To assess the contribution of the fundamental frequency of forced oscillations from radial fans and determine the natural frequency of the entire structure of the ventilation system, as well as its impact on the overall picture of oscillations of all structures combined into a single complex spatial system, spectral analysis of oscillations in the Z direction, the worst in terms of the magnitude of the amplitude of oscillations, was performed, with the installation of a vibration transducer on the fan frame above the damper inserts. The engine speed according to the passport data is 1450 rpm. The radial fan itself is driven by the engine by means of a belt drive with a coefficient of speed reduction equal to 2. The spectrogram (see Figure 8) visualizes two main peaks: at a frequency of ~ 12 Hz and 24 Hz, which are the frequencies of forced oscillations of the radial fan and its motor, respectively. The third small peak at a frequency of 17 Hz most likely corresponds to the frequency of natural oscillations of the entire ventilation structure.

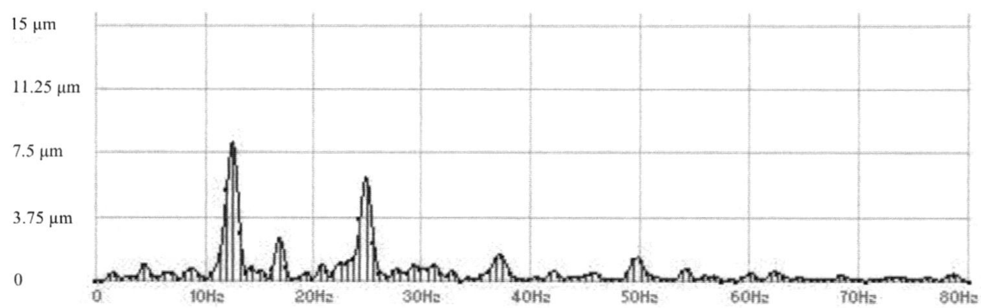

Figure 8. Spectrogram of oscillations of the support part.

To evaluate the efficiency of rubber spacers, as well as the damping effect of the foundation massif and soil base, the following measurements of similar parameters of forced oscillations below the rubber dampers were performed by placing the D21A vibration transducer on steel slides rigidly welded to the embedded structures in the foundation slab (see Figure 9).

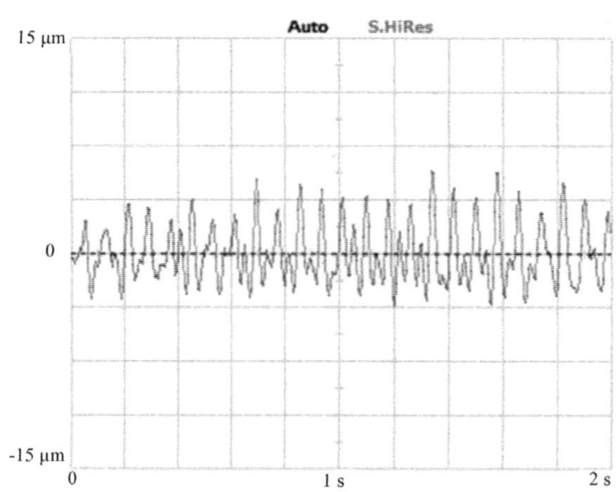

Figure 9. Oscillogram of oscillations of the foundation along the Z axis. (the division value along the Y axis = 3.75 μm, along the X axis = 0.2 s).

 doi:10.1088/1757-899X/1021/1/012010

During measurement of the parameters of the foundation oscillations, the following was established: due to the damping properties of the inventory rubber spacers and the foundation massif, the amplitude of forced oscillations is reduced by almost 2-3 times to 7-5 μm. In this case, any significant values of the amplitudes are registered only in the Z direction. In the X and Y directions, vibrations are almost completely absent and are not recorded by the device. The spectrogram of oscillations in this case shows a corresponding decrease in peaks at fundamental frequencies and almost does not affect the magnitude of the amplitude at the assumed natural frequency of ~ 17 Hz, which confirms the previous conclusions in this regard (see Figure 10).

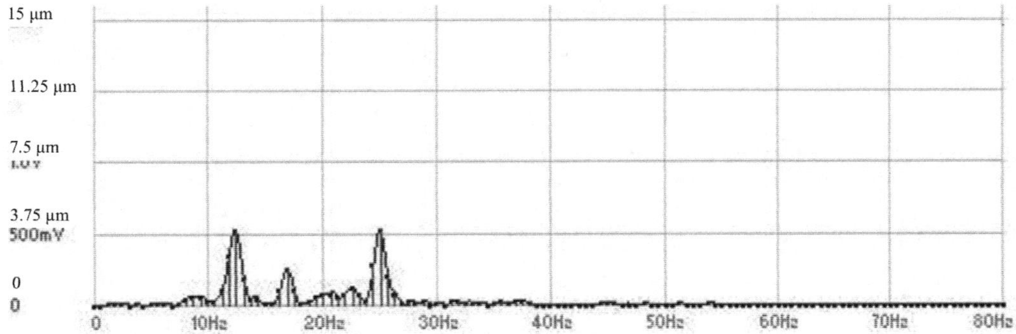

Figure 10. Spectrogram of foundation oscillations.

The study of possible transmission of oscillations to the foundations of the building did not show the occurrence of any vibrations on their surface due to good damping properties of the selected structure of the foundations and the base in the form of vibration damping sand pillow, as well as rather small dynamic forces compared to the building massif. According to studies [7] under the action of dynamic loads with an amplitude of oscillations of 10 - 15 μm on an artificial sand base with layer-by-layer compaction to the standard value of skeletal density $\gamma_d = 1.65 t/m^3$, which corresponds to the conditions at this site, additional subsidence may occur typical for Phase 1 or Phase 2. That is, with the rapid stabilization of deformations after compaction of the sand base, or with long-term deformations, the development of which is approximated by the logarithmic law.

6. Conclusions

The study of the parameters of forced oscillations of structures and foundations of the ventilation system of a new production building for yachts and boats in Kharkiv region demonstrates that determining the parameters of forced oscillations of complex engineering and geometric systems by analytical methods is often impossible or unreasonably time consuming. Given the potential risks to foundations and associated above-ground structures, provided they are laid on sandy natural or artificial foundations, this issue cannot be ignored. In this case, it is considered most rational to take all available measures to reduce the possible negative impact of vibration, such as balancing rotating and moving parts, installation of dampers between structures and foundations, and, if necessary, installation of anti-vibration screens in the soil mass to prevent vibration on the foundations of neighboring buildings.

In the framework of this study, it was found that the parameters of forced oscillations in the stationary mode of operation of the studied dynamic equipment are within safe limits for this type of basis. It should also be noted that the transmission of vibrations to the foundations of the building does not occur.

However, from time to time the condition of inventory factory dampers should be monitored, due to the deterioration of the properties of which the change in the dynamic mode of operation of the foundations in the direction of deterioration of their performance can occur.

IOP Conf. Series: Materials Science and Engineering 1021 (2021) 012010 doi:10.1088/1757-899X/1021/1/012010

References

[1] Savinov O A 1979 *Modern Structures of Foundations for Machines and Their Calculation* (Leningrad: Stroyizdat) p 200

[2] Kudryavtsev I A 1999 *The Effect of Vibration on the Foundations of Structures* (Gomel: Belarusian State University of Transport) p 200

[3] Vynnykov Yu L 2009 Influence of Vibration Regime of Rollers on Compaction of Low Cohesive Overburden *Industrial Machine Building, Civil Engineering* vol 25, ed S F Pichugin et al (Poltava: National University "Yuri Kondratyuk Poltava Polytechnic") pp 40–49

[4] Sawicki A and Mierczyński J 2015 Some Effects of Intrinsic Cyclic Loading in Saturated Sands *Journal of theoretical and applied mechanics* vol 53, ed W Kurnik et al (Warszawa: JTAM) pp 285–293

[5] Sawicki A, Mierczyński J and Sławińska J 2015 Structure and Calibration of Constitutive Equations for Granular Soils *Studia Geotechnica et Mechanica* vol 36, ed A Różański et al (Warszawa: De Gruyter) pp 35–46

[6] Aleksandrovych V A, Taranov V G, Luchkovskyi I I, Plashchev S A, Kornienko N V, Areshkovych O O 2013 Structure-Soil Massif System Behavior Features Under Static and Dynamic Loads *Proc. of the 18th Intern. Conf. on Soil Mechanics and Geotechnical Engineering* (*Paris*) vol 2, ed P Delage at al (Paris: Presses des Ponts) pp 1627–29

[7] Aleksandrovych V A 2012 Concerning the Vibrocreep Issue *Proc. of the 22nd European Yong Geotechnical Engineers Conf.* (*Gothenburg*) vol 1, ed V Svahn and T Wood (Gothenburg: Chalmers University of Technology) pp 173–178

[8] Lange D and Fanourakis G 2009 Comparing Vibratory and Impact Laboratory Compaction Methods *Proc. of 17th Intern. Conf. on Soil Mechanics and Geotechnical Engineering* (*Alexandria*) vol 1, ed M Hamza (Amsterdam: IOS Press) pp 93–96

[9] Kim S I, Park K B, Park S Y, Hwang S J, Lee J H and Choi J S 2005 Effects of Irregular Dynamic Loads on Soil Liquefaction *Proc. of the 16th Intern. Conf. on Soil Mechanics and Geotechnical Engineering* (Osaka) vol 4, (Amsterdam: IOS Press) pp 2673–76

[10] Areshkovych O, Boyko I and Sakharov V 2011 Determination of the Stress Strain State of Soil Base for the Structures at Static and Synamic Loads *Proc. of the 15th Europ. Conf. on Soil Mechanics and Geotechnical Engineering* (*Athens*) vol 2, ed A Anagnostopoulos (Amsterdam: IOS Press) pp 1225–30

Determination of rational locomotive operating modes and resource costs based on information from global positioning systems

Mykola Barybin[1,4], Anatoliy Falendysh[2], Victoria Hatchenko[3], Olha Kletska[2], Elena Kiritseva[2]

[1]Department of Heat Engineering, Heat Engines and Energy Management, Ukrainian State University of Railway Transport, Feuerbach square 7 61050, Kharkiv, Ukraine
[2]Department of Transport Technologies of Enterprises, Pryazovsky State Technical University, Universytets'ka str., 7 87555, Mariupol, Ukraine
[3]Department of Traction Rolling Stock of Railways, State University of Infrastructure and Technology, Kirillovskaya str., 9 04071, Kyiv, Ukraine

[4]kanifoli4.92@gmail.com

Abstract. The article is devoted to the study of energy-efficient use of traction rolling stock for driving trains. It deals with the optimal driving mode and the main parameters of traction calculations. Based on the use of GPS/GLONASS satellite navigation units, a method for determining the basic resistance of cars in real time is proposed. The possibility of using the results obtained on locomotives with analog devices and no movement tracking is analyzed. The "Traction Calculations" software package in the C# programming language was modified in order to reduce the time and facilitate calculations of the locomotive traction service. In the course of the study, the equation of the main resistance of the cars of train No. 3612 was determined. On the basis of the uncoupled rolling stock, the dependences of speed, time, heating and current of traction electric motors, fuel consumption by the CHME3 locomotive of the Kharkiv-Sortirovochny locomotive depot for the transportation of cars from the Kharkiv-Sortirovochny station to Merefa station were calculated, with recommendations for the driver to use the position of the driver's controller in order to reduce costs and fulfill the train schedule.

1. Introduction

The current state of railway transport in Ukraine is characterized by noticeable wear and obsolescence of fixed assets. The reform of railway transport is directly related to the development and implementation of a set of measures aimed at increasing the carrying capacity and reducing the cost of fuel and energy resources for train traction [7, 8]. During operation, there is a constant change in the parameters that characterize the quality of operation of the rolling stock. The decrease in operational characteristics is associated with the deterioration of the technical condition of individual equipment elements [1, 10]. This is due to increased wear of mechanical equipment parts and aging of insulation of electrical machines, which leads to a decrease in the service life of assembly units and parts. This results in an increase in drag forces, a decrease in braking and traction forces, with a consequent increase in the cost of pulling trains [4, 9]. Systemic analysis of traction calculations has been carried out throughout the existence of railway transport. The accuracy of the results obtained affects the cost of resources for traction and, as a result, reduces the cost of transportation [6, 8]. The definition of rational weight standards is the basis for increasing the carrying capacity of railways and cutting the repair cost of rolling stock. The development of microprocessor equipment and mathematical modeling of train movement parameters allows us to find hidden reserves in improving the efficiency of operation of traction rolling stock, efficient use of locomotive power and their rational modes of operation. The development of global vehicle positioning systems has become widespread in the world. It helps quickly get information about the movement and location of objects in space. Railway transport in Ukraine is strategically important for the functioning of the economy and social infrastructure, so the introduction of GPS/GLONASS systems in the locomotive facilities has led to the informative location of traction vehicles on the monitors of dispatching and other services [11, 12]. Improving the accuracy of determining the main parameters of movement, increasing the processing capabilities of electronic computers and automated systems, as well as the bandwidth of information transmission channels for processing and analyzing the received data, creation of guidelines on fuel economy in the operation of locomotives and train driving modes, their implementation in the transportation and training process of locomotive crews, will minimize the repair and operation cost [2, 3]. On the basis of analysis of the works [13-16], the relevance and main directions of development of traction calculations in the world are determined, the need for microprocessor control and the use of systems of increased reliability of information in the process of train movement is formed. Reducing the calculation time is the basis for algorithmization of processes. Software products for calculating traction parameters and optimizing energy consumption have been developed on the basis of algorithmization of processes. The most widely used specialized complexes are ISKRA-PTR, ERA, VECTRUM and MoveRW. The main parameters of the traction can be calculated and build a model of train coupling can be built to adapt the modes of reference when the timetable is changed. However, they have a number of disadvantages, such as high cost and technical requirements for electronic computing, insufficient accuracy of the impact of the track on the rolling stock, and the inability to respond flexibly to changing technical characteristics of the rolling stock. A material drawback of the operation is the lack of accounting for information received from the global positioning units GPS/GLONASS.

Since the parameters that characterize the quality of operation of the rolling stock are constantly changing during the operation, there is a need to create an adaptive train driving mode, the adaptation of which would include changes in the values of traction forces and fuel consumption. In [9], a software product is presented by optimizing the algorithm so that the set goals can be achieved. In order to establish the dependencies of the resistance forces to the movement of a real train, there is a need to diagnose them in real time. This is possible only when using an expression that can take into account all factors affecting the rolling stock-the law of conservation of mechanical energy. So the solution is to develop on the basis of the law of conservation of mechanical energy, using data blocks GPS/GLONASS, the equation of running resistance and determining the adaptive mode of the train with flexible adjustment of input parameters.

2. Formulation of the problem

The purpose of the work is to develop a model for determining the rational mode of driving a train, taking into account information from global positioning systems. To do this, on the basis of tracking (figure 1) and information from the global positioning system of train No. 3612, the equation of the main resistivity of cars should be determined. A mathematical model and software for determining the rational driving of the train in order to reduce the resources for traction should be developed.

Based on the results obtained, the resource consumption for traction should be optimized, the dependence of speed, time, heating and current of traction motors should be built for a CHME3 diesel locomotive for transporting cars from the Kharkiv-Sortirovochny station to Merefa. the minimum fuel consumption when running the train schedule should be determined, and the locomotive traction points should be set in the form of a table recommendation to the driver.

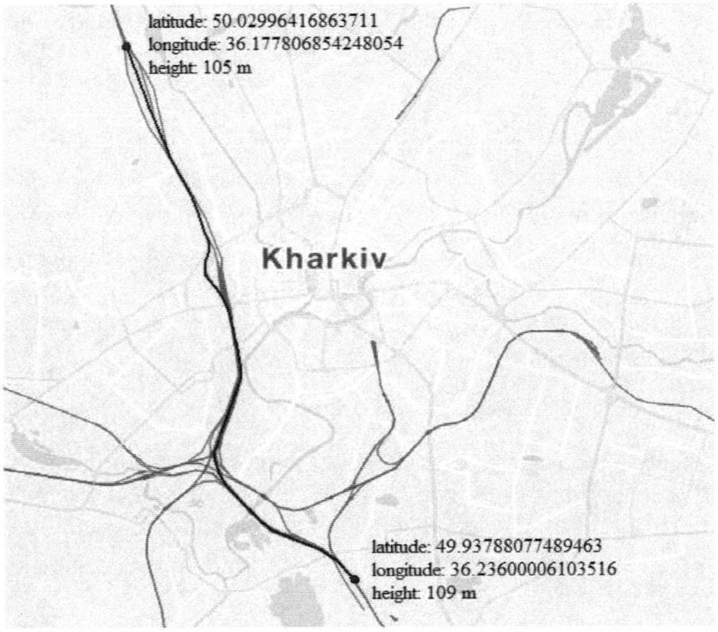

Figure 1. Tracking of train No. 3612 from Osnova station to Kharkiv-Sortirovochny station.

3. Solution

In general, the search for rational modes of locomotive operation (figure 2) is reduced to solving the integral type of equation (1) and establishing the time intervals of the stage run between stations. It is necessary to take into account the equal average speed in the time interval (t_1,t_3), but this leads to a decrease in the acceleration of the train and an increase in the time interval (t_1, t_2).

$$\int_{t_1}^{t_2} g_i(t)dt + \int_{t_2}^{t3} g_{xx}(t)dt > \int_{t_1}^{t_3} g_i(t)dt, \quad (1)$$

where $g_i(t), g_{xx}(t), g_j(t)$ accordingly, is a function of specific energy consumption on the position of controller of the driver "*i*", no load and on the position of controller of the driver "*j*" corresponding to the speed of movement; t_1, t_2, t_3 are time interval specific functions.

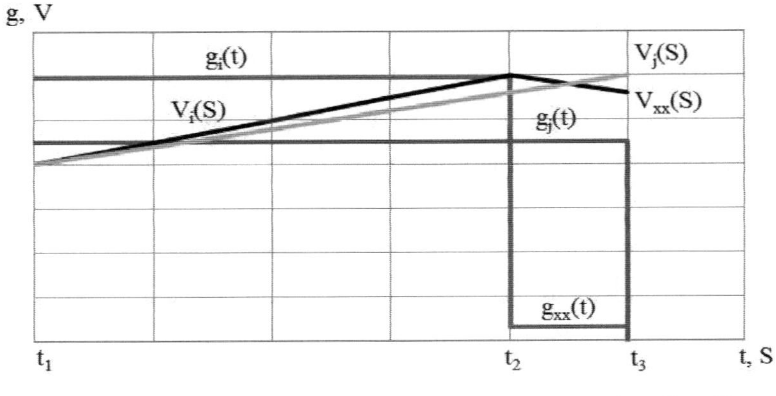

Figure 2. Graphical representation of search rationality (1).

It is proposed to use data on resource costs and hyperbolic characteristics of traction forces to install an energy-efficient position of the driver's controller. The logical interpretation of this statement is based on the need to determine the value of the traction force relative to fuel or electricity consumption. The mathematical representation of the coefficient η_{tb} is calculated using the expression

$$\eta_{tb} = F_{k_v} \cdot G_v^{-1}, \tag{2}$$

where F_{k_v} and G_v this is, respectively, the locomotive's traction force and energy consumption at the position of the driver's controller "*i*" and "*j*" corresponding to the speed of movement.

During operation, the technical parameters of locomotives change. We use the law of conservation of mechanical energy to determine the nature of the influence of the resulting sum of forces acting on the train during movement on the rails

$$\xi \cdot S \cdot (f_k - \varpi - b_T) = 0.5 \cdot (V_2^2 - V_1^2) + g \cdot (h_2 - h_1), \tag{3}$$

where V_1, V_2 are the speed at the beginning and end of the profile element, respectively; m is the mass of the train; g is the acceleration of gravity; h_1, h_2 are the heights at the beginning and end of the profile element, respectively; S is the length of the travel section; ξ is the coefficient of specific acceleration; f_K, w, b_T are, respectively, the specific force of traction, resistance to movement and braking.

The specific forces of resistance to movement are the sum of the forces of the main and additional resistance with the magnitude of the impact in accordance with the mass

$$\varpi = (P + Q)^{-1} \cdot (P \cdot \varpi'_O + Q \cdot \varpi''_O) + \varpi_D, \tag{4}$$

where P, Q are, respectively, the mass of the locomotive and the train; $\varpi'_O, \varpi''_O, \varpi_D$ are, respectively, the specific forces of resistance to the movement of the locomotive, the composition, and additional. Since the additional resistance ϖ_D includes the potential influence of the profile, we will not take into account the part of expression (3) $g \cdot (h_2 - h_1)$. The effect of ϖ_D for cars with rolling bearings is determined by the expression

$$\varpi_D = i + 200 \cdot R^{-1} + 1.5 \cdot \left| V^2 \cdot [12.96 \cdot R]^{-1} - h \cdot g \cdot S_K^{-1} \right| + 280 \cdot (q_0 + 70)^{-1} + \varpi_O \cdot (k_{HT} + k_B - 2), \tag{5}$$

where i is the steepness of the resulting ascent; R this is the radius of the curve; q_O is the axial load of the rolling stock; h is the difference between the heights of the rail heads; S_K is the distance between the rolling circles of the rolling stock wheelset; k_{HT} this is the coefficient of exposure to low temperatures; k_B is the coefficient of influence of air masses (wind).

In general, ϖ'_O, ϖ''_O are functions of a quadratic equation with corresponding coefficients [2, 3]. According to the requirements of [3] and the definition of data for the entire speed range, we present the forces of the main drag of cars in the form of the equation

$$\varpi'_O = a' \cdot V^2 + b' \cdot V + c', \tag{6}$$

$$\varpi''_O = a'' \cdot V^2 + b'' \cdot V + c'', \tag{7}$$

where $a', b', c', a'', b'', c''$ are, respectively, unknown coefficients of the locomotive and car resistivity equation. The coefficients of resistivity a', b', c' can be obtained during light running of the locomotive or from the data [2, 3]. On the basis of information from the GPS/GLONASS global positioning units integrated on the VL11 locomotive, we obtain accurate values of speed, time, and path with a minor error [11, 12]. Then the calculation of the real values of the main resistance of cars will be determined by the expression

$$\varpi''_O = Q^{-1} \cdot (P+Q) \cdot \left[f_k - b_T - \frac{V_2^2 - V_1^2}{2 \cdot \xi \cdot S} - \varpi_D - \frac{P}{P+Q} \cdot \left(a' \cdot \left[\frac{V_2 - V_1}{2}\right]^2 + b' \cdot \frac{V_2 - V_1}{2} + c' \right) \right], \quad (8)$$

We calculate the coefficients a'', b'', c'' using the least squares approximation of GPS/GLONASS block data

$$\begin{cases} a'' \cdot \sum_{i=1}^{n} V_i^4 + b'' \cdot \sum_{i=1}^{n} V_i^3 + c'' \cdot \sum_{i=1}^{n} V_i^2 = \sum_{i=1}^{n} V_i^2 \varpi''_{O_i} \\ a'' \cdot \sum_{i=1}^{n} V_i^3 + b'' \cdot \sum_{i=1}^{n} V_i^2 + c'' \cdot \sum_{i=1}^{n} V_i = \sum_{i=1}^{n} V_i \varpi''_{O_i} \\ a'' \cdot \sum_{i=1}^{n} V_i^2 + b'' \cdot \sum_{i=1}^{n} V_i + c'' \cdot n == \sum_{i=1}^{n} \varpi''_{O_i} \end{cases}, \quad (9)$$

where n is the number of data registrations by the GPS/GLONASS unit.

As a result of calculating the tracking of train No.3612 from Osnova station to Kharkiv-Sortirovochny station using the expressions (5, 8, 9), we obtained the equation of the main resistivity of a freight train weighing 4,483 tons, with 220 axles of gondola cars with rolling bearings, at a wind speed of 0.7 m/s and a temperature of 22 ^{0}C

$$\varpi''_O = 0.00009915 \cdot V^2 + 0.00420327 \cdot V + 0.87334037, \quad (10)$$

One of the tasks of this work is to prove that the system as a whole works with small volumes of global positioning blocks. Therefore, as an example of the interaction of information from the systems integrated in electrical locomotives VL11 and locomotives without GPS/GLONASS, let us consider the following case: upon arrival at the station Kharkiv-Sortirovochny, 7 cars from the end of the freight train weighing 571 tons, total 28 axles were coupled with the CHME3 diesel locomotive to transport cargo to the destination station Merefa. The temperature of traction motors is set to 37.3 ^{0}C before shipment. We use the software package "Traction calculations" developed by the authors of the article [5] in the Visual C# programming language and create a modification in the form of an energy-efficient train driving mode.

The essence of the modification will be as follows: to perform the train schedule, we use the balance method for calculating the time of passage of a section element. That is, the movement occurs according to the schedule and expression (2), which makes it possible to optimize fuel consumption in the time interval of the stage. Let us create an application that takes into account equation (10) for rational use of power and resources for traction. A graphical representation of the rationality of the driver's controller position, according to (2), CHME3 relative to the speed is shown in figure 3.

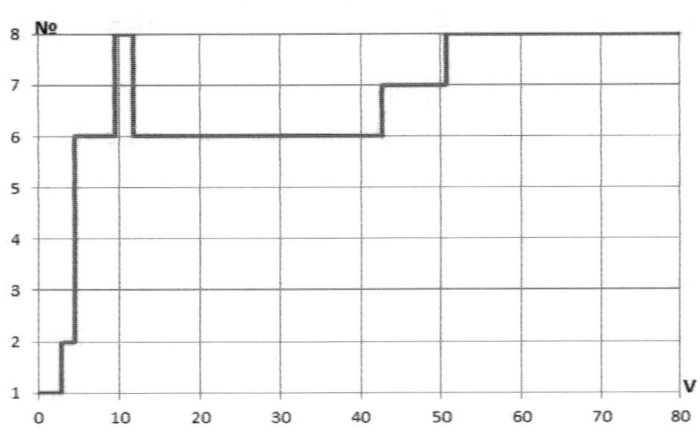

Figure 3. Rationality of the position of the driver's controller CHME3.

Determining the speed on the path interval using (3) will be determined by the expression

$$V_2 = \left[V_1^2 + 2 \cdot \xi \cdot S \cdot (f_k - b_T - \varpi)\right]^{1/2}, \tag{11}$$

On the other hand, V_2 speed can be calculated as

$$V_2 = V_1 + S \cdot \Delta t^{-1}, \tag{12}$$

where Δt is the growth time. Substitute (12) in (11) and perform the conversion, then the time increment is determined by the expression

$$\Delta t = S \cdot \left[\left[V_1^2 + 2 \cdot \xi \cdot S \cdot (f_k - b_T - \varpi)\right]^{1/2} - V_1\right]^{-1}, \tag{13}$$

The train schedule is essentially the sum of the driving time for races, then the relationship between Δt and the time of the stage run t_p must be directed by the expression

$$\sum_{i=1}^{n} \Delta t_i \cong t_p \; and \sum_{i=1}^{n} \Delta t_i - t_p \to 0, \tag{14}$$

If the left and right members of equation (14) do not match, increase the value of $S \cdot f_k$ in equation (11, 13) and repeat the calculation until condition (14) is met. The calculation results are shown in figure 4. Recommendations to the driver on the mode of operation of the locomotive on the route are shown in table 1

Table 1. Recommendations to the driver on the mode of driving the train.

Beginning of the thrust (km)	Position	End of thrust (km)	Fuel consumption (kg)	Note
774	1	774.006	0.042	
774.006	2	774.011	0.042	
774.011	6	774.02	0.157	
774.02	8	774.029	0.177	Sand
774.029	6	774.222	1.296	
775.014	6	775.946	4.008	
777.232	6	777.952	2.882	
779.495	6	781.092	7.524	
782.314	6	782.416	0.486	
784.03	6	791.5	36.2	
$\sum_{i=1}^{n} G_{xx_i} = 4.256 kg$		$\sum_{i=1}^{n} G_{T_i} = 52.814 kg$		$\sum_{i=1}^{n} G_i = 57.07 \, kg$

The results shown in table 1 fully correspond to the characteristics of the current load limitation of traction motors (Te-006) and the main generator (TD-802) [2-3], the fulfillment of the condition of wheel coupling with rails relative to speed [2-3], heating of electric machines in accordance with the insulation class (F) [2-3] of the CHME3 locomotive. The fuel consumption of a traction unit for moving trains in compliance with the traffic schedule and the weight mode map of a given section according to [4] is determined by the expression

$$G = \int_{\tau_1}^{\tau_k} G(\tau) \cdot d\tau , \tag{15}$$

where τ_1, τ_k are the time intervals for moving the train at the corresponding positions of the driver's controller; $G(\tau)$ is a function of fuel consumption relative to the speed and position of the driver's controller.

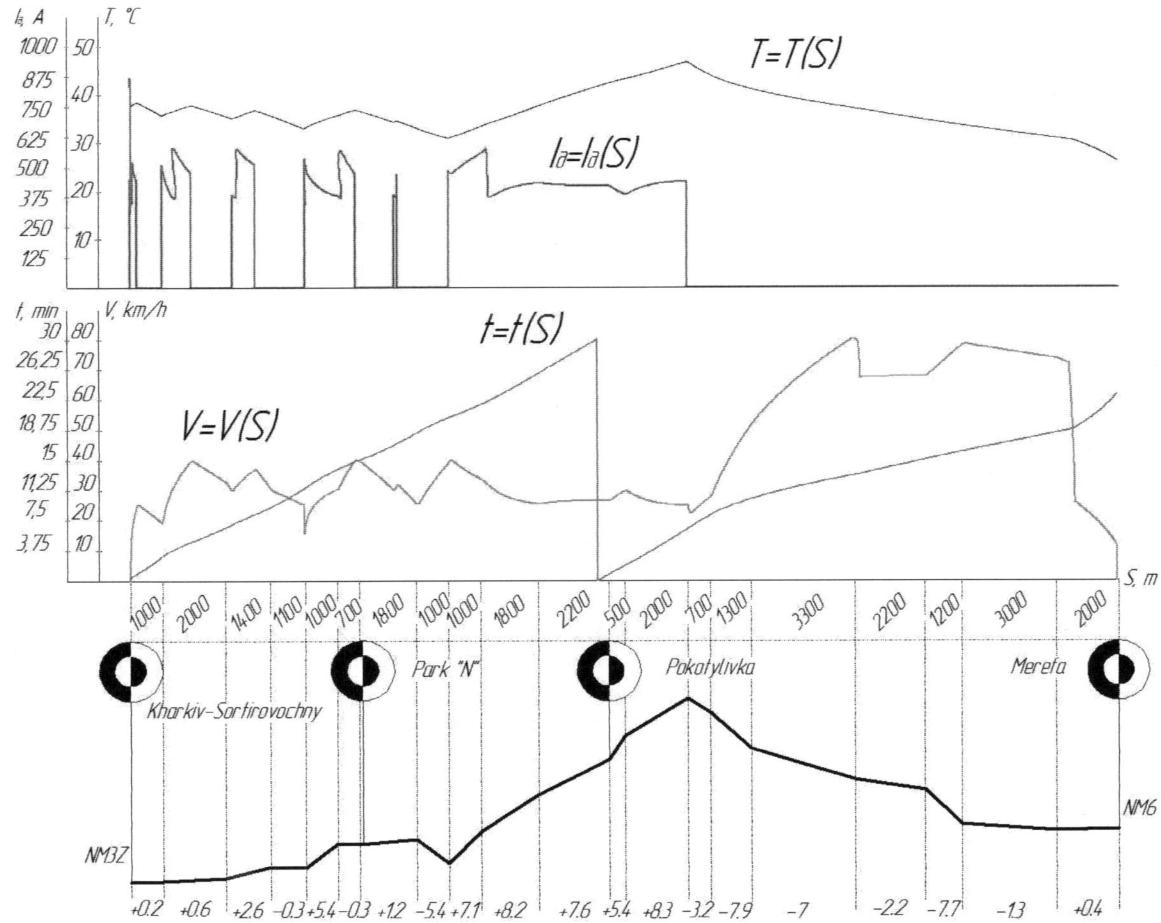

Figure 4. Graphical results of the calculation.

Based on the results of the calculations, the fuel economy is set at 12.75% compared to [4]. Using the balance mode of train driving will allow you to fulfill the schedule and reduce the consumption of fuel and energy resources for train traction. However, this requires diagnostic equipment and large-scale practical testing in the workplace.

4. Summary

This study solves the problem of determining the energy-efficient use of traction rolling stock for driving trains. The necessity of revising outdated methods of training locomotive crews in train driving is proved. Based on the law of conservation of mechanical energy and the use of GPS/GLONASS satellite navigation units, a method for determining the basic resistance of cars in real time is proposed. The possibility of using the results obtained on locomotives with analog devices and no movement tracking is analyzed. In the course of the study, the equation of the main resistance of the cars of train No. 3612 was determined. It is practically proved that resources can be saved with a few train tracking tools installed on mainline VL11 locomotives. The results of calculations of the main resistance are implemented in the optimization of the software complex algorithm in the C# programming language for calculating the uncoupled train at Kharkiv

Sortirovochny station. On the basis of the uncoupled rolling stock, the dependences of speed, time, heating and current of traction electric motors, fuel consumption by the CHME3 locomotive of Kharkiv-Sortirovochny locomotive depot on the transporting of cars from Kharkiv-Sortirovochny station to Merefa station were calculated with the issuance of recommendations to the driver to use the position of the driver's controller in order to reduce costs and fulfill the train schedule. The specified calculation algorithm allows implementing optimal solutions of resource-saving technologies in the process of cargo transportation. In the future, it is advisable to implement a mathematical model and software product in the locomotive economy in order to establish rational train driving modes, reduce traction resources and mass practical testing.

References

[1] Uyulan C, Gokasan M, Bogosyan S 2018 Readhesion control strategy based on the optimal slip velocity seeking method *Journal of Modern Transportation* **26(1)** pp 36-48

[2] Osipov S, Osipov S 2010 *Principles of traction* (Moscow: UMC MPS) p 592

[3] Order 867 2016 *Rules for traction calculations for train work* (Moscow: JSC Russian Railways) p 510

[4] Order 204-C 2014 *Regulations on inspection of energy efficiency control of Ukrzaliznytsia* (Kiev: Joint-Stock Company Ukrzaliznytsia) p 40

[5] Falendysh A, Hatchenko V, Voznenko S, Kletskaya O, Barybin M 2020 Mathematical modeling of basic parameters in traction calculations *Collection of scientific works of DUIT. Transport Systems and Technologies Series* **35** pp 102-112

[6] Cheremisin V 2017 The role of information technology in ensuring the reliability of the locomotive *Locomotive*, **9** pp 2-4

[7] Mikheyev V 2015 Estimation of the diesel locomotives operational efficiency at the given service area *Vesnik of the Siberian State Automobile and Road Academy* **1(41)** pp 91-96.

[8] Mostovych A, Cherniak A, Nozhenko O 2015 Application of the methods of correlation and spectral analysis for processing the results of dynamic testing of railway rolling stock *Railway transport of Ukraine* **4(113)** pp 20-24

[9] Cole C, Spiryagin M, Wu Q, Sun Y Q 2017 Modelling, simulation and applications of longitudinal train dynamics *Vehicle System Dynamics* **55(10)** pp 1498–1571

[10] Safna F, SunnyR 2015 Artificial Neural Network Based Data Mining *International Journal of Innovative Science, Engineering & Technology* **5(4)** pp 240-245

[11] Chen X, Lu C, Guo B, Guo F, Ge M, Li X, Schuh H 2018 GPS/GLONASS Combined Precise Point Positioning With the Modeling of Highly Stable Receiver Clock in the Application of Monitoring Active Seismic Deformation J*ournal of Geophysical Research* **123** pp 4025—4040

[12] Xingxing Li, Xiaohong Zhang, Xiaodong Ren, Mathias Fritsche, Jens Wickert, Harald Schuh 2015 Precise positioning with current multi-constellation Global Navigation Satellite Systems: GPS, GLONASS, Galileo and BeiDou *Scientific reports* **4** pp 1-14

[13] Ding J, Zhang P, Li J 2015 Fatigue life prediction of IGBT module for metro vehicle traction converter based on traction calculation *IEEE 11th International Conference on Power Electronics and Drive Systems, Sydney NSW*, pp 1116-1121

[14] Rassõlkin A, Gevorkov L, Vaimann T, Kallaste A, Sell R 2018 Calculation of the traction effort of ISEAUTO self-driving vehicle *25th International Workshop on Electric Drives: Optimization in Control of Electric Drives (IWED), Moscow*, pp 1-5

[15] Fu Q, Zhu J, Mao Z, Zhang G, Chen T 2018 Online Condition Monitoring of Onboard Traction Transformer Core Based on Core-Loss Calculation Model *IEEE Transactions on Industrial Electronics*, **65(4)**, pp 3499-3508

[16] Wang M, Kou B, Zhao X 2018 Analysis of Energy Consumption Characteristics Based on Simulation and Traction Calculation Model for the CRH Electric Motor Train Units *21st International Conference on Electrical Machines and Systems (ICEMS), Jeju*, pp 2738-2743

IOP Conf. Series: Materials Science and Engineering 1021 (2021) 012012 doi:10.1088/1757-899X/1021/1/012012

Calculation residual strength of reinforced concrete beams with damages, which occurred during loading

Yaroslav Blikharskyy[1], Rostyslav Vashkevych[2,3], Nadiia Kopiika[2], Taras Bobalo[2], Zinoviy Blikharskyy[2]

[1] Lviv Polytechnic National University, Institute of Building and Environmental Engineering, Department of Highways and bridges, Bandera Street 6, Lviv, Ukraine

[2] Lviv Polytechnic National University, Institute of Building and Environmental Engineering, Department of Building construction and bridges, Bandera Street 6, Lviv, Ukraine

[3] Email: rostyslav.v.vashkevych@lpnu.ua

Abstract. The article presents the development of a method of experimental study of reinforced concrete beams with damaged steel reinforcement. In addition, the purpose is to conduct experimental tests of undamaged samples with steel reinforcement of 20- and 16-mm diameter. Also, the 20 mm steel bars were damaged to the area equal to 16 mm samples, experimental tests of such samples were performed and corresponding analysis and comparison of the results of both tests was conducted. According to the study program, 6 beam samples were tested. Two of them were undamaged control samples with single load bearing reinforcement of Ø20 mm diameter; 2 samples with Ø20 mm reinforcement had about 36% damages (which corresponds to area of the 16 mm diameter samples) and 2 samples with Ø16 mm single reinforcement, which corresponds to 36% load bearing reinforcement damage of control samples (the area of 36% damages of 20 mm samples corresponds to 16 mm diameter). The experimental study identified the strength and strain parameters of reinforced concrete beams with damages in the load bearing stretched reinforcement due to the action of the bending moment, which causes the exhaustion of the bearing capacity.

1. Introduction

Reinforced concrete structures are the most widely used load-bearing elements of buildings and structures [1-4]. The most common are the structures, subjected to bending, which are used in the form of beams, slabs, frame crossbar, girders, etc. [5-7]. Reinforced concrete structures are often subjected to the action of aggressive environment [8]. In addition, according to the commonly used classification the considerable variety of environments could be identified, depending on the factors which cause their action [9]. The most intense aggressive environment is the one, caused by chemical industry. Herewith the reinforced concrete, as the construction material, is one of the best in terms of its exploitation in aggressive environments. Steel reinforcement, which is the most susceptible to corrosion, is securely protected by the concrete layer. If the concrete composition is properly selected, steel reinforcement will be relatively non-susceptible to corrosion [10]. However, significant number of buildings and structures has been in operation for decades, often without the selection of high-quality concrete and proper corrosion protection of reinforced concrete structures [11-14]. As the result, the corrosion or mechanical damages of concrete take place [15-17] and, subsequently, after destruction of protective concrete layer, the corrosion of steel bars begins. This process could result in decreasing of reinforced concrete elements bearing capacity and

IOP Conf. Series: Materials Science and Engineering 1021 (2021) 012012 doi:10.1088/1757-899X/1021/1/012012

serviceability [18-20]. If the corresponding strengthening and reconstruction measures are not implemented, the accidental destruction of the building or its separate elements could occur.

Corrosion could be the reason for decrement of steel bar cross-section in reinforced concrete elements. Therefore, as the result, the stress-strain state of construction changes and its serviceability characteristics are reduced. In addition, it is important to note that corrosion process take place under the load action and reinforced concrete elements have corresponding stress-strain state. These factors could cause certain specific features of the corrosion processes proceeding, which is related to initial existing stresses in both concrete and steel reinforcement, possible cracks, etc.

Nowadays the great number of studies is devoted to the issue of reinforced concrete structure strengthening with the use of composite materials [21], steel and reinforced concrete cages [22], various modern materials in order obtain better protection of such constructions [23-28].

In order to conduct the optimal calculations of such structure strengthening, the real stress-strain state should be identified, taking into consideration damages of reinforcement. Therefore, the study of influence of damages on stress-strain state of reinforced-concrete structures is really topical. In particular steel reinforcement damages need to be considered.

2. Purpose of the article

The main purpose of the article is to develop the method of experimental research of reinforced concrete beams with damaged steel reinforcement. In addition, the aim is to conduct experimental tests of undamaged samples with steel reinforcement of 20- and 16-mm diameter. Also, the 20 mm steel bars were damaged to the area equal to 16 mm samples, experimental tests of such samples were performed and corresponding analysis and comparison of both test results was conducted.

3. Materials and method of experimental research

The following legend and marking are used: C – control; B – beam; D – damaged. When the corrosion or mechanical damages of the reinforcement occurs, the reinforcement cross-section is reduced and, correspondingly the bearing capacity of reinforced concrete elements is decreased. In experimental studies the widely spread practice is to simulate the decrease of cross-section area of beam reinforcement by drilling holes in steel bars. Herewith, depending on the drilled hole diameter the level of reinforcement damage and cross-section area reduction of steel bars are simulated. In the tests the diameter of drilled holes was gradually increased, for example the area of Ø20-sample cross-section was reduced to the area of Ø16-samples.

The first number means the beam series. In our case there are two series: 1-st series- beams with steel reinforcement of 20 mm diameter; 2-d series- beams with steel reinforcement of 16 mm diameter. Second number is the sequence number of the sample.

Experimental samples used in the research had the length of 2100 mm, 100 mm weight and 200 mm height. The beam concrete composition: C: S: R =1:1.16:2, 5 with W/C=0.375. The cement used - M-500, sand quartz without impurities with a size module Mk=2.00, granite rubble of fractions 5 to 10 mm – 66%, 10 to 20 mm – 33%.

The reinforcement of samples: A500C steel bars, diameter was different, depending on series: Ø20 mm (1-st series, Figure 2.4), 16 mm (2-d series, Figure 1). Reinforcement of compressed zone and transverse reinforcement are made of Ø5 B 500 bars. Transverse reinforcement Ø5 B500C has 75 mm spacing. Connection of the reinforcement in the spatial frame was performed with the use of factory contact welding. To the longitudinal reinforcement were welded 4 metal holders with 200 mm spacing, intended for fastening mechanical devices for determining deformations. The samples were manufactured at the factory on the current production vibrating platforms.

During the experiment, samples were subjected to bending under short-term loading. Load level was controlled using K1 and K2 ring dynamometers, which simultaneously served as the hinge support on one side and a fixed support on the other side of the 1900 mm beam.

Load was applied in the form of two concentrated forces in the thirds of the beam span with the use of a hydraulic jack and the distribution beam. The scheme of device location is shown on Figure 3.

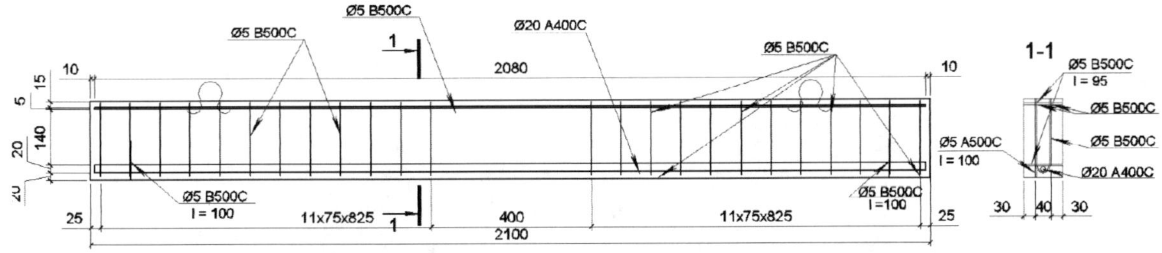

Figure 1. Construction and reinforcement of experimental samples of first series.

Deformations of the load-bearing reinforcement were measured with the use of clock-type micro-indicators Ai1 and Ai2 with the division value of 0.001 mm. The i1 and i2 micro-indicators with the division value of 0.001 mm were located on the beam upper face in order to measure the deformation of the most compressed concrete fibres. The i3 to i10 micro-indicators were located on the side face of the beam with 20 mm spacing increments and did not extend beyond the bending zone.

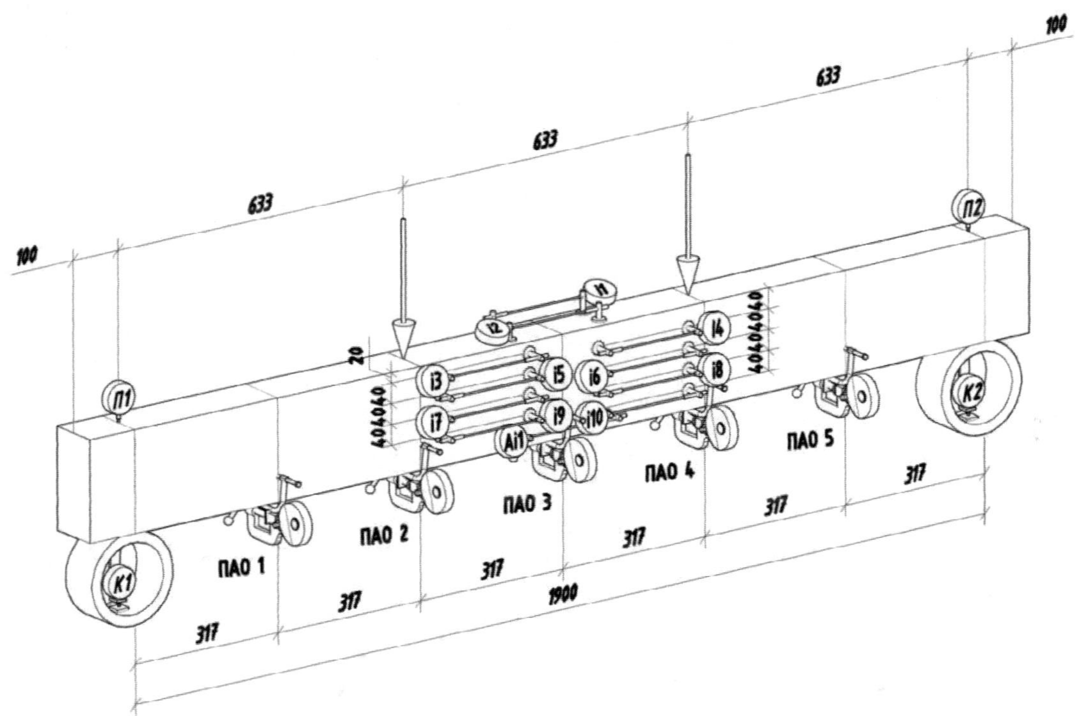

Figure 2. The scheme of location of devices.

The 1 series sample openings (load-bearing reinforcement of 20 mm diameter) were performed with their gradual incrimination: 1.0, 1.5, 2.0, 2.5, 3.0, 3.5, 4.0, 4.5, 5.0, 5.5 and 5.6 mm. For the 2 beams of the 1st series (BD-1.3 and BD-1.4), the openings were single (Figure 3).

Beam samples were concreted in single mix with appropriate control tests of cubes, prisms and cylinders. The compressive strength of concrete was determined by testing of cubes with 100 mm edge and cylinders with 100 mm diameter and 200 mm height. Prism strength and initial modulus of elasticity were determined with the use of prisms of 400 mm length and 100×100 mm cross-section and cylinders with 300 mm length and 150 mm diameter.

After the research the following results were obtained:

- prism strength fcd - 33.8 MPa;

IOP Conf. Series: Materials Science and Engineering 1021 (2021) 012012 doi:10.1088/1757-899X/1021/1/012012

- cubic strength of fcube - 48.7 MPa.

Figure 3. The general view of the single opening in load-bearing reinforcement of damaged samples.

Before the beginning of the experimental research, 3 samples of A500C class reinforcement were previously tested under tension in order to determine the mechanical properties of steel bars [17] (Figure 4).

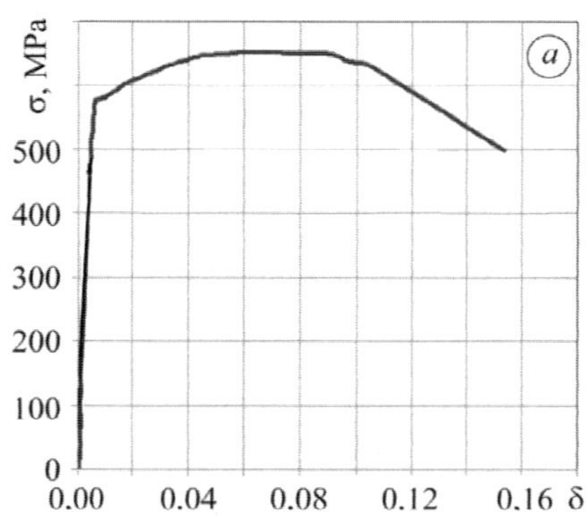

Figure 4. Strain diagrams of the continuous initial A500C steel bar [17].

The results of continuous load-bearing reinforcement of the 1st series beams are presented in Table 1.

Table 1. Mechanical properties of the A500C reinforcement [17]

Material	Yield strength $\sigma_{0,2}$, MPa	Ultimate strength σ_u, MPa	Specific elongation δ, %
A500C reinforcement of 20 mm diameter	570	650	15,3

4. Results of the experimental research

According to the research program 6 beam samples were tested. Among them were 2 undamaged control samples with single load bearing reinforcement of Ø20 mm diameter- BC-1.1 and BC-1.2; 2 samples with Ø20 mm reinforcement with damages about 36% (which corresponds to area of the 16 mm diameter samples)-BD-1.3 and BD-1.4 and 2 samples with Ø16 mm single reinforcement - BC-2.5 and BC-2.6, which corresponds to 36% load bearing reinforcement damage of control samples (the area of 36% damages of 20 mm samples corresponds to 16 mm diameter).

The strain characteristics of compressed concrete and stretched load bearing reinforcement of non-damaged control samples with Ø20 mm reinforcement (BC-1.1 and BC-1.2) and samples with load bearing Ø16 mm reinforcement (BC-2.5 and BC-2.6) are given in Figure 5. Herewith, it is important to note that graphs in Figure 5 and others show averaged values of both strains and bending moments of twin-beams.

(a)

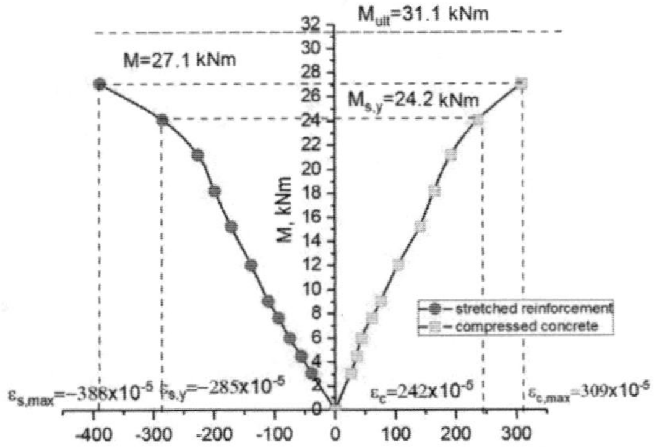

(b)

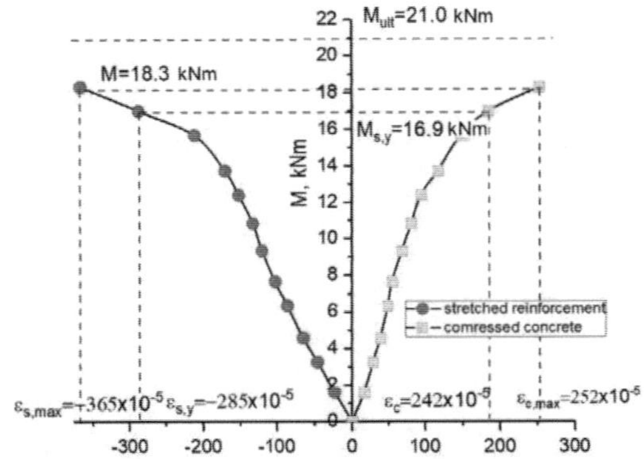

Figure 5. Averaged strain graphs of stretched reinforcement (left) and compressed concrete (right) of 1st series non-damaged beams BC-1.1 and BC-1.2 (a); 3d series beams BC-2.5 and BC-2.6 (b).

In samples with undamaged reinforcement, physical destruction occurs as the result of brittle fracture of the compressed concrete zone (Figure 6).

(a)

(b)

Figure 6. Brittle fracture of the concrete of the most compressed fibre of the BC-1.1 (a) and BC-3.5 samples

In the samples BP-1.3 and BP-1.4 with 20 mm reinforcement without applying a load in the in the steel bars was made a hole of 5.6 mm diameter, which resulted in a cross-sectional area of the steel bar equal to the sample of 16 mm diameter. However, it should be noted that hole mostly reduces the core zone of the steel bar, leaving the thermally-strengthened part of the rebar almost undamaged. It was identified according to the results of experimental testing of samples with holes. Therefore, no exact yield zone could be identified, which is shown by the experimental results [17]. The gradual increase in deformation of stretched reinforcement and the most compressed fibre of concrete could be clearly identified.

It is important to note the changes in the destruction characteristics of the samples, which are the result of damages mostly in non-strengthened steel-bar layer and insufficient damages in thermally-strengthened reinforcement layer. In addition, no destruction of compressed concrete and achievement of the strain limit values by most compressed fibre took place. Destruction was caused by the load-bearing reinforcement rupture. It could be explained by the fact that damages of the stretched armature were mostly local, which corresponds to the drilled hole. This could be affirmed by the absence on the graphs (Figure 7) exact yield zone of the steel bars.

Summary results of test samples without initial load level are shown in Table 2. In samples BD-1.3, BD-1.4, BC-2.5 and BC-2.6, the area of the load-bearing reinforcement, as well as all other parameters (concrete strength, frames location, etc.) are the same. However, according to table. 2, the strength of samples with 20 mm diameter damaged reinforcement (BD-1.3 and BD-1.4) is higher than the strength of samples with 16 mm load bearing reinforcement (BC-2.5 and BC-2.6).

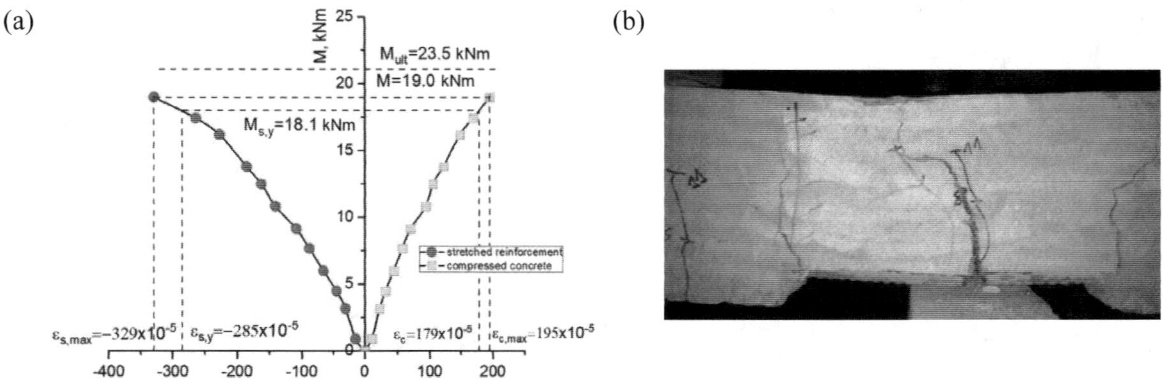

Figure 7. Averaged strain graphs of stretched reinforcement (left) and compressed concrete (right) of damaged beams of the 1st series BP-1.3 and BP-1.4 (a) and the destruction nature of the BD-1.3 sample (b).

Table 2 Bearing capacity of experimental samples

Sample mark	Bearing capacity, kNm		Physical destruction, kNm		Deviation in the bearing capacity values, %		Deviation in physical destruction, %	
	sample	average	sample	average	sample	average	sample	average
BC-1.1	24.9	24.2	32.9	31.1	-	-	-	-
BC-1.2	23.5		29.3		-		-	
BD-1.3	19	18.1	22.9	23.5	21.5	25.2	26.4	24.4
BD-1.4	17.2		24.1		28.9		22.5	
BC-2.5	16.3	16.9	20	21.1	32.6	30.2	35.7	32.2
BC-2.6	17.5		22.2		27.7		28.6	

Note: deviations were identified according to control samples KB-1.1 and KB-1.2

This could be explained by the fact that in the damaged samples the main load-bearing reinforcement zone is the thermally-strengthened outer layer. Therefore, the deviation of the bearing capacity of the damaged specimens was in average 24%, and of the non-damaged with the same reinforcement area - 31%.

5. Conclusion

In the article, by the experimental study the strength and strain parameters of reinforced concrete beams with damages in the load bearing stretched reinforcement due to the action of the bending moment, which causes the exhaustion of the bearing capacity were identified.

With the gradual reduction of the stretched reinforcement cross-section, certain increase in the deformation of both the reinforcement and the concrete of the most compressed fibre occurs; herewith if further increase in the load takes place the final bearing capacity of the reinforced concrete beams decreases in comparison with the samples with non-damaged reinforcement. The final bearing capacity decreases almost proportionally to the reduction of the load bearing reinforcement area.

Exhaustion of the load-bearing capacity of reinforced concrete beams without damages of reinforcement was due to reaching the yield strength of the steel bars and, after gradual load increase – reaching of the limit strain in the concrete fibre and its destruction. In reinforced concrete beams with damaged reinforcement if the load was increased the bearing capacity was exhausted due to reaching of the yield strength in the reinforcement. After further load increasing the rupture of the reinforcement occurs without fragmentation of the compressed zone concrete. It could be explained by the fact that the reinforcement damages were performed locally in one particular place by drilling of the hole, which, in addition, was the stress concentrator.

Reinforced concrete beams with ø20 mm load-bearing reinforcement, which area by damaging was decreased to the area of the ø16 mm steel bars had the final bearing capacity higher than reinforced concrete beams with non-damaged ø16 mm reinforcement. It could be explained by the fact that when of ø20 mm steel bars were damaged by hole drilling, the most damaged zone of the bar cross-section had lower mechanical properties; the outer thermally-strengthened layer with higher mechanical properties was damaged insignificantly.

References

[1] Vatulia G L, Lobiak O V, Deryzemlia S V, Verevicheva M A and Orel Y F 2019 Rationalization of cross-sections of the composite reinforced concrete span structure of bridges with a monolithic reinforced concrete roadway slab *IOP Conference Series: Materials Science and Engineering* **664(1)** 012014

[2] Vatulia G, Orel Y and Kovalov M 2014 Carrying capacity definition of steel-concrete beams with external reinforcement under the fire impact *Applied Mechanics and Materials* **617** 167-170

[3] Karpiuk V, Somina Y and Maistrenko O 2020 Engineering Method of Calculation of Beam Structures Inclined Sections Based on the Fatigue Fracture Model, *LNCE* **47** 135-144

[4] Vegera P, Vashkevych R and Blikharskyy Z 2018 Fracture Toughness of RC Beams with different shear span *Matec Web of Conferences* **174** 02021

[5] Pavlikov A, Kosior-Kazberuk M and Harkava O 2018 Experimental Testing Results of Reinforced Concrete Beams Under Biaxial Bending *IJST* **7(3.2)** 299-305

[6] Bobalo T, Blikharskyy Y, Kopiika N and Volynets M 2021 Influence of the Percentage of Reinforcement on the Compressive Forces Loss in Pre-stressed RC Beams Strengthened with a Package of Steel Bars *LNCE* **100** 53-62

[7] Vatulia G, Lobiak A, Chernogil V and Novikova M 2019 Simulation of Performance of CFST Elements Containing Differentiated Profile Tubes Filled with Reinforced Concrete *Materials Science Forum* **968** 281-287

[8] Selejdak J, Khmil R and Blikharskyy Z 2018 The Influence of Simultaneous Action of the Aggressive Environment and Loading on Strength of RC Beams *Matec Web of Conferences*. **183** 02002

[9] Lobodanov M, Vegera P and Blikharskyy Z 2021 Influence of Damages in the Compressed Zone on Bearing Capacity of Reinforced Concrete Beams, *LNCE* **100** 260-267

[10] Kovalchuk B, Blikharskyy Y, Selejdak J and Blikharskyy Z 2021 Strength of Reinforced Concrete Beams Strengthened Under Loading with Additional Reinforcement with Different Levels of its Pre-tension, *LNCE* **100** 227-236

[11] Cigada A and Zappa E 2014 Vision Device Applied to Damage Identification in Civil Engineer Structures *Structural Health Monitoring* **5** 195-206

[12] Koteš P, Brodňan M and Bahleda F 2016 Diagnostics of Corrosion on a Real Bridge Structure *Advances in materials science and engineering* 2125604.

[13] Khmil R, Tytarenko R, Blikharskyy Y and Vegera P 2021 The Probabilistic Calculation Model of RC Beams, Strengthened by RC Jacket Crack Resistance of RC Beams on the Shear *LNCE* **100** 17-24

[14] Orešković M, Klymenko Y, Aniskin A and Kozina G 2018 Analysis of Damaged Concrete Columns of Circular Cross-Section *Tehnički vjesnik* **25(2)** 337-343

[15] Karpiuk V M, Syomina Y A and Antonova D V 2019 Bearing Capacity of Common and Damaged CFRP-Strengthened RC Beams Subject to High-Level Low-Cycle Loading *Materials Science Forum* **968** 185-199

[16] Kos Z and Klimenko Y 2019 The Development of Prediction Model for Failure Force of Damaged Reinforced-Concrete Slender Columns *Tehnički vjesnik*, **26 (6)** 1635-1641

[17] Koteša P, Brodňana M, Ivaškováb M and Dubalac K 2015 Influence of Reinforcement Corrosion on Shear Resistance of RC Bridge Girder Subjected to Shear *Procedia Engineering* **111** 444-449
[18] Dmitrović L Kos Z and Klimenko Y 2019 The Development of Prediction Model for Failure Force of Damaged Reinforced-Concrete Slender Columns *Tehnički vjesnik* **26 (6)** 093612
[19] Tayeh B A, Naja M A, Shihada S and Arafa M 2019 Repairing and Strengthening of Damaged RC Columns Using Thin Concrete Jacketing *Advances in Civil Engineering* 2987412
[20] Kramarchuk A, Ilnytskyy B, Lytvyniak O and Famulyak Y 2019 Strengthening Prefabricated Reinforced Concrete Roof Beams that are Damaged by Corrosion of Concrete and Reinforcement *IOP Conference Series: Materials Science and Engineering* **708 (1)** 012060
[21] Vavruš M and Koteš P 2019 Numerical Comparison of Concrete Columns Strengthened with Layer of Fiber Concrete and Reinforced Concrete *Transportation Research Procedia* **40** 920-926
[22] Algburi A H, Sheikh M N and Hadi M N 2019 Analytical Investigation on the Behavior of Circular and Square RC Columns Strengthened with RPC and Wrapped with FRP Under Uniaxial Compression *Journal of Building Engineering* **25** 100833
[23] Algburi A H, Sheikh M N and Hadi M N 2019 New Technique for Strengthening Square-Reinforced Concrete Columns by the Circularisation with Reactive Powder Concrete and Wrapping with Fibre-Reinforced Polymer *Structure and Infrastructure Engineering* **15 (10)** 1392-1403.
[24] Blikharskyy Z, Bobalo T, Kramarchuk A, Ilnytksyy B and Vashkevych R 2021 Bearing capacity of stone beam reinforced by GFRP *LNCE* **100** 45-52
[25] Azizov T N, Kochkarev D V and Galinska T A 2019 New Design Concepts for Strengthening of Continuous Reinforced-Concrete Beams *IOP Conference Series: Materials Science and Engineering* **708 (1)** 012040
[26] Sanytsky M, Kropyvnytska T, Fic S and Ivashchyshyn H 2020 Sustainable low-carbon binders and concretes *E3S Web of Conferences*, **166** 06007
[27] Kramarchuk A, Ilnytskyy B, Lytvyniak O and Grabowski A 2019 The increase of seismic stability for existing industrial buildings *AIP Conference Proceedings* **2077** 145450
[28] Amran Y M, Alyousef R, Rashid R S, Alabduljabbar H and Hung C C 2018 Properties and applications of FRP in strengthening RC structures *A review Structures* **16** 208-238

IOP Conf. Series: Materials Science and Engineering 1021 (2021) 012013 doi:10.1088/1757-899X/1021/1/012013

Scientific approach to fire resistance calculation of reinforced concrete beams and columns

S L Fomin[1,2], Y V Bondarenko[1], S V Butenko[1], S M Koliesnikov[1]

[1] Department of reinforced concrete and stone structures, Kharkiv National University of Civil Engineering and Architecture, 40 Sumska Str., Kharkiv, 61002, Ukraine

[2]E-mail: sfomin@ukr.net

Abstract. Problems related to the fire resistance calculation of reinforced concrete beams and columns from a scientific point of view when using new design standards are considered. The first problem is the insufficient development of improved calculation methods, exactly their absence. Only basic principles and requirements are formulated for them. This article presents the principles and sequence of calculation. The second problem is the lack of an accurate mathematical nonlinear compression model of the stress-strain diagram for concrete. The article explains the imperfection of the existing one, the need to create a new model and presents the basic requirements that such a mathematical model must meet. The third problem of the discussed ones is the erroneous values of the relative ultimate deformation of concrete. The principle of formation of values of relative limiting deformation of concrete is specified and proceeding from it the additional requirement for mathematical nonlinear model of compression of the diagram of stress - strain for concrete is specified.

1. Introduction

Standards for calculating the fire resistance of reinforced concrete beams [1] and reinforced concrete columns [2] have been developed and adopted by NDIBK (State Research Institute of Building Constructions) in connection with the harmonization of national building codes with European ones. These standards apply to the design of reinforced concrete elements that are exposed to fire and in some cases complement the design of reinforced concrete elements at normal temperature. The norms describe the strength (compressive strength, tensile strength, relative deformation at maximum load, relative ultimate deformation, initial modulus of elasticity) and thermophysical (temperature elongation, specific heat, thermal conductivity) properties of concrete and reinforcing steel at elevated temperatures. The codes contain a description of tabular and simplified methods for calculating reinforced concrete beams and columns. These codes are a logical continuation of the previously adopted European standards [3], [4], but in turn they have also assimilated the shortcomings and inaccuracies [5] that took place in Eurocodes. There are a number of problems associated with the calculation of fire resistance and increasing its accuracy for these building elements when using these standards. The use of these standards is acceptable when assessing the fire resistance of reinforced concrete beams and columns as a separate structure and as part of the structure. With the accurate calculation of fire resistance and the global structural analysis when using these standards, there are a number of problems associated with improving its accuracy for these building elements.

2. Improved calculation methods

One of the main problems is the insufficient coverage of the improved calculation methods, their description and formulation of requirements to them. These standards [1] and [2] allow to carry out

IOP Conf. Series: Materials Science and Engineering 1021 (2021) 012013 doi:10.1088/1757-899X/1021/1/012013

calculations based on tabular data, simplified methods and improved calculation models, shown in Table 1, in accordance with the Codes [3].

It should be noted that only the improved methods allow to investigate the operation of the reinforced concrete structural element during a fire comprehensively and take into account possible options for the operation of all structural elements of the building as a whole like a part of the structural scheme of the building. Also, in real buildings reinforced concrete elements are interconnected and statically indeterminate, so the peculiarity of the structures during fire is that the calculated load effect during fire on the structural element E_ (d.fi) differs from the calculated load effect on structural element at normal temperature. However, depending on the type of fire, its location, scale of the fire and structural parameters of the elements (stiffness, curvature, cross section of the element), the calculated load effect during a fire on a structural element E_ (d.fi) for an individual element can both increase and decrease (Figure 1). This is due to the redistribution of forces between structural elements of the building on account of the nonlinear laws of deformation of concrete and reinforcing steel.

Table 1. Comparative table of fire resistance calculation methods.

	Tabulated data	Simplified calculation methods	Improved calculation models
Individual member analysis The individual member is considered as isolated. Indirect fire actions are not considered, except those resulting from thermal gradients	YES - Data given for standard temperature mode only; - Data could be developed for other fire temperature modes.	YES - standard fire and parametric fire, - temperature profiles given for standard fire mode only, - material models that consider change of material properties and apply only to heating rates similar to standard fire.	YES Only the principles are given.
Analysis of parts of the structure Indirect fire actions within the subassembly are considered, but no time-dependent interaction with other parts of the structure.	NO	YES - standard fire and parametric fire, - temperature profiles given for standard fire only, - material models that consider change of material properties are applied only to heating modes similar to standard fire.	YES Only the principles are given.
Entire structural analysis Analysis of the entire structure. Indirect fire actions are considered throughout the entire structure.	NO	NO	YES Only the principles are given.

The sequence of calculation by means of improved methods should include: setting absolute or equivalent temperature in each finite element, determining rigidity of structural elements by setting the law of nonlinear deformation and in case of reinforcement in the element, its percentage along all axes, modelling of nonlinear loads. The developed method of calculating the bearing capacity of structures and working loads in a fire [6] allows to predict changes in the condition of buildings, to develop scenarios of dangerous situations taking into account different combinations of heating, to develop proposals to ensure the required fire resistance.

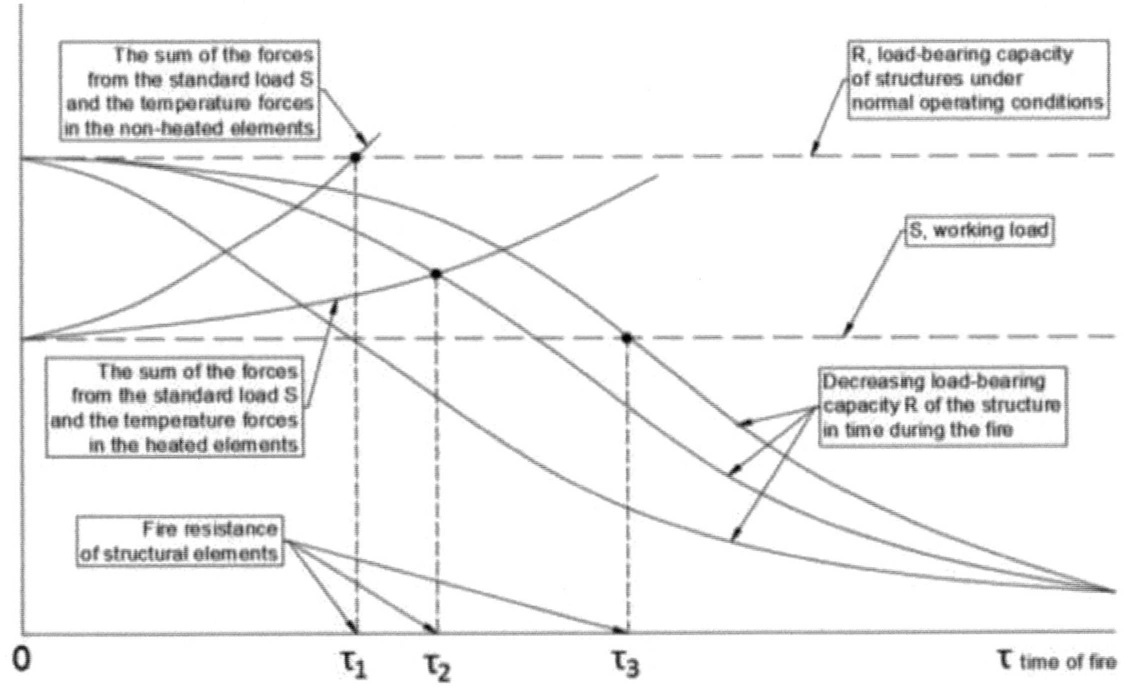

Figure 1. Scheme of calculation of fire resistance of structures.

3. Mathematical nonlinear compression model of stress-strain diagram for concrete at elevated temperatures

An important problem is the lack of clear mathematical nonlinear compression model of the stress-strain diagram for concrete, taking into account the action of temperature effects [5,6]. This is necessary for accurate calculations of fire resistance of building elements in software packages [7] such as Lira, Scad, Ansys and their analogues. In the standards [1,2] the ascending branch of the stress-strain diagram is described by the formula, in turn the descending branch of the stress-strain diagram for concrete has a generalized description for the use of linear and nonlinear models (Figure 2).

Such wording is unacceptable when calculating fire resistance by specified methods and when modeling the effect of fire on the structure. First, the mathematical nonlinear compression model of the stress-strain diagram for concrete at elevated temperatures when used at normal temperatures, given in norms [1,2], does not agree well with the compression model (1) of the stress-strain diagram for concrete at normal temperatures. given in the design standards [8].

$$\frac{\sigma_c}{f_{cm}} = \frac{k\eta-\eta^2}{1+(k-2)\eta} \tag{1}$$

$$\eta = \frac{\varepsilon_c}{\varepsilon_{c1}} \tag{2}$$

$$k = 1.05E_{cm}\frac{\varepsilon_{c1}}{f_{cm}} \tag{3}$$

IOP Conf. Series: Materials Science and Engineering 1021 (2021) 012013 doi:10.1088/1757-899X/1021/1/012013

where σ_c – compressive stress in the concrete; f_{cm} – mean value of concrete cylinder compressive strength; ε_c – compressive strain in the concrete; ε_{c1} – compressive strain in the concrete at the peak stress.

Table 2 shows the comparative calculations of the two models of concrete compression. Second, the arbitrary interpretation of the descending branch of the diagram leads to different results of calculations, and it is difficult to assess their reliability and accuracy. Third, the formula of the reduced nonlinear compression model of the stress-strain diagram for concrete does not contain the initial modulus of elasticity of concrete, in contrast to the model given in the design standards [8].

Table 2. Comparative calculations of the two models of concrete compression.

Compressive strain in the concrete, ppm	Compressive stress in the concrete by formula (1) in the standard [8], MPa	Compressive stress in the concrete by mathematical nonlinear compression model in the standards [1,2], MPa	Difference, %
0	0.0	0	
0.2	5.9	4.2	28.5
0.4	10.9	8.4	23.5
0.6	15.2	12.4	18.4
0.8	18.8	16.3	13.6
1	21.8	19.8	9.2
1.2	24.1	22.7	5.6
1.4	25.9	25.1	2.9
1.6	27.1	26.8	1.2
1.8	27.8	27.7	0.2
2	28.0	28.0	0.0

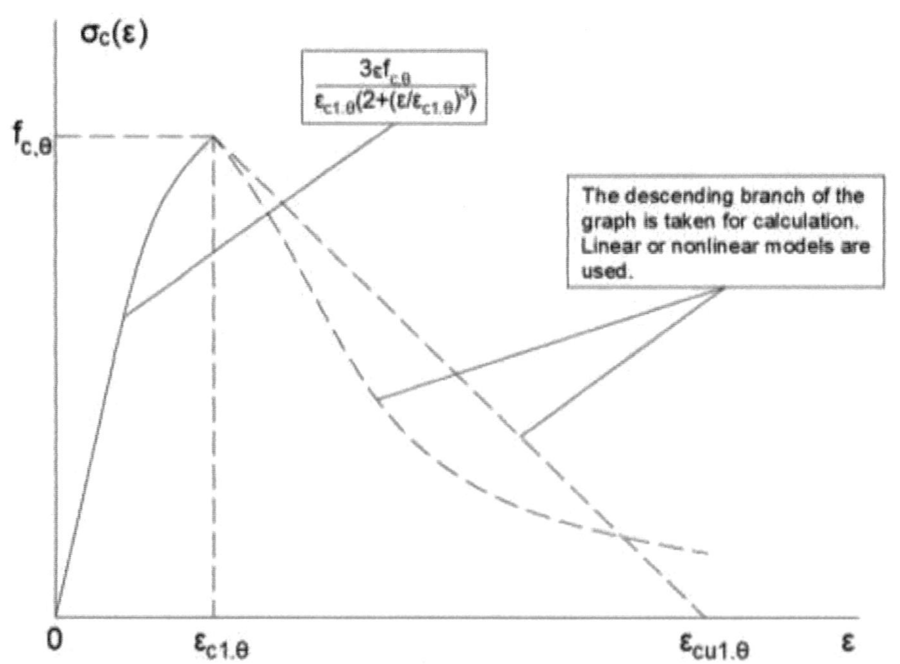

Figure 2. Mathematical model of the stress-strain diagram of compressed concrete at elevated temperatures.

Mathematical nonlinear compression model of the stress-strain diagram for concrete must be developed taking into account the following requirements:

- be a continuous mathematical function;
- have the extremum $f_{c,\theta}$ at the point $\varepsilon_{c1,\theta}$ε;
- at the starting point ε = 0 the differential of the function is equal to the modulus of deformation E_c
- be a nonlinear mathematical function;
- include the temperature variable θ.

In generalized form it will be obtained the following:

$$\sigma = F(f_{c,\theta}, \varepsilon, \varepsilon_{c1,\theta}, \theta) \tag{4}$$

$$\frac{d}{d\varepsilon}\sigma = 0 \quad \text{at } \varepsilon = \varepsilon_{c1,\theta} \tag{5}$$

$$\frac{d}{d\varepsilon}\sigma = E_c \quad \text{at } \varepsilon = 0 \tag{6}$$

where σ – compressive stress of concrete.

4. The value of relative ultimate deformation of concrete

Faultiness of values of relative ultimate deformation of concrete $\varepsilon_{c21,\theta}$ at elevated temperatures [9] in these standards [1, 2] should be noted. The increased value of the relative ultimate deformation of concrete $\varepsilon_{c21,\theta}$ at elevated temperatures leads to overestimated values of displacement of structural elements without their destruction [10]. This in turn increases the calculated value of the fire resistance of the structural element and the fire resistance of the building as a whole. This makes possible situations to arise when an element of the building should perform its functions [11], but in fact it is already destroyed according to the calculation.

To determine the relative ultimate deformation of concrete $\varepsilon_{c21,\theta}$ it is possible to use the criterion of specific potential deformation energy of deformation (fourth strength theory) [12]. As a criterion of strength in this case, the amount of specific potential energy of deformation accumulated by the deformed element is chosen. According to this theory, a dangerous state in the general case of a stress state occurs when the specific potential energy of deformation reaches its limit value.

Thus, this adds an additional requirement to the stress-strain compression diagram for concrete at elevated temperatures. The stress-strain diagram for concrete should give an accurate definition of the relative ultimate deformation of concrete $\varepsilon_{c21,\theta}$ as an extremum of the potential deformation energy at a given point (Figure 3).

$$\frac{d}{d\varepsilon}U = 0 \quad \text{at} \quad \varepsilon = \varepsilon_{cu1,\theta} \tag{7}$$

where U – potential deformation energy of concrete compression.

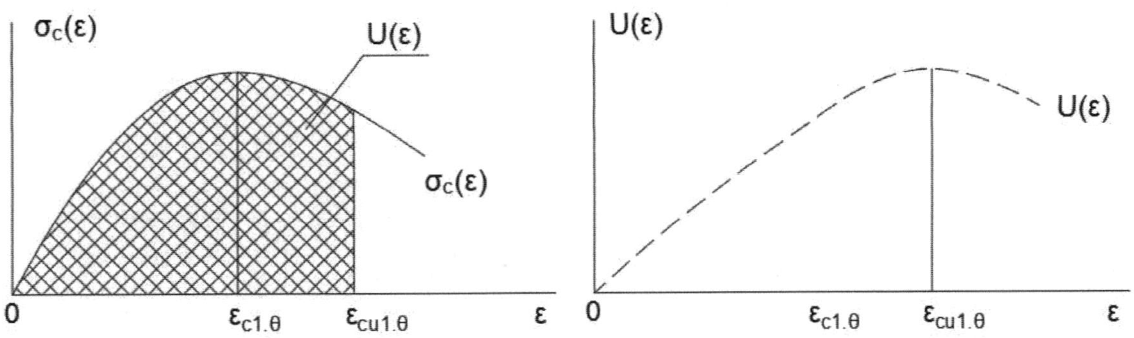

Figure 3. In terms of estimating the relative ultimate deformation of concrete.

Determination of real values of relative limit deformation of concrete $\varepsilon_{c2 1,\theta}$ at elevated temperatures is possible from experimental test data of samples (prisms, cylinders) of all classes of concrete at different values of elevated temperatures using a real stress-strain compression diagram [13,14]. Obtaining a real diagram of stress-strain with a descending branch is possible only in special presses that have a high rigidity of the structure compared to conventional ones [15]. In such installations the accumulation of potential deformation energy in the power elements of the unit (rods, traverses, frames) is minimized, which allows to fully investigate the descending branch of the stress-strain diagram [16].

5. Conclusions

1. The problems of fire resistance calculation of reinforced concrete beams and columns are scientifically grounded; they arise when the current standards are used.
2. The requirements and sequence of calculation are formulated applying the specified calculation methods.
3. The disadvantages for the existing model of compression of the stress-strain diagram for concrete at elevated temperatures have been stated. Requirements for formulating a more accurate chart model have been specified.
4. The faultiness of values of the relative ultimate deformation of concrete at elevated temperatures, which is given in the current standards, has been determined. An additional requirement for the stress-strain compression diagram for concrete at elevated temperatures has been formulated.

References

[1] DSTU-N B B.2.6-196:2014 National standard of Ukraine. Guidelines for design of reinforced concrete beams. Calculation of fire. (Kyiv: Min. of Reg. Devel. of Ukraine)
[2] DSTU-N B B.2.6-197:2014 National standard of Ukraine. Guidelines for design of reinforced concrete columns. Calculation of fire. (Kyiv: Min. of Reg. Devel. of Ukraine)
[3] EN 1992-1-2:2004 Eurocode 2: Design of concrete structures – Part 1-2: General rules - Structural fire design (Brussels: BSI)
[4] DSTU-N-P B V.2.6-159:2010 Structures of buildings and facilities. Design of composite steel and concrete structures. Part 1-2. General provisions. Structural fire design. (EN 1994-1-2:2005, Eurocode 4. MOD) (Kyiv: Min. of Reg. Devel. of Ukraine)
[5] Fomin S L, Bondarenko Y V, Butenko S V, Koliesnikov S M 2020 Major Issues of Theory and Practice of Fire Resistance of Reinforced Concrete Structures and Buildings *Materials Science Forum* 2020 (Kharkiv: National University of Civil Defence of Ukraine) **1006** 136-142
[6] Fomin S L, Butenko S V, Spirande K V, Iakymenko M V 2019 Mathematical model of concrete strain diagram under heating *IOP Conf. Ser.: Mater. Sci. Eng.* **708** 012048
[7] Bangash M Y H, Al-Obaid Y F, Bangash F N 2014 *Fire Engineering of Structures Analysis and Design* Springer-Verlag Berlin Heidelberg GmbH & Co. KG (Berlin, Germany)
[8] EN 1992-1-1:2005 Eurocode 2: Design of concrete structures - Part 1-1: General rules and rules for buildings (Brussels: BSI)
[9] Venkatesh Kodur 2014 Properties of Concrete at Elevated Temperatures *ISRN Civil Engineering* vol 2014 1-15
[10] Bastami M, Aslani F and Esmaeilnia Omran M 2010 High-Temperature Mechanical Properties of Concrete *International Journal of Civil Engineerng* **8** 337-351
[11] Purkiss John A and Long-Yuan Li 2013 Fire Safety Engineering Design of Structures Third Edition *Boca Raton: CRC Press* 452
[12] Pisarenko G S Resistance of materials: Textbook 2nd ed., add. and reworked. 2004
[13] Hachemi S, Ounis A, Chabi S 2014 Evaluating Residual Mechanical and Physical Properties of Concrete at Elevated Temperatures *World Academy of Science, Engineering and Technology International Journal of Civil, Architectural, Structural and Construction Engineering* **8(2)** 176-181

[14] Anunai R and Rakesh Siempu Behavior of high strength concrete subjected to elevated temperature 2019 *Rakesh Siempu International Journal of Innovative Technology and Exploring Engineering* **9(2)** 2997-3000
[15] John Newman and Ban Seng Choo 2003 Advanced concrete technology testing and quality *Butterworth-Heinemann* 1920
[16] Neville A M Properties of concrete Fifth Edition 2011 *Pearson Education Limited* 2866

Improvement of the method of probability evaluation of the failure-free operation of reinforced concrete beams strengthened under load

R Ye Khmil[1], R Yu Tytarenko[1,3], Ya Z Blikharskyy[2] and P I Vegera[1]

[1] Department of Building Constructions and Bridges, Lviv Polytechnic National University, S. Bandery street 12, 79013 Lviv, Ukraine
[2] Department of Highways and Bridges, Lviv Polytechnic National University, S. Bandery street 12, 79013 Lviv, Ukraine

[3] Email: roman.y.tytarenko@lpnu.ua

Abstract. The article is dedicated to the study of the load level effect on failure-free operation probability of undamaged rectangular cross-section reinforced concrete beams when their strengthening by the tensile steel reinforcement increasing. The presence of the load level simulates the real structure exploitation conditions at their strengthening. The validation of basic and improved reliability evaluation methods on real beam samples allowed receiving the recommended failure-free operation values *P(β)*. Both techniques are adapted to national standards for the reinforced concrete structure design. The difference between the techniques is the operating method of the random parameter of the load level at the moment of amplification. According to the results of approbation the comparative analysis of the received results is carried out – depending on additional steel bar diameter and the load level at the moment of amplification. The practical significance of the article results is in use of both existing and advanced methods of the probabilities of failure-free operation estimation of reinforced concrete bended elements, strengthened by addition of the stretched steel reinforcement under different load levels, in particular, for accepting higher design levels of reliability.

1. Introduction

Nowadays, probabilistic methods of building structure design are becoming more common as they make it possible to assign a guaranteed level of their reliability at the design stage [1, 2, 3, 4]. Moreover, taking into consideration the factors of high level of subjectivity in the selection of design random parameters of the load-bearing capacity reserve of structures, as well as the lack of sufficient regulatory base of calculation (multiple random parameters have large statistical variations); could be affirmed the relevance of the task of development of a reliability estimation method, including in terms of the failure-free operation ensuring. In turn, taking into account the rate of incensement of volumes of work during reconstruction of existing buildings and structures, the calculation of the elements that need reinforcement should be accompanied by a quantitative assessment of their reliability – by establishment of their failure-free operation probability. In addition, nowadays the problem of the load level effect (as a random parameter) in strengthening of reinforcement concrete structures on the probability of their failure-free operation has not been sufficiently studied and needs further research and analysis [5, 6]. Recently, many researchers have been devoted to the problems of corrosion analysis of reinforced concrete structure rebar [7-9], stress-strain state of bended and compressed-bended reinforced concrete elements [10-15], strengthened under loading with the use of traditional methods [16, 17] and various composite materials [18], as well as to work of reinforced concrete elements in the aggressive environment [19-21].

However, despite the novelty of the obtained experimental results in these studies no thorough analysis of the issue of reliability assessment of structures (especially those reinforced under loading) was provided. In turn, the reliability of reinforced concrete elements (as separate building structures) has been the subject of research for not a long time – since the end of the 20th century. It should be noted here that the vast majority of scientific works are dedicated to the calculation of the reliability of beams and slabs strengthened with composite reinforcement.

Among others, in some works [22, 23] analysis of bended elements (beams and slabs) of bridges strengthened with composite tapes of different cross-section was conducted, in addition to the method for safety factor calculation for such structures [24]. However, although in these studies indicators for reliability estimation were obtained [22, 23] and recommended safety factors were identified [24], they do not fully take into consideration the impact of load level at the moment of structure strengthening on their failure-free operation (as the random it was not considered at all). In studies [25, 26] was analyzed the influence of different size composite tapes, applied under the loading on the reliability of reinforced concrete beams with insufficient shear strength.

The advantage of article [25] is that the structural reliability analysis was performed with the use of simplified probabilistic algorithm based on the Monte Carlo method for a wide range of nominal load, under which the strengthening was conducted (from operational to limit values). Reliability estimation based on the statistical processing of experimental shear strength data for more than 250 beams reinforced under the load is given in the paper [26]. However, there is a number of comments regarding the studies [25, 26]: 1) only beams with insufficient shear strength were considered; 2) sufficient variation of obtained safety factor values was identified [26]; 3) only the strength parameters were accepted as the random parameters of the load-bearing capacity of the beams (the level of load at the moment of reinforcement was taken as a determinated value). In study [27] is proposed the method of reliability evaluation of reinforced concrete beams and slabs, strengthened with composite tape under loading.

Unlike works [25, 26], this research concerned elements with insufficient bending strength (normal cross-section of the structure), which makes it specially relevant. However, the sophisticated mathematical apparatus for reliability assessment, based on previous research statistics and adapted to US design standards, does not allow to fully use the proposed methodology in design practice. In addition, the calculation of reliability indicators was carried out only for one load level at the moment of amplification, which makes it impossible to plot a dependency graph of obtained β-indexes of this parameter – unlike the work [28].

Thus, in this work it is considered that it is appropriate to develop and improve approaches to the quantitative assessment of reliability (first of all, in terms of failure-free operation ensuring) of reinforced concrete bended elements strengthened under load. An important component of such an assessment is the study of the influence of load level at the moment of strengthening of such structures as a random parameter on their failure-free operation probability.

2. Purpose of the article

The purpose of the article is to investigate the influence of the load level on the probability of their failure-free operation when strengthening by addition of stretched steel reinforcement of reinforced concrete beams of undamaged rectangular cross-section. This purpose was achieved by approbation on the real samples of basic [28] and advanced methods for their failure-free operation probability. Both methods were adjusted to national standards for the reinforced concrete structure design. The difference between the techniques is the operating method of the random parameter of the load level at the moment of amplification.

3. The method of reliability evaluation of RC beams, strengthened under an action of load

Since the range of statistical variability of random parameters of building structures is within 5…25%, the evaluation of the reliability of the experimental samples of beams after reinforcement [15] will be conducted with the use of statistical linearization, using and developing the existing method, proposed previously [28]. The technique of assessing the reliability of new design structures has been used to find the reliability indicators. For reliability indicator identification, the method of estimation of reliability of newly designed structures [29], as well as well-known theses of probability theory [30] and recommendations for this theses application for building structures [31] were used.

According to previously developed [28] methodology for reliability estimation, the final equitation for calculating of the random value of the limite bending moment $\tilde{M}_{ult}$, which could be perceived by the beam after strengthening (maintaining the condition of $x \le x_R$), takes the following form:

$$\tilde{M}_{ult} = f\left(\tilde{E}_c, \tilde{E}_{s,aver}, \tilde{f}_y, \tilde{E}'_s, \tilde{\gamma}^{mid}_{s,inc}, \tilde{\gamma}^{add}_{s,dis}, \tilde{b}, \tilde{d}_{red}\right) = \left(A_s + A_{s,add}\tilde{\gamma}^{add}_{s,dis}\right)\tilde{f}_y \tilde{d}_{red} -$$

$$- \frac{0{,}5\lambda\varepsilon_c \tilde{d}_{red}\left[\left(A_s + A_{s,add}\tilde{\gamma}^{add}_{s,dis}\right)\tilde{f}_y - \varepsilon_c A'_s \tilde{E}'_s\right]}{\varepsilon_c + 0{,}5\tilde{\gamma}^{mid}_{s,inc}\left(-\varepsilon_c \pm \left(\varepsilon_c^2 + \dfrac{3{,}2\varepsilon_c^2 \tilde{b}\tilde{d}_{red}\tilde{E}_c \vartheta}{A_{s,tot}\tilde{E}_{s,aver}}\right)^{1/2}\right)} - \varepsilon_c A'_s \tilde{E}'_s a'\left(0{,}5\lambda + 1\right) + \tag{1}$$

$$+ \frac{A'_s \tilde{E}'_s a'^2}{\tilde{d}_{red}}\left[\varepsilon_c + 0{,}5\tilde{\gamma}^{mid}_{s,inc}\left(-\varepsilon_c \pm \left(\varepsilon_c^2 + \frac{3{,}2\varepsilon_c^2 \tilde{b}\tilde{d}_{red}\tilde{E}_c \vartheta}{A_{s,tot}\tilde{E}_{s,aver}}\right)^{1/2}\right)\right],$$

where $\tilde{X}_i(\tilde{x}_i)$, $Y_i(y_i)$ are, respectively, the random [$\tilde{X}_i(\tilde{x}_i)$] and determinate [$Y_i(y_i)$] parameter of bearing capacity reserve and structure loading level (all the random and determinate parameters are given in the study [28]).

The mathematical expectations for random arguments $\tilde{X}_i(\tilde{x}_i)$ are substituted in obtained above equitation (1) and the function $M_{ult} = f[X_1(x_1),\dots,X_n(x_n)]$ is linearized by identifying the coefficients $D_{X_i(x_i)}$ [31]:

$$D_{X_i(x_i)} = \partial\overline{M}_{ult}\left(\partial\overline{X}_i(\overline{x}_i)\right)^{-1}, \tag{2}$$

Formula for identification of standard of limit bending moment $\hat{M}_{ult}$ could be rewritten in following form [28]:

$$\hat{M}_{ult} = \left(\sum_{i=1}^{n}\left[D_{X_i(x_i)} \times \hat{X}_i(\hat{x}_i)\right]\right)^{1/2}, \tag{3}$$

where $\hat{X}_i(\hat{x}_i)$ is the standard of random value.

Probability of failure-free operation $P(\beta)$ (as a quantitative reliability indicator), could be identified by the following formula:

$$P(\beta) = 0{,}5 + \Phi(\beta), \tag{4}$$

where $\beta = (\overline{M}_{ult} - M_{ult})/\hat{M}_{ult}$ is the so-called "safety characteristics" according to works [29, 31] or the reliability index according to [24, 27]; M_{ult} is the design bearing capacity of the beam normal cross-section after strengthening; $\Phi(\beta)$ is the error function (Laplas function).

4. The results of theoretical research of influence of loading level during strenghthening of RC beams on the probability of their failure-free operation

The main difference to existing methods for failure-free operation probability evaluation is that in this study the load level is taken into account in calculations as the structure random parameter at the moment of strengthening. Coefficients of load level influence were identified experimentally and given in work [32]. The way of consideration of this coefficient in the method of estimation of failure-free operation probability is the topic of this study. The following approaches are proposed for this influence evaluation.

Variant 1 (basic). The values of standards ($\hat{\gamma}^{add}_{s,dis}$, $\hat{\gamma}^{mid}_{s,inc}$) and their variation coefficients ($V_{\gamma^{add}_{s,dis}}$, $V_{\gamma^{mid}_{s,inc}}$), given in [28] and based on [32] are taken for each load level and rebar diameter separately. Therefore, a separate standard value corresponds to each value in Table 1. Design values of material strength and deformability parameters, cross-section geometry, load level at the moment of strengthening and corresponding coefficients $\gamma^{add}_{s,dis}$, $\gamma^{mid}_{s,inc}$, as well as bearing capacity of strengthened beams M_{ult} are given in [28, 32]. Calculated numerical values of statistic characteristics (mathematical expectations, standards) of above-cited parameters, obtained on the basis of corresponding initial data are also given in [28] and based on [32]. Probability of failure-free operation of strengthened beams, derived from the assumption that load level differs in various strengthening conditions (in our particular case – for additional rebar diameter) is given in the table 1. Graphical representation of it is given in Figure 1 (a).

Table 1. Probability of failure-free operation *P(β)* (variant 1).

№	Additional rebar	Level of load at the moment of strengthening			
		$0.0\times M_{ult,0}$	$0.3\times M_{ult,0}$	$0.5\times M_{ult,0}$	$0.8\times M_{ult,0}$
1	Ø8 mm	0.999651	0.999635	0.999630	0.999602
2	Ø16 mm	0.999795	0.999767	0.999698	0.999672

Note: $M_{ult,0}$ – bearing capacity of unstrengthened beam

Variant 2 (advanced). In this case the values of standards ($\hat{\gamma}^{add}_{s,dis}$, $\hat{\gamma}^{mid}_{s,inc}$) and corresponding variation coefficients ($V_{\gamma^{add}_{s,dis}}$, $V_{\gamma^{mid}_{s,inc}}$) are taken regardless of the strengthening conditions, in this case they are the same for all load levels and additional rebar diameters. Thus, the following averaged values for standard coefficients $\hat{\gamma}^{add}_{s,dis}$, $\hat{\gamma}^{mid}_{s,inc}$ are obtained with the use of table 4 [28] and based on work [32]: $\hat{\gamma}^{add}_{s,dis} = 0.005$; $\hat{\gamma}^{mid}_{s,inc} = 0.002$. Therefore, taking into consideration changes of design bearing capacity of beam normal cross-section after strengthening M_{ult} (mainly towards underestimation) and standard $\hat{M}_{ult}$ (mainly towards overestimation), following changes of reliability indexes β and corresponding failure-free operation probability *P(β)* are obtained and given in table 2. Probability of failure-free operation of strengthened beams obtained due to assumption that load level does not depend on strengthening conditions is graphically represented in Figure 1 (b).

As could be seen from graphs the value of probability of failure-free operation obtained by the advanced method is lower at load level up to $0.3\times M_{ult,0}$ than those values obtained by the basic method. On the contrary, at load level over $0.3\times M_{ult,0}$ probability values of failure-free operation obtained by the advanced method exceed such values by the basic method. This tendency could be applied for both series of sample beams, strengthened under loading with the additional rebar of different diameter (Ø8 mm, Ø16 mm).

IOP Conf. Series: Materials Science and Engineering 1021 (2021) 012014 doi:10.1088/1757-899X/1021/1/012014

Table 2. Probability of failure-free operation *P(β)* (Variant 2).

№	Additional rebar	Level of load at the moment of strengthening			
		$0.0\times M_{ult,0}$	$0.3\times M_{ult,0}$	$0.5\times M_{ult,0}$	$0.8\times M_{ult,0}$
1	Ø8 mm	0.999646	0.999636	0.999624	0.999609
2	Ø16 mm	0.999782	0.999759	0.999730	0.999698

Note: $M_{ult,0}$ – bearing capacity of an unstrengthened beam

(a)
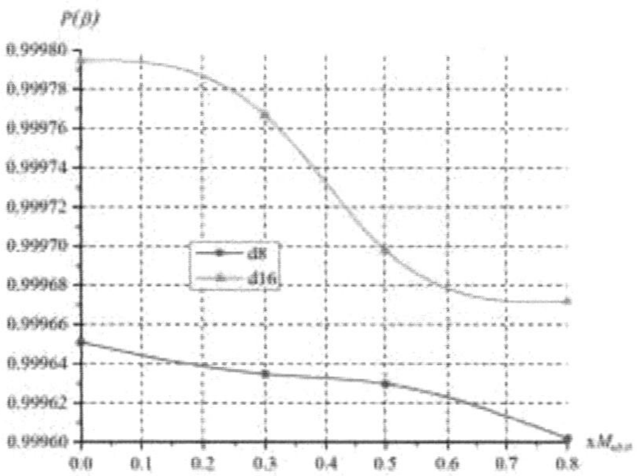

(b)
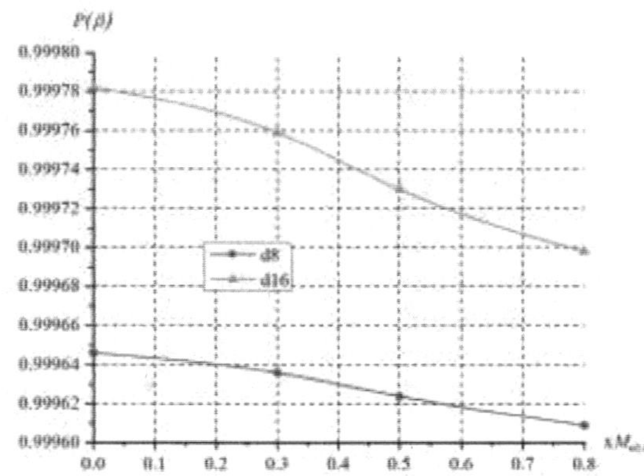

Figure 1. Dependency graph of the "load level ($x\times M_{ult,0}$) – probability of failure-free operation *P(β)*": (a) - (variant 1); (b) - (variant 2).

5. Scientific novelty

According to the results of this study, in order to more reliably design the strengthening of reinforced concrete beams by additional stretched rebar, as was proposed in this article, the advanced method should be used for probability of failure-free operation *P(β)* evaluation at load level less than $0.3\times M_{ult,0}$. The basic method should be used on the higher load levels.

This could be explained as follows: when using the method, which at reliability evaluation gives lower value of failure-free operation probability in order to obtain the desired (design) level of reliability, we need to increase the quantitative or qualitative parameters of other components that determine the bearing capacity of the reinforced element (in this case, the quantity or strength of additional reinforcement).

6. Practical significance

The practical significance of the obtained results is the possibility of differentiated assessment of the influence of the existing load level on the design process of reinforced concrete structures by the method of addition of the tensed rebar. This ranking of the influence of the magnitude of the load level through the use of basic (Variant 1) and advanced (Variant 2) techniques allows to achieve the design level of reliability with the necessary structural argumentation.

7. Conclusions

In this article was conducted the theoretical analysis of the effect of load level on the probability of failure free operation $P(\beta)$ of the reinforced concrete beams of undamaged rectangular cross-section, strengthened by addition of stretched steel bar.

On the basis of approbation of basic (variant 1) and advanced (variant 2) methods of reliability evaluation were obtained corresponding values of failure-free operation probability *P(β)* – in range from 0.999602 to 0.999795 (Variant 1) and from 0.999609 to 0.999782 (variant 2).

It was identified, that for design of such reinforced concrete beam strengthening type with higher reliability the advanced method (Variant 2) for estimation of failure-free operation probability *P(β)* at load levels less than $0.3\times M_{ult,0}$ should be used; otherwise the basic method (variant 1) at higher load levels is more appropriate.

References

[1] Klymenko Y, Grynyova I and Kos Z 2020 The Method of Calculating the Bearing Capacity of Compressed Stone Pillars *LNCE* **47** pp 161-7

[2] Vatulia G L, Lobiak O V, Deryzemlia S V, Verevicheva M A and Orel Y F 2019 Rationalization of cross-sections of the composite reinforced concrete span structure of bridges with a monolithic reinforced concrete roadway slab *IOP Conference Series: Materials Science and Engineering* **664(1)** 012014

[3] Karpiuk V, Somina Y and Maistrenko O 2020 Engineering Method of Calculation of Beam Structures Inclined Sections Based on the Fatigue Fracture Model *LNCE* **47** pp 135-44

[4] Kramarchuk A, Ilnytskyy B, Bobalo T and Lytvyniak O 2020 The Research Bearing Capacity of Crane Beams for Possible Establishment of Bridge Crane on Them *LNCE* **47** pp 202-10

[5] Blikharskyy Z, Khmil R and Vegera P 2017 Shear strength of reinforced concrete beams strengthened by P.B.O. fiber mesh under loading *Matec Web of Conferences* **116** 02013

[6] Khmil R, Tytarenko R, Blikharskyy Y and Vegera R 2021 The probalistic calculation model of RC beams, strengthened by RC Jacket *LNCE* **100** pp 182-91

[7] Lobodanov M, Vegera P, Khmil R and Blikharskyy Z 2021 Influence of damages in the compressed zone on bearing capacity of reinforced concrete beams *LNCE* **100** pp 260-7

[8] Blikhars'kyi Z Y and Obukh Y V 2018 Influence of the Mechanical and Corrosion Defects on the Strength of Thermally Hardened Reinforcement of 35GS Steel *Materials Science* **54(2)** pp 273-8

[9] Ma Y, Guo Z, Wang L and Zhang J 2020 Probabilistic life prediction for reinforced concrete structures subjected to seasonal corrosion-fatigue damage *Journal of Structural Engineering* **146(7)** 04020117

[10] Vatulia G, Lobiak A, Chernogil V and Novikova M 2019 Simulation of Performance of CFST Elements Containing Differentiated Profile Tubes Filled with Reinforced Concrete *Materials Science Forum* **968** pp 281-7

[11] Pavlikov A, Kosior-Kazberuk M and Harkava O 2018 Experimental testing results of reinforced concrete beams under biaxial bending *IJST* **7(3.2)** pp 299-305

[12] Bobalo T, Blikharskyy Y, Kopiika N and Volynets M 2021 Influence of the Percentage of Reinforcement on the Compressive Forces Loss in Pre-stressed RC Beams Strengthened with a Package of Steel Bars *LNCE* **100** pp 53-62

[13] Kovalchuk B, Blikharskyy Y, Selejdak J and Blikharskyy Z 2021 Strength of Reinforced Concrete Beams Strengthened Under Loading with Additional Reinforcement with Different Levels of its Pre-tension *LNCE* **100** pp 227-36

[14] Vegera P, Vashkevych R and Blikharskyy Z 2018 Fracture toughness of RC beams with different shear span *Matec Web of Conferences* **174** 02021

[15] Blikharskyy Z, Bobalo T, Kramarchuk A, Ilnytskyy B and Vashkevych R 2021 Bearing capacity of stone beam reinforced by GFRP *LNCE* **100** pp 42-52

[16] Azizov T N, Kochkarev D V and Galinska T A 2019 New design concepts for strengthening of continuous reinforced-concrete beams *IOP Conference Series: Materials Science and Engineering* **708(1)** 012040

[17] Blikharskyy Z, Brózda K and Selejdak J 2018 Effectivenes of strengthening loaded RC beams with FRCM system *Archives of Civil Engineering* **64(3)** pp 3-13

[18] Karpiuk V M, Syomina Y A and Antonova D V 2019 Bearing Capacity of Common and Damaged CFRP-Strengthened RC Beams Subject to High-Level Low-Cycle Loading *Materials Science Forum* **968** pp 185-99

[19] Kos Z and Klimenko Y 2019 The Development of Prediction Model for Failure Force of Damaged Reinforced-Concrete Slender Columns *Tehnički vjesnik* **26(6)** pp 1635-41

[20] Selejdak J, Khmil R and Blikharskyy Z 2018 The influence of simultaneous action of the aggressive environment and loading on strength of RC beams *Matec Web of Conferences* **183** 02002

[21] Kramarchuk A, Ilnytskyy B, Lytvyniak O and Famulyak Y 2019 Strengthening prefabricated reinforced concrete roof beams that are damaged by corrosion of concrete and reinforcement *IOP Conference Series: Materials Science and Engineering* **708(1)** 012060

[22] Okeil A, El-Tawil S and Shahawy M 2002 Flexural reliability of reinforced concrete bridge girders strengthened with carbon fiber-reinforced polymer laminates *Journal of Bridge Engineering* **7(5)** pp 290-9

[23] Pham H and Al-Mahaidi R 2008 Reliablity analysis of bridge beams retrofitted with fibre reinforced polymers *Composite Structures* **82(2)** pp 177-84

[24] Trentin C and Casas J 2015 Safety factors for CFRP strengthening in bending of reinforced concrete bridges *Composite Structures* **128** pp 188-98

[25] Alsayed S and Siddiqui N 2013 Reliability of shear-deficient RC beams strengthened with CFRP-strips *Construction and Building Materials* **42** pp 238-47

[26] Lima J and Barros J 2011 Reliability analysis of shear strengthening externally bonded FRP models *Proceedings of the Institution of Civil Engineers: Structures and Buildings* **164** pp 43-56

[27] Wang N, Ellingwood B and Zureick A 2010 Reliability-based evaluation of flexural members strengthened with externally bonded fiber-reinforced polymer composites *ASCE* **136** pp 1151-60

[28] Khmil R, Tytarenko R, Blikharskyy Y and Vegera P 2018 Development of the procedure for the estimation of reliability of reinforced concrete beams, strengthened by building up the stretched reinforcing bars under load *Eastern-European Journal of Enterprise Technologies* **5(7)** pp 32-42

[29] Pichugin S F 2018 Reliability estimation of industrial building structures *Magazine of Civil Engineering* **83(7)** pp 24-37

[30] Jaynes E T 2003 *Probability theory: the logic of science* (Cambridge: Cambridge university press) p 729

[31] Pichugin S and Severin V 2004 Reliability of structures under snow load in Ukraine *Snow Engineering* pp 67-72

[32] Blikharskyy Z Ya, Khmil R Ye, Rymar Ya V and Vasiliev I V 2008 Investigation of the stress state of the working reinforcement of reinforced concrete beams by building up its cross section under load *Collection of scientific works: Resource-saving materials, constructions, buildings and structures* **16** pp 311-7

IOP Conf. Series: Materials Science and Engineering 1021 (2021) 012015 doi:10.1088/1757-899X/1021/1/012015

Environmental assessment of small scale solar thermal electricity generation

Paul Koltun[1, 3], Vasyl Klymenko[2], Valentyn Soldatenko[2], Serhii Kononchuk[2] and Ruslan Teliuta[2]

[1] Victoria University, Melbourne, VIC 3001, Australia
[2] Central Ukrainian National Technical University, Universytetskyi Ave, 8, Kropyvnytskyi, 25006, Ukraine

[3] E-mail: pskoltun15@gmail.com

Abstract. Solar thermal electricity probably has the greatest potential of any single renewable energy process and it can be economically competitive with other technologies, if environmental costs are accounted for. In spite of the many investigations conducted so far there is no one dominant technology and no single life cycle assessment study conducted to compare different STE technologies. This life cycle study has been undertaken to identify potential materials flow, energy consumption and emissions for small scale STE options. The processes are described by the typical process inputs and outputs including energy, materials, solid wastes and other emissions. The results show sufficient efficiency for Australian conditions of small scale.

1. Introduction

Solar thermal electricity (STE) may be defined as the process by which collected solar energy is converted to electricity through the use of some kind of heat-to-electricity conversion device. STE probably has the greatest potential of any single renewable energy process and it can be economically competitive with coal-generated electricity if environmental costs are accounted for [1]. STE on grid was not achieved until the 1980s, although the basic technology for the production of mechanical energy had been under development for more than hundred years. During the last two decades, there has been rapid development in the basic technology, and prospects for rapid growth now appear to be very bright for newer approaches studied by a number of research groups [2-4]. In addition, approaches such as combined cycle power generation are being advanced by many to improve the overall energy conversion efficiency [4]. The world potential market places for STE technologies presented in Figure 1.

However, although many investigations have been conducted so far and some STE power plants are working more than two decades, unlike solar PV system, there is no one dominant technology for solar thermal power generation. And up to now there is no a single life cycle assessment (LCA) study conducted to compare different STE generation technologies, although some studies have been done to estimate current and future projection of costs based on accumulated experience [3]. Obviously, such analysis should be done before the large-scale implementation of STE technologies.

Taken into account the above mentioned situation, the whole LCA study has been divided in two parts: 1) design of most appropriate STE system in terms of its overall efficiency, cost effectiveness and environmental impact based on published in the literature results [6, 7] and 2) estimation of

IOP Conf. Series: Materials Science and Engineering 1021 (2021) 012015 doi:10.1088/1757-899X/1021/1/012015

potential environmental impact of the adopted system in the frame work of LCA methodology. To estimate environmental impact of the adopted STE the whole life cycle approach is adopted, viz. production and assembly of components, operation and disassembly and recycling. By chaining all processes, the assessment of the total impact on the environment and on energy and material resources demand can be done over the whole life cycle of the STE system.

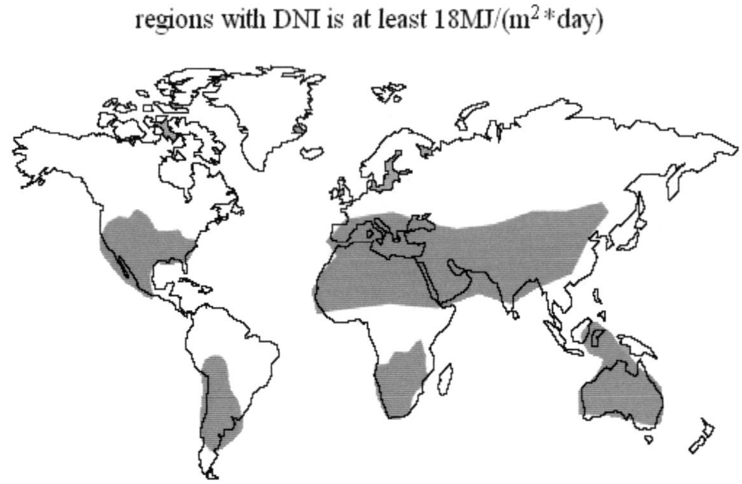

Figure 1. Potential solar thermal power marketplace [5].

2. Design of a small-scale STE system

There are three major types of concentrators: power tower, dish, and parabolic trough. The latter is most suitable for STE systems, as it is most efficient for medium temperature collectors (150–400°C). A north-south tracking axis orientation gives a high summer bias to annual electricity output, which is useful in Australia. Recently, considerable work has been undertaken to reduce structural cost, so it becomes the least expensive and most reliable solar concentrator technology. The performance of such concentrator can reach about 92% (the amount of absorbed solar radiation energy).

The major components of the system (shown in Figure 2) are briefly described below.

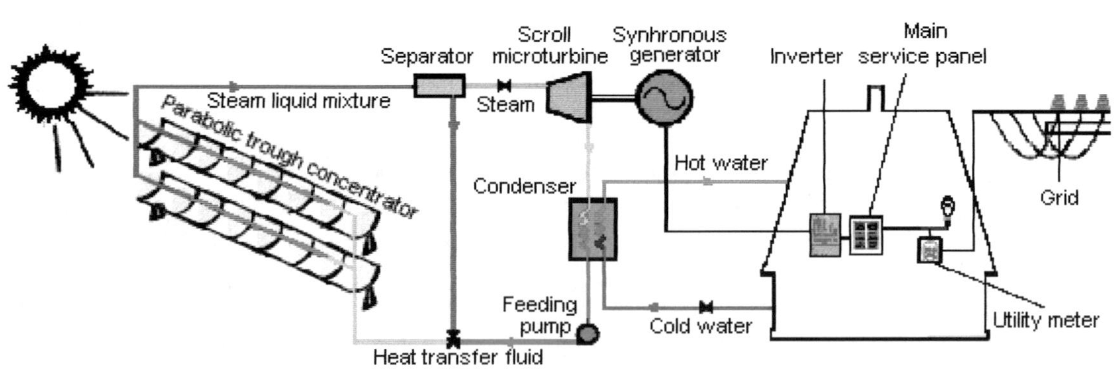

Figure 2. Schematic of the proposed solar thermal electricity generation system.

Solar collector: Two parabolic troughs working in parallel driven by tracking device govern by control unit.

Receiver and Heat Transfer Fluid: The main feature of the receiver is to covert the energy of solar radiation to heat energy using a heat transfer fluid (HTF), which transfers the absorbed energy further down to a device converting heat to mechanical energy. The absorbed solar radiation heats and vaporises HTF. A mixture of water and ammonia is adopted with the Kalina cycle [8]. The receiver employed for this system is based on works [9]. Steam created in the receiver is separated from the liquid in the separator. A T-junction separator is used [10]. Separated steam is then directed to micro-scroll turbine [11], where heat energy of the steam is converted to mechanical energy.

The steam from turbine is converted to liquid inside the heat exchanger. The heat from the steam recovered by cooling water inside heat exchanger can be used for household needs.

The feed pump is pumping working fluid (water-ammonia mixture) back to the receiver to close thermodynamic cycle.

The synchronous generator seating on the scroll micro-turbine shaft generates DC. The DC is converting to AC with voltage of 240V, which can be consumed on the spot or supplied to the grid.

To be workable, the STE system needs other balance-of- system (BOS) components. In material and energy analysis of STE system the BOS cannot be neglected as they represent rather significant part of overall system. Some complementary devices, which must be added to the proposing STE system, are: steam separator, heat exchanger, pump, inverter and sun tracking device.

A connection to the grid of small STE could be made at the node where voltage doesn't exceed 400V, i.e. it should be very close to consumers. Figure 3 shows a generalised scheme of the combined electric power system, which includes renewable energy sources.

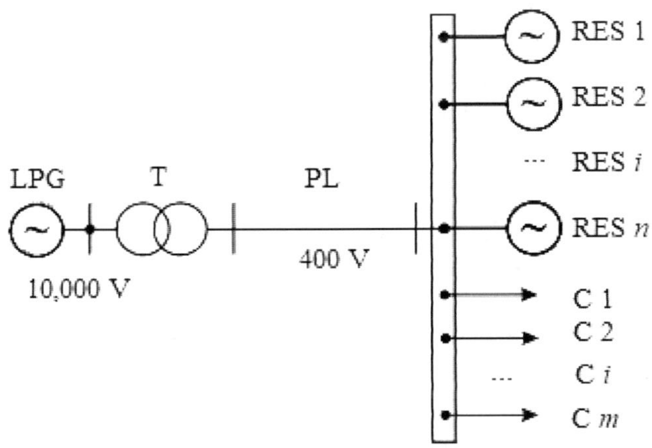

Figure 3. Generalised scheme of the combined electric power system.
LPG is a local power grid;
T is a step-down transformer: 10000/400V;
PL is a power line;
RES_1-RES_n are sources of renewable energy;
C_1-Cm are consumers

During use of such system the reverse generation modes may occur. Under such mode, the electrical energy from renewable energy sources (RESs) goes to electrical grid. Such mode is a positive situation for companies using "green tariffs" However, in the case of commensurate of the power of the transformer and RESs, the latter substantially influence working parameters of PL, which may exceed allowed voltage deviation [12]. Figure 3 presents simplified scheme for the reverse generation regime at the points of the joints of the consumers. If the transverse component of the voltage drop and power within power lines are not taken into account, then voltages at those points could be found by expression:

$$U = U_1 - \Delta U_{\mathrm{PL}} = U_1 - \frac{-P}{U_{\mathrm{NOM}}^2} r_{\mathrm{PL}},, \tag{1}$$

where, U is voltage at the points of SREs joints; U_1 is voltage at the beginning of PL; ΔU_{PL} is magnitude of voltage drop within PL (for the reverse generation mode ΔU_{PL} must be subtracted from U); P is power generated by SREs to the grid; U_{NOM} is nominal mode voltage; r_{PL} is active resistance of PL.

Expression (1) shows that increasing power generation by RESs increases PL voltage U at the points of joints of the consumers. Therefore, the mode of increasing RESs power generation may overcome allowed voltage deviation of PL at the points of joints of the consumers. Such extremal voltage deviation adversely effects reliability of electrical devices of the consumers, leads to reduction their service lives, and, as a consequence, to substantial economic loses [13].

Currently, there are several ways to solve this problem, the main of which are following: 1) measures to increase capacity of electricity distribution by grid elements; 2) measures to limit energy generation by RESs.

The first way requires reconstruction of the grid (replacing transformers with more powerful ones, replacing cables with those with bigger cross-section, etc.) Such approach requires significant capital investment by owners of RESs and as a consequence increase payback period of RESs and decrease their economic efficiency. The second way leads, on the one hand, to under receiving of money due to sale electricity by "green tariffs", although, on the other hand, it reduces economic losses due to extreme values of voltage deviation in the grid.

The most promising approach seems to be automatic control of power generated by RESs. Thus, the maximum level of power generation within the grid would be determined jointly by the profit from sale of electricity by "green tariffs" and reduction of economic losses from excessive voltage deviations. The implementation of such approach is possible through improvement of automatic control system (ACS) of power generation by RESs [14].

3 Life cycle assessment: goal, scope, and major assumptions

3.1 Purpose, scope and functional unit

The purpose of the LCA study is to investigate the environmental impact from the whole life cycle of the above STE generation system. A consequence of this objective is that the production level of the order of 100,000 STE units with capacity 3-5kW will be required to contribute 5% to current electricity supply (these figures are based on Australia electricity for the year 2017 [15]).

The scope of material flow analysis for the LCA study includes direct and indirect material inputs. The scope for the energy requirements, however, is broader and covers also the energy use for the production of materials and for the production of capital equipment for main materials.

The functional unit is assumed as the STE system itself, i.e. a system that produces an average of 3kW electrical energy from solar radiation. The overall efficiency of the STE system is 20%.

Hot water production (from the heat rejected within thermodynamic cycle) is not taken into account as it requires additional assumptions regarding its usage. The boundary of LCA study of the STE system is presented in Figure 4.

3.2 Major assumptions

The proposed STE system with overall efficiency 20% has been adopted from a detailed study of efficiencies of the main parts of the system.

The life cycle of each part of the STE system starts from the production of necessary materials (beginning from mining of natural resources) and then fabricating the different parts of the system (data for materials production and parts fabrication have been taken from SimaPro database [16] adopted for Australian conditions with taken into account all necessary transportations). Finally, the system is assembled and tested. At the use stage, the module produces electrical energy during 20 years (the lifetime of the system was assumed according [16]). At the end of the lifetime of the system, it will be decommissioned and the resulting waste will have to be disposed in a responsible way. As this event can occur at least 20 years ahead, we assumed that recycling rate for the major material will be improved and about 70% of steel and of aluminium and glass will be recycled. (As it is hard to obtain any data available on the recycling technology of auxiliary materials, we did not consider their recycling in our study).

Manufacturing and disposal of BOS such as inverter, separator, heat exchanger, feed pump, supporting structure and their accessories are also included in the LCA boundary (see Figure 4). The transportation of all necessary components to the final assembly site and transportation to the landfill

IOP Conf. Series: Materials Science and Engineering 1021 (2021) 012015 doi:10.1088/1757-899X/1021/1/012015

at the end of the life of the system have also been taken into account. It is assumed that average distance for components transportation is 300 km and to the final disposal is 100 km. (All transportations are assumed to be done by road diesel trucks with the capacity of 20 tons).

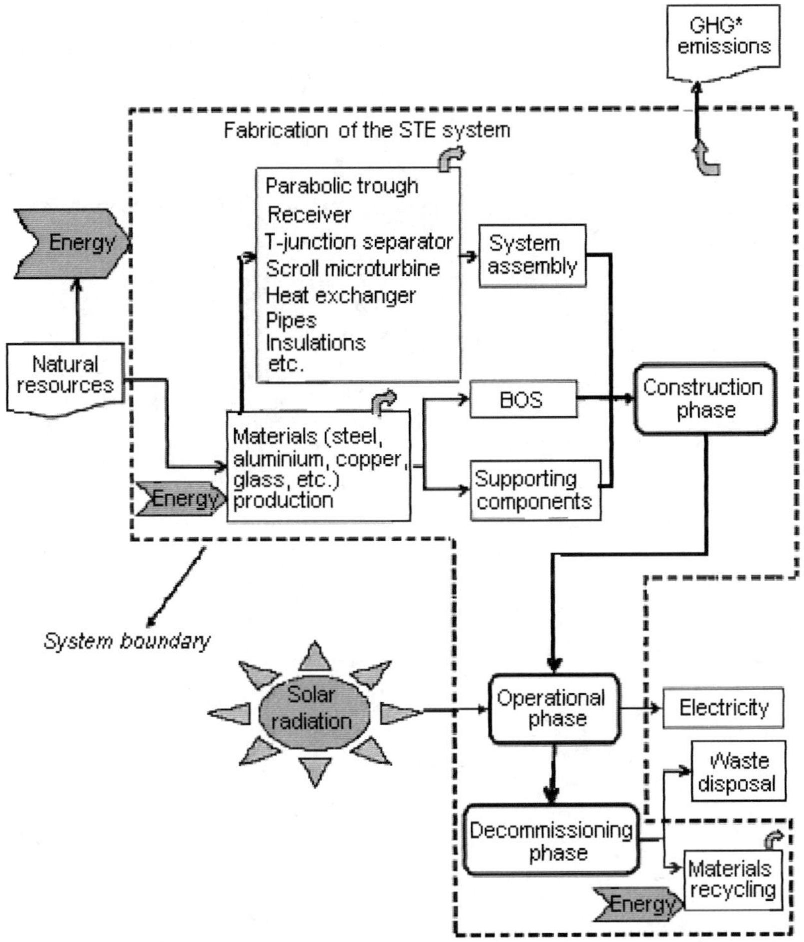

Figure 4. The LCA boundary of the small scale solar thermal electricity generation system.

At the use stage of the STE system, little so-called "parasitic" electric loads are necessary. The "parasitic" electric loads are the motor for the heat transfer fluid (HTF) pump and motor for PTC tracking mechanism. Additional parasitic load is a control unit. The estimated parasitic electric loads based on presented above data is 0.4 kW (0.1 kW – motor for HTF pump; 0.25 kW – for PC tracking device; 0.05 kW – for control unit). Also, to maintain proper reflectivity of PTC (about 0.95) washing water has to be used. According [7] mirror wash water consumption is 0.022 $m^3/(m^2 \cdot year)$. However, detergent use is not considered.

We also assume that 10% of HTF to be replaced each year due to some losses, which translates to an additional consumption of 6,4 kg of ammonia over the life time of operation of the STE system.

As the proposed STE system mostly made from metals, at the end of its life the STE system will be disabled and materials will be recycled based on the current recycling rate [17 - 19] and trends: aluminium – 60%; steel, copper – 70%; plastic – 50%. Recycling of glass module sheets represents some problems, because glass contaminated with plastics (more than 100 g per ton of glass) is normally not accepted. However, in this study we assumed that in 20 years the most problems of glass recycling will be solved, so 70% of glass is recycled, as well. We also assumed that 50% of metal and

IOP Conf. Series: Materials Science and Engineering 1021 (2021) 012015 doi:10.1088/1757-899X/1021/1/012015

plastic components from BOS will be recycled [17]. We assume the rest of materials after removal of the materials to be recycled are disposed of as a solid waste at landfill.

4 Materials and energy analysis and GHG emissions

The list of materials required directly or indirectly for the life cycle of the proposed STE system has been obtained through SimaPro8.0 modelling [16] based on inventory data. Table 1 shows the amount of required materials for the whole life cycle of the STE system. As can be seen from data presented in Table 1 dominating resources are metal ores (iron, copper, bauxite), fossil fuels (coal, natural gas, crude oil) and commodities for glass making (calcite, limestone, gravel). Other resources presented in Table 1 are mostly used as alloying metals for steel making, including stainless steel and for glass mirrors coating (silver). Obtained results for primary energy consumption and GHG emission through whole life cycle of the STE system presented in Table 2.

Table 1. Resources required for the full life cycle of adopted STE system

Substance	Unit	Total
Bauxite in ground	kg	25.7
Copper in ground	kg	34.6
Iron in ground	kg	592.1
Zinc in ground	kg	6.3
Lead in ground	g	100.3
Magnesium in ground	g	597.7
Manganese in ground	g	111.2
Molybdenum in ground	g	67.2
Nickel in ground	g	840.7
Silver in ground	g	60.0
Calcite in ground	kg	100.24
Limestone in ground	kg	8.336
Gravel in ground	kg	170.9
Sodium chloride in ground	kg	29.9
Coal hard in ground	kg	6143.0
Natural gas in ground	m^3	248.0
Oil crude in ground	kg	100.0
Water unspecified	m^3	3203.0

Based on the presented results and made assumptions and methodology adopted in [17], the energy and greenhouse gas (GHG) emissions payback, the energy and GHG emissions payback time have been calculated for where whole life cycle of adopted STE system:

The energy payback time is:

$$PBT_E = \frac{E_C \cdot p}{(E_P - E_L)/t} = \frac{32.5 \cdot 0.35}{(590 - 79)/20} = 0.45 \text{ (year),}, \tag{2}$$

where, E_C is total energy consumption through whole life cycle of the STE system; $p = 0.35$ is adopted within the study efficiency of conversion primary energy to electricity (including transmissions loss); E_P and t are total energy production of the system and the life time of the system; E_L is parasitic electric load (the energy consumed for the system's own needs).

The GHG payback time is:

IOP Conf. Series: Materials Science and Engineering 1021 (2021) 012015 doi:10.1088/1757-899X/1021/1/012015

$$PBT_{\mathrm{GHG}} = \frac{t \cdot T_{\mathrm{GHG}}/c}{E_{\mathrm{P}} - E_{\mathrm{L}}} = \frac{20 \cdot 10044/262}{590 - 79} = 1.95 \text{ (year)}, , \tag{3}$$

where T_{GHG} is total GHG emissions through whole life cycle of the STE system (see Table 2); c is the GHG average emissions per 1GJ of produced electricity in Australia (c = 262kg of CO_2 eq. per 1GJ [16]).

Table 2. The primary energy consumption and GHG

Processes (Components)	Primary Energy (MJt)	GHG Emissions (kg CO_2 eq.)
Parabolic Trough	16931	731.5
Solar Reflectors	192	131.2
Receiver	2952	177.2
Micro-turbine	6630	4944.5
Separator	91	61.0
Heat Exchanger	162	108.5
Water Pump	271	195.3
Pipes	369	75.8
Insulation	7	27.7
Inverter	1306	919.5
Control Unit	593	524.9
Tracking Mechanism	482	270.0
Woking Fluid	46	1.3
Miscellaneous	1613	1112.0
Use Stage	83	2.0
Disposal	261	19.9
Total	31989	9302.3

5 Conclusions

The LCA case study and results obtained for STE generation system for Australian conditions indicates a big potential of such systems in term of their environmental performance. It has been shown that proper design can bring about 20% efficiency of conversion of solar energy to electricity. Such rather high efficiency can be even substantially increased if integrated solar combined (heat-power or cooling-heat-power) generating is installed.

Although the presented case study has an intrinsic uncertainty and assumptions related to various factors (lack of detail information sources, data quality, etc.), however, it clearly shows the order of magnitude of energy consumption and GHG emissions ascribable to the whole life cycle of STE system.

An important insight from this study is the significance of the impacts at life cycle stages. Both energy consumption and GHG emissions are highest for manufacturing and assembly of the main components. Comparatively, impact from both use and disposal stages are negligible (less than 1%).

Further, the great energy and environmental advantage of the system are shown by very low values of payback times (less than a year for energy and less than 2 years for GHG). Even including the variability related to raw material eco-profiles and the uncertainties regarding different life cycle steps, it has been estimated that in pessimistic scenarios, the payback times are less than 4 years. Such results state a positive qualitative judgement for the environmental performances of the small scale solar thermal electricity production system that is not sensibly influenced by all the study uncertainties

IOP Conf. Series: Materials Science and Engineering 1021 (2021) 012015 doi:10.1088/1757-899X/1021/1/012015

References

[1] Clean Energy Australia Report 2018 *Clean Energy Council* p 62 URL: https://apo.org.au/node/173956

[2] Concentrating Solar Power Projects 2019 *National Renewable Energy Laboratory* URL: https://solarpaces.nrel.gov/khi-solar-one

[3] Gauché P at. al. 2017 System value and progress of CSP *Solar Energy* **152**, pp. 106-139

[4] Mills D 2004 Advances in Solar Thermal Electricity Technology *Solar Energy* **76** pp. 19-31

[5] Cost and Performance Characteristics of New Generating Technologies Annual Energy Outlook 2018 *U.S. Energy Information Administration*, URL: https://pdf4pro.com/view/cost-and-performance-characteristics-of-new-generating-460f47.html

[6] Ximenes A at. al. 2019 Life cycle costing as a bottom line for the life cycle sustainability assessment in the solar energy sector: A review *Solar Energy* **192** pp. 238-262

[7] Assessment of Parabolic Trough and Power Tower Solar Technology Cost and Performance Forecast 2003 *Report Department of Energy and National Renewable Energy Laboratory* **SL-5641** USA, p 344

[8] Wall G, Ishida M, Chuang C 1990 Exergy study of the Kalina cycle *Analysis and Design of Energy Systems: Analysis of Industrial Processes* **10-3** ASME pp. 73-77

[9] Flores V Almanza R 2004 Direct steam generation in parabolic trough concentrators with bimetallic receivers *Energy* **29** pp. 645-651.

[10] Yang B et. al. 2020 State-of-art of impacting T-junction: Phase separation, constituent separation and applications *International Journal of Heat and Mass Transfer* **148** pp. 119067-119116.

[11] Introducing the Plug and Play Micro-ORC URL: https://airsquared.com/news/introducing-plug-play-micro-orc/

[12] Pleshkov P G Garasova N Y Soldatenko V P 2018 Optimal control of the work of the hybrid electric energy system with renevable sources of energy *Bulletin of NTU "KhPI". Series: Problems of Electrical Machines and Apparatus Perfection. The Theory and Practice* **32 (1308)** pp. 64-70

[13] Plieshkov P Plieshkov S Zinzura V 2019 Automatic control of electrical distribution network mode with the voltage unbalance *Naukovyi Visnyk Natsionalnoho Hirnychoho Universytetu* **171** No 3. p.p. 83-91

[14] Plieshkov P Soldatenko V Zinzura V Plieshkov S 2020 Determining weight coefficients for an optimal system of control over electric energy generation in a combined electric power system *Eastern-European Journal of Enterprise Technologies* **103** No 1 p.p. 77-82

[15] Australian Energy Update, 2018 *Department of the Environment and Energy* p/ 39 *URL:* https://www.energy.gov.au/sites/default/files/australian_energy_update_2018.pdf

[16] Simapro 8.0 – the software tool to analyse and develop environmentally sound products *Pre Consultants B.V. Amersfoort, The Netherlands*, 2015

[17] Koltun P Tharumarajah A 2007 LCA of PV system included into distributed energy grid in Australia *Report CSIRO*

[18] Koltun P and Klymenko V 2020 Cradle-to-gate life cycle assessment of the production of separated mix of rare earth oxides based on Australian production route *Mining of Mineral Deposits* **14** N. 2 pp.1-15

[19] Victorian Recycling Industry 2013-14 *Annual Survey* p 47 URL: https://discover.data.vic.gov.au/dataset/victorian-recycling-industy-annual-survey/resource/2d3f6d61-6e8a-4177-b29b-c14ff45590dc

IOP Conf. Series: Materials Science and Engineering 1021 (2021) 012016 doi:10.1088/1757-899X/1021/1/012016

Physicochemical studies of the structure of energy-saving compositions based on slags

Tetyana Kostyuk[1], Varvara Vinnichenko[1], Andrii Plugin[2], Olga Borziak[2,3] and Artem Iefimenko[2]

[1] Kharkiv National University of Civil Engineering and Architecture, Sumska st. 40, 61002 Kharkiv, Ukraine
[2] Ukrainian State University of Railway Transport, Feuerbach sq. 7, 61050 Kharkiv, Ukraine

[3] Email: borziak@gmail.com

Abstract. Blast furnace slags are subdivided into granular and dump. Granulated slag is widely used in the construction industry to production of Portland cement, Portland slag cement, slag-alkaline cements, and concretes based on these cements. Dump slag is of less use, because it has low activity. The phase composition of granular and waste slag was investigated by methods of physicochemical analysis. The slags contain minerals that are present in Portland cement clinker. Concrete samples were made using granular, waste slag and sand. Unsatisfactory results in compressive and flexural strength were shown by a sample made of only two types of slag. The best strength results were obtained for a composition in which there are three components: granular slag, dump slag (filler) and sodium hydroxide. Electronic images have confirmed the appearance of low-basic calcium silicates, which provide the strength of the cementless composite. The resulting concrete composition is resource-saving, because made from production waste. And it can be called energy-saving, because no energy is expended to obtain components. The ecological feasibility of the production of such concrete is due to the absence of emissions of pollutants into the atmosphere (including carbon dioxide), since the firing of cement clinker is not needed.

1. Introduction

Energy-saving and resource-saving technologies in modern building materials science occupy one of the leading places of scientific research and programs. Cement binders in the production of modern building materials in the construction of buildings and structures are one of the main components of composite materials [1-3]. At the same time, the production process of binders is the most energy-intensive stage of the entire technological process of production of composite materials.

Development of research aimed at finding an alternative replacement of cement materials with production waste is relevant. The relevance is due to the fact that this area is resource-intensive due to the replacement of substances of natural origin with waste. Modern research shows that slag composites are the most effective substitutes for cement binders in concrete [4, 5]. Energy is used for cement production [6], and when cement is replaced by slag, this energy is not used at all, i.e. it is stored.

The aim of the study is to assess the possibility of replacing a traditional cement binder with a composite binder based on slags.

To achieve this goal, a study of the phase composition of slags was carried out; samples of composite materials based on slags were made and their physical and mechanical properties were evaluated.

2. Bibliography analysis

Blast furnace slag is a waste product from pig iron production in a blast furnace. Depending on the cooling rate, slag is divided into granular and dump. Granular slags include substances in an amorphous state and have high chemical and hydraulic activity. Granulated slag is widely used in the construction industry. It is used in the production of Portland cement, Portland slag cement, slag-alkali cements, as well as concretes and products based on these cements. Slag is also widely used in road construction. Crushed granular slag is also widely used as an additive of a mineral binder in the production of dry building and mortar mixtures. Slowly cooled slags - dump - are crystalline substances, most often they show low chemical activity. For this reason, dump slag is of less use. The founder of the scientific direction for the use of blast furnace slag in building materials and products is V.D. Glukhovsky [6].

In recent years, the world has seen a significant increase in the use of slags. One of the reasons for expanding the production of slag concrete is a large supply of thermal energy in the slag. In addition, concretes with aggregate from blast-furnace slag have a number of advantages over traditional concretes. As it was established in [7-9], the surface layer of slags reacts with calcium hydroxide released during the hydrolysis of alite. In this case, an additional amount of calcium hydrosilicates is formed. Calcium hydrosilicates create an extremely strong bond between the aggregate and the cementitious matrix. The number of capillary channels in the concrete stone decreases. This leads to a significant increase in the corrosion resistance of concrete in comparison with traditional compositions in most corrosive environments. In addition, due to the specific structure and the absence of micro-gaps at the interface between the binder and filler, such concretes have good physical and mechanical characteristics. The resulting concrete composition is resource-saving, because made from production waste. And it can be called energy-saving [10], because no energy is expended to obtain components. The ecological feasibility of the production of such concrete is due to the absence of emissions of pollutants into the atmosphere [11, 12], (including carbon dioxide), since the firing of cement clinker is not needed.

3. Materials and methods

Slag produced at the Dnieper Metallurgical Plant was used for experiments: granular and dump. To study the phase composition of slags and composite materials, physical and chemical research methods were used. To determine the quantitative phase ratio in both slag samples, petrographic analysis of samples was performed. The samples were examined in immersion preparations under transmitted light using a MIN-8 microscope at $\times$100 - 400 magnification. The sample of granular slag had a light gray color, boiled when exposed to 10% HCl with the release of hydrogen sulfide (with a sharp smell of H_2S). X-ray structural studies of the phase composition of slag samples were performed on an X-ray diffractometer DRON-1.5. Filtered (Ni β-filter) Kα 1.2 - radiation of Cu-anode, U = 42 kV, I = 19 mA. Range of scanning angles – $2\Theta = 6 - 100°$. Diffraction curves were recorded on the diagram tape of the potentiometer KSP - 4. The speed of the sample - 2°/min, the speed of the diagram tape - 10 mm/min. The JEOL LSM-840 scanning electron microscope was also used for the study.

4. Results of physicochemical research

The bulk of the sample is a colorless transparent isotropic glassy substance with a refractive index $N \approx 1.640 \pm 0.005$. On the surface of the grains of the glass phase there are films and smears of fine (particle size <0.004 mm) mixture of portlandite $Ca(OH)_2$ and calcite $CaCO_3$. The sample of granular slag had a light gray color, boiled when exposed to 10 % HCl with the release of hydrogen sulfide (felt a sharp odor of H_2S).

As impurities in the sample there are dense skeletal growths of gelenite $2CaO \cdot Al_2O_3 \cdot SiO_2$ and larnite (β - C_2S), oxermanite $2CaO \cdot MgO \cdot SiO_2$, fine aggregates with a predominance of portlandite with impurities of calcite. Calcite is presented in the form of individual crystals and aggregates of complex rhythmic-lobed shape, fine red-brown ferrites and, or calcium aluminoferrites, black non-magnetic grains of graphite.

The sample of slag dump had a gray color, boiled under the influence of 10 % HCl with the release of hydrogen sulfide (a sharp smell of H_2S).

Impurities include rectangular crystals of gelenite with a size of 0.015 mm (on the largest side), grains of β - C_2S, small (up to 0.008 mm) crystals of pseudovolastonite α - $CaO \cdot SiO_2$, growth crystals of belite β - C_2S and gelenite $2CaO \cdot Al_2O_3 \cdot SiO_2$. The impurities contain grains of rankinite $3CaO \cdot 2SiO_2$ and, or montmorelonite, or milelite, calcite in the form of individual crystals and aggregates of complex paw-shaped shape up to 0.01-0.02 mm. In the sample there are significantly more fine aggregates of ferrites - calcium aluminoferrites, which usually "cement silicates". There are fine red-brown ferrites and, or calcium aluminoferrites, black non-magnetic grains of graphite, single grains of quartz. In the glass phase, "stars" characteristic of the crystallization of calcium sulfide CaS are visible.

The phase composition of the slag samples of the blast furnace granulated Dnieper Metallurgical Plant is: glass phase 75-85 %, growths of gelenite and laryngitis 3-5 %, β - C_2S 2-3 %, Ochermanite 1-2 %, portlandite 3-5 %, calcite 2-3 %, calcium ferrites and aluminoferrites 1-2 %, black nonmagnetic particles 3-5 %.

The glass phase exists in significant quantities. To obtain the binding properties of the glass phase, the slag must be activated.

X-ray diffraction studies of the phase composition of the slag samples showed that sample 2 (granular slag) has very weak diffraction peaks in the crystalline phases due to the high dispersion of the material. The main phases are the phases of Portland cement clinker - $3CaO \cdot SiO_2$ (alite) and $2CaO \cdot SiO_2$ (belite): 3.03; 2.84; 2.72; 2.66; 2.29; 2.20Å. Possible content of calcite 3.03 Å. Sample 3 (slag dump) also has weak diffraction peaks. The following phases are observed in the sample: $3CaO \cdot SiO_2$ (alite) and $2CaO \cdot SiO_2$ (belite): 3.03; 2.68; 2.57; 2.29; 1.97 Å. You can also select the phase of calcite - $CaCO_3$: 3.86; 3.03; 1.99 Å. The sample contains quartz - SiO_2: 4.23; 3.32; 2.44; 1.824 Å; gelenite - $2CaO \cdot Al_2O_3 \cdot SiO_2$: 3.70; 3.06; 2.38; 2.09; 2.44; 1.753; 1.508 Å; anortite – $CaO \cdot Al_2O_3 \cdot 2SiO_2$: 3.22 Å; iron - Fe: 2.50; 2.04.

The slag was dried to constant weight at a temperature of 100 °C, cooled in a desiccator and examined on a scanning electron microscope. Electronic images of samples of the microstructure of granular and dump slag (figure 1-2) confirmed that the vitreous phase of granular slag (Figure 1) is fused with silicate splices and aluminate phase petals. In figure 2, the blast furnace slag sample is also represented by a glassy phase. Calcium aluminoferrites (in the upper part of the figure) as well as volostanite crystals, calcite cubes in the form of individual grains (in the lower part of the figure) are observed here.

Electronic images of ground samples of blast furnace granular and blast furnace slag showed that the main part is the vitreous phase. To obtain the binding properties of the slag, it is necessary to break the bonds of the silicate chains and release CaO from the vitreous phase of the slag.

5. Results of physical and mechanical research

A study was conducted to select the composition of heavy fine-grained concrete that does not contain Portland cement. They use a slag activator - sodium hydroxide. Blast furnace slag and quartz sand were used as active filler. Ground granular slag served as a binder. The test results are presented in table 1.

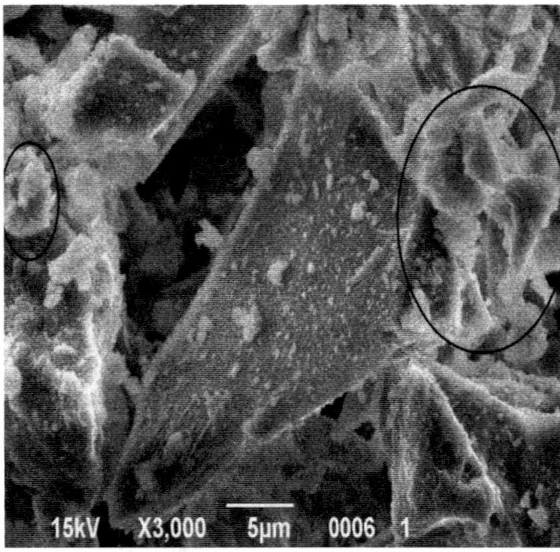

Figure 1. Microstructure of granular slag.

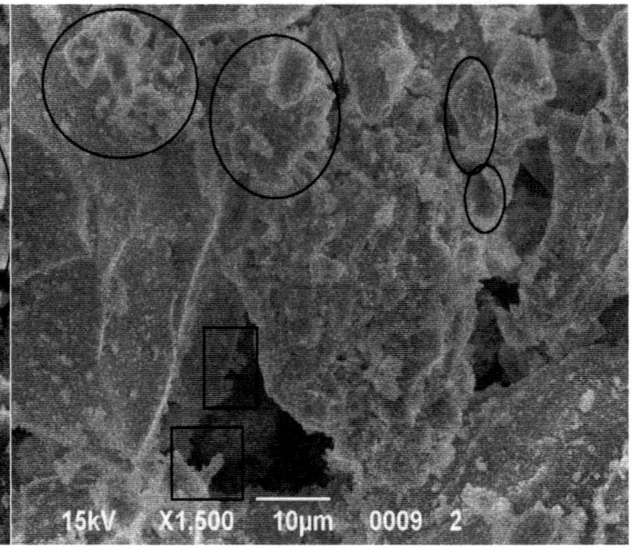

Figure 2. Microstructure of dump slag.

Table 1. The dependence of the strength of samples of slag mixtures on the composition

Composition	Composition of the composite	Compressive strength, MPa		Tensile strength in bending, MPa	
		3 days	28 days	3 days	28 days
1	The slag is granulated +NaOH + sand	3.0	7.2	2.06	4.2
2	The slag is granulated + NaOH + dump slag (as filler)	3.8	18.0	0.93	5.0
3	The slag is granulated + dump slag control	collapsed	13.9	collapsed	3.1

Table 1 shows that the best results in strength showed a sample where sodium hydroxide was used (as an activator), granular slag is a binder, dump blast furnace slag was used as a filler. Figure 3 shows that crystalhydrates are formed similarly to crystalhydrates of cement stone.

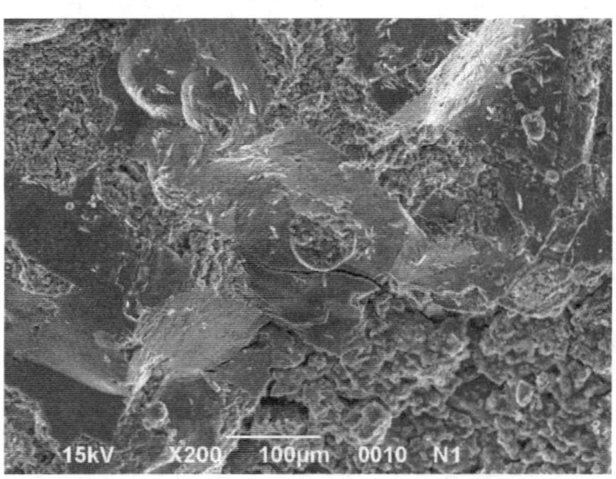

Figure 3. Microstructure of the sample 1.

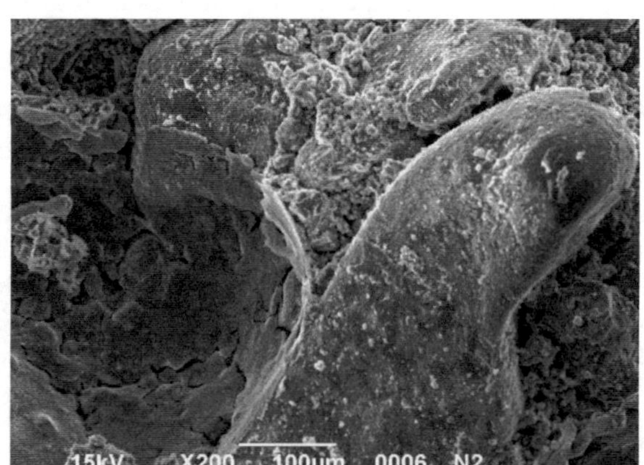

Figure 4. Microstructure of the sample 2.

Figure 4 shows the microstructure of a sample of cementless composite, where instead of dump blast furnace slag as a filler used quartz sand. The structure is inhomogeneous, to the left of the quartz particle is a half-destroyed shell of slag, there is no connection into a dense monolithic structure of tumors of crystal hydrates. Separate cubes of calcite are visible on the surface of quartz grain, there are accumulations of portlandite and petals of hydroaluminates with felt-like hydrosilicates of calcium.

Figure 5 clearly shows the structure of the control sample, which consists only of ground granular and dump blast furnace slag. At magnification in the picture, Figure 6-7, it is seen that all structural components are combined with tumors of low-basic calcium hydrosilicates, compacted by calcite cubes, Figure 7.

Thus, the results of the study showed that the cementless composite, which consists of blast furnace granulated slag and dump slag, has a compressive strength of 18 MPa.

Figure 5. Snapshot of the microstructure of the control sample 3.

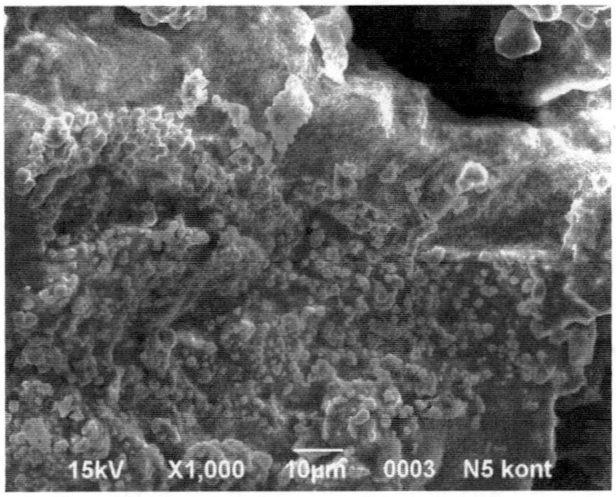

Figure 6. Snapshot of the microstructure of the control sample 3 at a magnification of x1000.

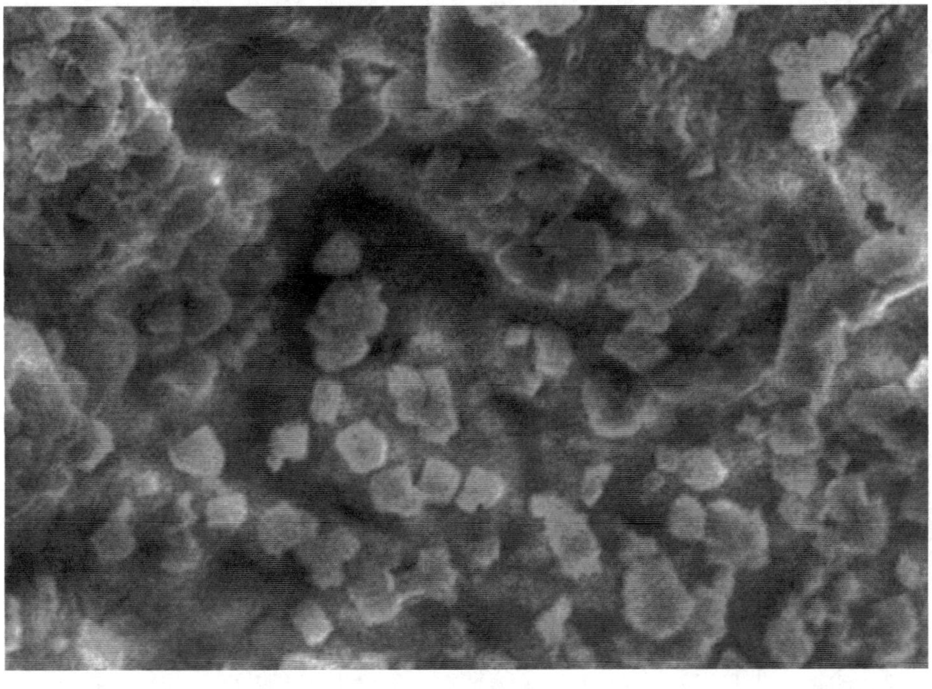

Figure 7. Microstructure of the control sample 3 at a magnification of x3000.

The use of blast furnace granular and dump slag in the production of concrete solves the problem [13-17]:

- conservation of natural and energy resources;
- improving the ecology of the environment (elimination of slag dumps, reduction of gaseous emissions into the atmosphere during cement production);
- solving a number of social problems (organization of additional jobs).

6. Conclusion

During the hardening of the composite, which consists of granular blast furnace slag, dump blast furnace slag and calcium hydroxide, low-basic hydrosilicates of calcium and calcite are formed. These crystal hydrates are insoluble and stable over time, which makes composites competitive with cementitious building materials.

References

[1] Mindess S 2019 *Developments in the Formulation and Reinforcement of Concrete* (Woodhead Publishing) p 442

[2] Marushchak U, Sanytsky M, Sydor N and Braichenko S 2018 *Proceedings of the 2018 IEEE 8th International Conference on Nanomaterials: Applications and Properties, NAP* (Zatoka; Ukraine) 8914835

[3] Borziak O S, Plugin A A, Chepurna S M, Zavalniy O V and Dudin O A 2019 *IOP Conf. Series: Materials Science and Engineering* (Bristol: IOP Publishing) **708** 012080

[4] Krivenko P, Petropavlovskyi O, Kovalchuk O, Rudenko I and Konstantynovskyi O 2020 *International Conference on Sustainable Futures: Environmental, Technological, Social and Economic Matters, ICSF 2020; E3S Web of Conferences* **166** 06001

[5] Ivashchyshyn H, Sanytsky M, Kropyvnytska T and Rusyn B 2019 *Eastern-European Journal of Enterprise Technologies* **4** 39-47

[6] Glukhovsky V D 1978 *Slag-alkali cements and concretes* (Kiev: Budivelnik)

[7] Moranville-Regourd M and Kamali-Bernard S 2019 Cements made from blastfurnace slag *Lea's Chemistry of Cement and Concrete* (Fifth Edition) chapter 10 pp 469-507

[8] Wang G C 2016 *The Utilization of Slag in Civil Infrastructure Construction* (Woodhead Publishing) p 442

[9] Netinger Grubeša I, Barišić I, Fucic A and Bansode S S 2016 The Indian experience of steel slag application in civil engineering *Characteristics and Uses of Steel Slag in Building Construction* (Woodhead Publishing) chapter 7 pp 141-160

[10] Vinnichenko V and Ryazanov A 2018 Energy Efficiency of Binder Application in Concrete *International Journal of Engineering and Technology* (*UAE*) vol 7 **4.3** 335-338

[11] Vinnichenko V, Ryazanov A and Krot O 2018 *MATEC Web Conf.* Vol 230 (EDP Sciences) 03020

[12] Vinnichenko V and Ryazanov A 2017 *MATEC Web Conf.* Vol 116 (EDP Sciences) 01020

[13] Holland R B and Kahn L F 2016 Effect of different concrete materials on the corrosion of the embedded reinforcing steel *Corrosion of Steel in Concrete Structures* ed A Poursaee (Woodhead Publishing) chapter 7 pp 131-147

[14] Black L 2016 Low clinker cement as a sustainable construction material *Sustainability of Construction Materials* ed Jamal M. Khatib (Woodhead Publishing) chapter 17 pp 415-457

[15] Mudersbach D 2004 Verringerung der Porosität von Hochofenstückschlacken *Report des FehS-Instituts* Vol 2 pp 5-6

[16] Steindl R, Baldermann A, Galan I, Sakoparnig M, Briendl L, Dietzel M and Mittermayr F 2019 *Constr. Build. Mater* **223** 55-68

[17] Cyr M and Pouhet R 2015 Resistance to alkali-aggregate reaction (AAR) of alkali-activated cement-based binders *Handbook of Alkali-Activated Cements, Mortars and Concretes* (Woodhead Publishing) chapter 15 pp 397-422

Strengthening of the reinforced concrete tank of anaerobic purification plants with the manufacture of biogas, damaged as a result of design and construction errors

A Kramarchuk[1], B Ilnytskyy[1], D Hladyshev[2] and O Lytvyniak[3, 4]

[1] Department of Building Constructions and Bridges, Institute of Civil Engineering and Building System, Lviv Polytechnic National University, S. Bandera str., 12 79013 Lviv, Ukraine
[2] Lviv Polytechnic National University, Institute of Architecture and Design, Department of Architectural Design and Engineering, S. Bandera str., 12 79013 Lviv, Ukraine
[3] Department of Civil Safety, Viacheslav Chornovil Institute of Sustainable Development, Lviv Polytechnic National University, S. Bandera str., 12 79013 Lviv, Ukraine

[4] E-mail: lytvyniak.oksana@gmail.com

Abstract. The defects and damages of bearing structures (walls, covering) of the solid-cast reinforced concrete tank were identified and detected using visual and instrumental testing. The actual reinforcement of the tank was determined and coordinated with designed reinforcement. Furthermore, the deformations were measured, the cracks opening in the walls and its depth in the concrete were detected and was also the strength quality of concrete was defined. During the year of the operation, the tank developed a considerable system of cracks in all lateral walls from technological loadings. At the same time, the width of the crack opening exceeded the limit values for reinforced capacitive structures (w_k), which is equal to 0.1 mm. The walls lost tightness, the use of the structure in the technological process was suspended. Checking calculations of the design solutions was carried out for the elements of the tank on technological loadings in the software package LIRA to find the causes of the cracks opening and water penetration in the walls of the reinforced concrete tank. It was determined that the existing design solutions did not ensure the bearing capacity and the fracture strength of bearing reinforced elements of the tank and its spatial rigidity from technological loadings. Necessary reinforcement was defined by ensuring the strength and possible limitation of the width of the cracks opening. This article presents the technical solutions concerning the strengthening of repair bearing elements of reinforced concrete spatial frame of the existing structure. The external bandages (reinforcing steel 4d32A400C) and inner tension bars (stainless steel round pipes 168 x 5 mm) were used as the strengthening of the reinforced concrete tank.

1. Introduction

During the operation of all types of buildings, a lot of the factors arise, which lead to the necessity of reinforcing the bearing structures that work on different types of loads. The main causes of the emergency technical situation of the structure include the increase of loading on bearing structures as a result of the change of their design map during the reconstruction and the modernization of technological processes [1 - 6], physical deterioration of materials and structures during intense and long operation, the aggressive impact of temperature and humidity on the concrete and reinforcement, and accidents at the enterprises [7 - 10]. In our case, two main factors lead to the emergency technical situation of the solid-cast reinforced concrete tank: carelessness during the development of working design documentation and considerable

IOP Conf. Series: Materials Science and Engineering 1021 (2021) 012017 doi:10.1088/1757-899X/1021/1/012017

defects and damages at the time of its erection. Therefore, only after years of the operation, the tank developed considerable operational abnormal damages, in the form of cracks of its lateral walls [5]. The considerable number of measured cracks with considerable depth in the concrete (some cracks went through the entire thickness) only confirms the work of the lateral walls of the tank as eccentric tension elements. The operation of a completely thick tank became impossible due to the insufficient reinforcement of its lateral walls and the errors of the design solutions about the reinforcement of its corners. Further operation of this structure is only possible after its reinforcing. Nowadays, a lot of reinforcements for the structures developed by the Department of Building Constructions and Bridges of Lviv Polytechnic National University are considered to be flexural elements and was presented in detail in scientific articles [11 - 16]. The strengthening methods for flexural and eccentric tension elements are similar and must ensure the necessary sectional area and the rigidity of strengthened bearing elements for the perception of necessary loadings. The scientific articles [11 - 16] present the strengthening of flexural elements with and without their unloading, depending on the operating conditions, a possibility of ensuring a joint action for the existing and additional elements of the strengthening taking into account the loss of physical-mechanical properties of concrete and reinforcement of the structures, which needed strengthening. In our case, the strengthening was necessary for all lateral walls that got considerable damages (cracks) along all their area. This strengthening was carried out with an empty tank, in other words with unloading that permits to use maximum existent physical-mechanical properties of concrete and reinforcement of the tank. The executed necessary instrumental tests and the calculations for the cross-section of the solid-cast reinforced concrete tank permit to use the most acceptable design arrangement and design diagram for the strengthening. The accepted steel elements of strengthening, such as inner tension bars and external bandages ensure the necessary spatial rigidity and tightness of the structure during the use of existent reinforcement that had a considerable negative reserve of bearing capacity at the moment of investigation. The introduced strengthening has to ensure the perception of all present loadings on the tank and it permits to use this structure in the future [1].

2. Materials and methods

The design map of the solid-cast reinforced concrete tank is a three-dimensional wall with the bearing slab roofs and foundation and four external reinforced concrete lateral walls, which are included in spatial work in different ways (figure 1). The thickness of the reinforced concrete walls is 400 mm, the thickness of the foundation plate is 500 mm, the thickness of the slab roof is 250 mm and the height of the tank from the bottom is 9000 mm. The barrier brick walls are foreseen from the contour of the structure above the mark +7.450. The reference mark ±0.000 m has accepted as the mark of the bearing floor of the workroom adjacent to the tank. At the time of the study, the tank was not used because of the emergency operational condition. [17]. The need for visual and instrumental investigations of the solid-cast reinforced concrete tank arisen due to the formation of multiple abnormal cracks and deformations of the lateral walls with the loss of their tightness, which is necessary for the capacitive structures. To conduct calculations for the structural elements of the tank with the determination of its prospects for the bearing capacity and the fracture strength, the investigations of the physical-mechanical properties of concrete and the defects and damages of the tank during the construction were carried out. During the investigation, the tank was not filled and the results of measurement for the deflections of three walls were as follows (in the central part of the walls): within axes A - E - f = 7.5 to 35 mm; within axes 1 - 3 - f = 9.3 to 22.3 mm; within axes E - A - f = 4.9 to 20 mm. The uncovered deformations along the walls of the tank from their tension at the height 3 m from the tank bottom were: +7.1 mm corner in axes 3/A along the axis 3; +8.3 mm – corner in axes 1/A along the axis 1; +8.7 mm - corner in axes 3/A along the axis A; +10.3 mm - corner in axes 3/EA along the axis E. The deflections for walls of the empty tank at the other levels are shown in figure 2.

IOP Conf. Series: Materials Science and Engineering 1021 (2021) 012017 doi:10.1088/1757-899X/1021/1/012017

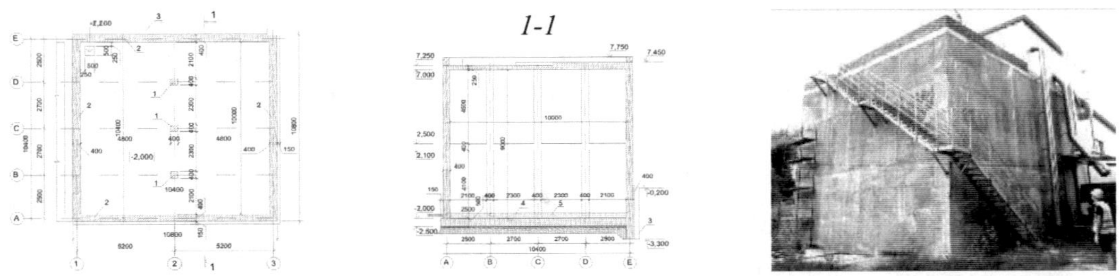

Figure 1. The scheme of the tank, the crosscut 1-1, and the total view of the structure on the crossing of axes 3 - A: 1 - existent supports for the covering of tank; 2 – existent precast reinforced concrete walls with thickness 400 mm, reinforced symmetrically by two mesh reinforcements d12A400C with step 150 x 150 mm; 3 -retaining wall under flotator; 4 – existent reinforced concrete foundation plate of the tank with reinforcement for the floor d14A400C with step 120 x 120 mm; 5 – sewage level at the time of investigation; 6 – existent reinforced concrete slab roof of the tank; 7 – contours of the foundation for tank

The deflections of the empty tank showed the values, which were less than their limit values fu=L/200=10000/200=50 mm [9] (figure 2, 3). However, the width of cracks opening and the depth of their permeation played a primary role in the decrease of capacity for the work of the investigated tank. The developed system of cracks for two lateral walls of the tank was deteceted after moistening by the water of the concrete surface. The cracks for front of axes E to A for the wall at the axis 1 are shown in the figure 4. When the tank is filled only to 300 mm, the measured width for the formation of the cracks varied in the range from 0.1 to 0.6 mm.

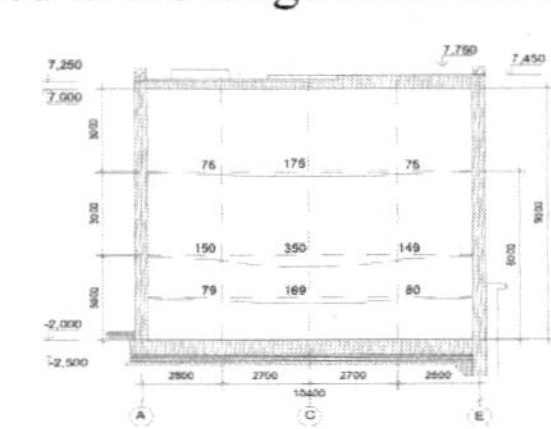

Figure 2. The results of deflection measurements for wall at the axis 3 for front axes A - E

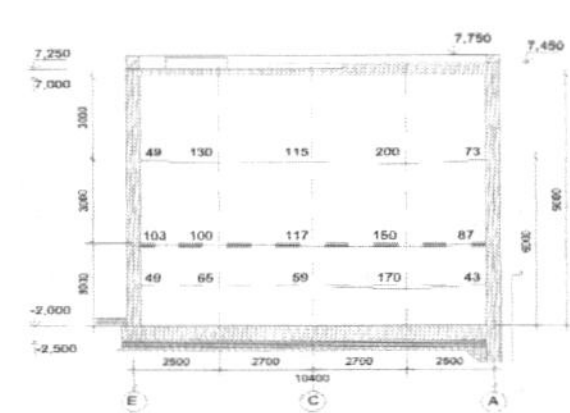

Figure 3. The results of deflection measurements for wall at the axis 1 for front axes E - A

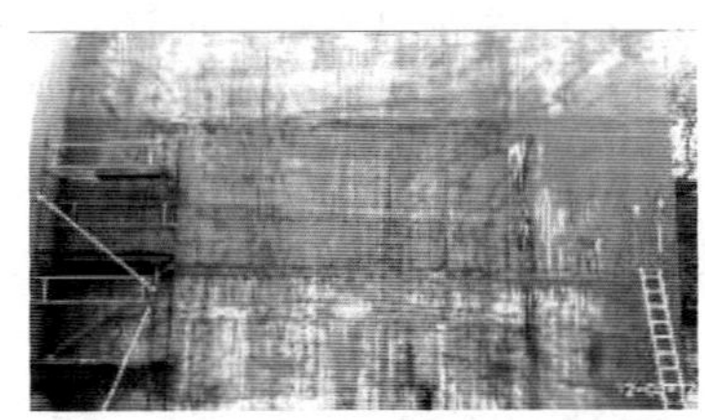

Figure 4. The cracks of front axes E - A for wall of axis 1 after the moistening by water of the concrete surface

It is worth noting that the width of cracks opening exceeded their limit values for reinforced concrete capacitive structures. The depth of cracks passing into the concrete walls of the tank is shown in figure 5. The step of cracks measurement was 120 mm (base of ultrasonic comparator UK-14PM). Because of that, the lateral walls of tank work as tension reinforced concrete elements with the flexure, accordingly most detected cracks formation are through the entire thickness in the central part on the outside lateral walls and near the corners of these walls. It indicates insufficient reinforcement for the lateral walls of the tank.

During the instrumental investigation of reinforced concrete structures of the tank, the determination of the concrete strength was carried out the non-destructive method using spring automatic machine A-1 by «Chmelnythprombud». At that, we discovered the porosity of concrete that was concerned with possible topping-up water and insufficient vibrocompaction of concrete, on which the smooth walls of air emptiness with diameter 2÷3÷4 mm indicate. The considerable height of the casing (over 3 m), two mesh reinforcement with small cells, which overflowed over the casing height, could influence on insufficient vibrocompaction of concrete. The age of reinforced concrete of the tank was approximately one year at the time of the investigation. IZS-10N machine was used for the determination of the location of the main reinforcement, the thickness of concrete cover and diameters of the main reinforcement for walls. By the results of researches was determined the actual guaranteeing strength quality of concrete with its 95 % ensuring. Also, the strength quality of concrete for lateral walls of axis front 1 - 3, A - E, E - A along the axes A, 3 and 3 was C12/15. The defined value of strength quality of concrete is in two times lower than the design strength quality of concrete, which must be equal to C25/30. The values W8 were not reached. The total coefficient of variation for the concrete strength (V = 11.89 %) does not exceed the normative coefficient of variation (V_n = 13.5 %) [5]. The insignificant variations of strength property of the concrete along all inspected area of the external surface of tank is evidence of stability for concrete strength.

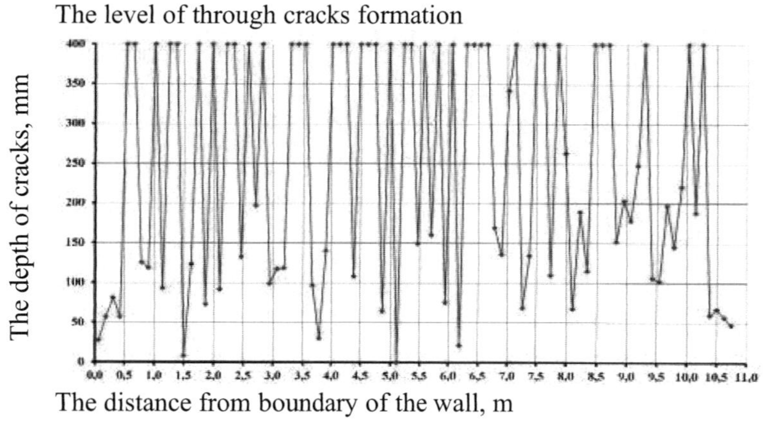

Figure 5. The depth of cracks passing into the concrete walls of the tank for axes 1 - 3 along the axis A

3. Results and discussions

The impossibility of operation of the tank by the purpose gave birth to the necessity to carry out the verifying calculations of its design solution as such that did not ensure the guaranteeing bearing capacity and fracture strength of bearing reinforced concrete elements. There were executed the verifying calculations of the tank as a spatial system in the software package LIRA on the design and technological loads for the possibility of the analysis of the main design solutions, accepted in the execution plan of the tank.

By the calculation the necessary reinforcement were determined according to the ensuring of the strength and possible limitation for the width of cracks opening till $w_k \leq 0.1$ mm for the capacitive structures. The comparison of the designed reinforcement of the tank elements and calculated reinforcement by the verifying calculation on the technological loadings are shown in table 1. By the calculation, the walls of the full tank were related to category 4 of technical state (hazardous) [17] by their insufficient bearing capacity that was concerned with insufficient reinforced of the slabs for the wall enclosure, the slab roofs and the bottom of the tank. Using the verifying calculations on the limit design loadings, we determined the characteristic zones in the walls, in which the stresses in main reinforcement of the meshes and inner girder greatly exceeded the conventional yield strength for the reinforcement that was used in the tank. There were also determined that the corners of the solid-cast joining of the wall elements of the tank between themselves and the joining the wall elements with the slab of the bottom were insufficient reinforced for the perception of the active tensile loads and the bending moments, which acted simultaneously.

Table 1. The comparison of the designed reinforcement of the elements of the tank with the received reinforcement by the verifying calculation

The elements of the tank	The location in the axis	The location's place of the area and the direction for the reinforcement	By the project Reinforcement A400C			By the verifying calculation			Compa-rison
			∅, mm	S, mm	As, running cm²/m	∅, mm	S, mm	As, running cm²/m	As,c / /As,p
Lateral walls	1 - 3 and 3 - 1	Bottom of the wall, vertical	12	150	7.53	25	150	32.7	4.34
		The middle, vertical			7.53	28		41.1	5.46
		The middle, horizontal			7.53	28		41.1	5.46
		Near the corners, horizontal			7.53	28 to 36		41.1 to 67.9	7.24
Lateral walls	A - E and E - A	Bottom of the wall, vertical, horizontal	12	150	7.53	28	150	41.1	5.46
		The middle, vertical			7.53	32		53.6	7.12
		The middle, horizontal			7.53	32		53.6	7.12
		Near the corners, horizontal			7.53	28 to 36		41.1 to 67.9	7.24
		The top of the wall, vertical, horizontal			7.53	22 to 28		25.3 to 41.1	4.41
Slab roof	1 - 3 - A - E	Perpendicularly to axis 6 (middle support)	14	150	10.26	22	150	25.3	2.46
Bottom	1 - 3 – A - E	Along the axes from the borders: along axis perpendicularly to axes 1, 3	14	120	12.83	20 to 28	120	20.9 to 41.1	2.42
		Perpendicularly to axis 2			12.83	32	120	53.6	4.18

The anchorage length for the reinforcement of the inner meshes for the wall enclosure in rigid joints of their connections on the combined action of the tensile loads and the negative bending moments were insufficient and the additional corner reinforcing elements were not effective enough. During calculation, the junction point for the wall elements of the tank with the slab roof must be accepted as a hinged joint due to insufficient length of the reinforcement dowels from the wall slabs.

For the possibility of further safe operation for the solid-cast reinforced concrete tank, it is necessary to carry out its strengthening for the prevention of further cracks opening and the ensuring of its spatial rigidity. Similarly to the calculation of the bearing capacity, we carried out the verifying calculations in the software package LIRA with regard to the design solution of the tank with technological loadings and accepted elements of the strengthening. The dead, changing, continuous, and variable changing loads were accepted similarly to the calculation of the design solution of the tank. Inasmuch as it is completely impossible to close the considerable formation of cracks even with the elements of the strengthening (through the cracks will act the hydrostatic pressure), then it is necessary to ensure the tightness for the reinforced concrete elements of the tank.

By the realized investigation, all walls of the solid-cast reinforced concrete tank (their thickness is 400 mm) are symmetrically reinforced by two meshes d12A400C (their step is 150 x 150 mm; the area of reinforcement is A_s = 7.53 cm²/m; the design resistance for the reinforcement is f_{yd} = 365 MPa). The distance between the axes of the bars of the reinforcement meshes is 250 mm. Taking into account the recommended deviation for the reinforcements meshes that is equal to 10 mm, the calculated distance between the axes of the reinforcement bars is $250 - 2 \cdot 10 = 230$ mm. We considered the reinforcement meshes as displaced to one of the walls. The linkage to one border was accepted 40 mm and the linkage to other border was accepted 400 – 230 – 40 = 130 mm.

The thickness of the tank's bottom was 500 mm. The protective layer was 50 mm that corresponded to the linkage of the bars for the reinforcement's diameter 14 mm: 50 + 2 · 14 + 2 · 10 = 98 mm. We accepted the linkage of the bars 115 mm as for the walls. The reinforcement for the bottom was d14A400C (the step is 120 x 120 mm; the area of reinforcement is $A_s = 12.82$ cm^2/m; the design resistance for the reinforcement is $f_{yd} = 365$ MPa). By the realized measurements, the concrete strength corresponds to the quality of the concrete C12/15. The cracks were in the walls of the tank with the residual opening after the unloading of the tank till 0.6 mm (more information in Materials and methods) which is evidence of the yield of the reinforcement during the filling of the tank.

The strengthening of the tank (figure 6, 7) was realized by means of the outer bandages (reinforcing steel 4d32A400C with $A_s = 32.17$ cm^2) and inner tension bars (stainless steel round pipes 168 x 5 mm). We accepted that the outer bandages had prestress 50 t and the inner tension bars without the prestress (table 2).

Table 2. The efforts in the inner tension bars (pipe 168 x 5 mm)

Stage	1 (lower)	2 (middle)	3 (upper)
	N	N	N
Extreme bars	+12.9	+12.3	+8.2
Middle bars	+22.1	+22.9	+16.2

The efforts in the wall:

$$M_{x,\max} = 80.61 \frac{kN \cdot m}{m}; N_y = -138.27 \frac{kN}{m}; \quad (1)$$

$$M_{y,\max} = 69.63 \frac{kN \cdot m}{m}; N_y = -133.37 \frac{kN}{m}; N_x = -358.92 \frac{kN}{m}; \quad (2)$$

$$M_x = 77.67 \frac{kN \cdot m}{m}; M_y = 14.32 \frac{kN \cdot m}{m}; N_{y,\max} = -160.83 \frac{kN}{m}; \quad (3)$$

$$M_x = 80.22 \frac{kN \cdot m}{m}; M_y = 15.1 \frac{kN \cdot m}{m}; \quad (4)$$

The tensile efforts:
in corners of the conjunction for two walls with the bottom

$$N_x = +161.81 \frac{kN}{m^2} \cdot 0.4 = +64.72 \frac{kN}{m}; M_x = 22.65 \frac{kN \cdot m}{m}; M_y = 16.48 \frac{kN \cdot m}{m}; \quad (5)$$

$$N_y = +84.37 \frac{kN}{m^2} \cdot 0.4 = +33.34 \frac{kN}{m}; M_x = 23.05 \frac{kN \cdot m}{m}; M_y = 16.08 \frac{kN \cdot m}{m}; \quad (6)$$

in the middle of the wall's height (in the direction X the wall is compressed)

$$N_y = +23.05 \frac{kN}{m^2} \cdot 0.4 = +9.22 \frac{kN}{m}; M_x = 22.85 \frac{kN \cdot m}{m}; M_y = 9.61 \frac{kN \cdot m}{m}. \quad (7)$$

The critical combination of the forces is the simultaneous action of the bending moments M and the forces N. During the calculation of the strength of the tank by the normal sections, we took into consideration the arrangement of the mesh reinforcement with the displacement, the thicknesses of the protective coatings (40 mm and 130 mm), the length (the span) of the element 1000 mm, the height of the cross-section 400 mm, the width of the cross-section 1000 mm and the values of the bending moments

IOP Conf. Series: Materials Science and Engineering 1021 (2021) 012017 doi:10.1088/1757-899X/1021/1/012017

M = 80.61 kN · m, the shear Q = 0 kN, the longitudinal force N = 138.3 kN. By the calculation the bearing capacity of the tank is ensured under the actual opening of the normal cracks in within 0.29 mm.

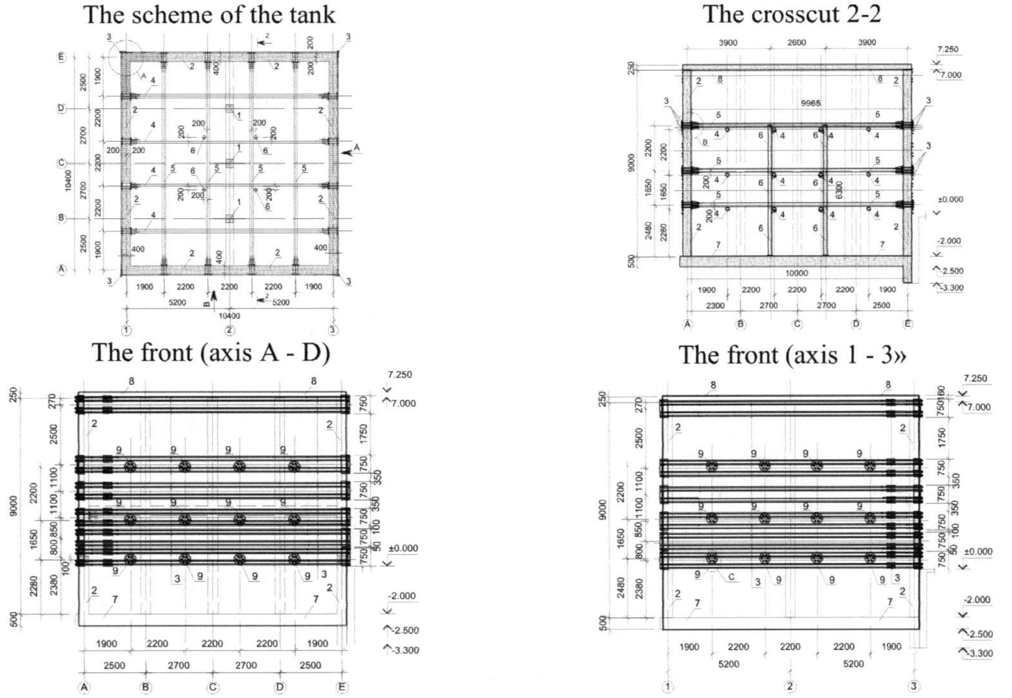

Figure 6. The scheme of the tank, the crosscut 2-2 with the arrangement of the inner tension bars and outer bandages of the strengthening, the front (axis A - D) and the front (axis 1 - 3): 1 – the existent supports for the covering of the tank; 2 – existent precast reinforced concrete walls with the thickness 400 mm, reinforced symmetrically by two reinforcement meshes d12A400C with the step 150 x 150 mm; 3 – outer bandages of the strengthening from the reinforcing steel 4d32A400C; 4, 5 – inner tension bars (steel round pipes 168 x 5 mm) in two directions; 6 – the support of the strengthening (pipe 168 x 5 mm); 7 – existent reinforced foundation slab of the tank (the reinforcement for the bottom d14A400C with the step 120 x 120 mm); 8 – existent reinforced concrete slab roof of the tank; 9 –supporting sheet of the inner tension bar with thickness t = 16 mm and with diameter d = 450 mm with the studs M24 (6 pieces). All elements are arranged inside of the tank and also the studs, which fix the inner tension bars, were executed of the stainless steel. Other elements of the strengthening was executed from the low-carbon steel C235

The strengthening of the tank was realized using the outer bandages and the inner tension bars, which ensured its spatial rigidity and, of course, the bearing capacity of all walls of the solid-cast reinforced concrete tank under the actual opening of the normal cracks in within 0.29 mm. Taking into consideration the limitation for the width of the opening cracks within 0.1 mm for this structure, the additional inner damp proofing to ensure the tightness of the wall enclosure and the bottom of the tank was realized after the executed strengthening. The material for the damp proofing must have considerable deformation features and good adhesion to the surface of the concrete. It is connected with the variation for the width of the cracks opening, which will not be bigger than the value 0.29 mm.

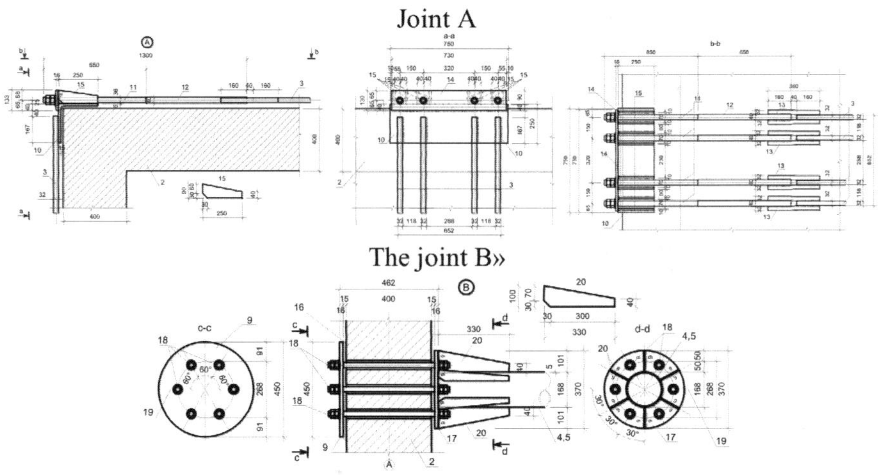

Figure 7. Joint A (the arrangement of the outer bandages of the strengthening) and the joint B (the arrangement of the inner tension bars of the strengthening): 2 – present precast reinforced concrete wall of the tank with the thickness 400 mm; 3 – outer bandages of the strengthening of the reinforcing steel 4d32A400C; 4, 5 – inner tension bars in two directions (pipe 168 x 5 mm); 9 – outer supporting sheet of the inner tension bar with thickness t = 16 mm and with diameter d = 450 mm; 10 – angle bar 250 x 250 x 16 mm; 11 – thread M36; 12 – thread shanks (reinforcing steel 4d40A400C); 13 - mounting joint (reinforcement steel 8d32A400C); 14 – supporting sheet of the outer bandages 16 x 730 x 130 mm; 15 - stiffening bar 10 x 90 x 250 mm; 16 – hard cement-sand grout; 17 – inner supporting sheet of the inner tension bar with thickness t = 16 mm and with diameter d = 370 mm; 18 – studs M24; 19 - symmetry line; 20 - stiffening rib 8 x 100 x 330 mm. The bays between flanges and the wall of the tank from inner face and also the openings under the studs were filled up sealing mastic. There was used the welding seams with leg 6 mm besides mentioned welding seams

Conclusions

The verifying calculations in the software package LIRA showed the considerable insufficiency of the reinforcement of the wall slabs of the solid-cast reinforced concrete tank. In all corners of the tank and the places, where the walls join with the slab of bottom, the simultaneous perception of the active tensile loads and bending negative moments is impossible due to that the reinforcement in corner joints of the tank elements is not ensured. Existing detected additional corner elements are not sufficiently effective and they do not ensure the necessary length for the anchorage of the reinforcement in inner meshes of the wall enclosure in rigid joints. The identified design errors caused formation of the large quantity of the abnormal cracks with considerable depth of their passing in the concrete that rendered impossible the further operation of the tank due to the loss of the tightness in the walls. The strengthening for the structures in accordance with the outer bandages and the inner tension bars was ensured the spatial rigidity and the bearing capacity of all walls of the solid-cast reinforced concrete tank under the actual opening of the normal cracks in within 0.29 mm. The outer bandages were executed from the reinforcing steel in quantity 4d32A400C and the inner tension bars from the round pipes 168 x 5 mm of the stainless steel. The operation of the tank is possible when the width of the cracks openings is in within 0.1 mm, therefore the ensuring of the tightness for the wall enclosure and the bottom of the tank was carried out inner damp proofing, which has considerable deformation behaviors and good adhesion to the surface of the concrete.

References

[1] Kramarchuk A P, Ilnytskyy B M, Bobalo T V and Lytvyniak O Ya 2020 The reinforcing for reinforced concrete tank of anaerobic purification plants with the manufacture of biogas, damaged as a result of the mistakes during the design and the construction *Actual problems of Engineering mechanics: proceedings of VII International conference* (Odessa) pp 126–130

[2] DBN V.1.2-2:2006 The loadings and influences (Kyiv: Min. bud-va, arkhit. ta zhytlovo-kom. hosp) p 75

[3] DBN V.1.2-14:2009 The basic principles for the reliability assurance and constructive safety of the buildings, structures, building constructions and foundations (Kyiv: Min. bud-va, arkhit. ta zhytlovo-kom. hosp) p 32

[4] DBN V.2.6-198:2014 Steel structures (Kyiv: Min. bud-va, arkhit. ta zhytlovo-kom. hosp) p 206

[5] DBN V.2.6-98:2009 The concrete and reinforced concrete structures (Kyiv: Min. bud-va, arkhit. ta zhytlovo-kom. hosp) p 71

[6] DSTU B V.1.2-3:2006 The Deflection and displacement. The design requirements (Kyiv: Min. bud-va, arkhit. ta zhytlovo-kom. hosp) p 15

[7] Kramarchuk A, Ilnytskyy B, Lytvyniak O and Grabowski A 2019 The increase of seismic stability for existing industrial buildings *AIP Conf. Proc.* **2077** 020029

[8] Ilnytskyy B, Kramarchuk A, Bula S and Bobalo T 2019 Study of the vibration influence on load-bearing floor structures in case of machinery operation *IOP Conference Series: Materials Science and Engineering* **708** 012052

[9] Kramarchuk A, Ilnytskyy B, Lytvyniak O and Famulyak Y 2019 Strengthening prefabricated reinforced concrete roof beams that are damaged by corrosion of concrete and reinforcement *IOP Conference Series: Materials Science and Engineering* **708** 012060

[10] Hladyshev D H and Hladyshev H M 2012 *The investigation of the technical state for the buildings, structures and their elements* (Lviv) p 304

[11] Khmil R, Tytarenko R, Blikharskyy Y and Vegera P 2018 Development of the procedure for the estimation of reliability of reinforced concrete beams, strengthened by building up the stretched reinforcing bars under load *Eastern-European Journal of Enterprise Technologies* **5/7** pp 32-42

[12] Krainskyi P, Blikharskyy Y, Khmil R and Vegera P 2018 Influence of loading level on the bearing capacity of RC columns strengthened by jacketing *MATEC Web of Conf.* **230** 02013

[13] Barabash V, Kramarchuk A and Ilnitskiy B 2018 Bearing capacity of steel-concrete bent elements reinforced with additional unstrained armature at different levels of strained state *The bulletin of Lviv Polytechnic National University. Series 'The theory and practice of construction* Volume **888** pp 3-11

[14] Kramarchuk A P, Ilnytskyy B M and Bobalo T V 2013 The experimental researches of the strength and deformability for the normal sections of the steel-aggregate granolithic concrete beams, reinforced by additional reinforcement bars without adhesion in tensile zone *The bulletin of Lviv Polytechnic National University. Series 'The theory and practice of construction'* Volume 755 pp 05-213

[15] D Gladyshev 2015 Variants of strengthening of shells of reinforced concrete cooling towers depending on constructional features and actual technical condition *Zeszyty Naukowe Politechniki Czestochowskiej 171: Budownictwo* Volume 21 pp 75-84

[16] Hladyshev H M and Hladyshev D H 2011 The experience of inspection, reinforcing and repair for the casing of the precast ribbed reinforced concrete biconical cooling tower *The bulletin of Lviv Polytechnic National University. Series 'The theory and practice of construction'* Volume **697** pp 68-75

[17] DSTU-N B V. 1.1 – 18:2016 The directions about the inspection for the buildings and structures for

IOP Conf. Series: Materials Science and Engineering 1021 (2021) 012017 doi:10.1088/1757-899X/1021/1/012017

the purpose the determination and appreciation of their technical state (Kyiv: Min. bud-va, arkhit. ta zhytlovo-kom. hosp) p 47

IOP Conf. Series: Materials Science and Engineering 1021 (2021) 012018 doi:10.1088/1757-899X/1021/1/012018

A study of bearing capacity of reinforced masonry beams with GFRP reinforcement

Andrii Kramarchuk [1], Borys Ilnytskyy [1], Taras Bobalo [1] and Oksana Lytvyniak [2, 3]

[1] Department of Building Constructions and Bridges, Institute of Civil Engineering and Building System, Lviv Polytechnic National University, S. Bandera str., 12 79013 Lviv, Ukraine
[2] Department of Civil Safety, Viacheslav Chornovil Institute of Sustainable Development, Lviv Polytechnic National University, S. Bandera str., 12 79013 Lviv, Ukraine

[3] E-mail: lytvyniak.oksana@gmail.com

Abstract. This article presents the results of experimental researches of bearing capacity and deformability of reinforced masonry beams with combined reinforcement. This study was carried out in the zone of maximum bending moment. The reinforcement A400C and A1000 were used for experimental samples. The combined action of ceramic masonry and reinforcement was ensured at the expense of a cement-sand grout in the joints of masonry and vertical reinforcing meshes in support zones of beams, which were welded to angle bars. Besides that, there was carried out a comparison of the results of experimental researches with calculated values by the active standards of Ukraine. This comparison permits to determine the expediency of the use of the existing design procedure of bearing capacity of reinforced masonry structures with combined double reinforcement. Finally, this scientific work has also presented the conclusions about the possibility of the efficient design of reinforced masonry structures in the effective area of normal forces.

1. Introduction

The masonry structures are widely used in civil and industrial building construction. These structures take a load from the dead weight, from ceilings, and from wind. Moreover, they execute the heat-insulating and soundproofing properties. The cost of masonry structures (foundations, walls) is 15…30 % of the total cost of the edifice, but their weight is equal to 60 % (approximately) from the weight of the edifice [1].

The reinforced masonry structures are used during the erection of foundations, outer and inner walls of buildings, ceilings, arches, chimneys, bridges, sewers, water towers, elevators, etc. (fig. 1, 2) [2]. The elements of masonry and reinforced masonry structures execute two functions: protective (these structures divide two environments with different physical properties) and bearing (they take and pass a load, which appears from operational factors (dead weight, payload, a change of temperature). The reinforced masonry structures must correspond to the operational requirements of bearing capacity, stiffness, life duration, reliability, reparability, and other properties, which will determine the quality and suitability of structure to the exploitation.

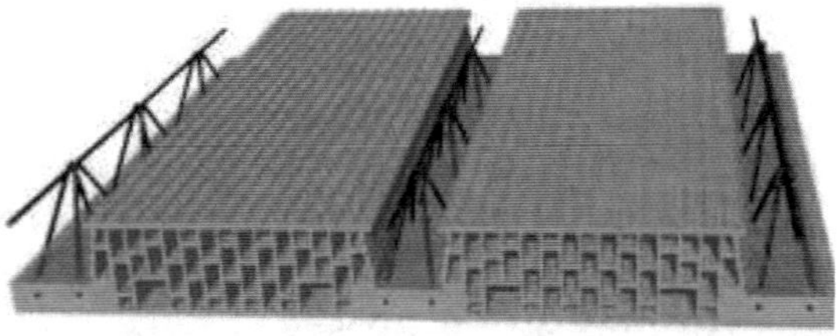

Figure 1. The reinforced masonry ceiling with reinforced concrete pieces (inserts)

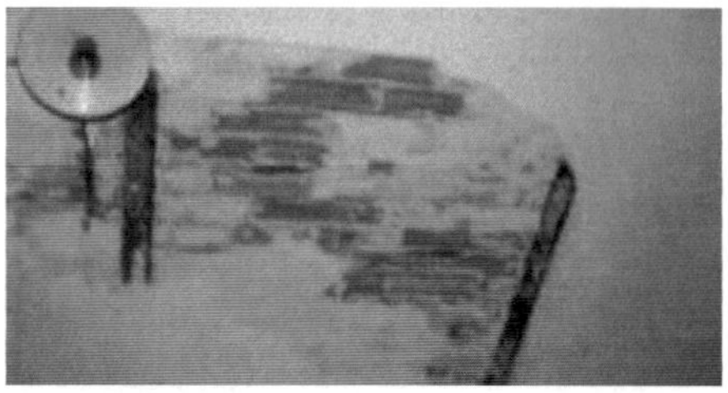

Figure 2. The reinforced masonry ceiling on the steel beams in housing building in Lviv (Ukraine)

The reinforced masonry structures are widely used not only in Ukraine but and in Europe since these structures have enough bearing capacity and deformability, they are environmentally friendly and simple in mounting [3 - 5].

However, now, reinforced masonry structures with additional longitudinal reinforcement and monolithic reinforced concrete insertions are widely used (complex structures). These solutions are necessary in cases of [6, 7]:

- an increase bearing capacity of structures, when the tensile loads appear in the cross-sections and these loads exceed the design tensile strength of masonry;
- the use of flexible elements, if their flexibility is equal more than 15;
- an application in thin walls and partitions for the increase of their firmness and bearing capacity during the action of lateral loads;
- the use in walls and columns for the increase of masonry rigidity, fracture strength, and seismic stability.

Moreover, there is recommended to use a netted reinforcement for the ensuring of better combined action of masonry with longitudinal reinforcement or reinforced concrete [8, 9].

This scientific work is devoted a study of reinforced masonry beams with combined reinforcement, in which are used reinforcement rods A400C and A1000C in a lower tensile zone. This reinforcement was arranged in the extreme borders of the cross-section.

2. Materials and methods

The four reinforced masonry beams with combined reinforcement (we used reinforcement rods A400C and A1000) and two reinforced masonry beams (with usual reinforcement by reinforcement rod A400C in the tensile zone) were designed, produced, and experimentally investigated. The cross-section of all beams was 140 x 250 mm. The length of these structures was 2300 mm and the design span was 2000 mm [1].

The construction of experimental samples is shown in fig. 3.

It should be noted that experimental samples of reinforced masonry beams have differed reinforcement in the tensile zone:

- beams B-1 – one reinforcement bar Ø8 mm A1000 and one reinforcement bar Ø10 mm A400C;
- beams B-2 - one reinforcement bar Ø10 mm A1000 and one reinforcement bar Ø8 mm A400C;
- beams B-3 were produced without combined reinforcement with the purpose of the comparison with the beams with combined reinforcement. However, the beams B-3 were reinforced by two reinforcement bars Ø10 mm A400C.

The combined reinforcement of reinforcement bars A400C and A1000 were used in the tensile zone of the beams B-1 and B-2. The adhesion of reinforcement and masonry was ensured using thick calking by cement-sand grout.

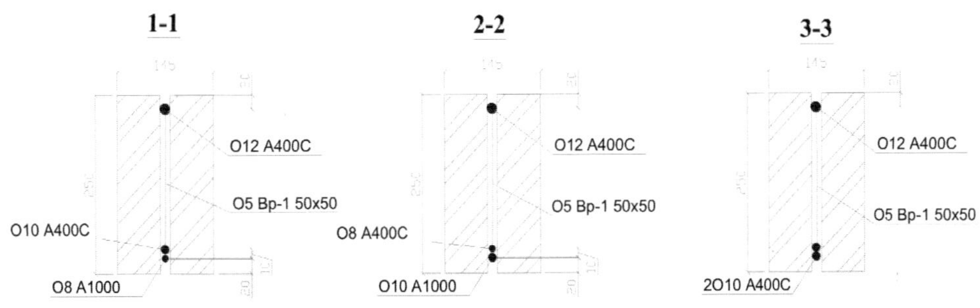

Figure 3. The cross-section of the investigated reinforced masonry beams

The mesh of wire was arranged in a zone of action of maximum shears. The diameter of the wire was 5 mm A240C. The size of the cell was 50 x 50 mm. The length of the mesh was 1/3 span of beams.

The feature of the production of experimental samples with combined reinforcement consists that the structure of beams was manufactured in the horizontal position in two stages. The experimental samples of beams are two horizontal rows of brick that are produced by means of layerwise laying.

Firstly, we put in the first row of brick. The length of the first row was 2300 mm and the medium thickness of the joints was 10 mm. After that, we put on the layer of the mortar, which was produced in accordance with the requirements [10], and then we arranged the reinforcement in tensile and compressed zones of beams with adherence of the height of the protective layer. Afterward, we located the reinforcement of support sections in the zone of action of maximum shears. For that, we used the reinforcing meshes of reinforcement bars Ø5 mm A240C with cells 50 x 50 mm. Next, we put on the second row of brick on a cement-sand grout. The medium thickness of the joints was 10 mm. The joints between bricks filled up by a mortar on 100 %.

The combined action of reinforcement and brick masonry was ensured using the insert of reinforcement on cement-sand grout (the grade of this mortar was M150). This action was ensured good adhesion of reinforcement and masonry. The arrangement of all elements of structures was executed under the level to ensure the straightness of structure.

In addition to the production of experimental samples, we produced prisms (the cross-section of prisms was 120 x 145 mm and length of prisms was 250 mm). Manufactured experimental samples and prisms were held by the temperature 20^0C and relative humidity 50…70 %. This action provided the growth of designed strength during 28 days.

After 28 days, the experimental beams were set up on the test bench strictly vertically by means of the level tube. After that, we plastered and fixed holders for microindicators with a base that was equal to 150 to 200 mm. The microindicators have measured the deformations of brick in the zone of pure bending.

The experimental samples of the reinforcement rod were produced for the purpose of determination of the physical-mechanical properties of longitudinal reinforcement, which was located in the tensile zone. The test was carried out stretching method of samples on the hydraulic tensile-testing machine with a simultaneous record of the data diagram of the tension of the reinforcement bar. The resistance strain gages were glued on experimental bars in places of the cavity. The deformations under the action of loading on the samples were measured using these devices. The strain-loading diagrams in reinforcement from the load for experimental samples of reinforcement bars A400C and A1000 are shown in fig. 4.

The characteristics of materials for experimental samples in accordance with building standards [11 - 13] are shown in table 1.

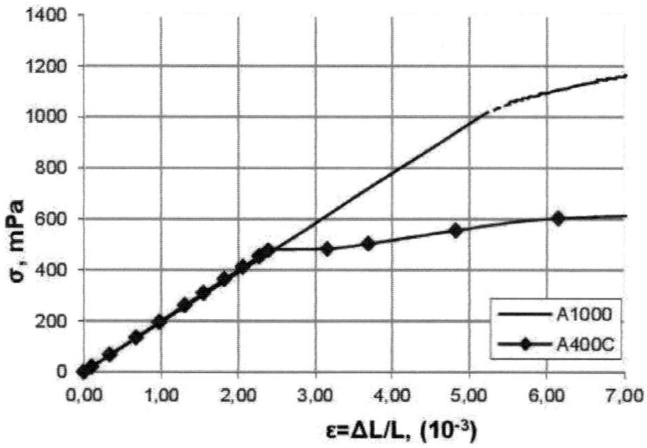

Figure 4. The diagrams of deformation from loading for reinforcement bars A400C and A1000

Table 1. The characteristics of materials for experimental samples

The type of beam	B-1		B-2		B-3
Geometrical parameters of cross-section					
b, [mm]	140		140		140
h, [mm]	250		250		250
A, [cm^2]	350.0		350.0		350.0
Ceramic blocks					
f_k, [MPa]	7.29		7.29		6.43
f_{vko}, [MPa]	0.9		0.9		0.8
E_{cm}x10^3, [MPa]	4.3		4.3		4.3
Reinforcement rod of the tensile zone					
d_s, [mm]	Ø8	Ø10	Ø8	Ø10	2Ø10
f_{yk} , [MPa]	1080	510	510	1080	510
A_s, [cm^2]	0.503	0.785	0.503	0.785	1.57
E_sx10^5, [MPa]	1.85	2	2	1.85	2
Class	A1000	A400C	A400C	A1000	A400C
Reinforcement rod of the compressive zone					
d_s', [mm]	Ø12		Ø12		Ø12
f_{yk} ', [MPa]	510		510		510
A_s', [cm^2]	1.13		1.13		1.13
E_s'x10^5, [MPa]	2		2		2
Class	A400C		A400C		A400C
Cross reinforcement					
d_s, [mm]	Ø5		Ø5		Ø5
f_{yd} ', MPa[MPa]	370		370		370
A_s', [cm^2]	0,126		0,126		0.126
E_s'x10^5, [MPa]	1.7		1.7		1.7
Class	A240C		A240C		A240C

The study of experimental beams (the calculated span was 2000 mm) was carried out using the test bench. The loading was created by means of hydraulic jack with power 1000 N that was set up on the distributive cross-arm. The effort was applied to the upper border of beams. This effort was divided into two concentrated forces, which were applied symmetrically relative to the middle of the experimental beam. The distance from the applied force to support was equal to 650 mm.

During the test, the load for beams was carried out with step, which was similar to 10 % from breaking load with the exposure 30 minutes on every stage. After the exposure of the load, we took a reading of all microindicators and deflection indicators, and also we fixed the creation of cracks and, of course, a width of the opening and development of cracks. The value of loading was controlled using standard pressure gage that was calibrated together with hydraulic pumping plant and jack, and also by the value of support reactions, which were fixed by two proving rings that were simultaneously fixed and movable supports. In hinged movable support, the supporting part was executed as a flat surface. At the same time, in the hinged fixed support, the supporting part of the proving ring was executed as cylindrical surfaces with a radius ring of the dynamometer.

The deformations were fixed using the microindicators that were set up with the base 200 mm. The division value of microindicators was 0.001 mm. All deformations were measured in the zone of pure bending. The width of the cracks opening and also the height of cracks were determined at every stage of loading and were measured using microscope МПБ-2М with division loading 0.005 mm.

The studies were carried out to an appearance of the yielding flow in the high-strength reinforcement or the destruction of the experimental sample. It permitted completely to use experimental samples.

3. Results and discussions

The executed experimental studies of reinforced masonry beams permitted to determine the features of their work.

On the first stage, the work of beams (to the moment of formation cracks in masonry) we observed predominantly elastic deformations.

Table 2 summarizes the maximum values of tensile and compressive deformations of longitudinal and transverse reinforcements along the height of beam's cross-section in the zone of the maximum moment, and also the value of the deflections to moment of the formation cracks that were obtained during the test studies of experimental samples.

Table 2. The main results of experimental studies of beams to the moment of the formation of the cracks

Type of beams	M, [kN · m]	$\varepsilon_{sA400} \cdot 10^5$	$\varepsilon_{sA1000} \cdot 10^5$	$\varepsilon_{в} \cdot 10^5$	f_{max}, [mm]
B-1	1.01	77	79	55	2.3
B-2	1.06	88	92	68	2.9
B-3	0.96	67	-	47	1.4

At the stage to the formation of cracks, the deformations of reinforcement and masonry of all experimental samples did not differ, because typical cracks for the loss of reinforcement adhesion were not observed, which permitted to draw a conclusion about that the reinforcement was deformed uniformly along its full length.

It should be noted that the tensile and compressive deformations of masonry in the zone of normal sections were distributed along the height of elements in accordance with flat cross-section hypothesis and these deformations arranged along the length of shear span according to the bending moment diagram.

The increase of loading conducted an appearance in tensile reinforcement inelastic deformations and the stress in the tensile zone of beams reached the tensile strength of masonry f_{xk1}. The next increase of loading on the beam conducted the appearance of cracks that was evidenced about the beginning of the second stage of deflected mode.

There was typical that the cracks in beams with the higher reinforcement percentage by high-strength reinforcement A1000 were formed later. The riffled reinforcement bar was deformed evenly according to the bending moment diagram.

The most deflections, till the creation of the cracks, were fixed in the second experimental sample B-2, where was designed high-strength reinforcement A1000 with the highest reinforcement percentage. The least deflections until the creation of the cracks were identified in experimental sample B-3, where high-strength reinforcement was absent.

In such a way, at the stage from the beginning of loading till the appearance of normal cracks, the reinforced masonry beams with combined reinforcement worked jointly as the monolithic cross-section.

When the normal cracks appeared, the homogeneity of masonry disturbed and the deflected mode changed along the span of beams. The deflected mode of bricks and joints in normal sections with cracks and between these cracks was different from one another.

The tensile and compressive deformations of masonry in sections between cracks were distributed to the linear law along the full height with maximum values near the tensile and compressive sides. Besides that, the tensile and compressive deformations of masonry in section with cracks were also distributed to the linear law, but not along the full height, yet only to the upper point of cracks in the tensile zone.

The most values of deformations for compressive and tensile zones of experimental samples, deflections of beams, deformations of tensile reinforcement till the moment of limit loading are presented in table 3.

Table 3. The main results of experimental studies of reinforced masonry beams under the loading $0,7M_{max}$ in the zone of maximum moment

The type of beams	$0.7M_{max}$, [kN · m]	$\varepsilon_{sA400} \cdot 10^5$	$\varepsilon_{sA1000} \cdot 10^5$	$\varepsilon_{b} \cdot 10^5$	f_{max}, [mm]
B-1	14.5	317	360	127	10,7
B-2	17,23	338	378	145	11,1
B-3	11,94	187	-	87	9,1

It should be noted that the beginning of the yielding flow for reinforcement A400C was happening under the achievement of the values of relative deformations $\varepsilon_{s,A400C}$=255 x 10^{-5}, and the yielding flow for reinforcement A1000 came under the values of relative deformations $\varepsilon_{s,A1000}$=584 x 10^{-5}Thus, the experimental sample was considered destroyed, if its deformations reached above the mentioned values [1]. The experimental values of the moment of crack formation and maximum permissible width of cracks opening are shown in table 4.

Table 4. The comparison the moments of crack formation and the maximum permissible width of cracks opening

The type of beams	The begin of crack formation		The maximum permissible cracks opening	
	M^d_{crc}, [kN · m]	test, a^d_T, [mm]	M, [kN · m]	test, a^d_T, [mm]
B-1	1.01	0.05	15.7	0.4
B-2	1.06	0.05	17.1	0.4
B-3	0.96	0.05	13.3	0.4

The cracks formation and the cracks opening caused the sharp growth of deformations in cross-sections of the experimental sample and, of course, the increase of deflections. Therefore, the value of the bending moment, which corresponded to an inflection on the chart of deflection, took for the moment of crack formation.

Further increase of loading was causing the growth of stresses in the compressive zone, the increase of the width of the crack opening, and the height of cracks. Accordingly, the experimental sample ceased to resist active forces and the stage of destruction was begun.

The beginning of the yielding flow in reinforcement A400C was happening earlier and under lesser values relative deformations than in reinforcement A1000. After the start of the yielding flow in reinforcement A400C, we observed a considerable increase of deformations' growth in high-strength reinforcement and, of course, the growth of deflections in beams during further load for samples and under fixed tensile load in reinforcement $f_{ykAs,A400C}$. All next increments of load were perceived high-strength reinforcement until its conventional yield strength was not reached. Even after the beginning of the yielding flow in reinforcement A400C, it continued to perceive part of the load, which was corresponded to its yield stress.

During the calculation of strength for the normal cross-section were determined such the stages of the work for reinforced masonry beams with combined reinforcement:

a) when the stress in reinforcement A400 C of reinforced masonry beams with combined reinforcement reached the yield point stress, there are observed the increase of the deformation growth and the deflection growth but the beams continued to perceive the growing loading;

b) during the beginning of the yielding flow in high-strength reinforcement A1000, the beam does not perceive the additional load, further loading of samples caused considerable deformations in reinforcement with the sharp excessive opening of cracks and certainly the growth of deflections;

c) physical destruction of beams, in particular the destruction of the compressive zone of beams.

During experimental studies, we carried out the analysis of the combined action of reinforcement and masonry with the formation of stress diagrams. The graphical correlations permitted to foresee the behavior of experimental samples under the next load and to determine the beginning of yielding flow in the reinforcement bar. The stress diagrams in reinforcement bars depending on the loading of the beam are shown in fig. 5.

The beams were designed in that way that the yielding flow of all tensile reinforcement bars of combined reinforcement began earlier than the destruction of the compressive zone of masonry. The destruction of experimental samples was plastic nature. Besides that, when the stress in reinforcement A1000 reached the yield point stress, the next loading caused the crush of the compressive zone of masonry. During the achievement of the yielding flow in reinforcement A400C the stress did not fall down in this reinforcement, in other words the reinforcement continued to perceive the corresponding maximum load and after its yielding flow.

The bending moment that corresponded to the yield point stress in reinforcement A400C was M_1, the value of maximum load, which was perceived by beam (under this load was happening the destruction of the compressive zone of the beam) was called – «limit (maximum) moment» - M_{max}.

The experimental and theoretical values of bearing capacity for beams and test loads of its physical destruction are presented in table 5.

We determined the change moment of the work for reinforced masonry beams with combined reinforcement from one stage to another by the stress diagrams in working reinforcement, taking into consideration the diagrams of experimental tests for the reinforcement and its deformations. During the test we carried out the analysis of the work of reinforcement and masonry using the stress diagrams. It was permitted to foresee the behavior of experimental samples under next loads, and also to determine the moment for the beginning of the yielding flow in reinforcement.

The determination of calculated values of bearing capacity of test reinforced masonry beams were carried out by the formulas [11]. The experimental and calculated values of limit (maximum) moments are shown in table 5.

The compressive strength of masonry for beams B-1 and B-2 was 9.06 MPa, and for beam B-3 – 7.15 MPa. The reinforcement percentage for beams B-1 and B-2 was 0.471 % and for beam B-3 – 0.956 %.

The destruction of beam B-1 was happened under the maximum bending moment that was equal to 21.9 kN · m. The theoretical maximum bending moment for this beam was 20.71 kN · m.

The beam B-2 was destroyed from bending moment, which was equal to 24.7 kN · m. The theoretical maximum bending moment for this beam was 23.61 kN · m.

The destruction of beam B-1 was happened under the maximum bending moment that was equal to 18.7 kN · m. The theoretical maximum bending moment for this beam was 17.06 kN · m.

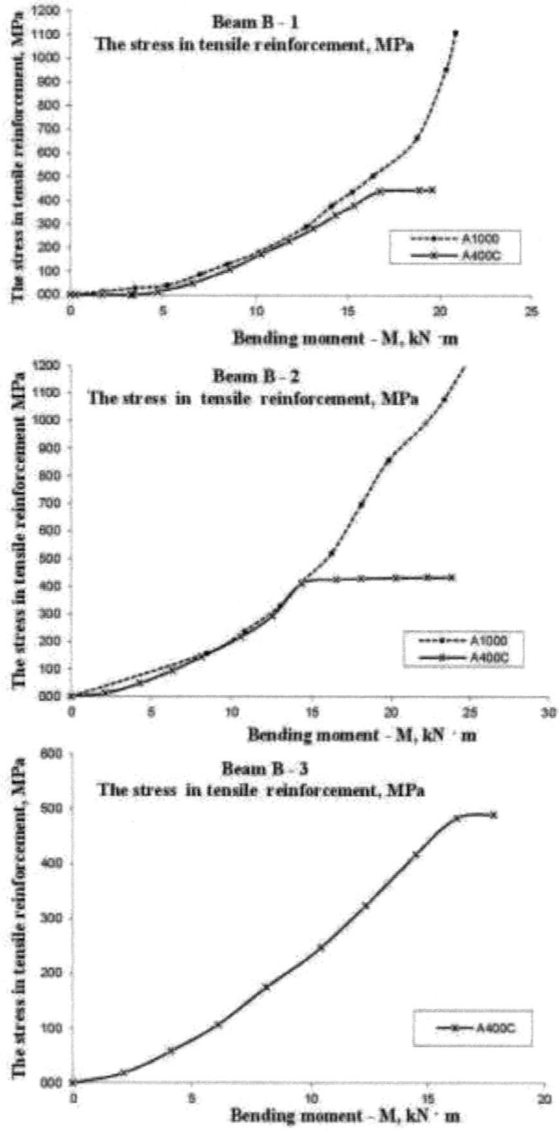

Figure 5. The stress diagrams in tensile reinforcement for beams B-1, B-2 and B-3

Table 5. The experimental and the theoretical values of bearing capacity of test beams and experimental loading of its physical destruction

The type of beams	Experimental moment M_1 during the beginning of the yielding flow of the reinforcement A400C, [kN · m]	The bearing capacity of reinforced masonry beams		
		Experimental moment M_{max} during under the physical destruction, [kN · m]	Theoretical moment by [10] M_N, [kN · m]	$(M_{max}-M_N)/M_{max}$, %
B-1	18.6	21.9	20.71	5.4
B-2	15.2	24.7	23.61	4,6
B-3	-	18.7	17.06	8.8

IOP Conf. Series: Materials Science and Engineering 1021 (2021) 012018 doi:10.1088/1757-899X/1021/1/012018

Conclusions

By theoretical and experimental studies were determined that the reinforced masonry beams with combined reinforcement by high-strength and normal steel continued to perceive the loading until the stress in the reinforcement A1000 do not reach the conventional yield strength.

Besides that, the physical destruction of the beam was happened after the beginning of the yielding flow of high-strength reinforcement A1000 and under the condition of sufficient reinforcement in the zone of action of maximum shears, in other words by total exhaustion of the summary bearing capacity of the reinforcements A400C and A1000.

The obtained experimental studies of bearing capacity for the reinforced masonry beams with combined reinforcement proved that the theoretical calculation of these beams by [10] is satisfactorily consistent with the bearing capacity of these structures (the difference is till 8.8 %).

The crack formation in reinforced masonry beams with combined reinforcement generally occurred in the joints between the bricks.

References

[1] Kramarchuk A P, Ilnytskyy B M, Bobalo T V and Lytvyniak O Ya 2020 Bearing capacity of reinforced masonry beams with combined reinforcement *Actual problems of Engineering mechanics: proceedings of VII International conference* (Odessa) pp 182–186

[2] IInes Fayala, Oualid Limam and Ioannis Stefanou 2016 Experimental and numerical analysis of reinforced stone block masonry beams using GFRP reinforcement *Composite Structures*. Volume 152 pp 994-1006

[3] IIlaria Cancelliere, Maura Imbimbo and Elio Sacco 2010 Experimental tests and numerical modeling of reinforced masonry arches *Engineering Structures*. Volume 32: Issue 3 pp 776–792

[4] Julia Nowak and Edyta Plebankiewicz 2018 Multicriteria assessment of selected types of ceilings *MATEC Web of Conf.* **219** 04011

[5] Storozhenko L I and Zaslavets O O 2017 Experimental research of stone constructions firmness reinforced with steel profiles *Collection of scientific works (branch machinery construction, building). (Yurij Kondratiuk Poltava State Technical University)* Issue 8 pp 38-41

[6] Babych V E and Karavan M S 2010 *Projecting of stone and stone reinforced constructions* (Rivne)

[7] Rotko S V, Uzhegova O A and Zadorozhnikova I V 2010 *Calculation of stone and reinforced stone constructions* (Lutsk) p 355

[8] Vakhnenko P F 1990 *Stone and reinforced stone constructions* (Kyiv) p 184

[9] Pershakov V 2009 *Reinforced Concrete and Stone Structures* (Kyiv) p 304

[10] DSTU B V.2.7-23-95 Building mortars. General technical conditions (Kyiv: Min. bud-va, arkhit. ta zhytlovo-kom. hosp) p 40

[11] DBN V.2.6-162:2010 Stone and reinforced stone constructions. Design rules (Kyiv: Min. bud-va, arkhit. ta zhytlovo-kom. hosp) p 100

[12] DSTU V.2.6-156:2010 Structures of buildings and edifices. Concrete and reinforced concrete structures (Kyiv: Min. bud-va, arkhit. ta zhytlovo-kom. hosp) p 123

[13] DSTU B EN 1052-1:2011 Testing methods of stone masonry. Part 1. Finding out the firmness by compression. Design rules (Kyiv: Min. bud-va, arkhit. ta zhytlovo-kom. hosp) p 19

IOP Conf. Series: Materials Science and Engineering 1021 (2021) 012019 doi:10.1088/1757-899X/1021/1/012019

Methodical approach in determining the reliability and efficiency of urban cargo transportation taking into account the congestion of street networks

O Kutiya[1,2], A Kravcov[1], T Gorodetska[1] and O Voytov[1]

[1]Department of Transport technology and logistics, Kharkiv Petro Vasylenko National Technical University of Agriculture, Alchevskyh street 44, Kharkiv, 61002

[2]E-mail: kafedrattl.2012@gmail.com

Abstract. A methodical approach has been developed in order to assess the reliability and efficiency of urban cargo traffic, taking into account the congestion of street networks. Modelling the transportation process is carried out online and allows one to determine the parameters of the transportation process, which take into account the presence of traffic jams on the city streets. A criterion for assessing the reliability of the logistics system of cargo urban transportation has been proposed – the reliability coefficient and the integral criterion of efficiency. The criteria take into account the travel time of the vehicle along the route and the time of delays in accepting the applications for services at the logistics centre (LC), as well as the time of delays at the transport company. It has been shown that in the absence of delays in the logistic chains, the reliability coefficient is equal to one, and the efficiency criterion tends to a minimum. The role of the LC capacity in the unit costs of transport services has been determined. Insufficient capacity of the LC increases the processing time of one application, which leads to an increase in the total unit costs.

1.Introduction

The urgency of the task of forming the routes of the urban transport network of cargo transportation is confirmed by the fact that at this stage there are Internet resources that allow the collection of information on the needs of cargo transportation. At the same time, for modelling and forecasting the cargo flows in the city transport network, it is necessary to have special modules of dynamic modelling of the real-time movement of cargo volumes taking into account the load (congestion) on transportation routes. Such modelling modules for urban cargo transportation should complement the existing Internet resources, and work on their basis, using the systems of information collection, the selection of the shortest distances, and the determination of congestion of highways and the presence of traffic jams on the routes.

Based on the above, the main argument of such scientific research is the search for solutions to improve the efficiency of cargo transportation in the city. The ways of such a search are the application of the mathematical model which interacts with the modern Internet resources and allows choosing a rational route taking into account the congestion of highways and streets of the city at different hours of the working day. The practical significance of such research is to increase the reliability and reduce the cost of transport services within the city.

2. Analysis of literature data and the problem statement

The authors of the work [1] the analysis of tendencies of development of city freight transportations is executed and the criterion of economic efficiency is offered. By means of the developed criterion the method of an estimation of cargo transportations of the city on the basis of the information on cargo flows is presented. Through the creation of an information matrix, tariffs for urban freight transportation are justified.

Analysis of works [2, 3] allows us to draw a conclusion about the prospects of the logistics approach in the development of freight models. In work [2] the results of formalization of the process of functioning of various production and transport systems are presented. In work [3] presents a mathematical model of the logistics system, which allows to rationalize the technology of interconnected enterprises based on the development of a set of technological and management solutions. In work [4] the analysis of the process of functioning of the transport and storage complex is carried out and the criterion of efficiency is formalized. This indicator represents the specific costs of cargo handling and takes into account the costs of the cargo owner associated with the downtime of cars waiting to be serviced.

Authors of works [5, 6] the results of forecasting the delivery time of goods in the city network are presented. The prospects of using IT systems for monitoring vehicles on the route are noted, which is substantiated in the work [7]. It is proved that this approach increases the informativeness of monitoring and allows to perform the forecast for transport services. This direction was further developed in the work [8], where an information model was developed, which was created on the basis of artificial neural networks. The simulation results show that the availability of information and communication systems is an important component of urban freight transport. A similar conclusion was made in the work [9], where it is emphasized that the exchange of information on the density and intensity of traffic on the streets of the city will increase the level of transport services.

Works [10, 11] devoted to assessing the reliability of transport services. Based on the analysis of transport networks, the authors concluded that it is advisable to evaluate them by comparing the criterion of reliability. Works are devoted to modelling of city motor traffic flows [12, 13]. They used the concept – "resistance to movement" on a separate section of the street. This approach allows you to develop a mathematical model of reducing the speed of vehicles in some parts of the city streets where there are traffic jams.

Methods of modelling dynamic traffic flows of urban freight traffic taking into account the congestion of streets in real time based on the use of available Internet resources need further development and improvement. The received forecast will allow to substantiate a rational route of delivery of cargo in real time thereby to increase reliability and efficiency of delivery.

3. The purpose and tasks of the study

The purpose of the study is to develop a methodical approach to assessing the reliability and efficiency of urban cargo transportation, taking into account the congestion of streets, online.

4. The methodical approach in conducting the research of urban cargo transportation

The mathematical models in the form of the second-order differential equations were chosen as a methodical approach in conducting the research. Such equations have proved themselves well in technical fields and have a well-founded apparatus of solution. The chosen method allows us to study the processes that are functions of time.

The target function of choosing rational routes for the urban cargo transportation is to increase the reliability of cargo delivery (in the exact time) and to reduce transportation costs. The factor that changes in the decision-making process is the congestion of the city streets, which will be taken into account by the parameter – the resistance of the route R_r. The method of calculating the resistance of the route is given in work [14, 15].

The difficulties in choosing rational urban routes taking into account the real congestion of city highways are stipulated by the fact that such tasks do not currently have formal methods of solution. For example, the classic transport problem allows one to determine the shortest route. However, in the city, the shortest route is not always rational owing to the uneven congestion of highways. There may be traffic jams on the shortest route, which will reduce the speed of vehicles and increase delivery time and delivery costs.

The process of "manual" route selection from those options that are considered and analysed in the logistics centre by the way of search and comparison, takes into account the quantitative rather than qualitative side of the process. Therefore, for the implementation of modelling, forecasting and quality management of urban transportation, the problem should be formalized in the form of developing a dynamic mathematical model.

5. The results of the research on transport processes of urban cargo transportations

The procedure of the methodical approach during the choice of rational routes of urban cargo transportations with the maximum reliability and the minimum expenses for transportation can be broken down into the following stages.

The first stage of modelling allows choosing a route with minimal resistance and determine the time of delivery t_d, i.e. the time of following the route, as well as the quality factor of the route Q_r. The physical content of the quality factor of the route and the formula for calculation are given in works [14, 15].

This allows choosing from several possible routes, using the Internet resource videoprobki.ua, the optimal value of $Q_r \rightarrow \max$. The value of the quality factor of the route $Q_м$ can serve as a criterion for choosing the optimal route of cargo traffic in the city network, as it takes into account the material, information and energy flows. The route is chosen on the basis of information from the Internet resources Google Maps and videoprobki.ua. The peculiarity of the proposed criterion Q_r, in comparison with the known ones is that it is determined online.

The second stage of modelling takes into account the inertia of the system in decision-making and the process of following the route, which allows determining the time of delays in the logistics system (LS).

The sum of delivery time and delay time determines the total time of executing the order for the transport service:

$$t_{\Sigma} = t_{dc} + t_d, \tag{1}$$

where t_{Σ} is the order execution time (hour); t_{dc} is the time of cargo delivery (hour); t_d is the time of delays in the LS (hour).

The author of work [14] obtained a second-order differential equation, which allows representing the transition process in the logistics system of urban cargo transportation from the moment of receipt of the application for transport services to the moment of the delivery of goods to the customer:

$$T_{LS}\frac{d^2m}{dt^2} + 2d_{LS}\cdot Q_r \cdot T_{LS}\frac{dm}{dt} + m\cdot[\,K_{LC}\cdot K_{TC}\,]^{-1} + m\cdot K_{LS} = J, \tag{2}$$

The time constant of the logistics system T_{LS}, which characterizes the inertia of the system, is determined by the expression:

$$T_{LS} = (T_{LC}\cdot T_{TC})^{1/2}\cdot(K_{LC}\cdot K_{TC})^{-1/2}, \tag{3}$$

where T_{LC} is the time constant of the logistics centre (LC), which takes into account the inertia (delay) in the processing of applications (hour); T_{TC} is the time constant of the transport company (TC), which takes into account the inertia (delay) in accepting applications for transport services (hour); K_{LC} is the coefficient of sensitivity of the logistics centre to the receipt of applications; K_{TC} is the coefficient of sensitivity of the transport company to the receipt of applications for transport services from the logistics centre; d_{LS} is the coefficient that characterizes the absence or presence of an oscillatory process in the LS.

The formulas for calculating the above time constants and coefficients are presented in work [14].

The right part of equation (2) is the order for the required productivity of the logistics system, t/h. The left part of the equation is the reaction of the *LS* to the order J.

The increase of the time constants of T_{LC} and T_{TC} makes the process less sensitive (less susceptible) to the order for transport service *J*. The process of processing the applications and the delivery of goods increases in time.

The gain coefficients of K_{LC} and K_{TC} characterize the sensitivity of *LC* and the transport company to the receipt of applications.

The solution for the above stated differential equation (2) is the expression:

$$m(t) = m \cdot [(1 - exp(-d_{LS} \cdot Q_r \cdot T_{LC}^{-1} \cdot t)) \cdot cos f \cdot t + A \cdot sin f \cdot t], \qquad (4)$$

where $m(t)$ is the function of the delivery of a given weight of cargo in time; t is the current time (hour); f is the oscillation frequency, dimension (1/hour), is determined by the expression given in work [14]; A is the amplitude of oscillations, measureless value, is determined by the expression given in work [14].

The big value of the time constant T_{LC} suggests that the logistics centre and the transport company have a great inertia in accepting applications for processing, and, as a consequence, will have a big delay time.

The dynamic model (3) allows determining the total time of the transport service taking into account various factors. The model works online and interacts with Google Maps and videoprobki.ua resources. This is the difference between the developed model and the previously known ones. The results of the dynamic modelling of the process of cargo delivery on urban routes, formula (4), with changing the weight of the cargo and the presence of congestion on the route, are presented in work [14].

The value of the total transport service time t_Σ is given by the joint use of the developed models.

The first allows one to determine the time of the vehicle on the route t_{dc}, taking into account the quality factor of the route Q_r. The second allows one to determine the delays in making decisions on the service t_d.

The values of t_{dc}, t_d, t_Σ obtained above allow determining the reliability of the logistics system of urban cargo transportation. According to works [10, 11], it can be stated that the reliability of the *LS* of cargo transportation is evaluated as the ratio of the following indicators of the process. In the numerator – of the mathematical expectation of the time spent on the delivery of goods along the route. In the denominator – of the mathematical expectation of the total time spent on the delivery of goods along the route and the time for delays that occurred during the delivery of goods along the route.

In accordance with the conclusions of work [15] let us write an expression for assessing the reliability of the *LS* when executing n is applications that were performed in the *LS*:

$$K_R = (\sum_{i=1}^{n} t_{dc,i} + \sum_{i=1}^{n} t_{d,i})^{-1} \cdot \sum_{i=1}^{n} t_{dc,i}, \qquad (5)$$

where K_R is the coefficient of reliability of the LS, a measureless value; n is the number of applications for transport services.

Based on expression (5), the reliability coefficient is less than one and becomes equal to one only when the total delays are zero. In the presence of even insignificant delays K_R is less than one. The obtained measureless parameter of the K_R, which has the physical content of the part of the execution of the application in the exact time, can be a criterion that assesses the reliability of the logistics system functioning. The present criterion will be used while determining the expenses for the urban transportation.

An important indicator of the work of the logistics system of urban cargo transportation is the value of expenses. The expenses characterize the effectiveness of the *LS*, and include the following components.

IOP Conf. Series: Materials Science and Engineering 1021 (2021) 012019 doi:10.1088/1757-899X/1021/1/012019

Based on works [1-4], which are devoted to the economic criteria, one can conclude that it is more appropriate to apply specific criteria or parameters that take into account the expenses per ton of the cargo transported, the measure of UAH/t.

The first component of the expenses is the operating costs, which are formed by the adopted tariff for the transportation of goods. This tariff is determined by the market and has a measure of UAH/km.

The expression using which one can determine the unit costs associated with the market value of the tariff, can be represented as follows:

$$B_1 = l_r^2 \cdot T_{tr} \cdot \omega \cdot [\, m \cdot \upsilon_{tech} \cdot K_R \,]^{-1}, \tag{6}$$

where B_1 is the expenses for transport services, which depend on the tariff (UAH/t); l_r is the length of the route (km); T_{tr} is the tariff for transportation (UAH/km); ω is the frequency of the receipt of applications for service in the logistics centre (LC) (1/hour); m is the weight of goods, t; υ_{tech} is the technical speed of the vehicle on the route (km/hour); K_R is the reliability factor, formula (5).

As expression (1) shows, the increase in the weight of the transported cargo m, the technical speed on the route υ_{tech} and the reliability factor K_R will help to reduce the unit costs of the transportation.

The second component of the expenses is the operating costs associated with the consumption of fuel by vehicles during the execution of the order. Based on works [1-4], we can write the expression:

$$B_2 = N_v \cdot \upsilon_{tech} \cdot t_\Sigma \cdot C_f \cdot 0.01 \cdot G_v \cdot [\, m \cdot K_R \cdot \beta \cdot \gamma \,]^{-1}, \tag{7}$$

where N_v is the number of vehicles which are on duty; t_Σ – the total time of the order execution taking into account delays (hour); C_f is the cost of one litre of fuel (UAH/l); G_v is the fuel consumption by the vehicle in the city movement cycle (l/100 km); β is the coefficient that takes into account the presence of idling (the coefficient of the mileage utilization); γ is the coefficient of utilization of the loading capacity of the vehicle.

The third component of the expenses is the expenses that take into account the drivers' salaries, the vehicle maintenance costs and the amortization, which depend on the initial cost of the vehicle.

The expression for the calculations of the third component of the expenses can be represented as follows:

$$B_3 = N_v \cdot K_{L-Unl} \cdot t_\Sigma \cdot R_{dr} \cdot [\, m \cdot K_R \,]^{-1} + 0.00041 \cdot C_v \cdot N_v \cdot [\, m \cdot K_R \,]^{-1}, \tag{8}$$

where $K_{L\text{-}UnL}$ is the coefficient that takes into account the increase in the total delivery time of cargo for loading and unloading operations, $K_{L\text{-}UnL}$=1.15–1.3; R_{dr} is the hourly rate of the driver's salary (UAH/hour); C_v is the initial cost of the vehicle (UAH).

The coefficient of 0.00041 takes into account the expenses for the vehicle maintenance, which are equal to 5% of the initial cost of the vehicle C_v and the amortization costs, which are equal to 10% of C_v per year, attributed to one day of operation. The total, integrated unit costs for urban cargo transportation will be expressed by the following formula:

$$B = B_1 + B_2 + B_3, \tag{9}$$

The analysis of the obtained formulas (6) – (9) allows concluding that the unit costs of urban transport services depend on the distance of the transportation, the cargo weight, the reliability factors, the mileage and load capacity utilization, the total time spent on duty, the tariff for the transportation, and fuel use. The presented integral value of unit costs, formula (9), can act as an economic criterion of urban cargo transportation in the process of solving optimization problems for the choice of routes. This criterion should be kept to a minimum.

6. The research results

We will model the impact of various parameters of the transport process on the total unit costs, which reflect the efficiency of urban cargo transportation.

The role of the *LC* in the general part of unit costs for transport services is represented by the dependencies in the figures 1–2.

The analysis of the dependencies presented in figure 1 allows estimating the degree of influence of the weight of the transported cargo and the time required to process one application in the *LC*, t_{LC}. With the increase of t_{LC} unit costs increase by 1.16-1.52 times, and the change in the weight of the transported cargo has an optimum that corresponds to $m = 10$ t and occurs at higher values of t_{LC} (low capacity of the *LC*).

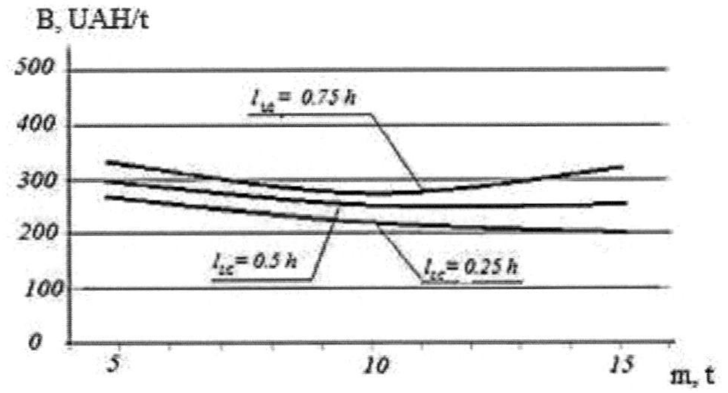

Figure 1. The dependencies of the change of the value of unit costs *B* of the logistic system at various weight of the transported cargo *m* and the time of registration of one application in the *LC*, t_{LC}

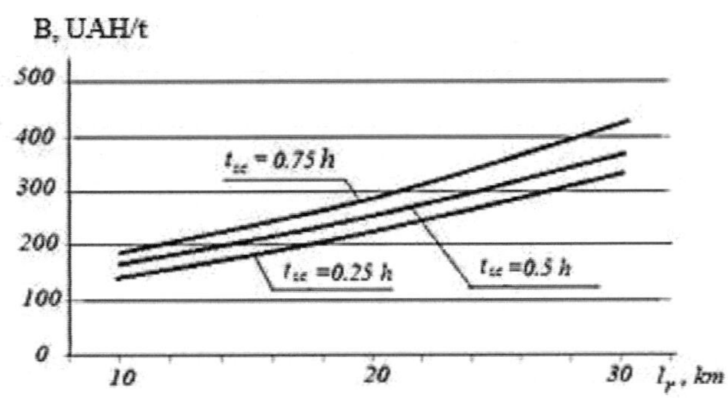

Figure 2. The dependencies of the change of the value of unit costs *B* of the logistic system at various length of the route l_r and the time of registration of one application in the *LC*, t_{LC}

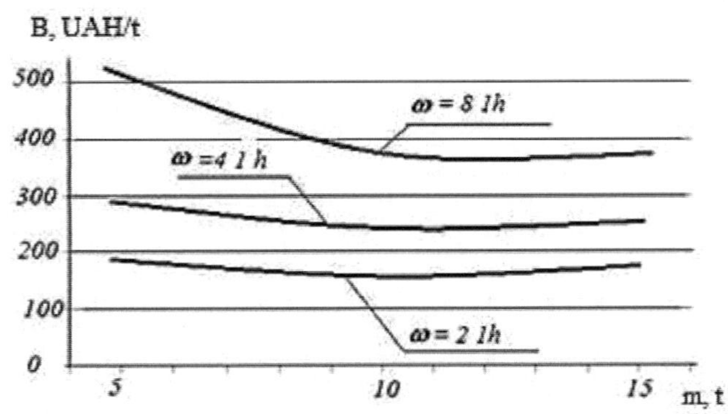

Figure 3. The dependencies of the change of the value of unit costs *B* of the logistic system at various weight of the transported cargo *m* and the frequency of receipt of applications in the *LC*, ω

The nature of the influence on unit costs of the route length l_r and time t_{LC} is presented in figure 2. From the dependencies it follows that the degree of the influence of t_{LC} on the value B is similar to the above dependencies, and the length of the route increases unit costs by 2.3-2.55 times.

The influence of the frequency of requests ω for service in the *LC*, which is represented by the dependencies in figure 3, allows concluding that when changing the weight of the transported cargo there is an optimum, which is characteristic of m=10t. The increase of the number of service requests per unit time increases B, because it requires an increase in the capacity of the *LC*.

The provided results of modelling the degree of influence of the *LC* on unit costs, Figures 1–3, suggest that the capacity of the logistics centre must be managed. The increase of the capacity of the *LC* will reduce the service time of one application t_{LC}, which will reduce unit costs for the transport services. In this case, the increase of the length of the route clearly increases unit costs, and the weight of the transported cargo has an optimum.

The degree of the influence on the value of unit costs of the coefficients of the mileage utilization β and the coefficient of the carrying capacity utilization γ, the change of which takes place during urban cargo transportation, is represented by the dependencies in the figures 4 –5.

The decrease of the mileage utilization coefficient from the value of Figures 1–2 β=1 to β=0.5 increases unit costs by 1.13-1.22 times. Thus, the influence of the weight of the transported cargo, Figure 4, has an optimum. It should be noted that the optimum exists for $\beta < 1$.

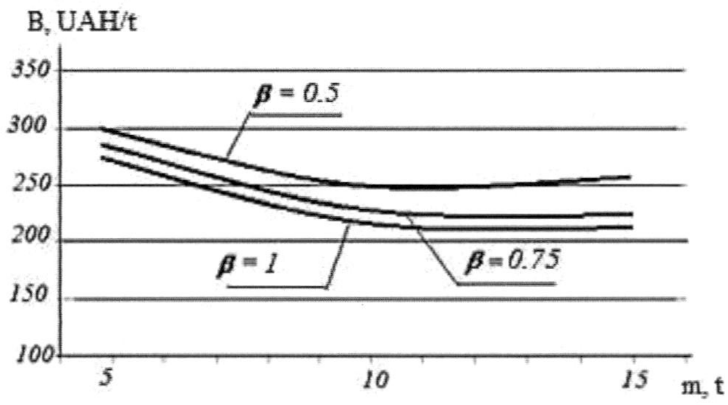

Figure 4. The dependencies of the change of the value of unit costs B of the logistic system at various weight of the transported cargo m and the mileage utilization coefficient β

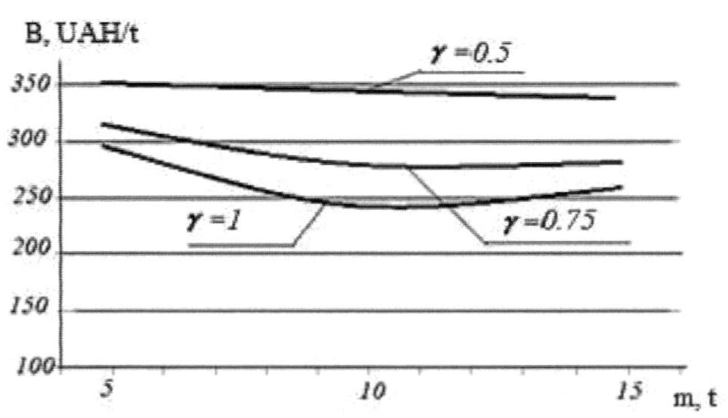

Figure 5. The dependencies of the change of the value of unit costs B of the logistic system at various weight of the transported cargo m and the coefficient of the carrying capacity utilization of the vehicle γ

The dependencies that characterize the degree of the influence of the carrying capacity coefficient on unit costs, Figure 5, are similar. The decrease of the value of γ from 1 to 0.5 increases the unit costs by 1.17-1.36 times. In this case, the increase in the weight of the transported cargo has an optimum, which is clearly expressed at γ=1 and becomes less expressed with the decrease of γ from 1 to 0.5, figure 5.

7. The analysis of research results

The dependencies given above allow one to estimate the degree of the influence of the weight of the transported cargo and the time required to service one application in the *LC*. With the increase of t_{LC} unit costs increase by 1.16-1.52 times, and the change in weight of the transported cargo has an optimum.

The nature of the influence on unit costs of the length of the route l_r suggests that the length of the route increases unit costs by 2.3-2.55 times.

The influence of the frequency of the receipt of service requests ω in the *LC* allows one to conclude that when changing the weight of the transported cargo there is an optimum.

The decrease of the mileage utilization coefficient from the value of β=1 to β=0.5 increases unit costs by 1.13-1.22 times. In this case, the influence of the weight of the transported cargo has an optimum.

The decrease of the value γ from 1 to 0.5 increases the specific unit costs by 1.17-1.36 times. In this case, the increase in the weight of the transported cargo has an optimum, which is clearly expressed at $\gamma = 1$.

8. Conclusions

The modelling of the influence of various factors and operating parameters of the transport process of urban cargo transportation within the accepted restrictions suggests that unit costs of transport services *B*, UAH/t, clearly increase with the increase of the route length, however, in this case, they have an optimum when changing the weight of the transported cargo.

It has been established that the existence of the optimum is influenced by the mileage utilization coefficient and the vehicle carrying capacity utilization coefficient.

The influence of the logistics centre (logistics centre capacity) on unit costs of transport services has been shown. The insufficient capacity of the *LC* increases the time of registration of one application, which leads to the increase in total unit costs.

References

[1] Zhang R M, Huang L 2017 Application of the freight rate on freight flow forecast. *Advances in Transportation Studies* **3** pp 61-68

[2] Shramenko N Y 2017 The methodological aspect of the study feasibility of intermodal technology of cargo delivery in international traffic. *Scientific Bulletin of National Mining University* **4 (160)** pp 145–150

[3] Shramenko N Y and Shramenko V O 2018 Mathematical model of the logistics chain for the delivery of bulk cargo by rail transport. *Scientific Bulletin of National Mining University* **5(167)** pp 136–141

[4] Shramenko N Y 2015 Effect of process-dependent parameters of the handling-and-storage facility operation on the cargo handling cost. *Eastern European Journal of Enterprise Technologies* **5/3 (77)** pp 43-47

[5] Yuan Y, Zhao Q Y, Wan X Y 2018 Evaluating transportation information service products through artificial neural networks. *Advances in Transportation Studies* **3** pp 59 – 68

[6] Sun S, Liu C 2016 Application of improved storage technology in Intelligent Transportation System. *Advances in Transportation Studies* **3** pp 51 – 60

[7] Wang Z, Goodchild A, McCormack E 2016 Freeway truck travel time prediction for freight planning using truck probe GPS data. *European journal of transport and infrastructure research* **16(1)** pp 76-94

[8] Combes F, Tavasszy L 2016 Inventory theory, mode choice and network structure in freight transport. *European journal of transport and infrastructure research* **16(1)** pp 38-52

[9] Meyer A, Sejdovic S, Glock K, Bender M, Kleiner N, Riemer D 2018 A disruption management system for automotive inbound networks: concepts and challenges. *EURO Journal on Transportation and Logistics* **7(1)** pp 25–56
[10] Gao J, Sun J, Shi Q Z, Liu F S 2015 A comparative reliability evaluation method for transportation network planning and design. *Advances in Transportation Studies* **3** pp 55–64
[11] Mishra S, Tang L, Ghader S, Mahapatra S, Zhang L 2018 Estimation and valuation of travel time reliability for transportation planning applications. *Case Studies on Transport Policy* **6(1)** pp 51-62
[12] Danchuk V D, Krivenko V I, Olijnik R V, Taraban S M 2010 Elektrotehnichna model' doslidzhennja transportnih potokiv. *Visnik NTU.* **21(2)** pp 28–32
[13] Danchuk V.D, Krivenko V I, Olijnik R V, Taraban S M 2015 Elektrichne modeljuvannja mis'kih avtotransportnih potokiv. *Visnik NTU «HPI»* **46(1155)** pp 109–114
[14] Kut`ya O 2019 Rozrobka dinami`chnoyi modeli` zatrimok prijnyattya ri`shen` u logi`stichnikh lanczyugakh mi`s`kikh vantazhnikh perevezen`, *Tekhni`chnij servi`s agropromislovogo, li`sovogo ta transportnogo kompleksi`v.* **16** pp 63-72
[15] Vojtov V, Kutiya O, Berezhnaja N, Karnaukh M and Bilyaeva O 2019 Modeling of reliability of logistic systems of urban freight transportation taking into account street congestion, *Eastern-European Journal of Enterprise Technologies.* **4/3 (100)** pp 15-21

IOP Conf. Series: Materials Science and Engineering 1021 (2021) 012020 doi:10.1088/1757-899X/1021/1/012020

Reconstruction of Shallow Foundations Using Peracetic Silicate Solutions

H M Levenko[1,2], V A Aleksandrovych[1]

[1] O.M. Beketov National University of Urban Economy in Kharkiv, 17, Marshal Bazhanov Street, Kharkiv, 61002, Ukraine

[2] Ganna.Levenko@kname.edu.ua

Abstract. The modern world is facing the acute issues of environmental pollution. Recently, they have begun to come out on top in the global relations and in the cooperation of various global business entities. The states allocate significant funds both for measures to prevent environmental pollution and to combat their consequences. The operating experience of many enterprises has shown that the issues of reconstruction of the foundations of existing buildings and structures are currently acute and urgent. One of the most effective types of soil consolidation is stabilizing with various binders, such as silication, bituminous grouting, tarring, cementation. As economic exploitation of soils intensively increases, various pollutants, both organic and nonorganic, affect them by deteriorating their composition, structure and properties. Different types of soils react differently to pollutants, some of them are more "sensitive" to them while others are less "sensitive". Pollution and environmental deterioration of soils due to the uncontrolled accumulation of various industrial wastes in the soils are the main problems. In this regard, under current conditions, along with the elimination of sources of pollution, it is necessary to develop new methods and technologies to prevent pollution. Physiochemical methods can solve these problems.

1. Introduction

As a rule, foundation consolidation is performed with vertical or inclined reinforcing elements. This is due to the technical characteristics of the construction equipment used, as well as the design requirements, while the consolidation of the soil layer is carried out at a sufficiently large depth. However, in cases of reconstruction of buildings and structures, additional superstructure, confined areas or prevention of the development of active deformations, it is reasonable to stabilize the soil at a shallow depth directly under the foundation in the zone of base active deformations [1–2].

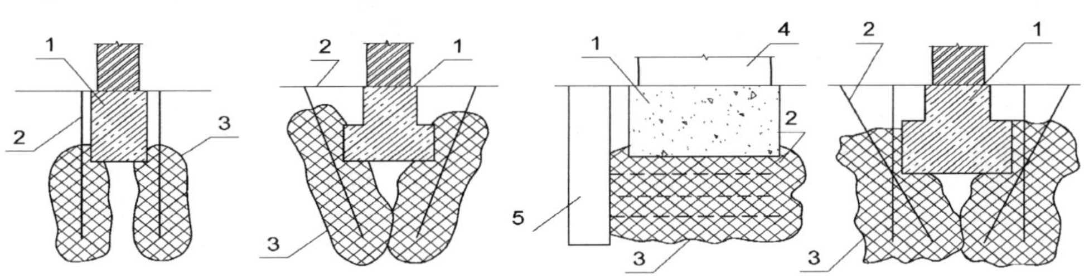

Figure 1. Schematic of the possible location of the injectors when stabilizing the foundations: 1 - reconstructed foundation, 2 - injector; 3 - stabilized soil body; 5 pit.

– When carrying out work on the stabilization of soil bodies, it is necessary to consider the environmental state of the plots to be stabilized, to strive to minimize the introduction of chemicals required for the work.

– The proposed stabilization method should ensure the stabilization of deformations as the bases of the system "base - foundation - structure".

– After carrying out the work, the mechanical and deformation indicators of the stabilized soil body should increase.

– The integrity of the stabilization the contaminated soil body should be ensured throughout its entire scope.

The condition for the choice of a certain stabilization technology is the condition for the fulfilment of all the listed requirements at a time. It is the most optimal method for finding and choosing the most effective method for chemical stabilization of soil bodies contaminated with industrial wastewater [3].

2. Basic material and results

Three methods of soil stabilization are most commonly used: boring and mixing method, jet silication, jet and mixing silication. The most widespread domestic method is boring and mixing, due to the simplicity of the mechanisms for its implementation.

The method consists in mechanical excavation and mixing of soil with a binder, which is fed in the form of a ready-made solution. Usually the binder is mixed with the soil to be stabilized directly in the well without excavation. The advantage of this method is the economy of the binding solution and the constant cross-section of the stabilized element in accordance with the diameter of the operating auger.

Regarding the jet silication method, it is possible to use one- and two-component binding solutions. It depends on the initial soil conditions and the required final stabilization conditions.

The studies on soil stabilization in two ways: by boring and mixing method and jet silication, were carried out in this paper.

The environmentally friendly compositions of silicate and peracetic gels were used in the studies, as can be seen in Table 1.

Table 1. Environmentally friendly compositions of silicate and peracetic gels.

Density of sodium silicate solutions, g/cm^3	Time intervals for the beginning of gelation, min	Boundaries of the volumetric ratios of the original components "sodium silicate/peracetic acid"	pH interval
1.10	1 - 60	4.5 - 8.0	6.0-8.0
1.15	1 - 60	6.5 - 10.5	
1.20	1 - 60	7.5 - 13.5	
1.25	1 - 60	12.5 - 16.0	

For the jet technology, the characteristic of the penetrating ability of the stabilization solution is the density of sodium silicate solutions. By reducing the density of the solutions used, it is possible to expand the range of soils to be stabilized in terms of permeability and particle size distribution.

As a result of testing the samples (see Figure 2) of the stabilized soil, it was found that with an increase in the density of sodium silicate and the concentration of peracetic acid, the calculated resistance of the soil, R, increases.

(a) (b) (c)

Figure 2. Soil compression testing.
(a) soil samples before testing; (b) sample during testing; (c) sample after testing.

The mechanical characteristics of soils after stabilization, which were determined as a result of laboratory tests, are given in Table 2.

Table 2. Mechanical characteristics of soils after stabilization

Mechanical characteristics of soils after stabilization			
φ, *deg*	*E*, *MPa*	*C*, *kPa*	R, kPa
23.9-39.8	37.9-42.0	38.2-39.2	220-560

After processing the results obtained, the diagrams of the dependence of the strength of the stabilized soil on the density of the applied sodium silicate solutions were plotted (see Figure 3).

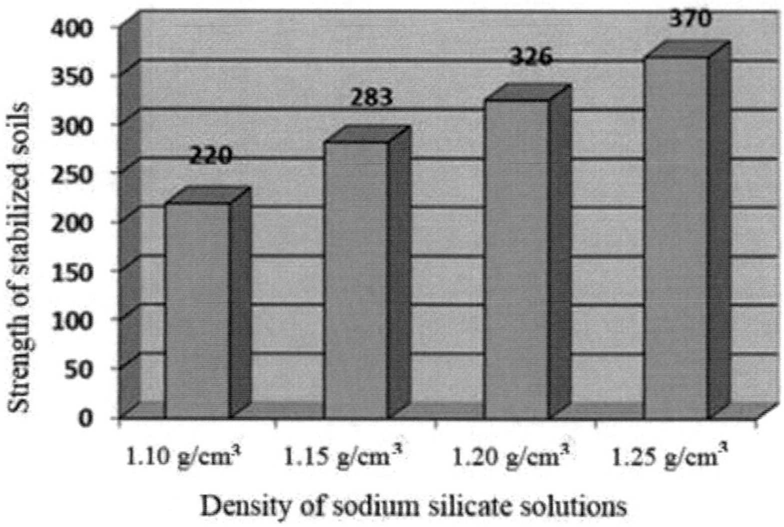

Figure 3. Dependence of the strength of the stabilized soil on the density of sodium silicate solution.

Regarding the radius of soil stabilization during jet silication, it depends on two components: soil density and stabilization solution density.

The physical and chemical process of soil stabilization by silication is based on the sufficient penetrating ability of a silicate solution, which has a low viscosity with the fast release of a silicic acid gel film, which cements the soil.

The research and comparative parallel analysis of the formulation of single-solution, two-component silication with solutions with a density of 1.05 g/cm^3; 1.10 g/cm^3; 1.15 g/cm^3; 1.20 g/cm^3 when stabilizing four groups of soil samples are carried out.

Based on the data obtained, the dependence of the stabilization radius on the density of sodium silicate solutions was plotted (see figure 4).

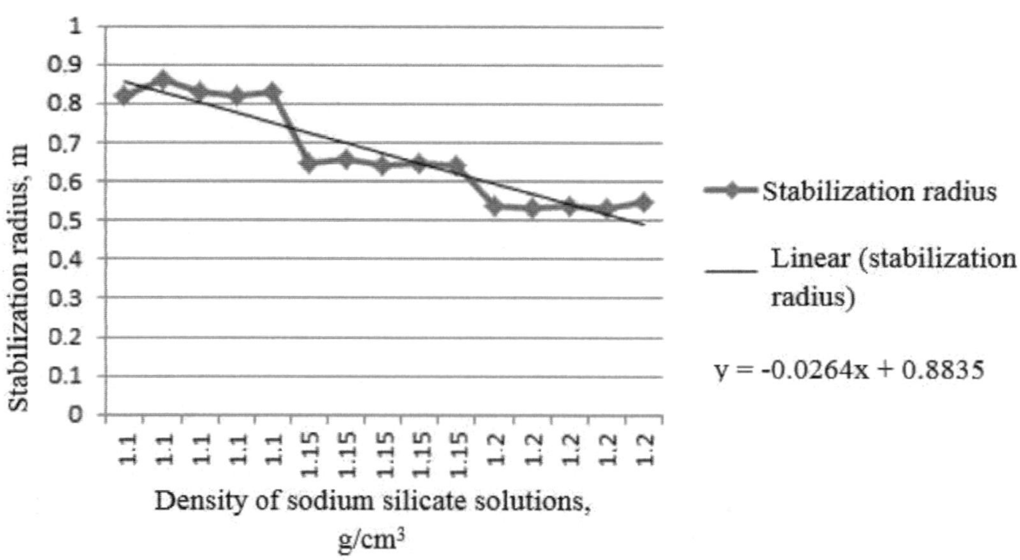

Figure 4. Dependence of the stabilization radius on the density of sodium silicate solutions.

The plot shows that with a decrease in the density of sodium silicate solutions, the stabilization radius increases.

Also, the stabilized soil samples were tested for durability and resistance to aggressive media.

Tap water with ion concentration of pH = 8.1-8.3 and 3% peracetic acid solution were used as an aggressive medium.

After 30 days and 90 days, the test samples were extracted and tested for compressive strength.

The results of the tests performed are shown in table 3.

Table 3. Compressive strength of stabilized soil after being exposed to aggressive media.

Studied soils	Aggressive medium	Ultimate compressive strength R_{compr}, kPa, when exposed to aggressive medium, days.	
		30	90
Soil I	Water pH = 8.3	288	238
	Peracetic acid 3%	301	337
Soil II	Water pH = 8.3	199	143
	Peracetic acid 3%	286	217
Soil III	Water pH = 8.3	196	140
	Peracetic acid 3%	216	185
Soil IV	Water pH = 8.3	342	290
	Peracetic acid 3%	348	335

The laboratory tests carried out indicate that the stabilized soil samples are resistant to aggressive media for a longer period.

This is primarily due to the resistance to acid aggression of the peracetic acid gel, which is the main binding material in the studied formulation.

The strength of the base soils increases, their mechanical characteristics improve after the stabilization by two methods, jet silication and boring and mixing. In some cases, the mechanical characteristics of the stabilized soil exceed their value in the natural state by several times. In clays, the soil strength R_{compr} increases by 1.81 to 3.01 times. The specific adhesion C of sandy loam increases by 9.6 times, and the deformation modulus E increases by 2.48 times. The angle of internal friction φ increases on average by 1.56 times.

3. Conclusions

Considering the above factors and based on the studies carried out, the authors recommend the use of a peracetic and silicate formulation for consolidation and reconstruction of the foundations of existing buildings and structures.

IOP Conf. Series: Materials Science and Engineering 1021 (2021) 012020 doi:10.1088/1757-899X/1021/1/012020

References

[1] Rzhanitsyn B A 1986 *Chemical stabilization of soil in construction* (Moscow: Stroyizdat) p 263
[2] Dombrovskyi Y *at al.* 2018 *DBN V.2.1-10-2018. Bases and foundations of structures. General design considerations* (Kyiv: Ukrarkhbudinform) p 107
[3] Holovko S I 2010 *Theory and practice of consolidating soil foundations by high-pressure cementation* (Dnipropetrovsk: Porohy) p 247
[4] Aleksandrovych V A, Taranov V G, Luchkovskyi I I, Plashchev S A, Kornienko N V, Areshkovych O O 2013 Structure-Soil Massif System Behavior Features Under Static and Dynamic Loads *Proc. of the 18th Intern. Conf. on Soil Mechanics and Geotechnical Engineering* (*Paris*) vol 2, ed P Delage at al (Paris: Presses des Ponts) p 1627–29
[5] Rogobete G, Tarau D, Bertici R 2011 Study of a soil with swelling and shrinking phenomena *Soil Forming Factors and Processes from the Temperate Zone* (Iaşi: Editura Universităţii "Al.I. Cuza") p 148–155
[6] Yahiro T, Yoshida H, Nishi K 1980 Sheet piles driving and h-steel pulling by high speed water jets *Proc.of the fifth international symposium on jet cutting technology* (Hannover: BHRA Fluid Engineering), p 237–250
[7] Patil B M, Patil K A 2013 Effect of industrial waste and RBI Grade 81 on swelling characteristics of clayey soil *Facta Universitatis Series: Architecture and Civil Engineering* vol 11, ed P Stanimirović (Serbia: University of Niš) p 231–236
[8] Amin E 2012 Solid-state NMR and FTIR studies of lime stabilized montmorillonitic and lateritic clays *Journal of Applied Clay Science* vol 67-68, ed F Bergaya at al (Elsevier B V) p 5–10
[9] Radhakrishnan G, Kumar A, Prasada R 2014 Swelling properties of expansive soils treated with chemicals and flyash *American Journal of Engineering Research* vol 3, ed M Sarker (Stamford: Natural State Research, Inc) p 245–250
[10] Muhanned Q W 2012 A laboratory evaluation of stabilization of silty clay soil by using Chloride Compounds *Engineering & Technology Journal* vol 30, ed R Dubey at al (Everant) p 3054–64
[11] Phanikumar B R, Radhey S S 2004 Effect of flyash on the Engineeering properties of Expansive Soil *Journal of Geotechnical and Geoenvironmental Engineering* vol 130, ed R Salgado at al (Reston: ASCE) p 764–767
[12] Zhao H, Ge L, Petry T, Sun Y 2014 Effects of Chemical Stabilizers on an Expansive Clay *KSCE Journal of Civil Engineering* vol 18, ed Lee D H (Cham: Springer Nature Switzerland AG) p 1009–17

IOP Conf. Series: Materials Science and Engineering 1021 (2021) 012021 doi:10.1088/1757-899X/1021/1/012021

Resource-saving steel hopper for loading of railway cars

Anton Makhinko[1] **and Nataliia Makhinko**[2,3]

[1]ETUAL LLC, 1, Bortnytska street, Petropavlivske, Boryspil district, Kiev region, Ukraine
[2]National Aviation University Ave Liubomyra Huzara 1, Kyiv, Ukraine

[3]Email: pasargada1985@gmail.com

Abstract. The article is devoted to the study of geometric characteristics of corrugated sheets of railway grain hopper. In particular, the problem of finding the optimal parameters of a corrugated sheet of a given length, such as the height and width of the corrugation shelves, is solved. The optimality condition in this problem is determined by the minimum mass of the corrugated sheet. Because of analytical calculations, it was found that, regardless of the angle of tip of the corrugation wall, there will always be such a value of optimum height at which the corrugation area will be the smallest. For the analysis of this parameter, a series of curves of dependence of the corrugation cross-section area on the height for different angles of tip were constructed. The study also provided concise and engineering-friendly equations to determine the optimum height and width of corrugation shelves for wall angles up to 60°. These expressions can be applied to the case of corrugated sheet with a limited number of erection joints. For other cases of jointing of the corrugated sheets with each other (complete or partial overlapping of the lower part), additional studies should be performed similar to the procedure presented.

1. Introduction

Grains are the main export products of our country. The length of transportation routes for transporting grain from different regions is quite large. Hence, the rational choice of delivery increases the overall level of profitability for this product. The freight rate of grain trucks is always high. This causes an increase in energy consumption for grain production, particularly when transporting over long distances. Consequently, it reduces the competitiveness of Ukrainian farmers, compared to foreign markets [1, 2]. In addition, road transport increases the burden on roads and their maintenance cost. Reducing the cost of the finished product is possible by reducing energy consumption and use of rail transportation. The effectiveness of grain terminals increases significantly when realizing the possibility of fast and uniform loading of cars (figure 1). Initially, it depends on the equipment and technological constructive solutions of constructions for loading. They have to meet the requirements for strength and rigidity and have an appropriate performance for temporary storage of grain. Loading structures with a loading capacity of 100 to 500 m^3 made of galvanized sheet steel are normally used as such constructions (Figure 2).

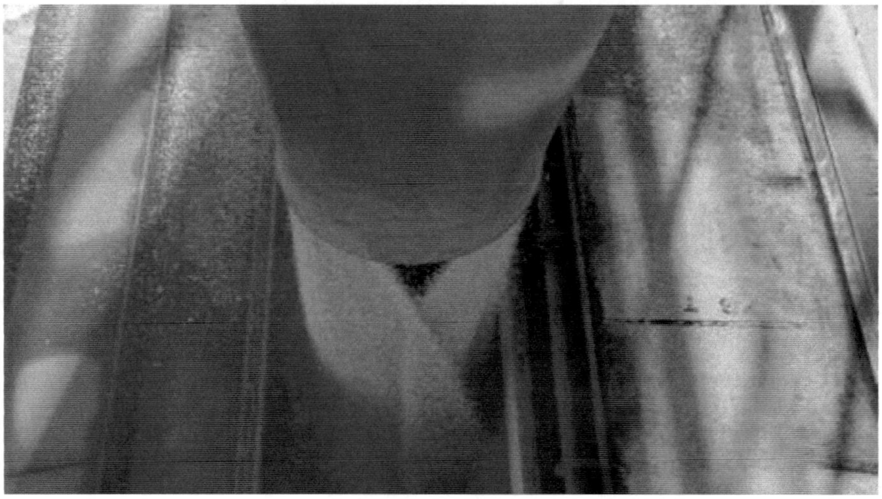

Figure 1. Loading of railway cars.

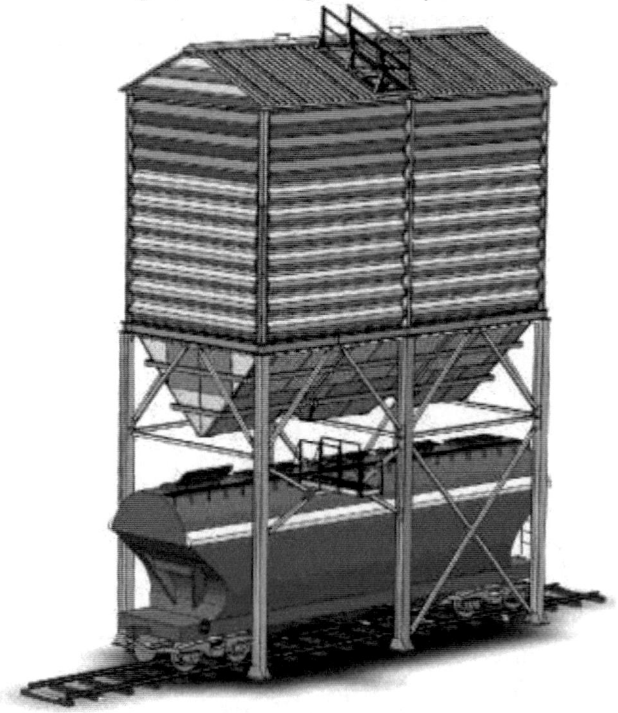

Figure 2. Steel hoppers with corrugated sheet body (ZEO Sokil LLC).

In order to decrease energy and financial losses, in the production of steel railway grain hopper, it is advisable to use standard elements in the form of corrugated steel panels. Corrugated steel panels have increased bending stiffness, therefore they less sensitive to local buckling due wind load and own local defects. The geometric characteristics of corrugated sheets of the hopper body (corrugation height and step) in different manufacturers acquire their own values because of production aspects. Quite often, these indicators are interpreted as advantages for some design solutions and are constantly compared to competitive ones. The method of simple analytical calculation of these characteristics for cylindrical silo was developed in the previous work of the authors [3].

It is worth noting that, as a whole, the cost of the hoppers is estimated by metal consumption of panels. Therefore, it is quite interesting to consider the problem of determining the minimum mass of grain hopper corrugated sheet. This will allow to obtain a resource-saving design solution for the loading hopper and increase the economic effect by reducing the metal content without losing load-bearing capacity.

The outlined direction of rational construction with corrugated sheets can be successfully applied also to other types of square hoppers [4-7], as well as to other structures, since the corrugated elements became considerable widespread in various areas of construction. They are used as part of load-bearing elements of building structures and as enclosing structures, and can combine these functions. Corrugated metal sheets are used for walls and roofs of buildings, culverts and overpasses, bridges, protruding elements of engineering structures, etc. [8-12].

2. Formulation of the problem

Corrugated sheet, which schematizes the wall of the hopper for the storage of bulk material, was loaded with pressure p_0. The cross-sectional dimensions of the sheet are determined by four independent parameters such as the height of corrugation h_p, the width of the corrugation shelves a_p, the distance between the corrugations b_p and the angle of tip of the corrugation walls β_p (figure 3).

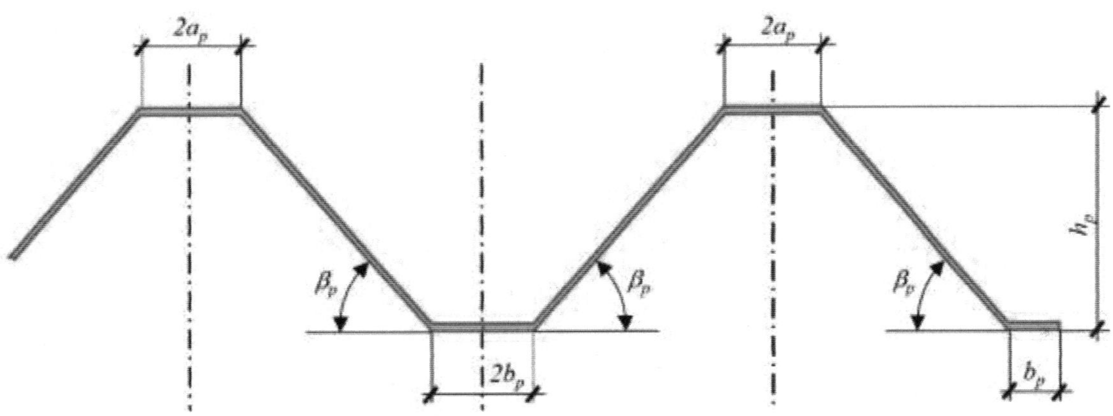

Figure 3. Geometric layout of the corrugated sheet.

Suppose that the sheet is fixed at the edges. It is necessary to find the size of corrugation h_p, $a_p = b_p$ and β_p, at which the mass of the sheet will be minimal.

3. Solution

We introduce the notation $\alpha_p = a_p/h_p$ and obtain the equations for the basic geometry of the corrugation, such as area A_p, length of horizontal projection ℓ_p, second moment of area J_p, and moment of resistance W_p. Given the symmetry of the profile relative to the vertical axis, the solution can be found for half the corrugation. Since $a_p = b_p$, the moments of resistance of the upper and lower shelves are equal. Considering previous studies [3] and skipping intermediate transformations, we write the result

$$A_p = h_p t_p \psi_A, \tag{1}$$

$$\ell_p = h_p \psi_\ell, \tag{2}$$

$$J_p = 0.333 h_p^3 t_p \psi_J , \tag{3}$$

$$W_p = 0.667 h_p^2 t_p \psi_J , \tag{4}$$

where ψ_A, ψ_ℓ and ψ_J are dimensionless factors whose values are given by equations

$$\psi_A = 2\alpha_p + \sin^{-1}\beta_p , \tag{5}$$

$$\psi_\ell = 2\alpha_p + \mathrm{ctg}\beta_p , \tag{6}$$

$$\psi_J = \left(2\alpha_p + 3\alpha_p^2 \sin\beta_p + 0.25\sin^{-1}\beta_p\right)\left(1 + 2\alpha_p \sin\beta_p\right)^{-1} . \tag{7}$$

To ensure the sheet strength, the equality should be maintained

$$M_p R_y^{-1} = W_p , \tag{8}$$

where M_p is the bending moment in the design section, which is calculated as for a single beam from linear load $q_0 = p_0 \ell_p$

$$M_p = p_0 \ell_p L_0^2 \alpha_0^{-1} = h_p p_0 L_0^2 \alpha_0^{-1} \psi_\ell = h_p p_0 L_0^2 \alpha_0^{-1}\left(2\alpha_p + \mathrm{ctg}\beta_p\right) , \tag{9}$$

where α_0 is the coefficient of the bending moment function. For example, with a hinged bearing $\alpha_0 = 8$, and with a fixed bearing $\alpha_0 = 12$.

Substituting (9) and (7) into equation (8) we obtain

$$p_0 L_0^2 \left(2\alpha_p + \mathrm{ctg}\beta_p\right)\left(\alpha_0 R_y\right)^{-1} = 0.667 h_p t_p \left(2\alpha_p + 3\alpha_p^2 \sin\beta_p + 0.25\sin^{-1}\beta_p\right)\left(1 + 2\alpha_p \sin\beta_p\right)^{-1} . \tag{10}$$

Given a number of simplifications and erection of similar ones, we get

$$p_0 L_0^2 \left(\alpha_0 R_y h_p t_p\right)^{-1} = \left(\alpha_p + 0.167\sin^{-1}\beta_p\right)\left(2\alpha_p + \mathrm{ctg}\beta_p\right)^{-1} . \tag{11}$$

Let us express the parameter α_p through the cross-section area

$$\alpha_p = 0.5 A_p \left(h_p t_p\right)^{-1} - 0.5\sin^{-1}\beta_p . \tag{12}$$

Simplifying the result, we get

$$p_0 L_0^2 \left(\alpha_0 R_y h_p t_p\right)^{-1} = 0.5\left(A_p \sin\beta_p - 0.667 h_p t_p\right)\left[A_p \sin\beta_p - h_p t_p (1 - \cos\beta_p)\right]^2 . \tag{13}$$

Subsequently, we write the expression for the cross-section area of the corrugation

$$A_p = 0.67 h_p t_p \sin^{-1}\beta_p \left[3 p_0 L_0^2 (1 - \cos\beta_p) - \alpha_0 R_y h_p t\right]\left(2 p_0 L_0^2 - \alpha_0 R_y h_p t_p\right) . \tag{14}$$

The geometric interpretation of this equation is shown in figure 4 in the form of graphs of dependence $A_p(h_p)$ at different angles of tip of the corrugation wall β_p.

The analysis of a series of curves of equation (14) shows that irrespective of the angle of tip β_p, there is always a value of height h_p, at which the corrugation area will be minimal. The height that satisfies this criterion is called the optimum corrugation height h_{opt}. Note that in this case the height

IOP Conf. Series: Materials Science and Engineering 1021 (2021) 012021 doi:10.1088/1757-899X/1021/1/012021

h_{opt} has two distinctive features. First, an increase in the angle of tip of the wall leads to a decrease in h_{opt}, and secondly, the function $A_p(h_p)$ in the vicinity of the abscissa h_{opt} is shallow in nature and has small deviations of the actual corrugation height from the optimum. This has a rather insignificant effect on the minimum steel intensity of the profile.

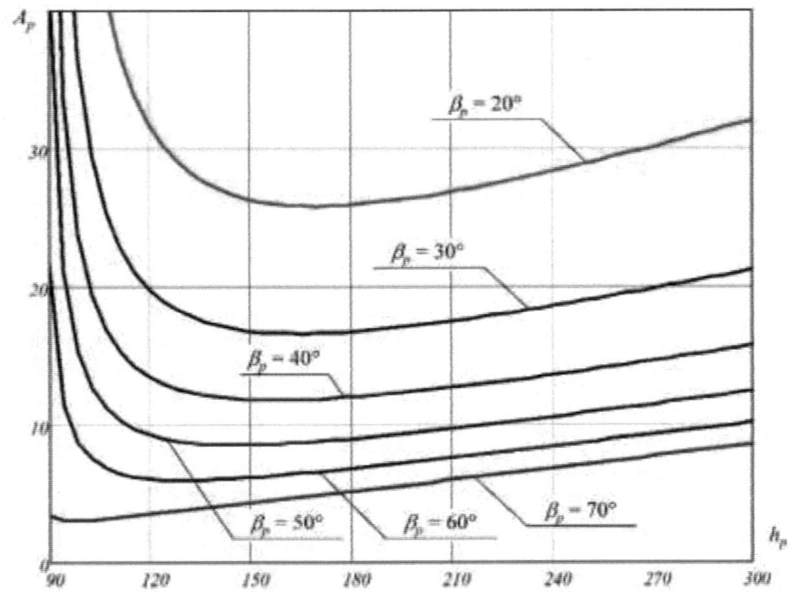

Figure 4. Geometric interpretation of equation (14).

To find the optimum height of the corrugation h_{opt} at a given angle β_p, we differentiate expression (14) by h_{opt} and equate the result to zero. After the transformations, we get

$$\frac{2}{3\sin\beta_p}\frac{12p_0^2L_0^2t_p\sin^2(0.5\beta_p)-\alpha_0h_{opt}t_p^2R_y\left(4p_0L_0^2-\alpha_0R_yh_{opt}t_p\right)}{4p_0L_0^2\left(p_0L_0^2-\alpha_0R_yh_{opt}t_p\right)+\left(\alpha_0R_yh_{opt}t_p\right)^2}=0. \quad (15)$$

Solving this equation with respect to the required height h_{opt}, we obtain a sufficiently concise and engineering-friendly equation

$$h_{opt}=p_0L_0^2\left(\alpha_0R_yt_p\right)^{-1}\left[2+(6\cos\beta_p-2)^{0.5}\right]. \quad (16)$$

Substituting the result into equations (14) and (13), we obtain an expression for the width of the corrugation shelf a_{opt}, which satisfies the criterion of optimality

$$a_{opt}=p_0L_0^2\left(3\alpha_0R_yt_p\right)^{-1}\sin^{-1}\beta_p\left[3\cos\beta_p+0.71\left(3\cos\beta_p-1\right)^{0.5}-2\right]. \quad (17)$$

Thus, given the length of the corrugated sheet, its thickness and the material strength, the size of the optimal cross-section (in terms of minimum steel intensity) will be determined only by the angle of tip of the corrugation wall to the horizontal plane.

There is one limitation. Given that the width a_{opt} cannot be negative, we find the limit value of the angle β_p at which this condition will be satisfied. Equating the expression (17) to zero and performing a series of trigonometric transformations, we obtain

$$\beta_p = 60° . \tag{18}$$

Thus, all the provisions obtained will be valid for angles of the corrugation wall up to 60°, which is sufficient for engineering calculations.

The use of equations (16) and (17) is quite convenient. For example, we calculate the optimal dimensions of the corrugated sheet of length $L_0 = 5000$ mm, made of steel with a yield point $R_y = 240$ MPa for two thicknesses $t_p = 4.0$ mm and $t_p = 3.0$ mm; for five load levels p_0 and four angles of tip β_p.

The ends of the sheet will be considered rigidly fixed at the ends. The results of the calculations are given in table 1

Table 1. Calculation of the optimal parameters of the corrugated sheet.

p_0, (kPa)	$\beta_p = 20°$		$\beta_p = 30°$		$\beta_p = 40°$		$\beta_p = 50°$	
	h_{opt} (mm)	a_p (mm)	h_{opt} (mm)	a_p (mm)	h_{opt} (mm)	a_p (mm)	h_{opt} (mm)	a_p (mm)
				$t_p = 4$ mm				
30	254.4	112.5	246.6	64.8	235.1	37.3	218.9	17.3
25	212.0	93.7	205.5	54.0	195.9	31.1	182.4	14.4
20	169.6	75.0	164.4	43.2	156.7	24.8	145.9	11.5
15	127.2	56.2	123.3	32.4	117.6	18.6	109.5	8.6
10	84.8	37.5	82.2	21.6	78.4	12.4	73.0	5.8
				$t_p = 3$ mm				
30	339.2	150.0	328.8	86.3	313.5	49.7	291.9	23.0
25	282.7	125.0	274.0	72.0	261.2	41.4	243.2	19.2
20	226.1	100.0	219.2	57.6	209.0	33.1	194.6	15.4
15	169.6	75.0	164.4	43.2	156.7	24.8	145.9	11.5
10	113.1	50.0	109.6	28.8	104.5	16.6	97.3	7.7

This problem is relevant for corrugated sheets with a limited number of erection joints, which will have little effect on its bending characteristics. In other cases, such as the relatively low sheets that face each corrugation overlapping, or the partial overlapping of the lower corrugation, the expressions for optimal performance will have a different representation, although the procedure for solving the problem will be the same.

4. Summary

1. In the framework of this study, the problem of determining the optimal geometric characteristics of corrugated sheet – the height, width and step of the corrugation shelves has been considered.

2. It has been found that regardless of the angle of the corrugation wall, there will always be a height at which the corrugation area will be minimal. This parameter will act as the optimum corrugation height.

3. Analysis of function of the corrugation cross-section area on the height showed that increasing the angle of tip of the wall would lead to a decrease in the optimum height.

4. We formulated concise and engineering-friendly equations to determine the optimum height and width of corrugation shelves for wall angles up to 60°. These expressions can be applied to the case of corrugated sheet with a limited number of erection joints. For other cases of jointing of the corrugated sheets with each other (complete or partial overlapping of the lower part), additional studies should be performed similar to the procedure presented.

5. The indicated calculation algorithm permits to implement optimal design solutions of resource-saving structures of rectangular steel hoppers for loading railway cars.

References

[1] Kingwell R, Elliott P, White P and Carter C 2016 *Ukraine: an Emerging Challenge for Australian Wheat Exports* (South Perth: AEGIC) p 92

[2] White P, Carter C and Kingwell R 2018 *Australia's Grain Supply Chains: Costs, Risks and Opportunities* (South Perth: AEGIC) p 106

[3] Lapenko O and Makhinko C 2018 Effect of profiled wall on the stiffness characteristics of structures for grain storage *Science and Construction* **16**(2) 40-5

[4] Makhinko A and Makhinko N 2019 To the calculation of the optimal level of reliability by using economic indicators *Lecture Notes in Civil Eng.* **47** 251-9

[5] Makhinko N 2018 Stress-strain state of the storage silos under the action of the asymmetric load *Matec Web of Conf.* **230** 02018

[6] Iwicki P, Rejowski K and Tejchman J 2017 Simplified numerical model for global stability of corrugated silos with vertical stiffeners *J. of Constr. Steel Res.* **138** 93-116

[7] Hajko P, Tejchman J and Wojcik M 2018 Investigations of local/global buckling of cylindrical metal silos with corrugated sheets and open-sectional column profiles *Thin-Walled Struct.* **123** 341-50

[8] Gallego E, Fuentes J M, Wiacek J, Villar J R and Ayuga F 2019 DEM analysis of the flow and friction of spherical particles in steel silos with corrugated walls *Powder Technol.* **355** 425-37

[9] Yossef N M 2020 A new approach to estimate the shear strength of curved corrugated steel webs *Struct.* **24** 400-14

[10] Abramczyk J 2016 Shaping innovative forms of buildings roofed with corrugated hyperbolic paraboloid sheeting *Procedia Eng.* **161** 60-6

[11] Kovalchuk V, Markul R, Pentsak A, Parneta B, Gajda O and Braichenko S 2017 Study of the stressstrain state in defective railway reinforcedconcrete pipes restored with corrugated metal structures *Eastern-European J. of Enterprise Technol.* **5/1** (89) 37-44

[12] Denga W, Liu D, Xiong Y and Zhang J 2019 Experimental study on asynchronous construction for composite bridges with corrugated steel webs *J. of Constr. Steel Res.* **157** 93-102

Management of the technical condition of hydraulic structures

B I Makovetsky[1], P M Sankov[1,5], R B Papirnyk[2], N O Tkach[3] and I V Trifonov[4]

[1]Architecture faculty, Prydniprovs'ka State Academy of Civil Engineering and Architecture Chernyshevskogo ave., 24a, 49005 Dnipro, Ukraine
[2]Faculty of Civil Engineering, Prydniprovs'ka State Academy of Civil Engineering and Architecture Chernyshevskogo ave., 24a, 49005 Dnipro, Ukraine
[3]Faculty of Civil Engineering and Ecology, Prydniprovs'ka State Academy of Civil Engineering and Architecture Chernyshevskogo ave., 24a, 49005 Dnipro, Ukraine
[4] Business Administration Department, Financial University under the Government of the Russian Federation, Leningradskiy prospect, 49, 125993 (GSP-3), Moscow, Russian Federation

[5] Email: petrsankov5581@gmail.com

Abstract. In the process of operation of hydraulic structures it is needed to study these structures for their technical condition, which affects the durability and other structures operational properties. Some defects are impossible or difficult to examine. This is due to the lack of open access for visual inspection. Such defects are cavities formed under reinforced concrete slabs, which protect the dam from erosion. Timely detection and elimination of cavities during repairs extends the service life of dams. Scientifically confirmed methods are proposed, by means of which the problem of finding the location and the volume of cavities under reinforced concrete of ground dams, dams and other hydraulic structures is solved. The research proves that with using the natural pulsed electromagnetic field of the Earth geophysical method it is possible to interpret in terms of the cavities which occur during the inspection of the reinforced concrete fastening of the upper slope of the dams. The authors propose to determine the depth of anomalies by the color of the equal level of the natural pulsed electromagnetic field of the Earth, which corresponds to the depth of the cavities obtained by the method of geoelectric tomography.

1. Introduction

In addition to the inspection of the technical condition of hydraulic structures, in some cases it is necessary to eliminate the causes that threaten the failure of these structures. This is due to the lack of open access for visual inspection. Sometimes such causes are harder to establish than to eliminate. Such reasons include the condition of the soil, which is in contact with the inner surface of the protective reinforced concrete slabs, which are usually laid on the slope of the summit pounds of the earth dam.

Under the influence of reservoir waves, such conditions as branching, erosion, and voids can be formed in the soil under the slabs. If measures to eliminate the cavities under the slabs are not taken, then eventually the dam may collapse under the pressure of pressure and waves of the reservoir. Currently, the problem is which method to determine the location of cavities and establish their volume rather quickly, with little labor efforts. Therefore, the purpose of this research is to develop an instrumental method of inspection of sections of reinforced concrete slabs, which protect the summit pounds of the dam from erosion, detection of voids and determination of their volume.

2. Materials and methods

The object of the inspection is a section of reinforced concrete slabs of the summit pounds of the dam of the cooling pond of Kryvyi Rih TPS of PJSC DTEK DNIPROENERGO within the limits of pickets 75-85 (length 1000 m).The purpose is to detect the location of the cavities under the reinforced concrete cover of the slope of the dam and to determine their geometry with the subsequent calculation of the volume.

2.1. Methods of inspection of reinforced concrete slabs

The most common method of inspection of the fastening of the slopes of dams is a visual method, which captures visible defects (cracks, potholes, concrete delamination, frost destruction, etc.) [1]. Methods of instrumental inspection are the method of non-destructive testing [2], which determines the strength of concrete and checks the integrity of the reinforcement. But what is the state of the system "reinforced concrete slab - bedrock" this method does not give an answer.

A common cause of defects and anomalies between the slabs is the destructive effect of water that occurs with the operation of the dam, when cracks appear, diverge temperature and deformation seams, reduces the layer of gravel under the slabs, and so on. First, the soil begins to be carried out from under the slabs through the seams and through the edges of the slabs. The upper part of the fastening near the crest of the dam is also washed away if there is no concrete parapet, which serves as a barrier to rain flows and no drainage. What is the danger of these phenomena? Over time, the reinforced concrete slab is covered with cracks and the destruction becomes avalanche-like. The earth dam becomes defenseless against pressure and waves of a reservoir.

2.2. Main provisions of research methods by different methods

The methods used in the inspection of reinforced concrete dams and described in scientific publications can be divided into two groups. The first is geophysical methods. These include georadar scanning, electrical exploration [2 - 5]. To the second - one known, it is a method of thermography [3, 6, 7]. When considering and analyzing the publications, we came to the conclusion that the first group uses the most common method is method of georadar [3]. Georadar is used for a wide range of technical tasks that are in one way or another related to the earth's crust. About a few dozen such problems are successfully solved using georadar [9, 12]. In the publication [3] it was found that georadar performed work to detect cavities under the protection slabs in earth dams. We witnessed the verification of the results of georadar by a company that performed a contract with the Kryvyi Rih TPS in 2013.

In research [3] V.V. Glazunov proves that georadar can overcome this trouble, but the interpretation in the article is not convincing enough. Desktop processing is complex, interpretations are ambiguous, the technical report is difficult for the customer to understand. In addition to these disadvantages, there is an economic one: the cost of the device is quite high. The thermal imaging method presented in the same article, which we assigned to the second group [6, 7, 12, 13], is based on distance measurements by a thermal imager. This method has more disadvantages than advantages and worthy results.

First, an expensive device, secondly, a long measurement procedure that requires separate requirements, thirdly requires unique weather conditions (only in summer, the presence of the sun, but when it is before sunset, so that there is no heat), fourthly, the results are such that not quite correct. Figure 1 [3] shows the cavities of the same color as the adjacent areas without anomalies. In general, thermogram analysis includes a proportion of subjective estimates. Any inaccuracy as a result leads to material losses of the customer of works. In research [14] it was also noted that this method is used in the United States as an approximate, evaluating in the early stages of the study, which shows in a fairly large area the affiliation of the anomaly. We completely agree with this conclusion.

One of the modern methods of geophysics is electrotomography as a component of a more general method of electrical exploration, and which is a more accurate and informative method than the georadar method [5 - 8]. This method solves many problems. Solutions to the problems of surveying the integrity of the dam are presented in figure 1. But the use of geophysical method of electrical

IOP Conf. Series: Materials Science and Engineering 1021 (2021) 012022 doi:10.1088/1757-899X/1021/1/012022

exploration is not suitable for our needs, because the method involves immersing steel electrodes in the soil, which cannot be done through reinforced concrete slabs.

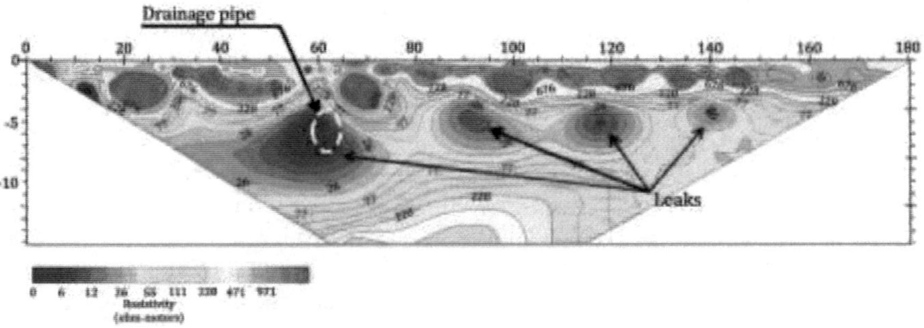

Figure 1. Geoelectric section of the body of the earth dam.

Another geophysical method is the method of the natural pulsed electromagnetic field of the Earth (NPEFE). It seems that the method of NPEFE [7] is also not used by anyone for our purposes because it does not answer the question: at what depth are the cavities. No researches were found on this topic. But this method is very attractive because it has one significant advantage over other geophysical methods. It consists in the fact that the radiation waves propagate in the opposite direction, ie not from the day surface of the Earth, as in georadar scanning, or in electrical exploration, but from the middle of the Earth. The power of these electromagnetic waves is not large enough, but they penetrate everywhere. Wave frequency from 1 to 50 KHz. Here is a graphical representation of the state of flooding among the urban development of the city made using the method of NPEFE.

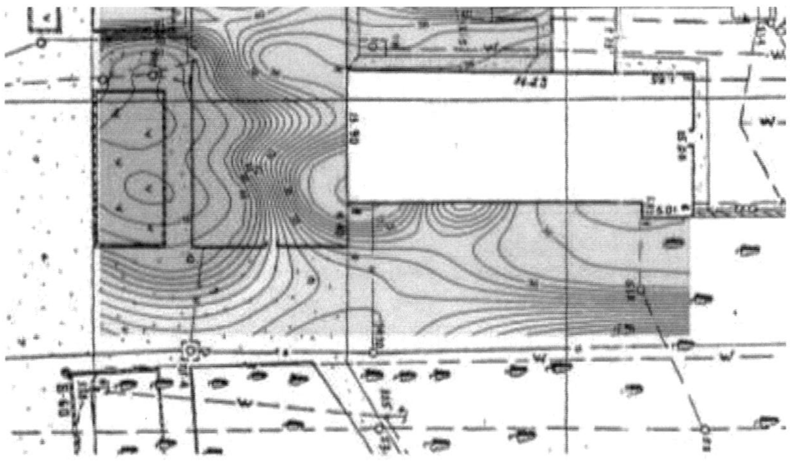

Figure 2. NPEFE scheme in urban development.

In order to find out at what depth underground the phenomenon appeared, which is reflected in figure 2, it is necessary to drill wells and take soil tests to determine its condition. In other words, the method gives a qualitative assessment of the situation that worries us. Quantitative assessment of soil condition can be obtained by applying to the heap another method that will show this condition by section in depth. Of the more acceptable methods, in our opinion, the method of geophysical tomography is suitable as a component of electrical exploration. In order to apply the two methods together, it would be advisable to overlay one graphic image on the other and as a result have the desired 3D image.

But as mentioned above, the method does not involve working in water, and it is difficult, and virtually impossible, to drive steel electrodes into the ground through reinforced concrete slabs. Given these circumstances, we came to the conclusion that it is necessary to make surveys by both methods on a separate experimental site with a pre-known location under the reinforced concrete slabs of the cavities, and with their known geometric dimensions. And then determine at what depth and what size the cavities are depicted and then compare how the image coincides with the natural location and size of the cavities. Finally, make a color image of the depth of the desired cavities using lines of equal level. To ensure the efficiency of the method, a measurement method was developed to examine the upper slope of the dam - cooler with subsequent verification of the coincidence of the results of NPEFE method with the natural location of cavities by local, adjustable drilling holes in the ground through reinforced concrete surface.

3. Theory and calculations

A particularity of electrical tomography is the measurement by the system of electrodes, preliminary placed along the same profile at equal distances from each other. Switching of electrodes is done by a special channel switch-board. The main purpose of this technique is the simultaneous observation of the section along the profile and depth. When working on a limited area, where there are no conditions for placing unboundedness, a dipole configuration is used. For this installation with the same dipoles, the second image can be obtained by displacements the measure point based on the principle of reciprocity, without additional measurements. Modern multi-electrode systems allow increasing the interval of shooting in depth. To visualize the picture, in other words the inversion of field measurements, the res2Dinv program is used, which is based on the quasi-Newtonian optimization method.

3.1 Methods of work performance

A section of the upper slope with two temperature seams, around which cavities usually occur, was found on the dam of the cooling pond of Kryvyi Rih CHP-plant, which is about 6 kilometres long. These cavities could be seen visually through a fairly wide control joint. Cavities are formed just near the joints, because through them the exhausted soil is washed away. The geometry of the cavities was measured in advance with measuring instruments. Surveying was carried out by the method of the natural pulsed electromagnetic field of the Earth, for which there were determined the profiles on which the operator with the RDVINS device will move and fix indicators in measuring points were determined (figure 3).

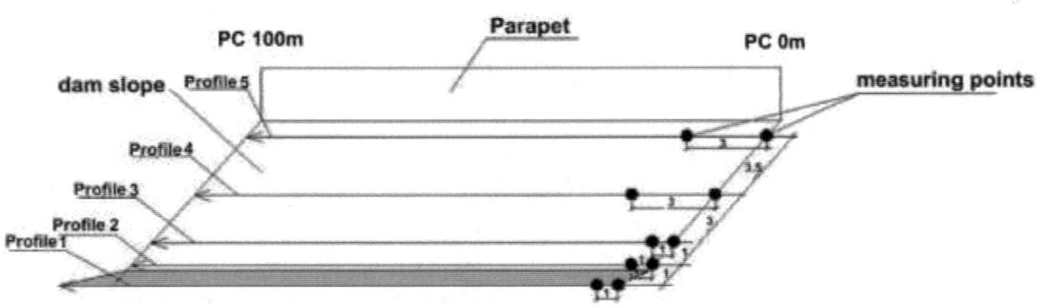

Figure 3. Diagram of the location of profiles and measurement points on them.

Operating conditions are satisfactory. The surface of the reinforced concrete pavement along the entire length of the site is smooth, not moistened. On profiles close to water the step of measuring points of 1 meter is accepted, on profiles of 3 and 4 it is 3 meters. Profile 1 is above the water surface. During the measurements, two horizontal components of the method of the natural pulsed electromagnetic field of the Earth were registered: Ex, which coincides with the direction of the profiles, and Ey, orthogonal to the extension of the dam. The results were processed independently for each measurement point. The standard program "surfer" was used for visualization. The results were presented in the form of isoline maps of components Ex and Ey in imp / s.

Zones of high inhomogeneity (flooded and accommodating cavities) are marked in bright yellow. The result is shown in figure 4. Anomalies indicate the belonging of cavities that are selected graphically. The slope width in meters starts from the profile of the 4th dam (figure 3).

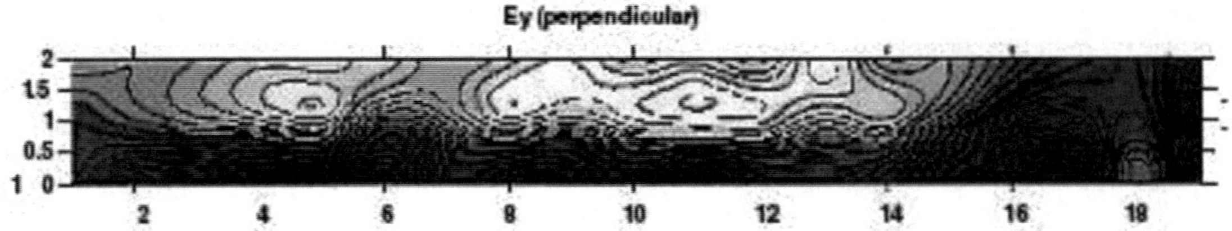

Figure 4. Map of lines of the same level of the natural pulsed electromagnetic field of the Earth on the experimental site.

To confirm the values of the depth of the anomalous zones, an experimental study was carried out in the same area by the method of electro geological tomography on the profile with a measurement step of 1 meter.

The measurement technique is as follows:

The EIH-209M meter was used as an electric field receiver at a frequency of 1.22 Hz in the dipole MN. The ГЕР -1/300 generator is used as a current source (figure 3).

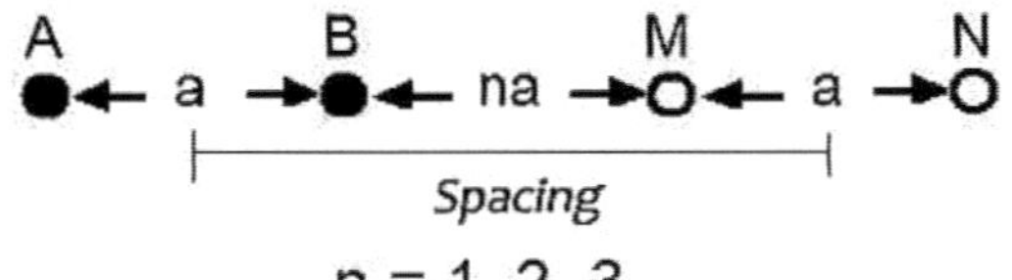

a)

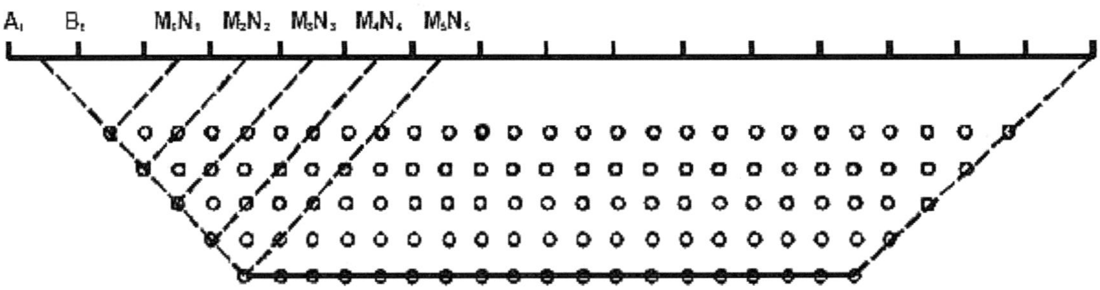

b)

Figure 5. a) Scheme of the field installation. AB - supply dipole; MN - receiving dipole. b) The layout of the recording points on the pseudo section.

According to the results of field measurements, pseudo sections are obtained, the construction features of which can be understood from figure 5 b). The extreme left dipole is used as a supplier, and located next to the right of it is as a recipient. The recording point refers to the middle of the installation and to the depth Z, which can be found when drawing lines from the centers of the dipoles at an angle of 45 ° to the intersection at the recording point.

The measurement technique involves the use of electrodes that are driven into the ground and connected to a conductive wire. We used metal cutter arms as electrodes, which rested on a foam substrate irrigated with aqueous saline solution. The general view of the location of the cutter arms electrodes is shown in the photo (figure 6).

Comparing the results shown in figure 4 and figure 7 we determine that the cavity in the middle of the section (between 8 and 12 meters) corresponds to the depth on the profile of the geoelectrical section of 0.5 meters, area - 0.35 m^2, volume - 0.175 m^3. The entire geometry of the cavity coincides with the previous measurements with an endoscope and measuring instruments. Thus, with the help of this study, we proved the possibility of using two methods of geophysics to accurately assess the state of the underlying layer of earth under reinforced concrete slabs.

IOP Conf. Series: Materials Science and Engineering 1021 (2021) 012022 doi:10.1088/1757-899X/1021/1/012022

Figure 6. General view of the location of the electrode cutter arms on the profile.

Figure 7 shows the tomography profile of the section of the reinforced concrete slab of the dam between the temperature seams with a pre-known location of large cavities under the slabs.

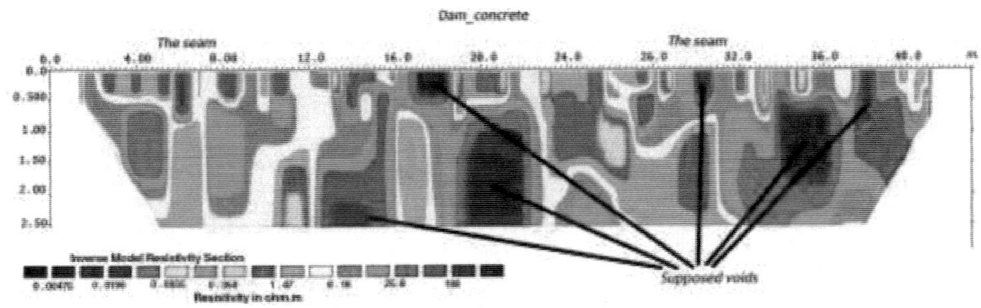

Figure 7. Geoelectrical section.

4. Study of the site of reinforced concrete pavement of the upper slope of the dam

The study was conducted according to the method of the natural pulsed electromagnetic field of the Earth described in the section above. According to the results of this study, the images of the natural pulsed electromagnetic field of the Earth are shown in Fig. 8 by segments of 25 m. Below these images are comments. Counting the width of the site in meters starts from the parapet of the dam.

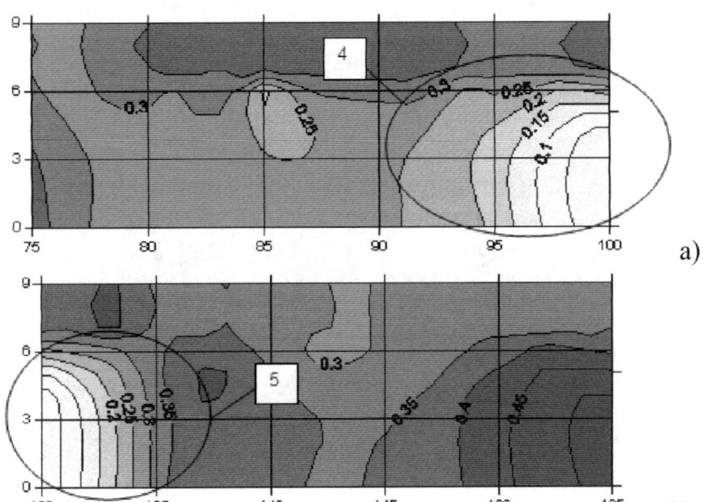

Figure 8. The image of the natural pulsed electromagnetic field of the Earth:
a) PC 75+75 – PC 76+00, b) PC 76+00 – PC 76+25.

Selected anomaly is 4, soil erosion at a depth of up to 0.5 m, anomaly 5 is the second part of the void from the previous section (it is located at the junction of sections).

4.1 The results of geophysical survey of the upper slope of the dam of the cooling pond by the method of the natural pulsed electromagnetic field of the Earth

The results of volumes on significant anomalies are shown in the diagram in figure 9.

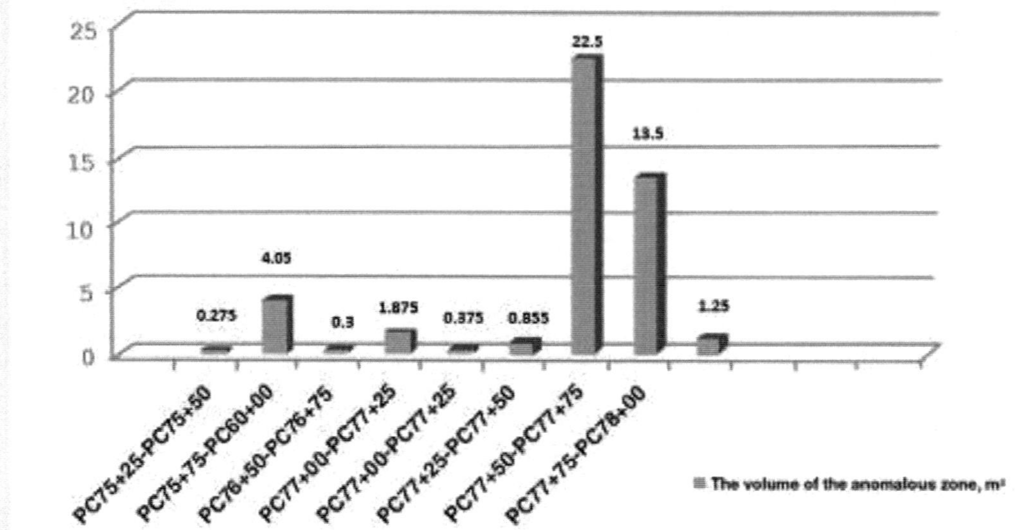

Figure 9. Diagram of the estimated volume of anomalous zones in the area from the picket PC 75+00 up to PC 85+00.

5. Conclusions

It is established that with the help of the geophysical method of the natural pulsed electromagnetic field of the Earth, it is possible to recognize in the plan of a branch cavity which meets at the study of reinforced concrete fastening of a top slope of dams. The depth of the anomalies is determined by the colour of the equal level of the natural pulsed electromagnetic field of the Earth, which corresponds to the depth of the cavities obtained by the method of geoelectrical tomography (figure 8). Thus the important economic task on timely detection and elimination of destructive defects of dams during their operation in the aggressive water environment is solved.

References

[1] N B V.1.2-18:2016 Guidelines for inspection of buildings and structures to determine and assess their technical condition (Research Institute of Construction Production)

[2] Klimenko E Technical operation and reconstruction of buildings and structures 2014 *textbook* (Kyiv: Training Center literature) p. 304 ISBN 966-8253-90-9

[3] V.V. Glazunov, V.G. Shtengel, V.S. Nedyalkov, N.N. Efimova and S.M. Danilev S 2018 Combined investigation by thermal imaging and georadar scanning for voids detection under reinforced concrete slabs of fastening soil slopes of hydraulic structures. Conference proceedings *Engineering and Mining Geophysics* **2018** p.1 – 11 doi.org/10.3997/2214-4609.201800474

[4] Kokinou E and A Sarris 2011 Detection of the near surface structure through a multidisciplinary geophysical approach **3 (4),** 349–357 doi.org/10.2478/s13533-011-0034-2

[5] Pigulevskyj P, Svystun V and Kyrylyuk O 2017 Geoelectric study of engineering-geological condition of southwestern kryvbas. Part 3. The results of application geoelectric methods in solving geotechnical problems *Geoinformatyka* **2 (62)** pp 55-632

[6] Vartanov A., Nabatov V. 2011 Thermographic control for construction *Mining information and*

analytical bulletin Moscow State Mining University **9** pp 193-200

[7] Belyaeva, LI B 44. Fundamentals of geophysics textbook - Ukhta: USTU ,. 2016 *Tutorial.* 182 p. ISBN 978-5-88179-969-4

[8] Ahmetov E, Asemov K and Zhumataeva M 2020 Research of accidents of hydraulic structures and safety control methods *Bulletin of the Tomsk Polytechnic University. Geo Assets Engineering* **331 (4)** pp 70-83 Almaty

[9] Glazunov V, Danilev S and Efimova N 2011 Application of the GPR method for assessing the state of fastening upper slope of land dams *Journal of Applied Geophysics* **74 (2-3)**, 131-141

[10] Eskin A, Dzhurin V, Serebryannikov S and Bryizhak U 2016 Regime studies of the dynamics of the physical characteristics of the rock fill dam *J. Success of modern natural science* **12 (2)** p.394

[11] Xingxin Xu, Qiaosong Z, Dong li and Jin wu 2010 GPR detection of several common subsurface voids inside dikes and dams *Engineering Geology* **111 (1)** pp 31-42 DOI: 10.1016 / j.enggeo.2009.12.001

[12] Pueyo A, Frongia P, Di Gregorio F, Kasas A and Pokovi A 2018 Internal characterization of embankment dams using ground penetrating radar (GPR) and thermographic analysis: A case study of the *Medau Zirimilis Dam Sardiniya, Italiya* **237** pp 129-139 doi:10.1016/j.enggeo.2018.02.015

[13] Glazunov V, Shtengel V, Nedyalkov V, Efimova N and Danilev S 2018 Combined investigation by thermal Imaging and georadar scanning for voids detection under reinforced concrete slabs of fastening soil slopes of hydraulic structures *Conference Proceedings Engineering and Mining Geophysics* Apr. pp 1-11

[14] Xingxin Xu, Qiaosong Z, Dong li and Jin Wu 2010 GPR Detection of several common subsurface voids inside dikes and dams Engineering Geology **111 (1)** pp. 31-34 DOI: 10.1016/j.enggeo.2009.12.001

IOP Conf. Series: Materials Science and Engineering 1021 (2021) 012023 doi:10.1088/1757-899X/1021/1/012023

Determining the Dependence of Fire Parameters in a Cable Tunnel on its Characteristics

Oleksandr Nuianzin[1,6], Stanislav Sidnei[2], Petro Zayika[3], Svitlana Fedchenko[4], Bogdan Alimov[5]

[1]Science department of Cherkasy Institute of Fire Safety named after Chernobyl Heroes, National University of Civil Defence of Ukraine, 94 Chernyshevska Str. Kharkiv 61023, Ukraine
[2]Department of Safety of Construction Facilities and Labor Protection of Cherkasy Institute of Fire Safety named after Chernobyl Heroes, National University of Civil Defence of Ukraine, 94 Chernyshevska Str. Kharkiv 61023, Ukraine
[3]Department of Fire Prevention Work of Cherkasy Institute of Fire Safety named after Chernobyl Heroes, National University of Civil Defence of Ukraine, 94 Chernyshevska Str. Kharkiv 61023, Ukraine
[4]Post-graduate department of Cherkasy Institute of Fire Safety named after Chernobyl Heroes, National University of Civil Defence of Ukraine, 94 Chernyshevska Str. Kharkiv 61023, Ukraine
[5] Department of Substances and Materials of the Scientific and Testing Center, Institute of Public Administration and Research on Civil Defence, 18 Rybalska Str. Kyiv, 01001, Ukraine

[6]alexandrnuyanzin@gmail.com

Abstract. In order to study the parameters of fires in cable tunnels both experimental methods and numerical simulations are used. The purpose of this article was to determine the dependences of the maximum temperature inside the cable tunnel during a fire, the duration of the fire in a local area of the cable tunnel and the time of reaching the maximum temperature inside the cable tunnel during a fire on its characteristics (cross-sectional area of the cable tunnel, fire load) inside the tunnel. In the course of the research a complete factorial experiment was developed and conducted. According to its results, the required regressions were obtained. Thus, there are reasons to believe that the obtained results of the study are the basis for the creation of mathematical models that describe fires in cable tunnels and can be used for engineering assessment of fire resistance of cable tunnel structures. In addition, computer simulation of heat and mass transfer in case of fire in cable tunnels was further developed.

1. Introduction

Despite significant progress in solving the problem of improving the fire safety of cable products, there are now many problematic issues in the field of safety of cable structures.

Combustion of electric cables is accompanied by the release of a significant amount of heat, which is determined by the specific heat of combustion of insulation materials, protective sheaths of cables and the mass of these materials contained in a unit length of cable. Experiments on combustion of cable streams in a cable tunnel have shown that the temperature in the combustion zone of cables with PE insulation or with paper impregnated insulation reaches 1000-1200 °C [1]. At the same time, there

IOP Conf. Series: Materials Science and Engineering 1021 (2021) 012023 doi:10.1088/1757-899X/1021/1/012023

is a significant amount of black smoke and other gaseous products, which leads to reduced visibility and complicates the actions of fire fighters and evacuation personnel.

2. Analysis of recent achievements and publications

In order to study the parameters of fires in cable tunnels both experimental methods [1-4] and numerical simulations [5-10] are used.

In the work [2], changes in the temperature during the propagation of flue gases during a fire in a cable tunnel are considered. However, this paper describes only the initial stage of the fire. The temperature during the entire time of the fire was not determined. In the work [3], field tests were conducted in a communal tunnel. A fire during the spillage of flammable liquid was simulated. This work can be the basis for the study of violations of the modes of operation of cables with oil filling. The work [4] describes the temperature distributions in tunnels with different ventilation conditions during full-scale experiments. The conducted study enabled analyzing the influence of air velocity on the specifics of fire development.

In the work [5] the causes of fires in tunnels were analyzed by numerical simulation. However, computational experiments on fire simulation were not performed. The work [6] is devoted to the analysis of the parameters of the burnout rate of PVC cable insulation. The linear velocity of fire propagation for different cable laying types was mathematically calculated. In the work [7], the temperature regime of the fire in the cable tunnel was determined depending on its shape, size and fire load. This made it possible to determine the parameters that most significantly affect the temperature of the fire inside the tunnel. The study [8] shows that the temperature in the real fire reached 800-900 °C. Based on this information, the type of thermocouples was selected to measure temperatures during the experiment. In the work [9] it was found that the width of the tunnel has little effect on the rate of burnout of the fire load. The influence of two ventilation systems on the temperature distribution in a small cross-section tunnel model was investigated in the work [10]. This allows us to conclude that the aerodynamics in the space of the tunnel affects the temperature of the fire.

Based on the above, conducting additional study is required to establish the mathematical dependences of the temperature of the fire in the tunnel on its parameters.

3. Purpose

To determine the dependence of the maximum temperature inside the cable tunnel during a fire, the duration of the fire in a local area of the cable tunnel and the time of reaching the maximum temperature inside the cable tunnel during a fire on its characteristics (cross-sectional area of the cable tunnel, fire load) inside the tunnel.

4. Method

The scheme of the tunnel with geometric parameters that have changed is shown in figure 1.

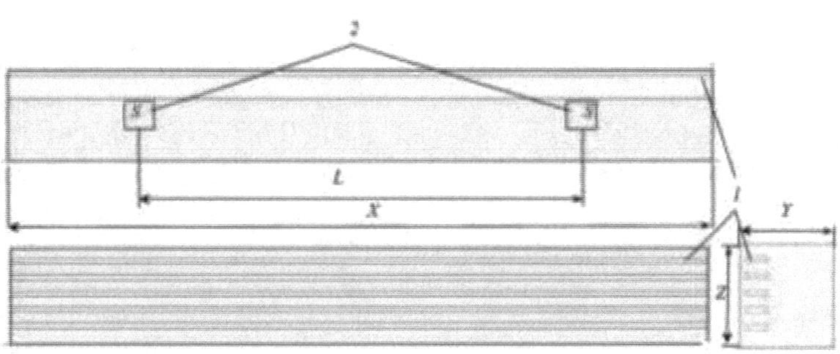

Figure 1. The scheme of the tunnel with geometric parameters that have changed: 1 – Model of cable lines; 2 – Model of ventilation and inspection hatches area S, m^2; X, Y, Z, L – the distance between the holes

Table 1. Thermophysical characteristics of materials and conditions of convection and radiation heat exchange.

Characteristic	Units of measurement	Value
Thermal conductivity coefficient	W / (m·K)	0.159
Specific heat capacity	J/(kg·K)	1320
Density	kg/m^3	1400
Degree of blackness	-	0.85
Thermal conductivity coefficient	W/(m·K)	390
Specific heat capacity	J/(kg·K)	420
Density	kg/m^3	8900
Degree of blackness	-	0.7
Stefan–Boltzmann constant	W/(m^2·K^4)	$5.67 \cdot 10^{-8}$

Theoretical studies were carried out on the basis of systems of differential equations of continuous media of the type of Navier-Stokes equations and Fourier thermal conductivity equations. To solve the equations in the work, the method of finite or boundary elements, non-relation methods, Galerkin method, and optimization methods are used. Computational experiments were carried out in the Fire Dynamics Simulator (FDS).

According to the calculation scheme of the mathematical model of the tunnel (fig. 1) in order to conduct a full analysis of the parameters that affect the temperature of the fire, the following parameters varied:

- Fire load: cable lines were modelled on two and one side; the number of cable lines was from 1 to 10; the insulation material of cables and wire cores changed. Due to this, the rate of heat output from 1 m^2 of cable lines varied (table 1).
- Geometric parameters of the cable tunnel: varied as the cross-sectional area, using a combination of parameters Y and Z, the general parameter – the cross-sectional area of the cable tunnel (table 1).
- Aerodynamic parameters: the horizontal component of air velocity was set as a parameter that characterizes the excess or lack of oxidant and affects the intensity of development and spread of fire.

To take into account the specific features of the heterogeneity of the material of which the cable lines are made, as well as the conditions of conductive heat transfer (thermal conductivity in the solid material), cable lines are given as a multi-component material.

5. Consideration on methods and results

To build a mathematical model of the temperature of a fire in a cable tunnel, a complete factor computational experiment should be conducted. There are three independent factors – the cross-sectional area of the cable tunnel, fire load, as well as the horizontal component of air movement inside the tunnel. In the table 2 shows the parameter intervals in the experiment, which are selected as factors.

Table 2. Intervals of variation of factors in a computational experiment.

Factor 1. Fire load in terms of 1 m^2 of cable tunnel, mJ/m^2 (hereinafter – x_1)	Factor 2. Cross-sectional area of the cable tunnel, m^2 (hereinafter – x_2)	Factor 3. The horizontal component of air velocity, m/s (hereinafter – x_3)
224.7–2247	2.88–4.4	0–5

The selected mathematical model is a linear dependence of the maximum temperature inside the cable tunnel on the selected factors, which has the form.

$$y = b_0 + b_1x_1 + b_2x_2 + b_3x_3 + b_4x_1x_2 + b_5x_1x_3 + b_6x_2x_3 + b_7x_1x_2x_3 \quad (1)$$

where b_0, b_1, b_2, b_3, b_4, b_5, b_6, b_7 – regression coefficients.

To construct a regression according to formula (1) it is necessary to conduct 8 numerical experiments according to the accepted planning matrix, which is written in the form of the tab. 3.

Table 3. Typical matrix for planning a complete factorial experiment to determine the temperature of a fire in a tunnel.

No.	x_1	x_2	x_3	x_1x_2	x_1x_3	x_2x_3	$x_1x_2x_3$
1	+	+	+	+	+	+	+
2	-	+	+	-	-	+	-
3	+	-	+	-	+	-	-
4	-	-	+	+	-	-	+
5	+	+	-	+	-	-	-
6	-	+	-	-	+	-	+
7	+	-	-	-	-	+	+
8	-	-	-	-	-	-	-

To determine the initial data of the complete factorial experiment, 8 computer models were calculated [7] in which the parameters of the maximum and minimum intervals in different combinations were laid down. The results of experiments are shown in figure 2.

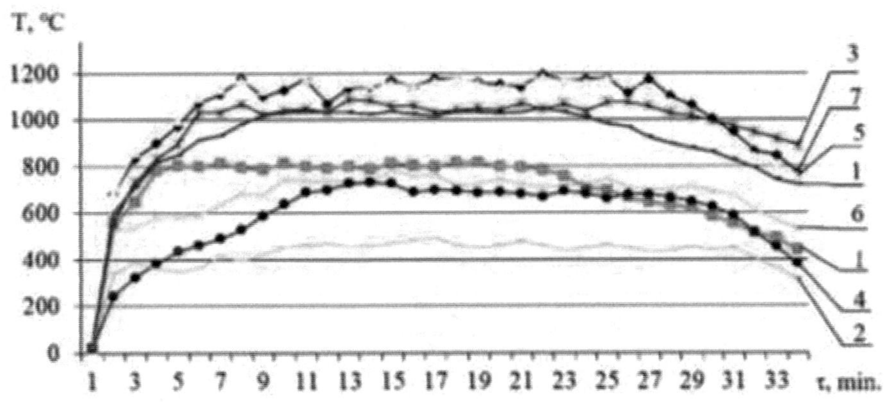

Figure 2. Summary graph of calculation of 8 computer models, the data of which are the input for a complete factorial experiment: 1–8 – experiment number according to Table 3.

The graphs in figure 2 show that the temperature regime of the fire in cable tunnels with different sizes, aerodynamic parameters and fire load, as well as the dependence of the temperature regime of the fire on these parameters are determined. With 10 laid cable lines and fire load 2247 MJ/m^2 the maximum temperature exceeded 1200℃, when there is 1 line and fire load 224.7 MJ/m^2 – 500℃.

With the smallest cross-sectional area of the tunnel and a decrease in air flow velocities, the temperature inside increases 50% faster than the average parameters. In addition, excess fresh air reduces the combustion temperature by 50–70℃, although it contributes to the faster spread of fire along the cable lines.

Further, in figures 3 – 5 the results of a complete factorial experiment to determine the temperature of the fire in the tunnel are presented.

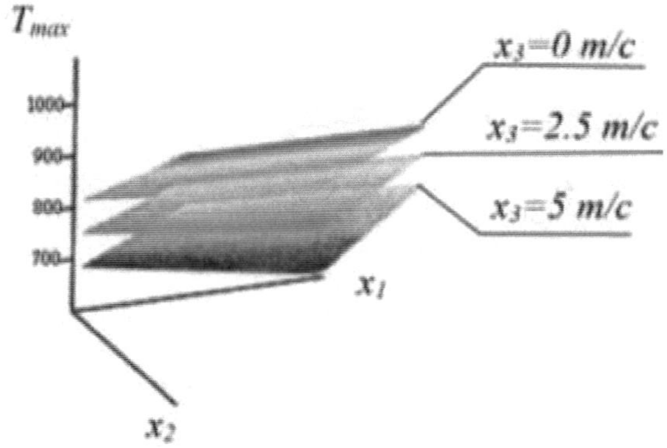

Figure 3. Maximum temperature inside the cable tunnel during a fire (T_{max}).

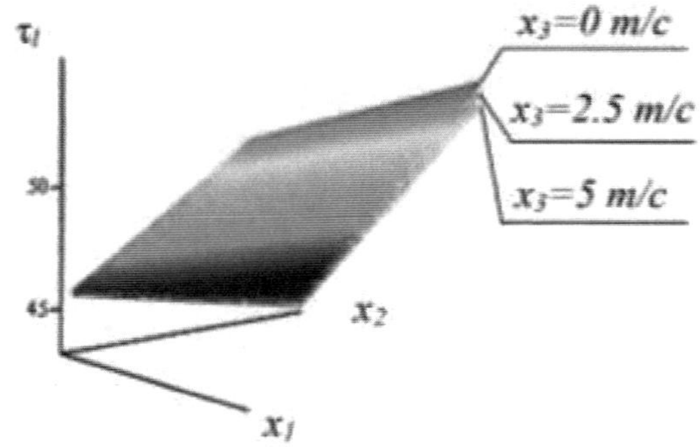

Figure 4. The duration of the fire in a certain area of the cable tunnel (τ_l).

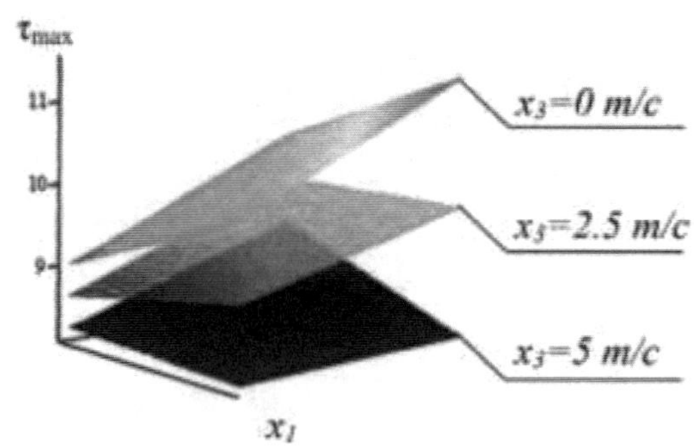

Figure 5. Time to reach maximum temperature inside the cable tunnel during a fire (τ_{max}).

According to the results of a complete factorial experiment, the regression of the maximum temperature inside the cable tunnel during a fire was obtained (T_{max}), duration of the fire in a certain local area of the cable tunnel (τ_l) and the time to reach the maximum temperature inside the cable tunnel during a fire (τ_{max}) are represented by expressions (2) – (4):

$$T_{max} = 0.097 \cdot x_1 - 27.92 \cdot x_2 - 11.391 \cdot x_3 + 0.01 \cdot x_1 \cdot x_2 - 0.001 \cdot x_1 \cdot x_3 - 5.279 \cdot x_2 \cdot x_3 + 0.001 \cdot x_1 \cdot x_2 \cdot x_3 + 870.594 \quad (2)$$

$$\tau_l = 0.002 \cdot x_1 - 1.439 \cdot x_2 + 0.0125 \cdot x_3 + 0.001 \cdot x_1 \cdot x_2 - 0.016 \cdot x_2 \cdot x_3 + 48.969 \quad (3)$$

$$\tau_{max} = 0.001 \cdot x_1 + 0.596 \cdot x_2 + 0.05625 \cdot x_3 + 0.025 \cdot x_2 \cdot x_3 + 6.55 \quad (4)$$

After obtaining regression dependences, it became possible to create a mathematical model of the behaviour of enclosing building structures of cable tunnels and to estimate their limit of fire resistance at different fire temperatures.

6. Conclusions

1. A complete factorial experiment was developed and conducted. According to its results, the regression of the maximum temperature inside the cable tunnel during a fire, the duration of the fire in a certain local area of the cable tunnel and the time to reach the maximum temperature inside the cable tunnel during a fire are represented by expressions (2) – (4).

2. The temperature regime of the fire in cable tunnels with different sizes, aerodynamic parameters and fire load, as well as the dependence of the temperature regime of the fire on these parameters are determined. With 10 laid cable lines and fire load 2247 MJ/m^2 the maximum temperature exceeded 1200℃, when there is 1 line and fire load 224.7 MJ/m^2 – 500℃.

3. With the smallest cross-sectional area of the tunnel and a decrease in air flow velocities, the temperature inside increases 50% faster than the average parameters. In addition, excess fresh air reduces the combustion temperature by 50–70℃, although it contributes to the faster spread of fire along the cable lines.

4. It is expedient to direct further work on research of fire resistance of enclosing building designs of cable tunnels at the temperature modes of fire defined in this section of work, using regression dependences (2) – (4).

References

[1] Nuyanzin O Pozdieiev S Samchenko T (et al) 2018 Experimental study of temperature mode of a fire in a cable tunnel *Eastern European Journa of Enterprise Technologies* **No 3/10 (93)** p 21–27

[2] Niu Y Li W 2012 Simulation Study on Value of Cable Fire in the Cable 3 *Tunnel Procedia Engineering* **T 43** pp 569–573

[3] Zhao Y Zhu G Gao Y 2018 Experimental Study on Smoke Temperature Distribution under Different Power Conditions in Utility Tunnel *Case Studies in Thermal Engineering*

[4] Tian X (et al) 2017 Full-scale tunnel fire experimental study of fire-induced smoke temperature profiles with methanol-gasoline blends *Applied Thermal Engineering* **T 116** pp 233–243

[5] Modic J 2003 Fire simulation in road tunnels *Tunnelling and underground space technology* **T 18 № 5** pp 525–530

[6] Vaari J (et al) 2012 Numerical simulations on the performance of water-based fire suppression systems *VTT Technol* **T 54**

[7] Nuianzin O (et al) 2018 Investigation of the regularities of temperature regime of fire in cable tunnels depending on its parameters *MATEC Web of Conferences EDP Sciences* **T 230** p 02022

[8] Hsu W S (et al) 2017 Analysis of the Hsuehshan Tunnel Fire in Taiwan *Tunnelling and Underground Space Technolog* **T 69** pp 108-115

[9] Ji J (et al) 2016 Influence of aspect ratio of tunnel on smoke temperature distribution under ceiling in near field of fire source *Applied Thermal Engineering* **T 106** pp 1094-1102

[10] Brahim K (et al) 2013 Control of Smoke Flow in a Tunnel Journal of Applied Fluid Mechanics **T 6 № 1**

IOP Conf. Series: Materials Science and Engineering 1021 (2021) 012024 doi:10.1088/1757-899X/1021/1/012024

Research of Integrity of Fire Insulation Cladding with Mineral Wool of Steel Beam under Fire Impact

Serhii Pozdieiev[1,4], Oleksandr Nuianzin[1], Olena Borsuk[2], Inna Nedilko[2]

[1]Science department of Cherkasy Institute of Fire Safety named after Chernobyl Heroes, National University of Civil Defence of Ukraine, 94 Chernyshevska Str. Kharkiv 61023, Ukraine

[2]Adjunctura of Cherkasy Institute of Fire Safety named after Chernobyl Heroes, National University of Civil Defence of Ukraine, 94 Chernyshevska Str. Kharkiv 61023, Ukraine

[3]svp_chipbbk@ukr.net

Abstract. In the article, the processes of integrity loss of fire-retardant cladding from mineral wool of a steel I-beam are considered. The relation between fire-retardant cladding integrity and its thermal insulating ability was studied. Computer model of a steel beam with fire-retardant cladding of mineral wool, which takes into account its own weight, current mechanical load and temperature mode of heating under the heat-insulating effect of the cladding, using the computer system LS-DINA, is developed in this article. It is shown that the fire-retardant cladding loses its integrity under the thermal influence of fire long before the onset of the limit state of loss of fire resistance in terms of bearing capacity. The conditions under which fire-retardant cladding loses its integrity and fire-retardant ability are investigated. Having considered the time of fire-retardant cladding integrity loss the fire resistance limit can be significantly clarified and allows to avoid early collapse of building structures cause by fire.

1. Introduction

Due to the high thermal conductivity of the metal and small cross-sectional dimensions, steel structures heat up quickly in a fire. Using fire-retardant cladding remains an effective means of ensuring the design fire resistance of steel structures. Promising material for fire-retardant cladding are mineral wool boards, as their using provides advantages when performing installation work, their cost, as well as the ability to increase fire resistance without dismantling the main structures.

When fire-retardant cladding with mineral wool is used, an important question arises of preserving its integrity to ensure appropriate fire-retardant functions during the entire time of fire exposure until the onset of the limiting state of loss of fire resistance in terms of bearing capacity. Taking this into account, it is important to study the behavior of a steel beam with a fire-retardant mineral wool cladding and to establish the conditions under which the destruction of the cladding is possible with a possible loss of fire-retardant ability. Considering such conditions allows us to more accurately take into account the thermal effect of a fire and obtain updated data on the fire resistance of steel structures with a similar fire-retardant cladding, in particular, I-beams, as one of the most common elements of such structures. This can be achieved by switching at the appropriate time from the calculation of the temperature in condition of the fire-retardant cladding existing to the calculation when it is absent. Obviously, the limit of fire resistance will be less value. Taking this feature into account, the risk of erroneous assessment of compliance with the required fire resistance class can be reduced and in this way fire safety level of such type building structures can be significantly improved. The above material is distinguishing actual of studies in this area.

2. Analysis of recent achievements and publications

Both experiments and calculations are devoted to the study of the thermal conductivity coefficient of building materials [1–8].

In [1], four options for assessing the fire protection of steel structures are described. Alternatives to the use of unprotected steel by the method of geometric average methodological and analytical hierarchy are also shown. However, using of mineral wool as a fire protection is not considered.

The work [2] is devoted to the study of realistic behavior of steel structures under the influence of fire. However, the behavior of fire-retardant structures has not been studied.

In [3-4] the results of application of the finite element method to obtain the necessary database for assessing the fire resistance of building structures are shown.

The paper [7] presents the essence of the published Brazilian standards for the design of fire-retardant steel structures. One of the options for fire protection is the use of mineral wool cladding. However, the requirements for the thickness and quality of the cladding are not given.

In the study [8], according to international recommendations, the results of the experimental study and the numerical calculation of the temperature inside the thermally protected steel structures under the action of fire were compared. However, the models were not verified.

3. Purpose

The purpose of this work is to identify the conditions under which the fire-retardant coating of mineral wool for a steel beam loses its fire-retardant ability due to loss of integrity under the thermal influence of the standard temperature of the fire. To do this, a numerical experiment was performed using a computer model created using the computer system LS-DYNA.

4. Method

To simulate the stress-strain state in a steel beam with fire-retardant cladding of mineral wool, we used a generalized engineering approach, which is based on the following provisions.

To calculate the stress-strain state of a deformed body, a general theoretical approach is used, which is based on the initiation of displacements of point of the system of deformed bodies using a system of generalized dynamics equations and stress-strain equations as a response to the displacement data, which is approximated by the finite element method, implemented in the computer system LS-DYNA.

To model a steel beam, two-dimensional shell finite elements (FE) of a rectangular shape of the Belichko-Tsay type [9-10] with four nodal points and five points of integration over thickness are used.

Three-dimensional massive hexahedral FE with eight nodal points are used for modeling mineral wool plates of fire-retardant facing.

As a model of steel beam material, a thermoelastic material with the possibility of plastic deformations is used, the deformation diagrams of which correspond to the recommendations of the second part of Eurocode 3, the shape of which includes only the growth section and the horizontal section. The descending branch is not taken into account, since its presence does not significantly affect the nature of the deformation of the steel beam, due to the prevalence of transverse displacements under conditions of section stability loss.

The Blatz-Co model is used to describe the nonlinear behavior of the material of mineral wool plates of fire-retardant facing [9-10].

It is assumed that the disturbance of the connection tightness between the mineral wool fire-retardant plates and the surface of the beam occurs under the condition of mineral wool delamination, because according to the technical conditions of the manufacturer, the glue has good adhesion and is resistant to high temperatures, while the tensile strength at separation and shear of the mineral wool itself is much lower [7].

To describe the performance of the connection between the mineral wool plates of fire retardant cladding and the surface, the model of contact interaction with its violation is used, provided that the shear strength and abruption strength between the contacting surfaces are equal to the corresponding strength of the mineral wool itself as the least durable material.

The application of the load has a dynamic history and occurs gradually with the initial application of its own weight, the actual load and the application of the temperature load as calculated in the last stage.

The structural scheme of a beam with a protective mineral wool cladding is shown in Figure 1.

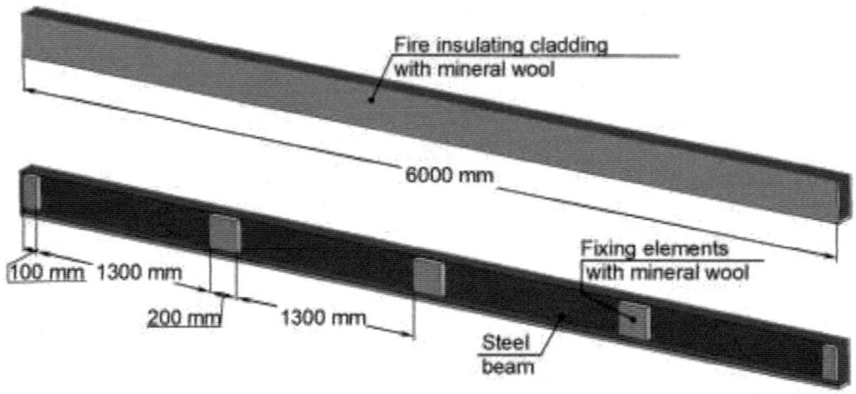

Figure 1. Structural scheme of a steel beam with fire-retardant facing on the basis of mineral wool.

For the calculation, it was assumed that the likely separation of the fire retardant cladding occurs not along the adhesive seam, but along the adjacent layers of mineral wool, that is, the destruction is cohesive. This assumption is made taking into account the fact that the strength of the adhesive seam is much higher than the tensile strength and shear strength of the mineral itself. This assumption is taken into account by means of the LS-DYNA software environment when describing the contact interaction between the elements of the model. The parameters of contact interaction are shown in Figure 2.

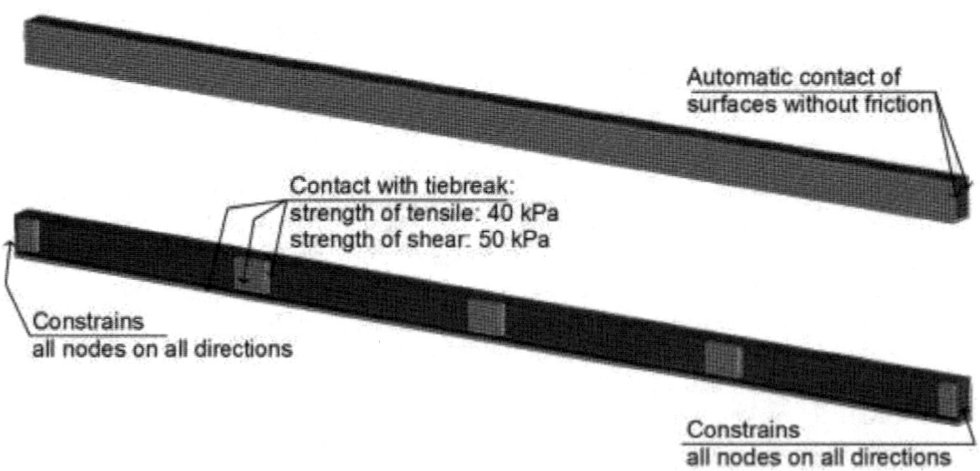

Figure 2. Finite element scheme of a steel beam with fire-retardant mineral wool cladding with boundary conditions and conditions of contact interaction

IOP Conf. Series: Materials Science and Engineering 1021 (2021) 012024 doi:10.1088/1757-899X/1021/1/012024

The scheme of load application of a steel beam with fire-retardant cladding from mineral wool, which is resulted as shown in Figure 3, shows type and a direction of the applied loading.

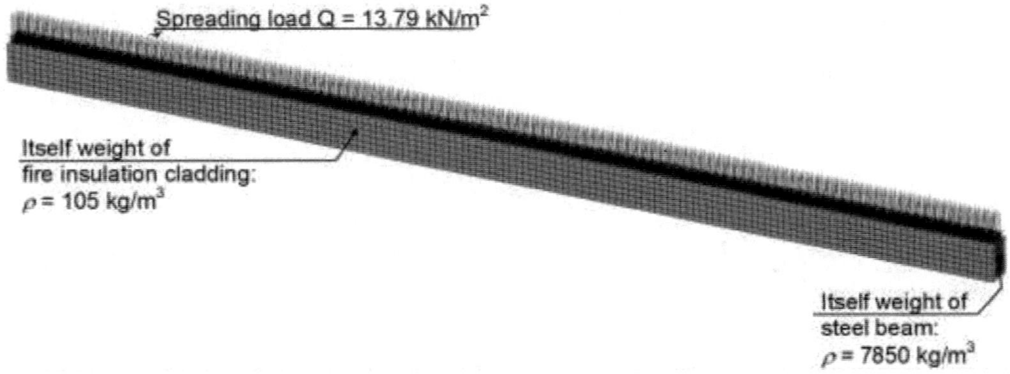

Figure 3. Scheme of load application to a steel beam with fire-retardant cladding on the basis of mineral wool with boundary conditions and conditions of contact interaction

5. Consideration on methods and results

After the calculation, the results were obtained on the behavior of a steel beam with fire-retardant cladding of mineral wool under the influence of the fire standard temperature.

Figure 4 shows graphs of changes in the maximum deflection of the beam and the rate of increase in the maximum deflection depending on the time of exposure to the standard temperature mode of the fire.

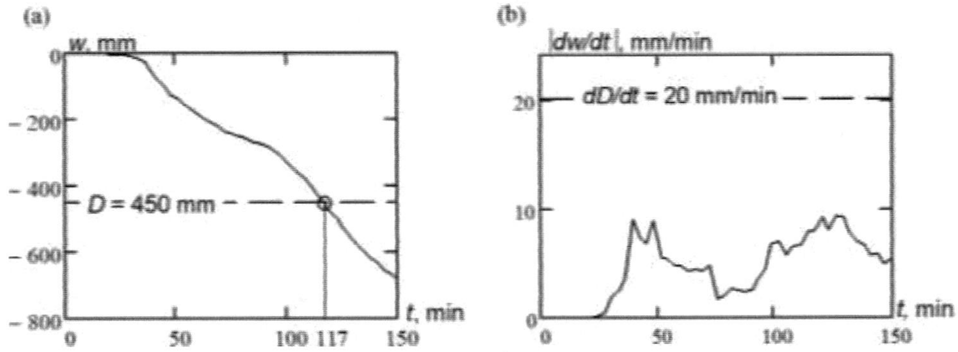

Figure 4. Graphs of changes in the maximum deflection of the beam (a) and the rate of increase of the maximum deflection (b) depending on the time of exposure to the standard temperature mode of the fire.

Figure 4 shows the critical values of the maximum deflection D of the beam and the rate of increase of the maximum deflection of the beam, above which the onset of the state of loss of bearing capacity is recorded. Critical values are determined by the formulas used in [11-12]:

$$D = \frac{l^2}{400H}, \quad \frac{dD}{dt} = \frac{l^2}{9000H}, \tag{1}$$

where l is beam span length (mm), H is height of beam cross-section (mm).

According to the graphs given in Figure 4 it can be concluded that the onset of the state of loss of fire resistance by bearing capacity occurs at 117 min under the influence of the standard temperature mode of the fire.

In order to study the integrity of the fire-retardant mineral wool cladding of the tested steel beam, the positions and geometric shape of the cladding at different points in time of exposure to the standard temperature mode of the fire, shown in Figure 5.

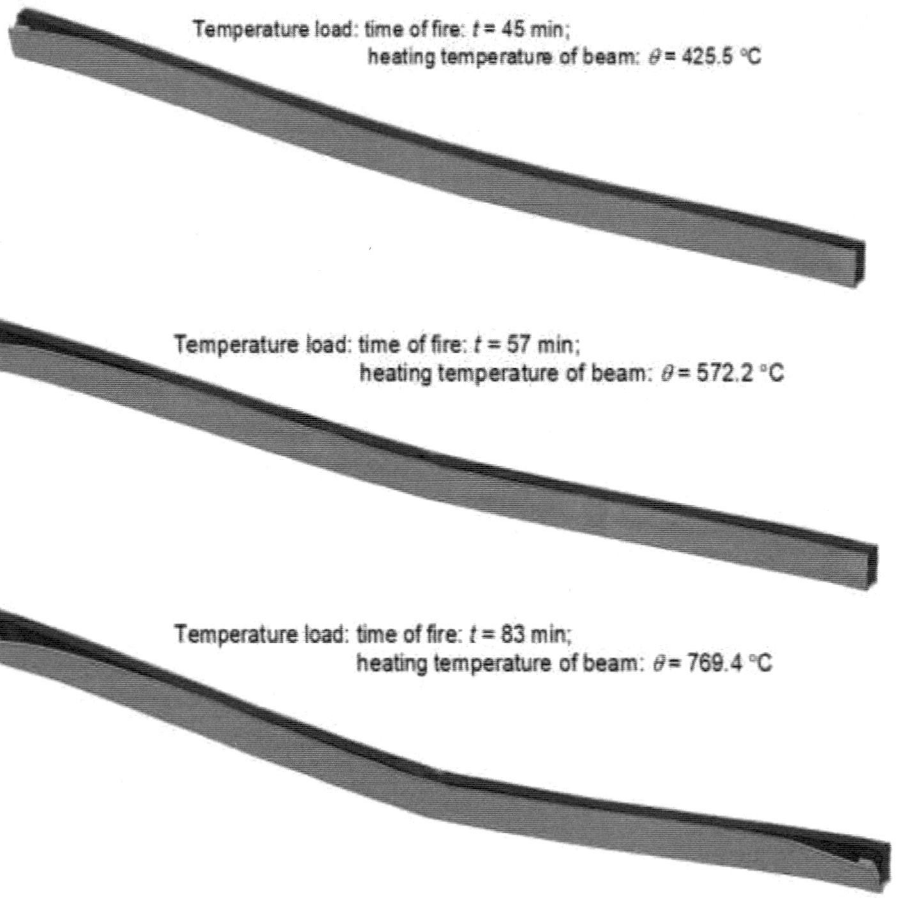

Figure 5. Position and geometric shape of the cladding at different points in the time of exposure to the standard temperature of the fire.

The graph of the gap depending on the time of exposure to the standard temperature mode of the fire is shown in Figure 5.

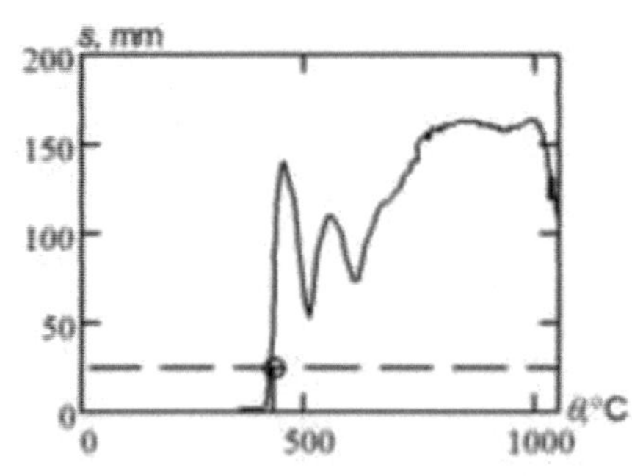

Figure 6. Graph of the gap between the cladding and the edge of the steel beam depending on the heating temperature.

These graphs indicate the moment of opening the gap with a delamination of 25 mm. It was found that the opening of the gap between the beam and the cladding of 25 mm in size is observed at a time of 44.5 min at a heating temperature of the steel beam 425°C.

6. Conclusions

Conducted research lets to make such conclusions.

1. The behavior of a steel beam with fire-retardant mineral wool cladding under the influence of the standard temperature mode of the fire in the period from 0 to 150 minutes was studied.

2. During fire exposure of 60 minutes at a temperature of 433°C, detachments of fastening mineral wool elements are observed, and at a temperature of 967°C, signs of local loss of stability are observed in the form of folds in the shelves and the wall of the I-beam in the middle and at its fixed ends it was found that the onset of the state of loss of fire resistance by the bearing capacity of the investigated steel beam with mineral wool fire-retardant cladding occurs at 117 minutes of exposure to the standard temperature mode of the fire.

3. It was found that the opening of the gap between the beam and the mineral wool fire-retardant cladding 25 mm in size is observed at a time of 44.5 min at a heating temperature of the steel beam 425°C.

4. It was found that the loss of fire-retardant capacity of mineral wool on the basis of opening the gap between the beam and cladding 25 mm in size occurs earlier by 73 min at the onset of the limit state of fire-resistance loss, which must be taken into account in the calculations.

References

[1] Akaa O U Abu A Spearpoint M Giovinazzi S 2016 A group-AHP decision analysis for the selection of applied fire protection to steel structures *Fire Safety Journal* **86** pp 95-105

[2] Wang Y C & Kodur V K R 2000 Research toward use of unprotected steel structures *Journal of structural engineering* **126 (12)** pp 1442-1450

[3] Nekora O Slovynsky V Pozdieiev S 2017 The research of bearing capacity of reinforced concrete beam with use combined experimental-computational method *MATEC Web of Conferences* **116** p 02024

[4] Pozdieiev S., Nuianzin O., Sidnei S., Shchipets S. 2017 Computational study of bearing walls fire resistance tests efficiency using different combustion furnaces configurations *MATEC Web of Conferences*, 116 p 02027.

[5] Akaa, O. U., Abu, A., Spearpoint, M., & Giovinazzi, S. 2016 A group-AHP decision analysis for the selection of applied fire protection to steel structures. *Fire Safety Journal* **86** pp 95-105.

[6] Bilotta, A., de Silva, D., & Nigro, E. 2016 Tests on intumescent paints for fire protection of existing steel structures. Construction and Building Materials **121** pp 410-422.

[7] Silva V P Fakury R H 2002 Brazilian standards for steel structures fire design *Fire Safety Journal* **37(2)** pp 217-227

[8] Silva V P 2005 Determination of the steel fire protection material thickness by an analytical process – a simple derivation. *Engineering Structures* **27(14)** pp 2036-2043

[9] Hallquist J O LS-DYNA 2005 Theory Manual *Livermore Software Technology Corporation California USA*

[10] Belytschko, T., Chiapetta, R.L. & Bartel, H.D. 1976 Efficient Large Scale Non-Linear Transient Analysis by Finite Elements. *International Journal for Numerical Methods in Engineering* **10**, p 579-596.

[11] DSTU B V.1.1-4-98. 2005 Building structures. Fire resistance testing methods. General requirements. Fire safety. *Kyiv Ukrarhbudinform* 36 p.

[12] DSTU B V.1.1-13:2007. 2008 Fire protection. Beams. Fire resistance testing method. *Kyiv Ukrarhbudinform* 12 p.

IOP Conf. Series: Materials Science and Engineering 1021 (2021) 012025 doi:10.1088/1757-899X/1021/1/012025

Research of Disclosure of Relief Venting Structures with Polycarbonate Fencing in Conditions of Explosion

Serhii Pozdieiev[1,5], Vadym Nizhnyk[2], Yuriy Pidhoretskiy[3], Andriy Shvydenko[4]

[1]Science department of Cherkasy Institute of Fire Safety named after Chernobyl Heroes, National University of Civil Defence of Ukraine, 94 Chernyshevska Str. Kharkiv 61023, Ukraine

[2]Fire Protection Research Center of Institute of Public Administration and Research in Civil Protection, 18, Rybalska St., Kyiv, Ukraine, 01011

[3]Adjunctura of Cherkasy Institute of Fire Safety named after Chernobyl Heroes, National University of Civil Defence of Ukraine, 94 Chernyshevska Str. Kharkiv 61023, Ukraine

[4]Department of Civil Defense of Cherkasy Institute of Fire Safety named after Chernobyl Heroes, National University of Civil Defence of Ukraine, 94 Chernyshevska Str. Kharkiv 61023, Ukraine

[5]svp_chipbbk@ukr.net

Abstract. The article considers the processes of release of the openings of relief venting structures based on honeycomb polycarbonate sheets under action of an explosion. The use of honeycomb polycarbonate sheets enables saving transparent fence against destruction, decreasing risks of human traumatism with glass shards, reduce weight of relief venting structures. There were investigated the conditions of the ratio of the opening sizes of the sections of relief venting structures with honeycomb polycarbonate sheets, under which this opening is reliably released to reduce the pressure during the explosion to safe values. A mathematical model was proposed that establishes a relationship between the value of the explosion pressure and the opening size at which the opening of relief structures with honeycomb polycarbonate sheets. The article proposes a calculation method for designing the sections of relief structures with honeycomb polycarbonate sheets using nomograms. The use of developed method lets significantly simplify calculating for design of relief structures with honeycomb polycarbonate sheets.

1. Introduction

Premises of categories A and B, as to the explosion and fire hazard, in accordance with the norms in force in Ukraine DSTU B B.1.1-36: 2016 [1], should be equipped with external relief venting structures (RVS). In this case, this document states that the required area of RVS should be determined by calculation. The use of flexible fencing elements installed in standard window profiles made of polyvinyl chloride polymeric material is promising for the RVS construction, as they can be used repeatedly after their removal from the frames in case of an explosion.

The use of polycarbonate honeycomb sheets is promising as flexible fencing elements [2]. When using such a design of RVS it is necessary to study the main features of its behavior in case of an explosion, and to determine the conditions of releasing the edges of flexible elements from the locks of window frames of standard profile. At the same time, it is important to ensure that the design dimensions of

RVS openings are such that their release under the action of excess pressure from explosion with a boundary value would be reliable and guarantee a rapid decrease of the pressure to safe values.

2. Analysis of recent achievements and publications

As a result of the analysis of scientific studies it was found that ensuring safety from an explosion in industrial buildings with certain technological processes is quite important and the main technical solution for this is the installation of relief structures [2-10].

When installing relief venting structures to protect buildings and structures from explosion, there are methods of their designing with the calculation of explosion resistance in the deflagration explosion of gas-air mixtures in premises [1]. As an example of such effective methods of RVS designing can be considered the method given in the Technical Code of Established Practice, which is valid in the Republic of Belarus TKP 45-2.02-38-2006 (02250) [11], the methods given in other standards in this field, for example, the US standard NFPA 68 [12].

British standard BSEN 14491: 2012 [13], there are also a number of monographs, publications, theses [14], etc., which address the issues of explosion protection of explosion hazardous facilities with the use of safety structures.

A study [2] of the behavior of RVS fencing based on cellular polycarbonate sheets (CPS) proposed a mathematical model describing the behavior of CPS under explosion conditions, but did not propose methods for designing such RVSs based on it.

3. Purpose

The purpose of this work is to develop a method for calculating the design dimensions of the openings in relief structures for their reliable release for the boundary values of excess explosion pressure by developing special nomograms.

4. Method

When considering the CPS fixed in the frames of the window profile, it is considered as a rectangular plate, hinged with the possibility of free movement of the CPS ends in the longitudinal direction, loaded with an evenly distributed load. figure 1 shows a calculated scheme of application of the load on the CPS under the influence of the explosion on it, in accordance with the approach developed in the study [2].

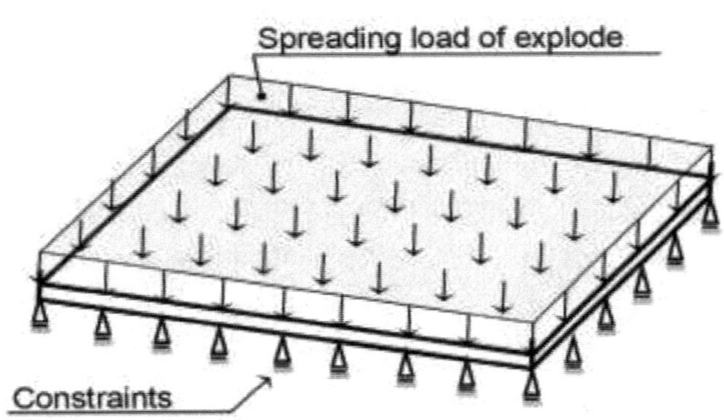

Figure 1. Scheme of load application on the cellular polycarbonate sheet in the section of the relief structure

According to the approach, developed in [2], at the moment of release of the CPS edges from the locks of the standard window profile, their behavior can be described by the equilibrium equation under the action of excess explosion pressure (P_e), distributed friction forces along the mounting unit in the element frame (q_{fl}) and internal elastic forces arising in the middle of the CPS due to deformation. Given that the deformation of the CPS occurs in the elastic zone, the equilibrium equation for the CPS appears as a following expression [2]:

$$P_e \cdot a \cdot b - \frac{w_{max}\pi^6 D \cdot a \cdot b}{16} \times \left[\sum_{m=0}^{\infty} \sum_{n=0}^{\infty} \frac{(-1)^{\frac{(2m+1)+(2n+1)}{2}-1}}{(2m+1)(2n+1)\left(\frac{(2m+1)^2}{a^2} + \frac{(2n+1)^2}{b^2}\right)^2} \right]^{-1} -$$

$$-2q_{fr}\left(\frac{b\pi w_{max}}{\sqrt{a^2+\pi^2 w_{max}^2}} + \frac{a\pi w_{max}}{\sqrt{b^2+\pi^2 w_{max}^2}}\right) = 0. \quad (1)$$

here a and b are the height and width of the opening in the RVS section, according to Figure 2; D is the effective parameter of CPS stiffness, m = 1, 3, 5, 7,…; n = 1, 3, 5, 7,… are odd whole numbers, w_{max} is the maximum deflection of the CPS, at which its edges release from the locks of the window profile of the RVS sections.

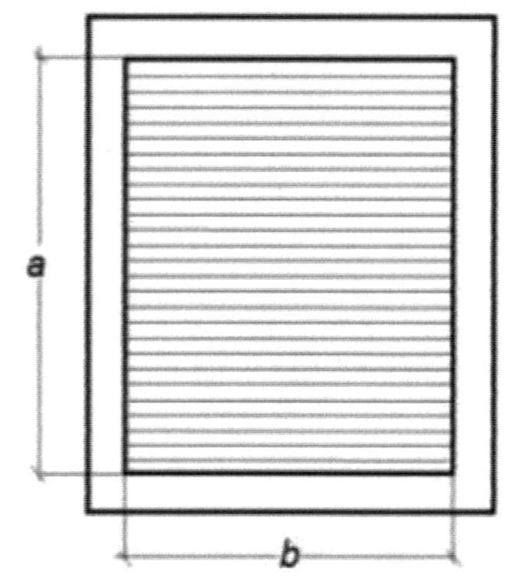

Figure 2. Scheme of the opening in the section of the relief venting structure with honeycomb polycarbonate sheets

The maximum deflection of the CPS is calculated by the following expression:

$$w_{max} = \frac{9a^2}{\pi(a-2\Delta x_{max})} - \frac{12a}{\pi} + \frac{3}{\pi}(a-2\Delta x_{max}) \quad (2)$$

here x_{max} is the maximum displacement of the CPS edge when it is released out of the locks of the standard window profile of the RVS section.

The parameters of the effective stiffness of the CPS D and the maximum displacement x_{max} do not have a specific physical meaning in this case, but are the effective characteristics of the CPS, which are determined experimentally using the setup shown in Figure 3.

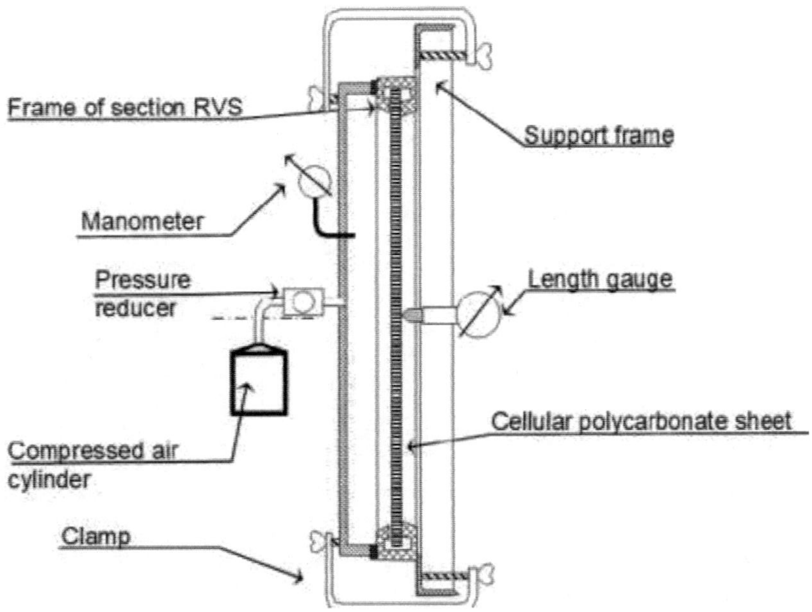

Figure 3. Scheme of an experimental setup to determine the effective stiffness

It was found that the effective stiffness and critical displacement depend only on the thickness of the CPS and their values are given in Table 1.

Table 1. Effective stiffness and critical displacement characteristics of cellular polycarbonate sheets

Effective parameter	Thickness of honeycomb polycarbonate sheet		
	4 mm	6 mm	8 mm
Stiffness, D, N·m	289.064	244.196	206.044
Maximal displacement, x_{max}, mm	2.915	7.708	9.773

Using these characteristics and expression (2), the equation takes the form, which is given below:

$$P_{виб}ab-\left[\frac{9a^2}{\pi\left(a-2\Delta x_{кр}\right)}-\frac{12a}{\pi}+\frac{3}{\pi}\left(a-2\Delta x_{кр}\right)\right]\frac{\pi^6 abD_h}{16}\left[\sum_{m=0}^{\infty}\sum_{n=0}^{\infty}\frac{(-1)^{\frac{(2m+1)+(2n+1)}{2}-1}}{(2m+1)(2n+1)\left(\frac{(2m+1)^2}{a^2}+\frac{(2n+1)^2}{b^2}\right)^2}\right]^{-1}-$$

$$-\frac{2q_{тр}\left(\frac{9a^2}{\pi\left(a-2\Delta x_{кр}\right)}-\frac{12a}{\pi}+\frac{3}{\pi}\left(a-2\Delta x_{кр}\right)\right)b\pi}{\sqrt{a^2+\pi^2\left(\frac{9a^2}{\pi\left(a-2\Delta x_{кр}\right)}-\frac{12a}{\pi}+\frac{3}{\pi}\left(a-2\Delta x_{кр}\right)\right)^2}}-\frac{2q_{тр}\left(\frac{9a^2}{\pi\left(a-2\Delta x_{кр}\right)}-\frac{12a}{\pi}+\frac{3}{\pi}\left(a-2\Delta x_{кр}\right)\right)a\pi}{\sqrt{b^2+\pi^2\left(\frac{9a^2}{\pi\left(a-2\Delta x_{кр}\right)}-\frac{12a}{\pi}+\frac{3}{\pi}\left(a-2\Delta x_{кр}\right)\right)^2}}=0 \quad (3)$$

The obtained equation can be used to determine the design dimensions a and b of the openings of the sections of the RVS with CPS, by solving it with respect to these parameters as unknown.

5. Consideration on methods and results

Using equation (3), the nomograms were built to determine the design parameters of the openings of the CPS sections in accordance with Figure 2 for certain boundary values of excess pressure from explosion. The built nomograms are shown in Figure 4.

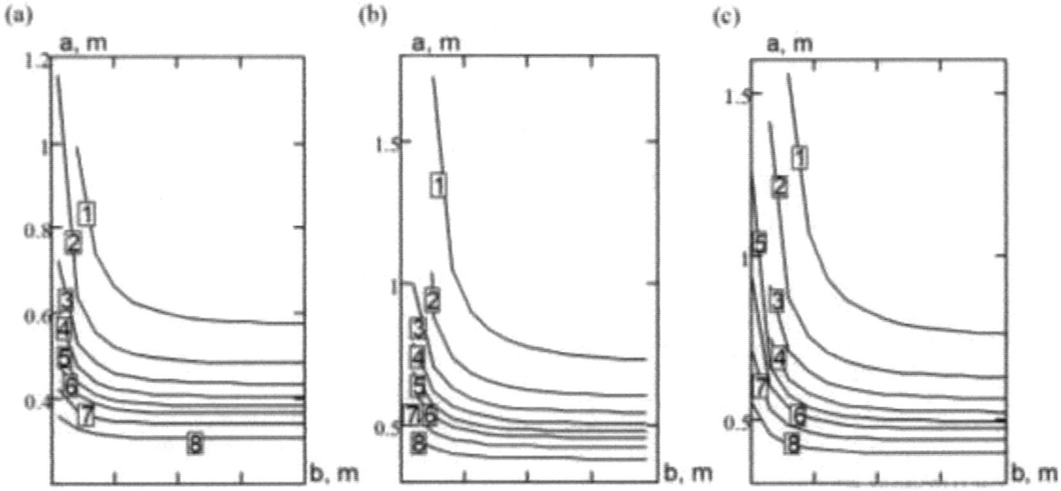

Figure 4. Nomograms to determine the parameter a, depending on the parameter b of relief venting structures based on cellular polycarbonate sheets with a thickness of 4 mm (a), thickness of 6 mm (b) and thickness of 8 mm (c) at a certain excess pressure for which it must be provided the release: 1 – 2,500 Pa; 2 – 5,000 Pa; 3 – 7,500 Pa; 4 – 10,000 Pa; 5 – 12,500 Pa; 6 – 15,000 Pa; 7 - 20,000 Pa; 8 - 30,000 Pa.

When determining the design parameter a, according to the previously accepted values of the excess pressure from explosion and the height of the opening of the RVS section, one can use the nomograms shown in figure 4. Excess pressure is determined from the preliminary calculation, and the height of the opening of the RVS section can be determined basing on the design of the premises and the convenience of the location of the RVS sections in the premises. When choosing the height of the RVS sections, there is a restriction: the height of the opening of the RVS section must be at least 500 mm. When using nomograms from Figure 4, linear interpolation can be used to determine the design parameter a from the intermediate values of the excess pressure from explosion and the height of the opening of the RVS section.

Thus, the method of selecting the design parameters of the RVS sections on the basis of CPS consists of performing the following procedures.

The minimum height of the RVS section a is selected on the basis of the design and architecture requirements for construction of the relevant floor. That is, the height can be set by the designer of the building.

The value of excess pressure from explosion is set by means of preliminary calculation.

Depending on the RVS location, the thickness of the CPS is selected. If the RVS is indoors, a CPS with 4 mm thickness should be used. If the RVS is installed in the fence of the building from the outside space, there should be installed CPS with 8 mm thickness. The installation of 6 mm thick CPS requires additional justification, taking into account wind loads in the area.

According to the values of excess pressure and thickness of the CPS, the appropriate curve is determined to determine the design parameter of the opening width of the RVS section. According to this curve, the value of the opening height of the RVS section and the corresponding values of the opening width are selected. When finally determining the width of the opening, the design or architectural solutions can be taken into account by increasing the width of the opening from a certain minimum value, but not more than 1.75 m.

6. Conclusions

As a result of our research we can point following.

1. To develop a method for determining the dimensions of the opening of the RVS section with CPS for their guaranteed release after reaching the critical value of the excess pressure from the explosion, an equation was proposed in which these dimensions are included as independent variables.

2. According to the obtained equation, the nomograms were built to determine the geometric dimensions of RVS openings with CPS for their guaranteed release after reaching the critical value of excess explosion pressure; tables were developed to determine the design dimensions of RVS openings.

3. A method for designing RVS out of CPS on the basis of the offered nomograms was developed.

4. It was shown that the developed method has a simple implementation, cost-effective calculation algorithms and allows designing effective RVS based on CPS.

References

[1] DSTU [National Standardization System of Ukraine] B V.1.1-36: 2016 Definition of categories of premises, buildings, installations according to explosion and fire danger.

[2] Pozdieiev S., Nuianzin O., Sidnei S., Shchipets S. 2017 Computational study of bearing walls fire resistance tests efficiency using different combustion furnaces configurations *MATEC Web of Conferences*, **116**, 02027.

[3] Pozdieiev, S., Pidgoretskiy, Y., Nekora, O., Sidnei, S., Tyshchenko, O. 2018. Research of Explode Exposure at the Relief Vent System Structures with Soft Transparent Material. *International Journal of Engineering & Technology*, **7** (4.3), 298.

[4] Danchenko Y., Andronov V., Teslenko M., Permiakov V., Rybka E., Meleshchenko R. 2018 Kosse A. Study of the free surface energy of epoxy composites using an automated measurement system *EasternEuropean Journal of Enterprise Technologies* **1** (12-91), pp. 9-17.

[5] Pospelov B., Andronov V., Rybka E., Popov V., Romin A. 2018 Experimental study of the fluctuations of gas medium parameters as early signs of fire *EasternEuropean Journal of Enterprise Technologies* **1** (10-91), pp. 50-55.

[6] Boris Pospelov, Vladimir Andronov, Evgeniy Rybka, Vadym Popov, Oleg Semkiv Development of the method of frequency-temporal representation of fluctuations of gaseous medium parameters at fire 2018 **Vol 2**, **10** (92), pp. 64-66.

[7] Hall, R. W. and Board, S. J. 1979 The Propagation of Large Scale Thermal Explosions, *Int. J. Heat and Mass Trans.* **22** pp 1083-1093.

[8] Bradley D., Mitcheson A 1978 The Venting of Gaseous Explosion in Spherical Vessels *Combustion and Flame* **32** pp 221-236.

[9] Board, S. J., Farmer, C. L., and Poole, D. H. 1974 Fragmentation in Thermal Explosions, *Int. J. Heat Mass Transfer* **17** pp 331-339.

[10] Rabie, R. L., Fowlen, G. R., and Fickett, W. 1979 The Polymorphic Detonation *Physics of Fluids* **22** pp 222-235.

[11] TCP 45-2.02-38-2006 (02250) 2006 Relief panels. Calculation rules. *Minsk: Ministry of Architecture and Construction of the Republic of Belarus*, 27 p.

[12] NFPA 68. Standard on Explosion Protection by Deflagration Venting. 2013 Edition.

[13] BS EN 14491:2012. Dust Explosion Venting Protective Systems.

[14] L. Pilyugin. 1988 Structures of buildings of explosion-hazardous production. *Moscow: Stroizdat*, 315 p.

IOP Conf. Series: Materials Science and Engineering 1021 (2021) 012026 doi:10.1088/1757-899X/1021/1/012026

General model and the mechanics of concrete elements and structures deformation

V M Romashko

Architectural Designing Bases, Constructing and Graphics Department, National University of Water Management and Nature Recourses Use, Soborna str. 11, 33028 Rivne, Ukraine

Email: romashkovasyl@gmail.com

Abstract. The article presents a critical analysis of the model development of reinforced concrete elements and structures resistance by force influences. Their relationship with the defining provisions of the mechanics of reinforced concrete elements deformation was evaluated. The main disadvantages of the models considered are identified. A universal deformation and force model of deformation of reinforced concrete elements and structures is proposed, which is free of most of the disadvantageous characteristic of other models. The main advantage of the deformation-force model of resistance is indicated, due to the expansion of the known system of static, geometric and physical relationships to the level of statically definable. It becomes noticeable when calculating reinforced concrete elements and structures for deflection and crack resistance. This is one of the main features of the deformation-force model, which makes it possible to quite simply "embed" the technical theory of reinforcement to concrete adhesion into the general theory of reinforced concrete deformation. Due to the averaged design cross-section and the averaged materials deformations in the block between cracks, all parameters of cracks formation and propagation in the deformation-force model are directly related to the parameters of reinforcement to concrete adhesion.

1. Introduction

Reinforced concrete theory is not just a set of provisions and statements about the regularities of concrete and reinforcement deformation and their joint interaction. As a true scientific theory, it must simultaneously describe the properties of these materials, explain the holistic picture of the reinforced concrete elements and structures deformation laws and provide opportunities to determine their true stress-strain state at any stage of deformation. Therefore, its development has always been accompanied by the improvement of the following structural elements:

- prerequisites – mainly in the form of fundamental concepts and principles, laws and axioms, assumptions and hypotheses specific to reinforced concrete;
- an idealized object, mainly in the form of abstract models of the definitions, properties and interconnections of reinforced concrete as a solid body;
- logic – in the form of a set of specific relationships (rules, methods and means) aimed at clarifying the structure of knowledge about concrete and aimed at their development
- conclusions – as a set of laws and statements, methods and techniques capable of predicting the condition of reinforced concrete elements and structures at any stage of their deformation.

The degree of mutual coherence of these structural elements continues to affect the completeness and effectiveness of the developed methods of calculating reinforced concrete elements.

IOP Conf. Series: Materials Science and Engineering 1021 (2021) 012026 doi:10.1088/1757-899X/1021/1/012026

2. Analysis of recent research and publications

In the early stages of its development, concrete was modeled as a completely solid (elastic) body. His condition was described by a system of static, geometric and physical relations, called the corresponding equations of solid-state mechanics. These relationships were based on the following assumptions, hypotheses and preconditions:

- the reinforcement and concrete of the compressed zone are deformed elastically by Hooke's law;
- the concrete does not accept the extended area of participation in the work of the reinforced concrete element;
- in sections normal to the longitudinal axis of the element, the flat section hypothesis holds.

All this together allowed us to form a methodology for calculating reinforced concrete elements and structures by “allowable stresses”. However, it was later found that the actual stress distribution in concrete over the height of the cross-section of the element does not correspond to the triangular stress plot, and therefore concrete should not be considered as elastic material.

At the second stage of the reinforced concrete theory development, since the early 1930's, it began to be modeled as a deformed solid. Since taking into account the elastic-plastic properties of concrete required correction of the corresponding ratios (equations) of the mechanics of deformed solid (MDS), it was implemented due to the following assumptions:

- the hypothesis of boundary equilibrium has been advanced, according to which the structure material is in the limiting state before destruction and the cross-section of the reinforced concrete element is in equilibrium [1];
- the performance of the compressed zone concrete was described by nonlinear deformation diagrams;
- the performance of the stretched concrete in deforming the reinforced concrete element was not taken into account.

It is on their basis in the late 1930's, a new method of calculating cross-sections of reinforced concrete structures by “destructive efforts” was formed and introduced into the normative documents of many countries of the world. Since then, the development of general theory of reinforced concrete has been divided into two separate directions.

In most countries of the world, it continued to develop on a conditionally “deformation” model, which involved the use of additional prerequisites:

- the operation of the compressed zone concrete in the cross section of the element is described by idealized but nonlinear deformation diagrams;
- in cross sections normal to the longitudinal axis of the reinforced concrete element, the flat section hypothesis remains valid throughout the deformation process.

In the USSR and some other countries, a “power” model of reinforced concrete deformation was adopted. According to her, it was assumed that the destruction of structures in the limiting stage is due to the formation of so-called “hinges of plasticity”. It was suggested to use in the calculations a rectangular plot of stresses in the concrete of the compressed zone [2] without involving the hypothesis of flat sections. This simplification made it possible to significantly simplify the calculations and achieve a significant economic effect for many types of reinforced concrete slabs, non-cut beams and frames compared to the allowable stress method. However, simplified calculations of reinforced concrete structural elements sections by “destructive forces” could not cover the issue of their rigidity and fracture toughness. In addition, possible deviations of actual loads and strength characteristics of materials from their calculated values were taken into account by one synthesizing factor of safety margin. And this made it very difficult to adjust it. But the biggest methodological disadvantage of simplified calculations for “destructive efforts” was that the use of the “plastic hinge” model removed the theory of reinforced concrete from the methods of construction mechanics.

The further development of the theory of reinforced concrete structural elements deformation by the model of “plastic hinge” was the calculation method of structures cross sections by boundary states, proposed in the late 1950's [3, 4]. Its implementation was associated with a system of coefficients, which ensured that such conditions would not follow either in the most unfavorable combination of all loads or in the lowest values of the strength materials characteristics. The load bearing strength of reinforced concrete elements remained virtually unchanged. The introduction of the system of the above calculated reliability coefficients allowed to develop "empirical" methods of calculating reinforced concrete elements for rigidity and fracture toughness, but, unfortunately, did not ensure the methodological unity of structures calculations by boundary states.

By the beginning of the 21st century, it became clear to everyone that the "power" model had completely exhausted itself. Since then, the theories of reinforced concrete, in general, and the method of calculating cross-sections of structures by boundary states, in particular, have been oriented to the "deformation" model. With the development of computer technology, this model is not only finally established in the Western countries, but also gained recognition in the countries of Eastern Europe and the former USSR. However, now it seems that the "deformation" model already needs significant refinement, since often "idealized objects" (abstract models of reinforced concrete) are rather poorly linked to the "logic" of his research, and therefore:

- technical theory of reinforcement with concrete coupling is extremely poorly integrated into the general theory of reinforced concrete;
- the question of methodological (dialectical) unity of structures calculations by boundary states is almost never solved;
- the internal static uncertainty of reinforced concrete elements, even in the load bearing strength calculations, is in most cases revealed by numerous iterations;
- in the calculations of fracture toughness and deflection, there is a need not only for numerous iterative operations, but also for the use of various empirical parameters, dependencies and coefficients;
- the lack of universal calculating methods of reinforced concrete elements and structures that could be implemented both programmatically and by engineering express methods, leads not only to the loss of engineering complexity of the calculation results, but also to the physical nature of the deformation processes.

3. The purpose and objectives of research

These studies are aimed at forming a universal theory of deformation of reinforced concrete, which is based on:

- defining hypotheses and preconditions not only of deformation, but also of power models;
- an idealized object of study in the form of an abstract model of a deformed rigid body with elastic-plastic properties;
- sufficiently substantiated logic of research in the form of the generally recognized "extended" system of MDS equations;
- ability to calculate or predict the condition of reinforced concrete elements and structures at any stage of their deformation.

The construction of a universal theory of reinforced concrete on these principles will help to solve all the above problems related to the "logic" of research.

4. Results and discussion

Modern analysis of the research state in the field of reinforced concrete suggests that one of the most successful idealized objects of the general (universal) theory of its deformation can be considered a model of deformed solid. This model allows not only to reproduce the elastic-plastic properties of the concrete itself, but also to offer their variant solutions in the form of force, deformation and universal deformation-force models [5].

The variants themselves or the ways in which the general theory of concrete is developed depends to a great extent on the basic initial hypotheses and assumptions. And in this the decisive role should be given to the hypotheses of boundary equilibrium, flat sections and "nonlinearity" of rigidity [6].

In the universal deformation-force model, whose development is initiated by the author, the limit equilibrium hypothesis ($M_{Ed,max} = M_u$) is one of the most important. It is not only the basis for static relationships characterizing the reinforced concrete deformation, but also serves as a general criterion for the exhaustion of the load-bearing strength of a reinforced concrete element or structure [7].

Equally important is the hypothesis of flat sections, by which geometric relationships of reinforced concrete elements and structures deformation are formed. In general, there are a number of studies where the validity of this hypothesis is generally denied [8-10]. Indeed, in some sections of a reinforced concrete element with cracks, the flat section hypothesis does not work. But V. Ya. Bachinsky [11, 12] substantiated why the general stress-strain state (SSS) of a reinforced concrete element cannot be estimated by deformations and stresses of materials measured only in a separate section. It is not necessary to consider even the average, but the "focal" section of the element in the block between the cracks (Figure 1). And for him, in such circumstances, the conjecture of flat sections generally remains valid throughout the deformation process of an element. Perhaps this is why the vast majority of alternative solutions [8, 10, 13] also boil down to the "hidden" use of the flat cross-section hypothesis, rather than being constructed by the "deplanation" of normal cross sections.

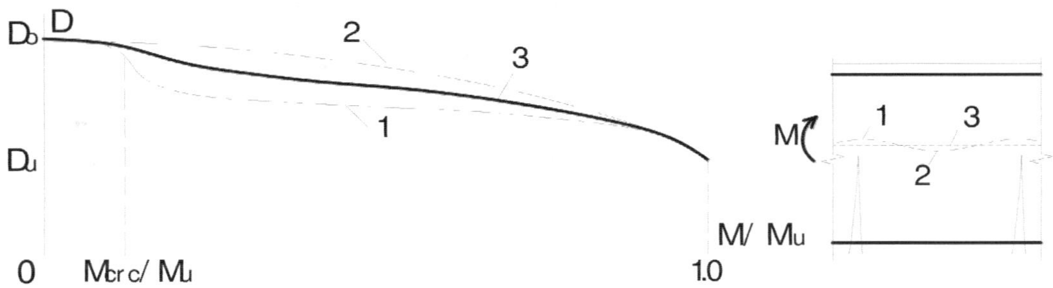

Figure 1. Stiffness diagrams of bending reinforced concrete element in section: 1 - with crack; 2 - between cracks; 3 - focused.

The decisive role in the formation of general (universal) theory of reinforced concrete is given to the hypothesis of rigidity "nonlinearity". It underlies the "deformation-force" model of reinforced concrete elements resistance to force influences [5]

$$M \cdot r = A - B \cdot r^{-1} \cdot r_u - C \cdot M \cdot M_u^{-1}, \qquad (1)$$

where A, B, C are the parameters that reflect the strength and deformation characteristics of the element section, which are directly related to its initial stiffness (D_o), curvature (r_u^{-1}) and bearing strength (M_u).

According to this hypothesis, the generalized diagram of the reinforced concrete element condition (Figure 2) takes on a universal appearance

$$M = \left[D_o \cdot r^{-1} - M_u \cdot (r_u \cdot r^{-1})^2\right] \cdot \left[1 + (D_o \cdot M_u^{-1} - 2 \cdot r_u) \cdot r^{-1}\right]^{-1} \qquad (2)$$

and is transformed, under certain initial conditions, into a generally recognized concrete deformation diagram

$$\sigma_c = f_{ck} \cdot \left[E_{co} \cdot \varepsilon_c \cdot (E_{cu} \cdot \varepsilon_{cu})^{-1} - (\varepsilon_c \cdot \varepsilon_{cu}^{-1})^2\right] \cdot \left[1 + (E_{co} \cdot E_{cu}^{-1} - 2) \cdot \varepsilon_c \cdot \varepsilon_{cu}^{-1})\right]^{-1}, \qquad (3)$$

where f_{ck} - the characteristic value of the concrete compressive strength; E_{co} - initial value of modulus (deformation) of concrete; E_{cu} - limit value of the concrete deformation module; ε_{cu} - value of compressed concrete boundary deformation.

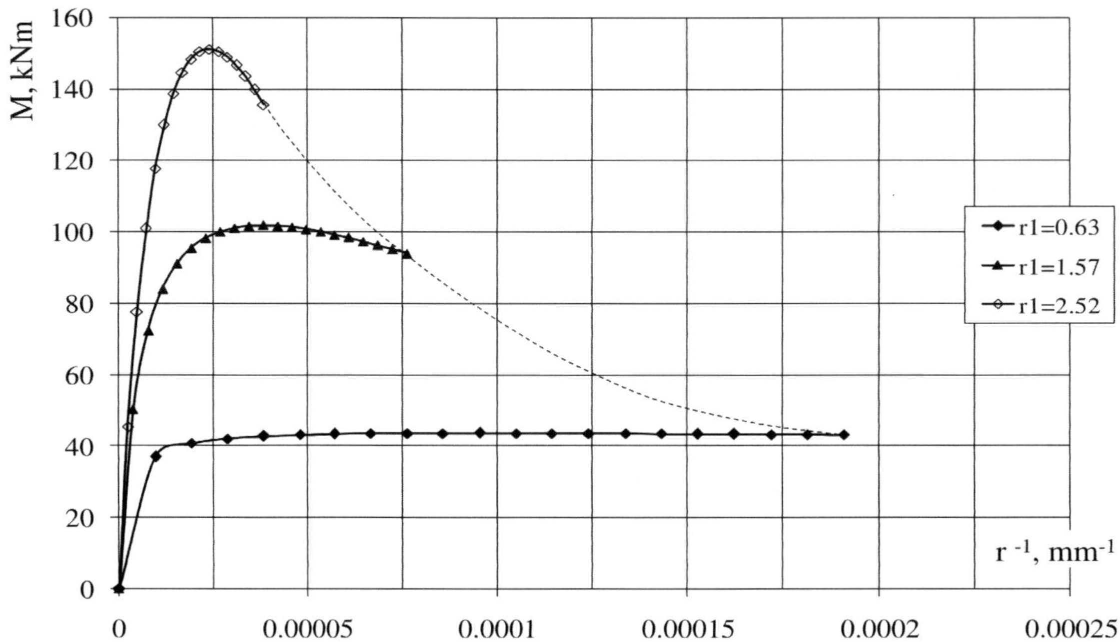

Figure 2. Forms of the state diagrams of the bending reinforced concrete element depending on the percentage of its reinforcement ($\rho_1 = 0.63 - 2.52$).

Thanks to the above transformation of diagrams $M - r^{-1}$ and $\sigma_c - \varepsilon_c$ methodological unity of reinforced concrete elements and structures calculation is provided for bearing strength, crack resistance and deflections. Moreover, the universal state diagram of the reinforced concrete element (2) acts as a link that not only links but also complements the MDS ratio system [5, 6]:

$$\left.\begin{array}{l} \bullet\ \text{static } M = f(\varepsilon_c, \varepsilon_{ct}, \varepsilon_s),\quad N = f(\varepsilon_c, \varepsilon_{ct}, \varepsilon_s); \\ \bullet\ \text{geometric } r^{-1} = f(\varepsilon_c, \varepsilon_{ct}, \varepsilon_s); \\ \bullet\ \text{physical } \sigma_c = f(\varepsilon_c),\ \sigma_{ct} = f(\varepsilon_{ct}),\ \sigma_s = f(\varepsilon_s). \end{array}\right\} \quad (4)$$

In general, the system of MDS equations (4) is complemented not only by the state diagram (2) but also by the function of compressed concrete boundary deformations [5]. It was obtained from the same diagram (2) by the application of the extreme criterion of bearing strength (the Fermi criterion). As a result, the “logic” of reinforced concrete elements and structures studies by deformation-force model allows to extend the mentioned system of MDS equations (4) to the level of statically significant. In such circumstances, it is possible to significantly reduce the number of iterative operations in the calculations and even completely eliminate them. And this, in turn, contributes to the development of engineering express methods of calculating reinforced concrete elements and structures. In the power model, similar methods have already been used. When using the “plastic hinge” system (4) became statically significant, and the need for the use of geometric and physical relations in general was eliminated [3].

The advantages of the deformation-force resistance model [5-7] with the extended relations system (4) become especially noticeable in the calculations of reinforced concrete elements and structures in the deflections and fracture toughness. According to the vast majority of existing methods [8, 14-16], the calculation of bends and cracks is based on the fact that the concrete element or structure SSS is first determined separately in the sections with cracks and in the middle sections between them. Subsequently, on the basis of already averaged parameters or characteristics of an element SSS, its averaged settlement cross-section is formed. The characteristic difference between the deformation-force model is that it allows to determine directly the curved curvature of the reinforced concrete element at any deformation stage

$$r^{-1} = (2M_u \cdot r_u)^{-1}\left[(1 - M \cdot M_u^{-1}) \cdot D_o \cdot r_u^{-1} + 2M - \sqrt{((1 - M \cdot M_u^{-1}) \cdot D_o \cdot r_u^{-1} + 2M)^2 - 4M \cdot M_u}\right] \quad (5)$$

and to relate it, in the flat section hypothesis, directly to the relative materials deformations in the central section of the block between cracks

$$r^{-1} = (\varepsilon_{c2} + \varepsilon_{s1}) \cdot d^{-1}, \quad (6)$$

where ε_{c2} - current values of the most compressed face concrete relative deformations; ε_{s1} - current values of elongation of reinforcement relative deformations; d - the working height of the section of the rectangular profile element, respectively.

Another peculiarity of the deformation-force model is that it allows to quite simply "embed" the technical theory of reinforcement with concrete coupling into the general theory of reinforced concrete deformation [17-19]. Due to the averaged cross section and the averaged materials deformation in the block between the cracks, all the crack formation parameters are directly related to the parameters of the reinforcement with concrete coupling (Figure 3).

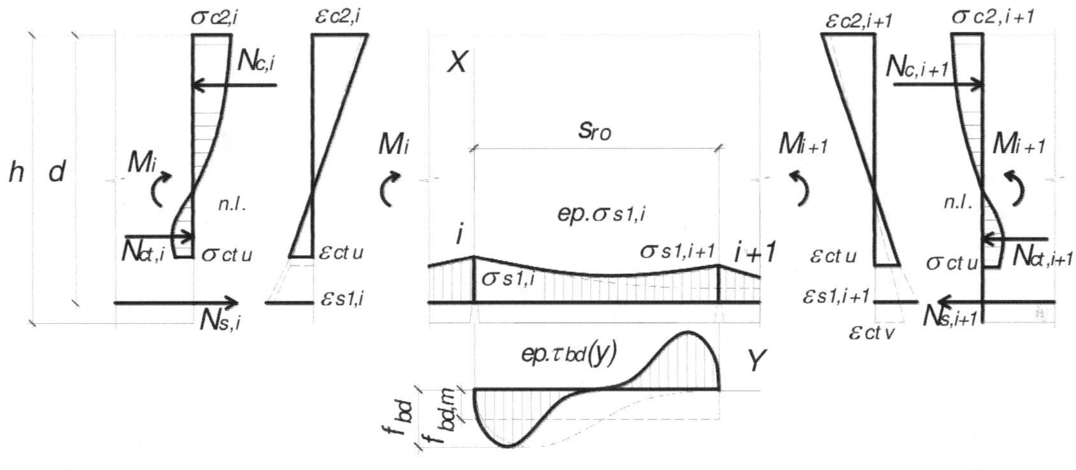

Figure 3. Calculation scheme of bending reinforced concrete element stress-strain state in the block between two cracks stretched element with corresponding plots of adhesion stresses.

In this case, the solutions obtained are much simpler than those offered in the framework of fracture mechanics by models of “double console elements” or folded rods. The stated advantages of the mentioned models [8, 13, 20] in the general reinforced concrete theory are generally debatable. The fact is that the fracture mechanics can fairly effectively describe the laws of origin and heterogeneities development and defects in the structure of relatively homogeneous materials under different loading types. The application of the destruction mechanics to reinforced concrete, in which even before the loading start there are already many heterogeneities and defects in the form of pores and sinks, micro- and macro-cracks and dislocations of various kinds, leads to solutions whose implementation is extremely difficult, even with the help of special software complexes. In these circumstances, it is premature to speak about the fracture mechanics introduction into the practice of designing reinforced concrete structures.

The effectiveness of the proposed deformation-strength model of the reinforced concrete to the forces resistance and the methods of reinforced concrete elements and structures calculation developed on its basis were evaluated by comparing the theoretical calculations with the experimental data results (Table 1). At the same time, similar comparisons were made with the results of calculations performed according to the current norms methods [15, 16]. All of them showed that the priority in the accuracy of the bearing strength estimation, deflections and width of the reinforced concrete elements normal cracks opening belongs to the calculation methods, which underlie the deformation-strength model of reinforced concrete resistance.

Table 1. Comparison of theoretical and experimental values of the studied parameters.

Authors and research parameters	Deviation of the experimental data from the calculated methods								
	[15]			[16]			[5]		
	Δ	σ	v ,%	Δ	σ	v ,%	Δ	σ	v ,%
	Bearing strength of bending elements								
Pam H. J. [21]	1.020	0.083	7.83	1.016	0.080	7.83	1.013	0.073	7.25
Sarkar S. [22]	1.052	0.170	16.00	1.048	0.174	16.65	1.028	0.133	12.93
	Deflections								
Burns N., Siess C. [23]	1.019	0.1553	15.24	1.022	0.1585	15.51	0.978	0.1104	11.29
Ernst G. C. [24]	1.011	0.1798	17.78	1.009	0.1881	18.64	1.018	0.1421	13.96
	The width of the opening of normal cracks								
Pundinaitė M. [25]	1.289	0.3002	23.29	1.334	0.3209	24.06	1.106	0.1988	18.31
Gilbert R., Nejadi S. [26]	1.095	0.3321	30.33	1.103	0.3766	34.14	1.087	0.2102	19.34

Δ and σ - mean and standard deviation, v - coefficient of variation

5. Conclusions

Thus, suggested by the authors direction of reinforced concrete general theory development by deformation-force model of deformed solid body allows:

- to achieve the methodological unity of the reinforced concrete elements and structures calculation by bearing strength, fracture resistance and deflections due to the dialectical transformation of the elements state diagrams into material deformation diagrams;
- significantly reduce the static uncertainty level of the generally recognized system of equations (ratios) of MDS or in general transform this system into statically significant due to its expansion by the element state diagram analytical dependence and the function of the compressed concrete boundary deformations;
- effectively predict not only the overall rigidity, but also any other parameter of the stressed-stain state of reinforced concrete elements and structures at any stage of their deformation with minimal use of empirical parameters and coefficients, or even without their involvement;
- due to its relative simplicity, to create engineering express methods of calculating reinforced concrete elements and structures by boundary states without the use of special software;
- it is quite simple to "integrate" the technical theory of reinforcement with concrete coupling into the general theory of reinforced concrete deformation due to the use of parameters and characteristics of the reinforced concrete elements and structures cross-section in the areas between normal cracks.

References

[1] Loleit A F 1933 *Instructions for the selection of sections of reinforced concrete elements for critical efforts* (Moscow: VNIITOB)
[2] Pasternak P L 1944 Remarks to the draft of new design standards for reinforced concrete structures *Construction industry* **7** 20-3
[3] Gvozdev A A 1949 *Calculation of the bearing capacity of structures by the method of limiting equilibrium: The essence of the method and its justification* (Moscow: Gosstroyizdat)
[4] Goldenblat I I 1955 *Basic principles of the method of calculating building structures based on design limit states and loads* (Moscow: Gosstroyizdat)

IOP Conf. Series: Materials Science and Engineering 1021 (2021) 012026 doi:10.1088/1757-899X/1021/1/012026

[5] Romashko V M 2016 *Deformation-force model of resistance of concrete and reinforced concrete* (Rivne: O Zen)

[6] Romashko V and Romashko O 2017 The construction features of the deformation and force model of concrete and reinforced concrete resistance *MATEC Web of Conf* **116** 02028

[7] Romashko V and Romashko O 2019 Fundamentals of the general theory of resistance of reinforced concrete elements and structures to power influences *Materials Science Forum* **968** 534-40

[8] Bondarenko V M and Kolchunov V I 2004 *Calculated models of force resistance of reinforced concrete* (Moscow: ACS)

[9] Kolchunov V I, Yakovenko I A 2011 On the use of the hypothesis of flat sections in reinforced concrete *Construction and Reconstruction* **6 (38)** 16-23

[10] Kochkarev D V, Babich V I 2011 Prerequisites for calculating and calculating the deflections of reinforced concrete elements experiencing bending, taking into account the nonlinear deformation of materials *Building structures* **74 (1)** 406-13

[11] Bachinsky V Ya 1982 About the loss of deformation stability of a bending bar *Building structures*: *rep. interdepartmental scientific and technical sat. NIISK Gosstroy of the USSR* **35** 55-8

[12] Golyshev A B, Bachinskii V Ya 1985 On the development of an applied theory for the calculation of reinforced concrete structures *Concrete and reinforced concrete* **6** 16-8

[13] Kolchunov VI, Yakovenko IA 2009 Development of a two-cantilever element of fracture mechanics for calculation of crack opening width of reinforced concrete structures. *Bulletin of Civil Engineers SPbGASU* **4 (21)** 160-3

[14] Bambura A M 2005 *Experimental bases of the applied deformation theory of reinforced concrete* Doctoral thesis, NDIBK

[15] DSTU B B.2.6-156:2010 Designs of buildings and structures - Concrete and reinforced concrete structures made of heavy concrete

[16] EN 1992-1-1:2004 Design of Concrete Structures - Part 1-1: General rules and rules for buildings

[17] Romashko O and Romashko V 2018 Evaluation of bond between reinforcement and concrete *MATEC Web of Conf* **230** 02027

[18] Romashko V and Romashko O 2018 Calculation of the crack resistance of reinforced concrete elements with allowance for the levels of normal crack formation *MATEC Web of Conf.* **230** 02028

[19] Romashko O V and Romashko V M 2019 Model of multilevel formation of normal cracks in reinforced concrete elements and structures *IOP Conf. Ser.: Materials Science and Engineering* **708** 012069

[20] Iakovenko I A 2018 *Models of deformation of reinforced concrete on the basis of fracture mechanics* Doctoral thesis, Poltava National Technical University

[21] Pam H J , Kwan A K H, Islam M S 2001 Flexural strength and ductility of reinforced normal- and high-strength concrete beams *Structures & Buildings* **146 (4)** 381

[22] Sarkar S, Adwan O, Munday J G L 1997 High strength concrete: an investigation of the flexural behavior of high strength RC beams *The Structural Engineer* **75 (7)** 115-21

[23] Burns N H, Siess C P 1966 Plastic Hinging in Reinforced Concrete *Proceedings ASCE* **92 (ST5)** 45-64

[24] Ernst G C 1957 Plastic hinging at the intersection of beams and columns *ACI Journal* **28 (12),** (*Journal Proceedings* **53**) 1119-44

[25] Pundinaitė M 2010 *Experimental and theoretical studies of cracking of precast concrete elements* Master's thesis, Vilnius Gediminas Technical University

[26] Gilbert R I, Nejadi S 2004 *An Experimental Study of Flexural Cracking in Reinforced Concrete Members under Sustained Loads: UNICIV Report № R-435* (Sydney: University of New South Wales)

IOP Conf. Series: Materials Science and Engineering 1021 (2021) 012027 doi:10.1088/1757-899X/1021/1/012027

Systems and subsystems of track, control and management of high-speed railway

A A Shevchenko[1, 5], **O O Matviienko**[2], **V A Lyuty**[3], **V G Manuylenko**[4] **and N A Murygina**[4]

[1] Railway Survey and Design Department, Ukrainian State University of Railway Transport, Feierbakh Square 7, 61050, Kharkiv, Ukraine
[2] Engineering systems "El-term". Vosstaniya Square 7/8, 61000 Kharkov. Ukraine
[3] Department of Building Materials and Structures, Ukrainian State University of Railway Transport, Feierbakh Square 7, 61050, Kharkiv, Ukraine
[4] Department of track and track facilities, Ukrainian State University of Railway Transport, Feierbakh Square 7, 61050, Kharkiv, Ukraine

[5] Email: annshevc@gmail.com

Abstract. The article discusses the following issues: according to the current level of development and equipment of the existing network of railways in Ukraine, it is necessary to state the determination of the gauge of the high-speed line; the main characteristics of the high-speed rail system and the general requirements for laying the line; track systems and subsystems, as well as including their movement and control on free terrain, ensuring the safety of operation and transportation of passengers with the declared conditions of comfort and speed; the distance between the centres of the tracks increases in proportion to the speed of movement of passenger flows, which requires additional arrangement of the roadbed, taking into account the radii of the curves, the elevation of the outer rail, the angle of descent and ascent of the track, the equivalent taper, the slope of the rail, the profile of the rail head, such as rail and sleepers; lateral and vertical forces acting on the track. In order to ensure traffic safety at the maximum permissible speeds, it is necessary that the indicated vertical and lateral forces (static and dynamic) meet the requirements, which in turn depend on the above factors.

1. Introduction

The geographical position of Ukraine is traditionally considered to be the crossing point between the Baltic and Black Seas, the border between Europe and Central Asia. With an area of 603,549 square kilometers, about 1300 km from west to east and 900 km from north to south, Ukraine is the largest country in all of Europe, slightly larger than France. Based on the successful introduction of high-speed rail in Europe and around the world, the following conclusions can be drawn:

- Countries that have established high-speed rail systems have significant national goals that contribute to their achievement;
- These countries were ready for huge investment costs, as they were convinced of the need for these projects;
- In Europe, the possibility of unification and interaction of transport systems is considered as a factor for joint work of all countries of the European Union;
- In Japan, high-speed rail is the main means of transportation that contributes to the development of the country's economy and tourism.

Based on the foregoing, Ukraine needs to put into operation high-speed railways that will meet all the requirements for the design, maintenance and operation of systems [1, 2, 4].

2. Track subsystem

To maximize the operation of trains, the lines of the high-speed network of Ukraine should be designed in such a way that trains with a length of 400 m, with a maximum weight of 1,000 tons, moving at a speed of 350 km/h or more, correspond to the dimensions of the track presented in this article [3, 5, 8].

The track subsystem of a high-speed railway system includes all permanent stationary devices that serve to perform the following functions in accordance with the basic requirements:

- Keeping vehicles on track, including moving and steering them in clear areas, and installing safety equipment to ensure this function is performed. Boarding and disembarking passengers from trains at stations.
- Hence, permanent stationary devices include:
- Track sections without directional switches and crossings,
- Turnouts and level crossings used to change the direction of movement of vehicles,
- Structures such as bridges and tunnels that allow obstacles to be crossed under special conditions,
- Necessary safety equipment as well as protective equipment to maintain the integrity of the subsystem,
- Associated structures at stations (platforms, driveways, etc.).

3. Infrastructure

3.1. Line location

The design of high-speed lines should meet the following general requirements aimed at satisfying the main characteristics of the high-speed railway system presented in Table 1

Table 1. Main characteristics of the high-speed rail system.

Main characteristics of the high-speed rail system	General requirements for laying lines
Straight line between big city centres	Using the existing infrastructure of the main stations of large cities
Connection to urban/suburban and other rail services at hubs	Use of existing long-distance train equipment for high-speed trains, if the project does not allow, the shortest system of passenger movement, delivered by local or regional modes of transport
Reduce trip duration as much as possible	Design as straight as possible line lay between main stations with the greatest possible length of track sections with high-speed railway parameters. Design with the smallest distance between platform and standard train door placement
Minimize infrastructure costs	Design of all infrastructure subsystems with parameters standard for high-speed lines, which excludes the use of road and railroad crossings, except for high-speed overpasses and switches at junctions. Minimize line lengths by leveraging standard high-speed train performance as well as optimizing maximum allowable slopes and curves

Minimize operating costs	Designing a depot for maintenance of infrastructure and rolling stock, taking into account the location of the routes
Minimize the cost of project implementation stages	Design compatibility solutions high speed and existing lines, so that high speed trains can run on existing lines

These principles for the introduction of high-speed rail in Ukraine should be carried out taking into account the relevant criteria and parameters for the location of the track, which are described below [6, 7].

Compatibility of high-speed lines with traditional lines is part of the operational requirements for the gradual implementation of a high-speed rail network. Possible scenarios should cover all subsystems to achieve the required compatibility [9].

3.2. Distance between track centers

An important condition is the location of the tracks of high-speed lines at a considerable distance from each other, since with the passage of two trains, the possible sum of speeds will be 750 km/h. Too close passage of two trains can lead first to a sharp increase and then to a sharp decrease in pressure between the cars. Repetitive pressure on windows can cause glass fatigue, which can cause windows to break.

Thus, to maintain safety, the two tracks in each direction must be farther apart than on conventional lines.

Adequate space must be provided along each track used for high-speed trains; this space should allow passengers to disembark from the train on the side opposite to the adjacent tracks, if the latter are operated during the disembarkation of passengers. In places where tracks are located on engineering structures, a protective parapet should be installed on the side of the tracks so that passengers can safely exit.

At the design stage, the minimum distance between the centres of the tracks of lines specially built for high speeds up to 350 km/h should be 5 m.

3.3. Earth bed

Due to the conditions described in the previous paragraphs, the width of the top layer of the subgrade for a line with two tracks should be at least 14.5 m. The subgrade for high speed lines is located deeper than for conventional lines. Typically, a layer of concrete and a tarmac of crushed stone (like on a highway) is laid, followed by ballast to prevent soil movements affecting the alignment of the railroad. For drainage purposes, the subgrade must be constructed with a 4% cross slope from the centre of the base of the straight track. For the feasibility study, embankments and branches should be constructed in a 3/2 (base/height) ratio [10, 16].

The sub-ballast layer should be made of granular materials, 0/31.5 mm thick, 0.35 m. This will avoid puncturing the sub-layer for water drain and distribute the load evenly along the subgrade. The subgrade should be made of granular materials (0/125 mm, 0.50 m thick), you should also pay attention to the effect of deep soil freezing in winter.

In areas with unsatisfactory natural soil conditions, to increase the stability of the embankment, it is necessary to add a layer of geotextile, as well as a geogrid layer in the lower part of the embankment. To strengthen the embankment in areas prone to soil settlement, pile driving is necessary to create transitional structures to adjacent bridges and culverts. Attention should be paid to the effect of deep freezing of the soil in winter.

3.4. Curve radius

When designing high-speed lines, the minimum radius of the selected curve should be such that the failure of the outside rail to exceed the curve for the curve under consideration does not exceed the level planned for the passage of trains at maximum speed [4, 11, 15]. Thus, the minimum radius of the horizontal bend of the curve for a high-speed line designed to travel at a speed of 350 km/h should be at least 7,200 m (in exceptional cases - 5,600 m). The standard minimum for a vertical curve should be around 25,000 m which is also confirmed by the condition (1) of comfortable ride of passengers and prevention of overloading of the outer rail:

$$h_{min} = 12.5V_{max}^2(R)^{-1} -115, \tag{1}$$

where V_{max} −maximum speed developed by passenger trains on this curve (km/h); 115 −the value of the maximum understeer (mm), calculated under the conditions of not exceeding the allowable outstanding acceleration.

Condition (1) is performed in compliance radii design and operation.

On track sections where trains run at low speed (station and siding tracks, tracks in depots and at parking lots), the minimum horizontal design radius for an isolated curve should be at least 150 m. Allowing in practice various options for laying the track, the minimum effective radius is not should be less than 125m.

Vertical laying of parking and service tracks should not include curves with a radius of less than 600m in hills and 900m in lowlands.

The sub-ballast layer should be made of granular materials, 0/31.5 mm thick, meter this will avoid piercing the sub-layer for water drain and evenly distribute the load along the subgrade. The subgrade should be made of granular materials (0/125 mm, 0.50 m thick), you should also pay attention to the effect of deep soil freezing in winter.

In areas with unsatisfactory natural soil conditions, to increase the stability of the embankment, it is necessary to add a layer of geotextile, as well as a geogrid layer in the lower part of the embankment. To strengthen the embankment in areas prone to soil settlement, pile driving is necessary to create transitional structures to adjacent bridges and culverts. Attention should be paid to the effect of deep freezing of the soil in winter [12].

3.5. Elevation and failure to exceed the outer rail in a curve

The minimum radius of the curved track, as well as the elevation of the track, determine the maximum inadequacy of the outer rail in the curve for a given speed. Insufficient excess of the outer rail in the curve is one of the elements that determine the load on the track.

Infrastructure design should take into account the performance capabilities and technical constraints of rolling stock. Acceleration and train braking and stopping requirements are very important issues to ensure potential speeds.

The recommended elevation for high speed lines should be limited to 180 mm. The lack of excess of the outer rail on straight track sections at the design stage of high-speed lines should be limited to 65mm. Higher indicators of insufficient excess of the outer rail in the curve are possible if the construction of the lines takes place in difficult topographic conditions.

3.6. Track width

The article [10, 11] describes the advantages and disadvantages of using 1.520 and 1.435 mm gauge for the Ukrainian high-speed network. The final decision on the track gauge will be made by the Ukrainian authorities.

In Europe and around the world, high-speed lines with a rail spacing (track gauge) of 1.435mm are used. This standard has been adopted by the European Union for all high-speed lines.

The track gauge corresponds to the distance between the active sides of the rail heads, measured at a height of 14.5 mm (± 0.5 mm) below the rolling surface. At the design stage, as well as during the construction and operation of the track, the track gauge must be maintained within the specified limits. It should also be noted that the figures presented in the article should be considered as "standard or standard values" with a standard base gauge of 1.435 mm.

3.7. The angle of descent and ascent of the track

Higher slope and rise angles are allowed for high-speed lines than for traditional railways. This fact is due to two main reasons. First, the line should be as straight as possible and not contain curves in order to thus reduce its length. Secondly, the faster the train goes, the less it will slow down on the rise, due to the kinematic energy accumulated by the train. The angle of descent and ascent of new high-speed lines should be limited to 35 mm/m. The average slope of the cross-sectional slope on a track section longer than 10 km should be 25 ‰ or less. The maximum length of a continuous track with an inclination angle of 35 ‰ should not exceed 6.000 m.

3.8. Equivalent taper

The interaction of the wheel and rail is a major determinant of the dynamic behavior of a moving rail vehicle. Equivalent taper plays a significant role, since it allows you to evaluate the interaction of the wheel and rail on a straight track section, as well as in a curve with a large radius.

Taking into account the axle characteristics (wheel profile and distance between the active surfaces of the wheel), compliance with these values for the equivalent taper for high-speed rail should be obtained by a suitable, well-documented selection of three components for track sections and switches: track gauge and tolerances, rail slope and rail head profile [12, 13].

3.9. Rails and sleepers

The rail curtain is the angle between the axis of symmetry of the profile of the new rail installed on the support and the perpendicular to the rolling surface. For sections of the track, the speed of which is more than 280 km/h, the rails, as a rule, are laid with a slope of 1/20, which guarantees the preservation of the design parameters of the taper of the wheel profile of the rolling stock.

However, the track may be laid with a different slope than 1/20, and it may be necessary to adopt different parameters for the “rail head profile”, track gauge and axle characteristics of the rolling stock. In this case, the compatibility of the new system with the rolling stock wheel profile in terms of its equivalent taper must be ensured.

In the design drawing of a rail, the rail head profile is defined as a sequence of circles forming a continuous curve. This curve changes as a result of rail wear and generally tends to maintain a constant shape, which must be measured using highly accurate methods to determine the equivalent taper.

The profile of the rail head must include a lateral slope from the side of the rail head at an angle of 1/20–1/17.2 with respect to the vertical axis of the rail head. This slope in the direction of the upper surface should be followed by a sequence of curves with radii of 12.7 and 13 mm, and then 80 and 300 mm, with respect to the vertical axis of the rail head.

Rail type. It is recommended to use rails of type 60 E1 (UIC 60), made of steel grade 900 A. Draft European legislation defines the flatness of the surface of class "A". It is necessary to use a jointless track, with the only exception in the areas of bridges and overpasses [4, 10, 11].

Sleepers. Based on the experience of European countries, it is necessary to use monolithic or double-block reinforced concrete sleepers with the following characteristics (these characteristics are presented as an example):

- Sleepers per kilometer: 1.666 /km
- Weight: 250 to 290 kg
- Length: 2415/2500 mm
- Width: 290 mm

3.10. Impacts on the track

Transverse and vertical forces acting on the track are important factors that determine both the dynamic characteristics of the rolling stock on the track and the fatigue characteristics of the track superstructure. In order to ensure road safety at the maximum permissible speeds, it is necessary that the specified vertical and lateral forces (static and dynamic) meet specific requirements. The infrastructure should be designed to withstand the maximum axle loads specified for the rolling stock.

IOP Conf. Series: Materials Science and Engineering 1021 (2021) 012027 doi:10.1088/1757-899X/1021/1/012027

The coefficient of the force of the dynamic lateral and vertical action of the wheel on the rail should not exceed the coefficient of derailing of the car from the rails. The geometrical characteristics of the wheel-rail contact area must ensure stable movement of the bogies (equivalent taper). The braking system of the rolling stock must not cause shear of the rails in the rail fastening system and/or slip of the wheels [17, 18].

The forces of influence exerted by the rolling stock on the track are determined both by the conditions associated with traffic safety and the characteristics of the track's ability to resist these forces. These forces arise exclusively as a result of contact between wheels and rails, as well as as a result of direct action on the rail of the braking equipment. These forces include:

- Vertical impact forces: static, resulting from the distribution of the weight of the rolling stock on the wheelsets; quasi-static in curves arising from the transfer of vertical loads as a result of lateral acceleration, not compensated by the elevation of the track; as well as dynamic ones arising from the geometric characteristics of the track and movement of the rolling stock.
- Track and turnouts must be designed to withstand a minimum of 16 tonnes per axle, which is the maximum dynamic axle load permissible for high-speed trains at 350 km/h.
- Lateral forces of action: quasi-static in curves, arising as a result of lateral acceleration, not compensated by the elevation of the track; as well as dynamic ones arising from the geometric characteristics of the track and movement of the rolling stock.

The track and turnouts must withstand at least the following lateral forces:

- Maximum total dynamic lateral force of the wheelsets on the track:

$$(\sum Y)_{max} = 10 + P(3)^{-1}, \tag{2}$$

where P –is the maximum static axle load (expressed in kN) of the rolling stock allowed on the given line (maintenance rolling stock, high-speed and other trains).

This limit is specific to the risk of displacement of the track on the ballast due to shear dynamic forces. The ratio of shear forces to vertical forces of the wheel:

$$Y(Q)^{-1} = 0.8, \tag{3}$$

where Y and Q are, respectively, the dynamic transverse and vertical forces of the wheel on the rail.

This limit characterizes the risk of wheel rolling on the rail and is shown in Figure 1 for the existing railways of Ukraine for different speeds.

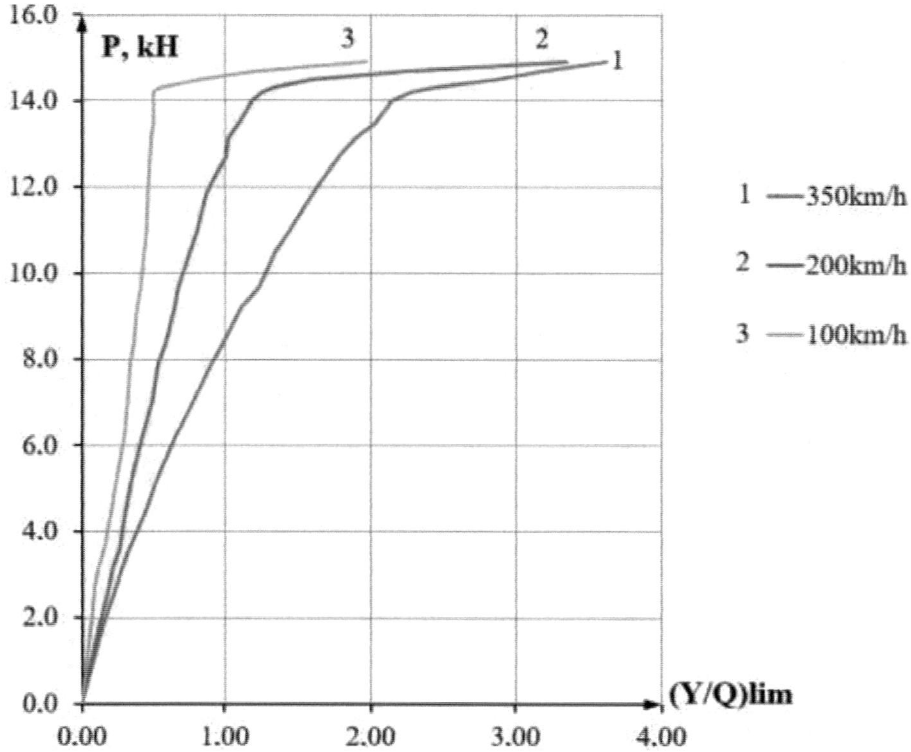

Figure 1. Limit characterizing the risk of wheel rolling on the rail from speed

- Lateral forces resulting from the acceleration and deceleration of the rolling stock during braking and acceleration.

Tracks and turnouts must withstand the transverse forces of the rolling stock, corresponding to a maximum acceleration and deceleration of 2.5 m/s^2, as well as the accompanying effects of temperature rise. The track superstructure must also withstand the lateral forces exerted by other trains that will use the line (maintenance rolling stock as well as other trains) that meet the above acceleration and deceleration rates [14].

3.11. Structural loads

Vertical, lateral horizontal and longitudinal loads on structures should be calculated using loads from load models as well as calculation methods specified in the standards for all types of rolling stock expected to operate on the line (maintenance rolling stock, high speed and other trains). Consideration must be given to dynamic effects on bridges [19].

3.12. Pointer transfers

The points of the turnouts and the marks of the cross on the turnouts and diamond-shaped crossings must be equipped with devices for securing and blocking.

Usually, it is required to comply with UIC standards 60 A 74. It is possible to use turnouts with stationary fixed switch points.

The permissible speed of movement of trains along the main and branch line of the turnout switch depends on the characteristics of the permissible speed of movement and has the following assessment conditions: strength of the elements of the turnout switch; resistance to permissible stresses and deformations; rolling stock resistance to rollover and loading; ride comfort for uncanceled accelerations and permissible impulse accelerations [20].

Dependence of the permissible train speed on the strength of the switch elements:

$$(V) = f([\sigma], \Delta x, \Delta y) \tag{4}$$

where $[\sigma]$ −permissible voltages on the limited elements of the turnout switch; $\Delta x, \Delta y$ −permissible deformations in the horizontal and vertical plane of the limited elements of the turnout switch.

Dependence of the permissible train speed on the condition of ensuring the durability of the rolling stock:

$$(V) = f(R, n, Y) \tag{5}$$

where R - the smaller of the 2 radius of the wit or the transfer curve; n - the coefficient of safety factor against overturning (sliding) of the wheel flange on the elements of the turnout switch.

Dependence of the permissible train speed on the condition of ensuring the comfort of passengers' ride:

$$(V) = f([a_{unc}], [F]) \tag{6}$$

where $[a_{unc}]$ - allowable uncanceled horizontal centered acceleration; $[F]$ - permissible vertical damping of the crew.

Taking into account these speeds, the necessary geometric parameters and design solutions of the turnouts are established.

For example, on high-speed railways in France, special turnouts are used, which allow you to maintain high speed when crossing them:

- Tg 0.0154, (length - 226 m),
- Tg 0.0218 (length - 160 m).

4. Conclusions

And this is not the whole list of necessary implementations and innovations for the successful commissioning and maintenance of a high-speed railway.

Having considered and analyzed the issues raised in the article, we can draw the following conclusions:

- According to the current level of development and equipment of the existing network of railways in Ukraine, it is necessary to state the determination of the gauge of the high-speed line, and as a result, the choice of a scenario for the implementation of standards.
- The main characteristics of the high-speed railway system and the general requirements for laying the line are significantly different and require more detailed study, taking into account the specific conditions of the terrain and operation.
- Systems and subsystems of the track, as well as including their movement and control on free terrain, ensuring the safety of operation and transportation of passengers with the declared conditions of comfort and speed.
- The distance between the centres of the tracks increases in proportion to the speed of passenger traffic, which requires additional arrangement of the roadbed, taking into account the radii of curves, the elevation of the outer rail, the angle of descent and ascent of the track, equivalent taper, rail slope, rail head profile, rail type and sleepers.
- Transverse and vertical forces on the track are very important factors that determine both the dynamic characteristics of the rolling stock and the fatigue characteristics of the track superstructure. In order to ensure road safety at the maximum permissible speeds, it is necessary that the specified vertical and lateral forces (static and dynamic) meet specific requirements. Which, in turn, depend on the above factors of these conclusions.

The studies presented in this article are analyzed taking into account investment and the results are presented in Table 2.

Table 2. Investments in high-speed transport in Ukraine.

Invested object	The size of the cost, million euros	%
High speed lines	17451.25	57.87

Stations	2173.5	7.21
Including:		
- new	897	2.97
- modernized	1276.5	4.23
Depot and repair company	693.45	2.30
Rolling stock	7665.9	25.42
Total	30157.6	100

The research presented in this article is an intermediate one and should be modified for specific operating conditions and the chosen option for the implementation of a high-speed railway.

Acknowledgments
The authors would like to thank their families and colleagues of the Ukrainian State University of Railway Transport for their support and patience, who helped in writing this article.

References

[1] *Regulations on carrying out planned preventive repair and road works on the railways of Ukraine* (Kiev: Ministry Transport of Ukraine, 2004) p 278

[2] Kurhan M B and Kurhan D M 2016 *Theoretical basis for the introduction of high-speed trains in Ukraine* (Dnipro, DNUZT) p 254

[3] Svintsov E S and Poletaev V I 2007 Geodesic works when shooting railways of industrial enterprises. *Proceedings of Petersburg Transport University* (Kharkov, Izvestia PGUPS) pp 84-91

[4] Volodymyr Kuzyo, Oksana Zatvornytska, Dmytro Yablonovskyy 2019 Liberalization of the railway market in Ukraine. Lessons learned from the EU countries *Juan Montero, Professor, Part-time professor* (Florence School of Regulation, RSCAS, EUI, Italy) pp 55

[5] Nemets K A and Nemets L M. 2014. Theory and methodology of geographical science. *Methods of spatial analysis. Tutorial.* (Kharkiv, VN Karazin KhNU) p 172

[6] Konversky A E 2010 Fundamentals of methodology and organization of scientific research. *A textbook for students, cadets, graduate students and adjuncts.* (Kyiv: Center for Educational Literature) p 352

[7] Kichirok I P and Onishchak R M. 2014. Creating digital maps for the use of auto navigation. *Bulletin of geodesy and cartography* (Kiev, Research Institute of Geodesy and Catrography) **2 (89)** pp 49-51

[8] Kirpa G M 2004 *Integration of Ukraine's railway transport into the European transport system* (Dnipro, DNUZT) p 268

[9] Kurgan M, Baydak S, Khmelevska N 2014 Prerequisites for the introduction of accelerated train movement in the direction of Kuma – Dnipropetrovsk, *Ukrainian Railways* **10 (16)** pp 51-64

[10] Shevchenko A A and Matviienko O O 2018 Problems of accuracy of curve measurements on international transport corridors of Ukraine *Scientific and Production Journal "Metrology and Devices"* **4 (72)** pp 56-60

[11] Anna Shevchenko, Oleksander Matviienko, Vitalii Lyuty, Vladimir Manuylenko and Mykhailo Pavliuchenkov 2018 Ways of introduction of the high-speed movement of passenger trains in Ukraine *Matec Web of Conferences* **230** 01014 pp 1-9

[12] Nikitin A B 2018 Theory and methods of geodetic support road transport infrastructure *Thesis for the degree of doctor of technical sciences* (Novosibirsk, Siberian State University of Geosystems and Technologies) p 230

[13] Shevchenko A A, Matviienko O O, Lyuty V A, Manuylenko V G and Murygina N A 2019 Digital models and the effect of error when shooting terrain for high-speed traffic *Conf. Series: Materials Science and Engineering* **708** (2019) 012028

[14] Bugaec N V, Shevchenko A A, Murygina N A, Shevchenko O S 2019 Development of a mathematical model of joint work of the track and rolling stock in the zone of rail joints *Proceedings of the XX International Scientific and Practical Conference International Trends in Science and Technology* (Warsaw, Poland) Vol. **1** pp 30-38

[15] Kitov Yu P, Verevicheva M A, Vatulia G L, Orel Ye F and Deryzemlia S V 2017 Design solutions of the structures with the optimal internal stress distribution *MATEC Web of Conf.* **133** 03001

[16] Railway Applications-Communication, Signaling and Processing Systems-Safety Related Electronic Systems for Signaling. IEC. IEC62425: 2013

[17] Armin Z, Gunter H 2015 Towards Modeling and Evaluation of ETCS Real-time Communication and Operation. *The Journal of Systems and Software*. 2015 pp 761-767

[18] Bei Wang 2018 High-Speed Railway Train Operation Control System *IOP Conf. Series: Materials Science and Engineering* 452 (2018) 042109 pp 1-5

[19] Bin Ning, Member, Tao Tang, Hairong Dong, Ding Wen, Senior Member, Derong Liu, Fellow, Shigen Gao, and Jing Wang 2011 An Introduction to Parallel Control and Managementfor High-Speed Railway Systems *IEEE Transactions on intelligent transportation systems*, Vol. **12** No. **4** December 2011 pp 1473-1483

[20] H R Dong, B Ning, B G Cai and Z S Hou 2010 Automatic train control system development and simulation for high-speed railways *IEEE Circuits Syst. Mag.*, Vol **10** No. **2** pp 6-18

Theoretical basis of bearing capacity calculation of the statically indeterminate reinforced concrete beams and its experimental research

A A Shkurupiy[1], P B Mytrofanov[1,2], Yu O Davydenko[1] and A G Horb[1]

[1]Structures from Metal, Wood and Plastics Department, National University "Yuri Kondratyuk Poltava Polytechnic", Pershotravneviy sq. 24, 36011, Poltava, Ukraine

[2] Email: mytrofanov.p@gmail.com

Abstract. New normative documents of Ukraine for the design of reinforced concrete and their elements contain a number of fundamentally new approaches to the calculation of their bearing capacity. Deformation model with an empirical criterion of durability is accepted as a basis of calculations of concrete and reinforced concrete elements on action of bending moments and longitudinal forces. Statically indeterminate reinforced concrete structures, in particular fixed beams, are widely used in various buildings and structures as load-bearing structures. Therefore, the research of their work and the direct calculation of such structures on the basis of the deformation model is a very important issue in the current trend of development of reinforced concrete both in our country and abroad. Theoretical calculations of bearing capacity of reinforced concrete continuous beams on the basis of deformation model with extreme strength criterion with the use of optimization and numerical methods and the limiting equilibrium method are performed. Their experimental studies are carried out. The results of theoretical calculations of such constructions are compared with experimental data obtained by the authors of the article and given in the works of other researchers.

1. Introduction

Specific features of reinforced concrete make additional requirements to the methods of calculating the strength, rigidity and stability of reinforced concrete structures (RCS) and their reinforced concrete elements (RCE), which must take into account the deformation of the material, i.e. use deformation models.

Today, one of the main tasks in the design of reinforced concrete is to clarify existing and develop new methods of calculation, including on the basis of deformation methods, which is proposed in the new normative documents [1, 2, 9]. Continuous beams are widely used in various buildings and structures as load-bearing structures. Therefore, the study of their work taking into account high-strength concrete and improving the method of calculation of such structures is an important issue.

The use of a deformation model (DM) makes it possible to fairly accurately take into account the stress-strain state (SSS) at different stages of the RCE, including in the limit state. As noted in [3, 4, 18], DM with extreme strength criterion (ESC) has significant advantages over existing DM with empirical criterion [1, 2, 9, 18, 20-22], especially for the calculation of statically indeterminate RCE, and is a significantly new approach to the calculation their strength and load-bearing capacity.

DM with ESC necessitates careful control of all values of SSS limit parameters of reinforced concrete elements, especially stress in stretched and compressed reinforcement, as their design dependences differ significantly in the ranges of elastic and inelastic work, especially in reinforced structures, and limited control of SSS parameters can lead to results that do not correspond to reality. Based on the developed optimization technique [4-6, 16-18], a wide range of problems in the calculation of reinforced concrete (beams, columns, frames, etc.) can be solved using a wide range of classes of concrete, including high-strength [18, 19]. Therefore, theoretical and experimental studies to study the operation of continuous beams taking into account the redistribution of forces (in the formation of characteristic cross sections of conventional plastic hinges [7]) and determine the limits of bending moments and ultimate load using the limiting equilibrium method [8, 16] is an urgent task.

2. Analysis of recent research and publications

A number of literature sources are devoted to calculations of bearing capacity of reinforced concrete and their elements on the basis of DM [4-6, 9, 10, 11, 16, 17, 19-22]. In particular, more detailed calculation of the RCE bearing capacity on the basis of DM with ESC is paid attention in [3, 4].

DM with ECM is an alternative model to existing DM with an empirical criterion of strength [1, 2, 9, etc.]. At present, none of the existing DM, except for DM with ESC, makes it possible to calculate the bearing capacity of reinforced concrete and their elements working on different types of deformation (bending, central and off-center compression or tension, etc.) without a predetermined value of the most compressed concrete fiber of normal cross section ε_{cu1}. They need experimental data to determine ε_{cu1}, and their values are assumed to be constant for a particular class of concrete. In fact, ε_{cu1} depends on many factors, the main of which are: the compressive strength class of concrete, the shape of the cross section, the class of reinforcing steel and its location, and so on. Only DM with ESC makes it possible to analytically obtain the value ε_{cu1} as a result of calculation according to the method described in [3, 4, 18]. Clarification of the existing methods of calculation of reinforced concrete allows to identify reserves of material savings and is an urgent task.

3. The purpose of this work

Calculation of bearing capacity of continuous reinforced concrete beams based on DM with ESC using high-strength concrete with different classes of rod reinforcement and the percentage of reinforcement loaded by concentrated forces in the beams, and comparison of calculation results with experimental data obtained by the authors and given in other works.

4. The main material

Theoretical research was limited to the task of calculating the bearing capacity of reinforced concrete continuous beams using DM with ESC, optimization and numerical methods [4, 5] and the limiting equilibrium method [8].

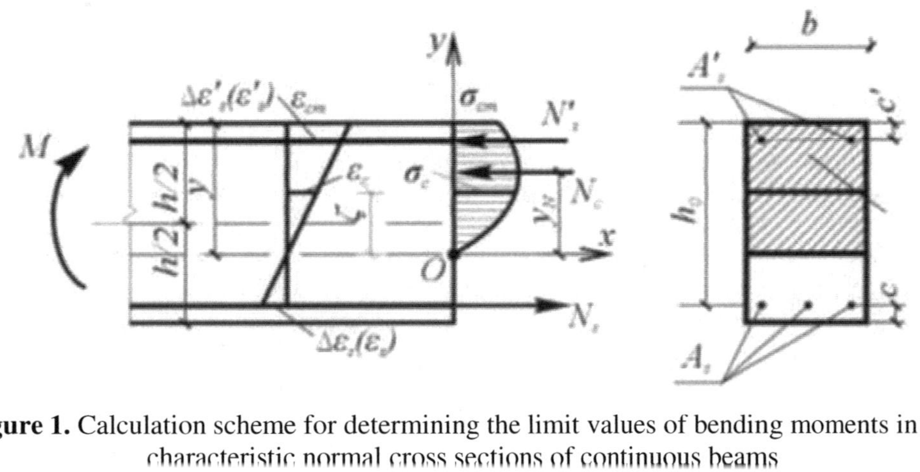

Figure 1. Calculation scheme for determining the limit values of bending moments in the characteristic normal cross sections of continuous beams

The calculated positions and physical dependences are given in [4, 5, 6]. The state of destruction of the normal section is described by the calculated schemes of forces and deformations, which are shown in Figure1.

According to the method described in [4, 5, 12], analytical calculations of the bearing capacity of continuous reinforced concrete beams of high-strength concrete using numerical and optimization methods and the method of boundary equilibrium. The algorithm for calculating such structures by this method is implemented in a specially developed software package for a PC based on the Windows operating system (Figure 2).

The experimental specimens were rectangular in length and 4.000 mm long. Their height and width were constant and equal to 180×120 mm. All experimental beams had a ratio of span length (l) to section height (h) greater than 10 (Figure 3 and Figure 4). This condition makes it possible to neglect the influence of transverse forces when calculating their bearing capacity. A total of 18 samples were tested. Concentrated forces were applied in the middle of the spans - a=b=l/2 (Figure 3 and Figure 4).

Reinforcement of continuous double-beam beams was constant in length and was performed by longitudinal reinforcing rods of class A400C with a diameter of 12 and 14 mm, as well as welded clamps from reinforcement of class A240C with a diameter of 6.5 mm with a 50 mm step in the support zones and with a 100 mm step in the middle zone.

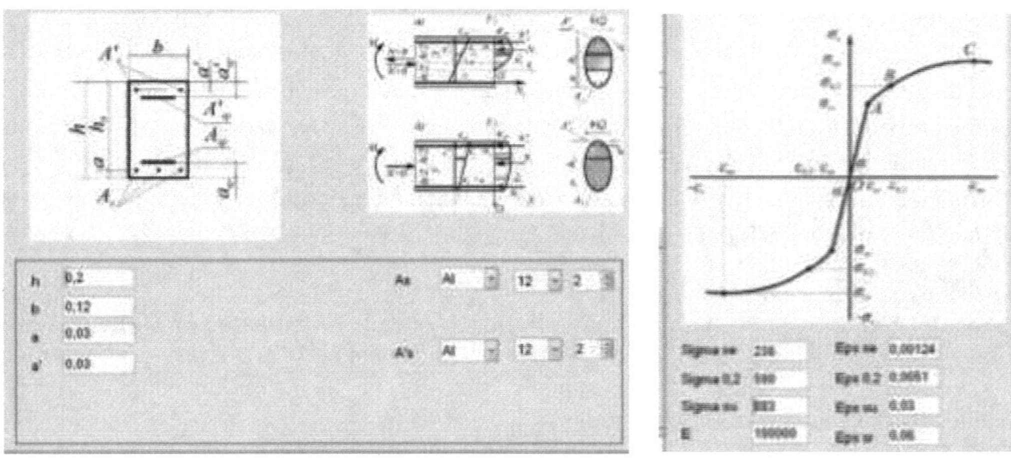

Figure 2. The software package CRC-12

The results of calculations are obtained in comparison with experimental data obtained personally by the authors and other researchers [13, 14, 15]. The results of experimental researches of two-span continuous reinforced concrete beams made of high-strength concrete of different compressive strength classes, made it possible to obtain new data on their strength, deformability and bearing capacity, as well as to analyze the obtained values ε_{cu1} of ultimate deformation of compressed zone concrete in such normal cross-sections.

The nature of the development of cracks and the destruction of experimental samples are shown in Figure 5. Comparison of theoretical calculations of bearing capacity of such beams with experimental data obtained by the authors and their statistical processing are given in Tables 1 and 2. Geometrical and mechanical characteristics of experimental samples of continuous reinforced concrete beams tested by the author are given in [12].

The bearing capacity of similar structures and their comparison with experimental data of other researchers are calculated (S. Krylova, Yu. Gushchi, M. Abakanova, A. Oatula [13, 14, 15]). The results of their comparison are shown in Tables 3 and 4.

Beams with spans of 2.7 *m* reinforced with high-strength longitudinal bars of A-IV class and 80C grade with proof load were used in the research [13, 14].

Beams, used in the study [15], had spans of 3 *m* and the cross-sectional dimensions bxh=170x220 mm. Bars of A-III class (35GC grade) with a minimum value of proof load $\sigma_y = 400\,MPa$ were used as the longitudinal reinforcing. The prismatic strength of concrete for reinforced concrete continuous beams of this series is $f_{ck,prizm} = 29.2\,MPa$, and the initial concrete modulus of elasticity is $E_{cm} = 30.4\,GPa$.

The design scheme of beams, that were considered in the articles [13, 14, 15], is shown in Figure 6, where *m* is the concentrated forces ratio.

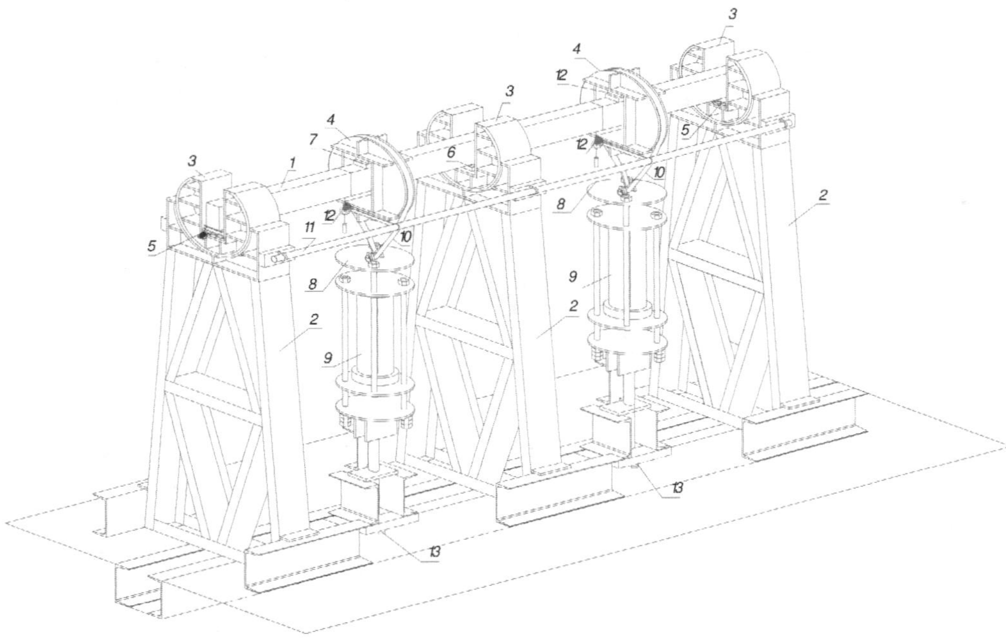

Figure 3. Scheme of installation and placement of measuring instruments:
1) prototype of the beam; 2) support frame; 3) support ring; 4) loading ring; 5) power dynamometer; 6) articulated fixed support; 7) the hinge which transfers loading to a sample; 8) the compressor of compressive forces into tensile forces; 9) hydraulic jack; 10) metal cable; 11) bar for fastening of devices; 12) deflector; 13) power floor anchors

Figure 4. General view of the installation for testing continuous beams

Table 1. Experimental and theoretical limit values of bending moments of normal sections of continuous beams

Sample code	Limit values of bending moments of normal sections, kNm					
	Section under concentrated force in span			Section on the middle support		
	$M_{u1,test}$	$M_{u1,theor}$	$\frac{M_{u1,theor}}{M_{u1,test}}$	$M_{u2,test}$	$M_{u2,theor}$	$\frac{M_{u2,theor}}{M_{u2,test}}$
B-1-12-1	13.150	13.327	1.0135	10.751	13.327	1.240
B-1-12-2	13.177	13.327	1.0114	9.746	13.327	1.367
B-1-12-3	13.211	13.327	1.0088	8.966	13.327	1.486
B-1-14-1	17.435	17.554	1.0068	14.085	13.366	0.949
B-1-14-2	17.180	17.421	1.0140	10.290	13.366	1.299
B-1-14-3	17.214	17.421	1.0120	9.746	13.366	1.371
B-2-12-1	13.244	13.616	1.0281	11.988	13.616	1.136
B-2-12-2	13.341	13.616	1.0206	11.080	13.616	1.229
B-2-12-3	13.817	13.616	0.9855	12.265	13.616	1.110
B-2-14-1	17.366	17.883	1.03	11.581	13.863	1.182
B-2-14-2	17.147	17.734	1.0342	10.594	13.683	1.292
B-2-14-3	17.277	17.734	1.0265	10.571	13.683	1.294
B-3-12-1	13.893	13.756	0.9901	10.214	13.616	1.333
B-3-12-2	13.897	13.9	1.0002	12.107	13.616	1.125
B-3-12-3	13.861	14.196	1.0242	11.703	13.616	1.163
B-3-14-1	17.993	17.883	0.9939	11.277	13.683	1.213
B-3-14-2	17.506	18.349	1.0482	13.439	13.683	1.018
B-3-14-3	17.374	18.191	1.0470	12.798	13.683	1.069
Mean value			1.023			1.215
Standard deviation			0.009			0.315
The coefficient of variation			0.030			0.178

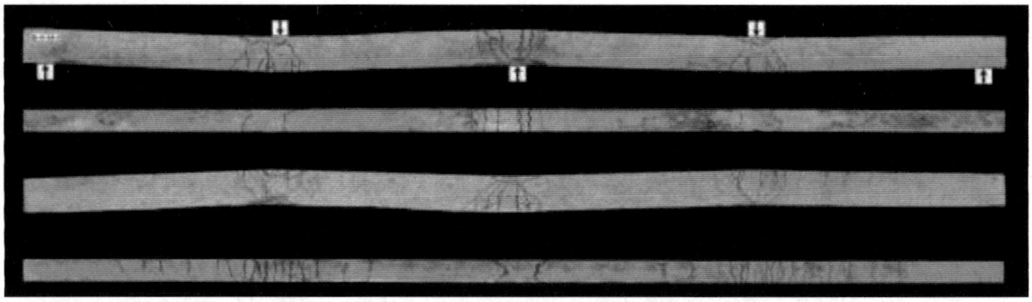

Figure 5. The nature of the crack development and the destruction of experimental samples

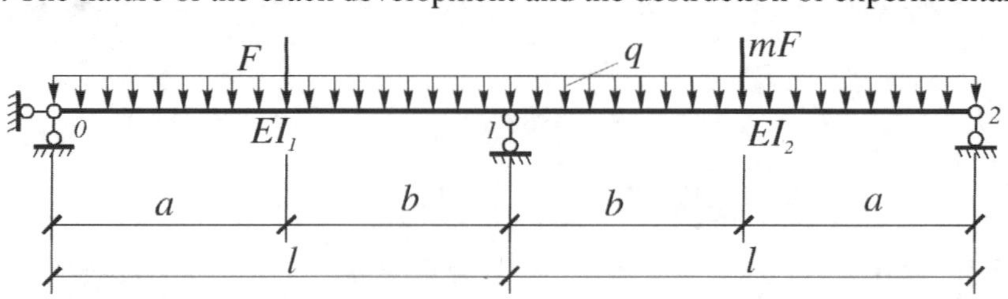

Figure 6. Calculated scheme of the beam (m – ratio between the concentrated forces)

Table 2. Experimental and theoretical values of deformations of normal sections of continuous beams in the boundary stage

Maximum relative deformation of the most compressed facet of concrete, ‰						
	Section under concentrated force in span			Section on the middle support		
	$\varepsilon^{r}_{cu1,test}$	$\varepsilon^{r}_{cu1,theor}$	$\frac{\varepsilon^{r}_{cu1,test}}{\varepsilon^{r}_{cu1,theor}}$	$\varepsilon^{mid}_{cu1,test}$	$\varepsilon^{m}_{cu1,theor}$	$\frac{\varepsilon^{mid}_{cu1,theor}}{\varepsilon^{mid}_{cu1,test}}$
B-1-12-1	-2.869	-3.061	1.067	-2,641	-3.061	1.159
B-1-12-2	-2.78	-3.061	1.101	-2.669	-3.061	1.147
B-1-12-3	-2.94	-3.061	1.041	-2.707	-3.061	1.131
B-1-14-1	-2.962	-2.934	0.990	-2.631	-3.073	1.168
B-1-14-2	-2.95	-2.923	0.991	-2.365	-3.073	1.299
B-1-14-3	-2.987	-2.923	0.979	-2.661	-3.073	1.155
B-2-12-1	-3.115	-3.261	1.047	-2.97	-3.261	1.098
B-2-12-2	-3.043	-3.261	1.072	-2.972	-3.261	1.097
B-2-12-3	-3.026	-3.303	1.092	-2.646	-3.261	1.232
B-2-14-1	-2.998	-3.13	1.044	-2.919	-3.266	1.119
B-2-14-2	-2.969	-3.116	1.049	-2.515	-3.266	1.299
B-2-14-3	-3.056	-3.116	1.019	-2.625	-3.266	1.244
B-3-12-1	-2.991	-3.275	1.095	-2.978	-3.261	1.095
B-3-12-2	-3.006	-3.29	1.094	-2.472	-3.261	1.319
B-3-12-3	-2.874	-3.317	1.154	-2.356	-3.261	1.384
B-3-14-1	-3.088	-3.13	1.014	-3.094	-3.266	1.056
B-3-14-2	-3.048	-3.171	1.040	-2.68	-3.266	1.219
B-3-14-3	-2.992	-3.158	1.055	-2.452	-3.266	1.332
Mean value			1.053			1.197
Standard deviation			0.034			0.160
The coefficient of variation			0.059			0.127

Table 3. Comparison of experimental and theoretic values of limit bending moments and loads for beams described in the articles [13, 14].

#	Sample	Supporting			Middle					
		M_{exp} kNm	M_{theor} kNm	$\frac{M_{\exp}}{M_{theor}}$	M_{exp} kNm	M_{theor} kNm	$\frac{M_{\exp}}{M_{theor}}$	$F_{\exp}$ kN	F_{theor} kN	$\frac{F_{exp}}{F_{theor}}$
I	BOI-1A	47.1	36.74	1.28	34.6	36.42	0.95	110.5	101.7	1.09
	BOI-1B	42.8	36.11	1.19	38.7	37.88	1.02	112	103.5	1.08
	BOI-2A	41.1	32.02	1.28	31.5	31.53	1.00	100	88.3	1.13
	BOI-2B	40.6	31.6	1.28	34.5	33.21	1.04	103.1	90.6	1.14
II	BVI-3A	59	51.08	1.15	51	51.64	0.99	150.6	143.1	1.05
	BVI-3B	60.3	51.28	1.18	49	52.4	0.94	148.7	144.6	1.03
	BVI-4A	35.4	30.56	1.16	30.9	30.96	1.00	93.7	85.7	1.09
	BVI-4B	33	30.64	1.08	29.3	32.19	0.91	88.3	87.8	1.00
III	BVU-5A	54.5	44.14	1.23	41	46.02	0.89	136	126	1.08
Mean value										1.08
Standard deviation										0.039
The coefficient of variation %										3.61

IOP Conf. Series: Materials Science and Engineering 1021 (2021) 012028 doi:10.1088/1757-899X/1021/1/012028

Experimental samples considered in [13, 14] have a=1.8 m, d=0.9 m, m=1. Beams of various series considered in [15] have d=1.5 m, m=1.45 – 1.49. The values of limit bending moments in the spans and on supports (Figure 7) were calculated by the DM with ESC using special application for the PC developed by the authors. Theoretic limit loads were calculated using the limiting equilibrium method taking into account the layout of the loads and equality of beam bending moments.

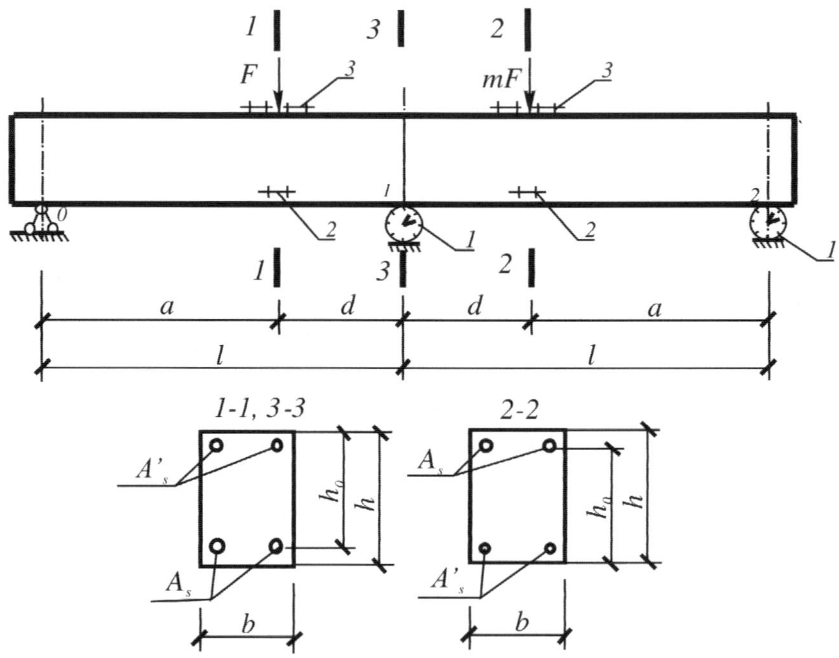

Figure 7. Calculation scheme, nature of reinforcement and scheme of arrangement of measuring devices and calculated cross-sections of beams: 1 - ring dynamometer; 2 - strain gauge for fittings (base 100 mm); 3 - strain gauge for concrete (base 100 mm)

The values of limit bending moments in sections 1 and 2 for the both series of beams (B-4 i B-5) obtained in the calculations are equal to $M_{u1} = 26.39\,kNm; M_{u3} = 40.24\,kNm$ respectively. The estimated values of limit bending moments on the support 1 (section 3) are different, because longitudinal reinforcements of beams in series B-4 and B-5 are different in this section. Their values are equal to $M_{u2} = 20.72\,kNm$ (B-4) and $M_{u2} = 43.76\,kNm$ (B-5).

Table 4. Comparison of theoretical and experimental values of limit loads according to the data given in [15]

A series of experimental beams [15]	Experimental values of the ultimate load, kN		Estimated values of the maximum load, kN		$\frac{F_{u1,exp}}{F_{u1,theor}}$	$\frac{F_{u2,exp}}{F_{u2,theor}}$
	$F_{u1,exp}$	$F_{u2,exp}$	$F_{u1,theor}$	$F_{u2,theor}$		
B-4	46.87	68.14	49.00	67.47	0.96	1.01
B-5	66.37	98.60	64.36	82.83	1.03	1.19
The average relative error %						4.70

5. Conclusions.

Experimental and theoretical study of continuous reinforced concrete beams with different classes of concrete, including high strength ones, led to a comprehensive reliable, qualitative and quantitative assessment of the impact of structural factors and factors of influence on their carrying capacity, redistribution of internal forces and determination of limit load. It is established that the proposed engineering technique based on ESC combined with numerical and optimization methods allows to analyse with sufficient accuracy the full range of the limiting parameters of normal sections of reinforced concrete structures in phase of destruction, makes it possible to calculate the value of bending moment in the sections where conventional plastic hinges are formed. Also, the technique allows to take into account redistribution of internal forces in statically indeterminate systems and to determine the limit load using the limiting equilibrium method. Assessment of the reliability of the results obtained by the developed engineering methods based on deformation techniques with extreme strength criterion and its comparing with experimental data obtained by the methods of mathematical statistics shows good consistency of experimental and theoretical data.

References

[1] *DBN B.2.6-98:2009 Reinforced concrete structures. Design rules* (K.: Minregionstroy of Ukraine) p 71
[2] *DSTU Б B.2.6-156: 2010 State Standard of Ukraine. Buildings and facilities structures. Concrete and reinforced heavy concrete structures* (K.: Minregionstroy of Ukraine) p 168
[3] Mitrofanov V P 2004 Practical application of the stress-strain model with the extreme strength criterion of the reinforced concrete elements *Municipal services Series: Architecture and technical sciences* **60** pp 29 – 48
[4] Shkurupiy O A 2011 Strength of concrete structures and their elements on the basis of a stress-strain model with the extreme strength criterion *Building structures: Interdepartmental research engineering collection of studies* **74** pp 605–614
[5] Rzhanitsyn A R 1984 *The theory of calculating building structures for reliability.* (Moscow. Stroyizdat) p 294
[6] Kudzis A 1985 *Reliability assessment of reinforced concrete structures.* (Vilnius, Moxlas) p 156
[7] Tykhyi M, Rakosnyk I 1976 Reinforced concrete structures calculation at the plastic phase *Force redistribution*; transl. from Czech (M.: Stroyizdat) p 198
[8] Gvozdyov A A 1949 Bearing capacity calculation by means of the limiting equilibrium method *The method essence and substantiation* (M.: Gosstroyizdat) p 280
[9] Eurocode 2 2004 *Design of Concrete Structures General Rules and Rules for Buildings* (Brussels: CEN) p 226
[10] Bambura A N, Gurkovskiy A B 2003 On the problem of deriving the strains theory for reinforced concrete of bar systems on the experimental basis *Building structures: Collection of studies* pp 121–130
[11] Bambura A M, Nemchinov Yu I, Gurkivskyi O B, Bezbozhna M S, Dorogova O V 2010 Design fundamentals of the bearing capacity calculation for concrete and reinforced concrete structures according to the national standardized document DBN B.2.6 – 98: 2009 *Building structures: Interdepartmental research engineering collection of studies (building construction)* **73** pp 724–736
[12] Shkurupiy A, Lazaryva E 2012 Bearing capacity calculation for continuous reinforced concrete beams using the strains model with the extreme strength criterion *Engineering sciences: theoretic and applied aspects: proceedings of the International research and practical teleconference* (Novosibirsk) pp 143–153
[13] Krylov S M, Gushcha Yu P, Abakanov M C 1980 Bearing capacity of statically indeterminate reinforced concrete beams made of concretes with strength up to 90 MPa, reinforced with high-strength bar reinforcement *Concretes and reinforced concrete structure elements subjected to different duration affects* (Moscow) pp 102–119
[14] Krylov S M, Gushcha Yu P, Abakanov M C 1979 Force redistribution in statically indeterminate reinforced concrete structures reinforced with steels without the yield point *Strength, rigidity and crack safety of reinforced concrete structures* pp 171–186

[15] Oatul A A, Chirkov V P 1965 Studying natural force distribution in continuous reinforced concrete beams with non-tensioned reinforcement at durable operation loads affect *Studies of concrete and reinforced concrete: Collection of studies* **34** pp 116–135
[16] Shkurupiy A A, Mytrofanov P B, Davydenko Yu O, Horb A.G. 2019. Statistical characteristics of strength distribution of normal sections of bended reinforced concrete elements and their analysis *Materials Science and Engineering*, **708** (1) doi:10.1088/1757-899X/708/1/012072
[17] Shkurupiy O, Mytrofanov P, Masiuk V. 2018. Calculation of the stability of the form of equilibrium of discrete systems *International Journal of Engineering and Technology (UAE)*, **7**(3), 401-407. doi:10.14419/ijet.v7i3.2.14561
[18] Mitrofanov V P, Pinchuk N M, Mitrofanov P B 2014 The primary design conceptions of concrete and reinforced concrete structures in the ADM model code *Paper presented at the 4th International Fib Congress 2014: Improving Performance of Concrete Structures, FIB Proceedings*, pp 259-261
[19] Olsson K, Dahlblom O 2016 *Structural Mechanics: Modelling and Analysis of Frames and Trusses* (Wiley) p 344
[20] Smith P 2001 *An Introduction to Structural Mechanics* (Palgrave Macmillan) p 368
[21] Nageim H, Durka F, Morgan W, Williams D 2010 *Structural Mechanics* (Palgrave Macmillan) p 456
[22] Zalesov A, Chistyakov E, Laricheva Yu 1996 Deformation design model of reinforced concrete elements under the action of bending moments and longitudinal forces *Concrete and reinforced concrete* **5** pp 16–18

Features of assessing the evenness of road surfaces using laser road scanners

R V Smolianiuk[1,3], N V Smolianiuk[2] and I V Kiashko[1]

[1] Department of Automobile Road Construction and Maintenance, Kharkiv National Automobile and Highway University, Yaroslava Mudrogo St. 25, 61002 Kharkiv, Ukraine
[2] Department of Bridges, Structures and Structural Mechanics, Kharkiv National Automobile and Highway University, Yaroslava Mudrogo St. 25, 61002 Kharkiv, Ukraine

[3] Email: rovlsm@yahoo.com

Abstract. Road pavement evenness is one of the basic parameters characterizing road quality. Evenness influences car wear, fuel consumption, transportation costs, emotional state and comfort of drivers and passengers, and many other parameters. Assessment of road pavement evenness, which is widely used all over the world, has a number of features that do not allow to correctly assess an evenness at bridges, viaducts and other structures. Moreover, particularities of laser scanners operation do not allow to compare the obtained data with data obtained with the help of three-meter straight-edge. The paper suggests ways to solve problems in question, which enable significantly improving quality of evenness assessment at such important transport infrastructure objects as bridges, overpasses and viaducts. There is also a solution to compare the data of transverse evenness obtained by laser scanners with data obtained with the help of a three-meter straight-edge that will permit to use these data later for comparison with standard values.

1. Introduction

Road pavement evenness (roughness) influences car wear, fuel consumption [1], transportation costs, emotional state and comfort of drivers and passengers, and many other parameters. In current practice related to pavement evenness assessment, the most widely used evenness parameter is the IRI (International Roughness Index) [2], which is calculated by summing fluctuations of prediction model of the “golden car” [3]. Fluctuations are simulated by special software according to the measured profile of the road pavement. Profile can be measured with the help of equipment of 4 classes.

The 2nd class equipment is used most widely; it is generally a measuring complex based on laser sensors that measure the distance between the level at which the sensor is located and the pavement surface. After adjusting for the angle of measurement, the profile of road pavement is obtained. Laser sensors have shown the best resistance to atmospheric phenomena and other disturbances and today are actually the standard for such equipment. A feature of all laser sensors is a very small diameter of laser trace on pavement. Another feature of the general equipment operation is that the distance between the measured points ranges from 50 to 300 mm. It depends on equipment design determined by the manufacturer. To calculate the IRI, it is sufficient to take measurements with a step of 300 mm. However, with such step, a number of important points can be lost that can have a significant impact on car traffic comfort [4]. To calculate the IRI, RoadRoof or ProVAL software can be used.

2. Investigation of evenness at bridges and other objects

Laser sensors used in the 2nd class equipment for assessing evenness give a laser trace on pavement surface within 2-5 mm (Figure 1) that enable actually determining pavement texture.

Figure 1. Laser trace on pavement.

Frequency of laser sensors operation is (according to the manufacturers' data) in the range from 2 to 16 kHz. Such operating speed can be achieved under ideal conditions. In real practice, taking into account necessary storage time of laser sensor and speed of ports and other equipment operation, measurement frequency is from 200 Hz. However, even such speed is excessive and most of the data is simply ignored or not obtained, since they are not needed to calculate the evenness of road pavement according to IRI. But this data can be especially valuable for analyzing such objects as bridges, overpasses and others. As a rule, such places have defects influencing evenness (Figure 2).

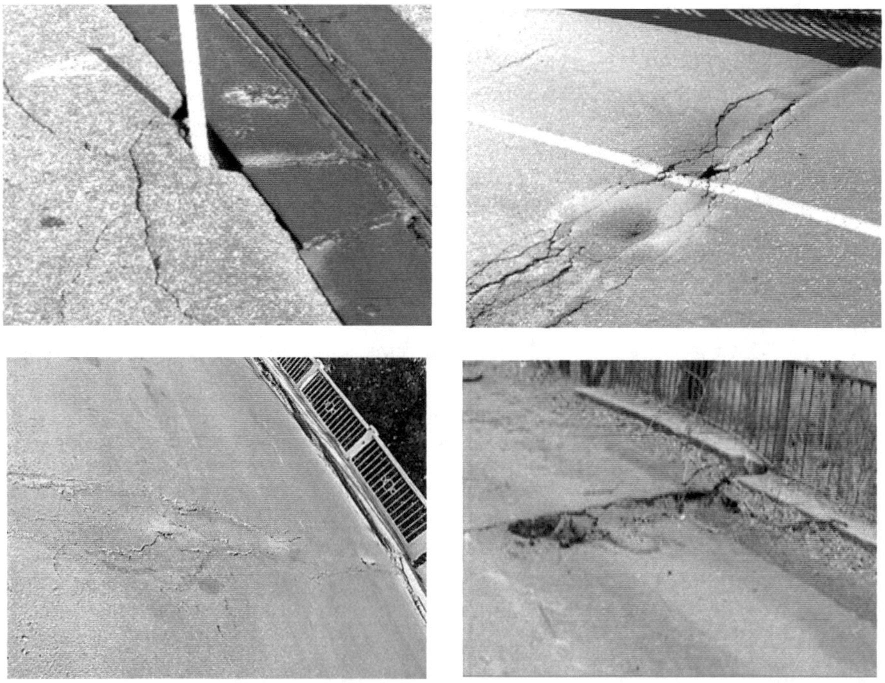

Figure 2. Defects on the bridges of Ukraine.

A lot of defects of bridges are caused by presence of joints of reinforced concrete slabs [5-8]. These places require constant maintenance and very often are in poor condition. Typical profiles corresponding to such types of defects were analyzed as well as models of such profiles were developed in this work. Based on the developed profiles, vehicle movement was simulated to calculate the IRI. Simulation and calculation were carried out by ProVAL software.

Figure 3 shows one of typical profiles and the result of movement simulation along it.

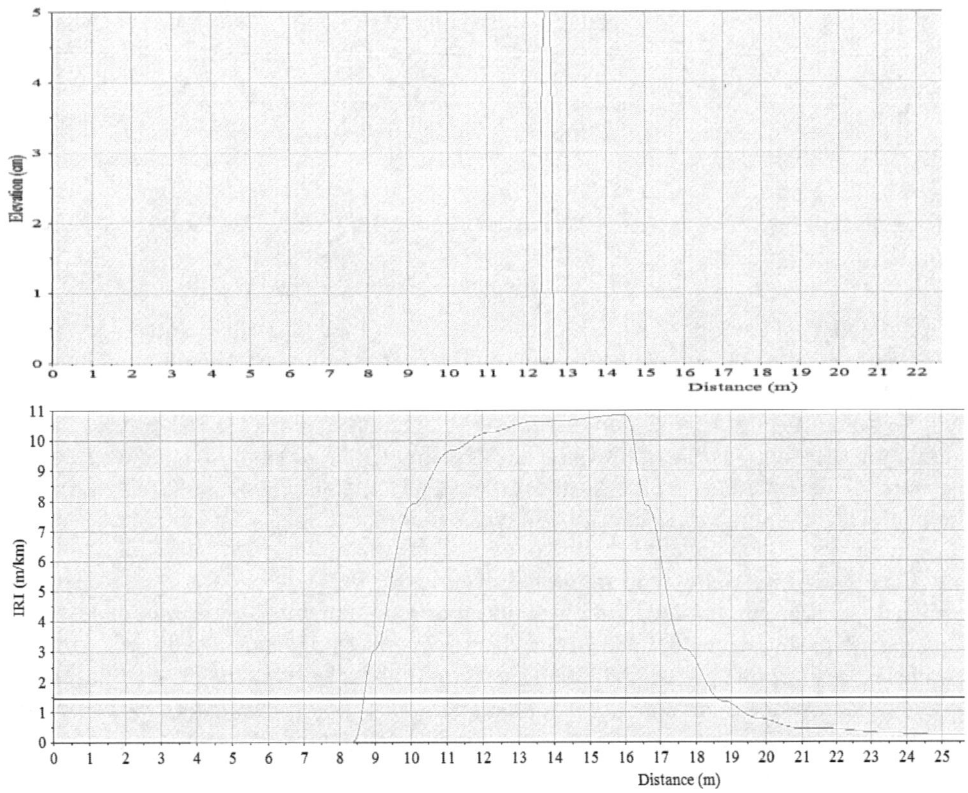

Figure 3. One of typical profiles and the result of movement simulation along it.

As it is shown at the diagram, the instantaneous IRI values are quite high. Fluctuations of the vehicle caused by such defects continue for a long time. Calculation of average IRI values for the section of 10 m length containing a single defect was also carried out using ProVAL software. Calculation results show that for a number of defects the average IRI value can exceed 4 m/km that is the boundary value of evenness in terms of safety.

Figure 4. Laser scanning complex LVS-4-KNAHU

Laser scanning complex LVS-4-KNAHU (Figure 4) has been used for laser scanning of the pavement and calculating evenness according to IRI at 4 experimental sections, which contain bridges. Table 1 presents the basic results of the experiment.

IOP Conf. Series: Materials Science and Engineering 1021 (2021) 012029 doi:10.1088/1757-899X/1021/1/012029

Table 1. Results of experiment.

Number of the object	Evenness according to IRI, m/km for 10 m length section containing a defect		
	Profile measured by 20 cm step	Profile measured by 2 cm step	Evenness value obtained for the defect model
1	2.57	3.86	3.56
2	3.39	4.82	4.74
3	1.86	2.42	2.66
4	3.62	4.64	3.99

The first column of Table 1 shows evenness values obtained under pavement surveying with usual step. Low values of evenness can be explained by the fact that the laser beam of a sensor may not fall or partially fall into the joint between the beams of the bridge. Therefore, this defect does not influence the value of evenness. Column 2 presents evenness values for the profile measured with a significantly smaller step. These values are usually slightly larger than the theoretical ones obtained with the help of defect model. This is due to the fact that along the section length there may be additional defects that increase the overall value of evenness.

3. Using a laser scanner to determine lateral evenness (rut)

The peculiarity of scanning system laser sensors operation, namely, small area of laser trace on road pavement, causes problems under determining lateral evenness of roads [9,10]. In Ukraine, as in many other countries, a three-meter straight-edge (in some countries, two-meter straight-edge) is used to determine lateral evenness or tracing rut (presence of rut). A straight-edge is put down across pavement and the distance between the straight-edge surface and pavement is measured using special device (Figure 5).

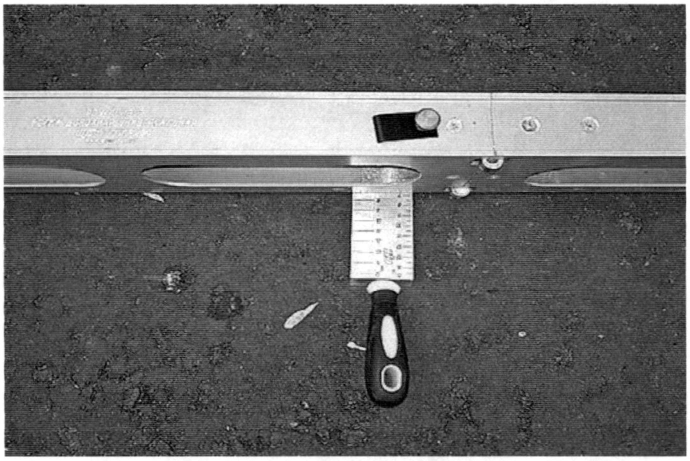

Figure 5. Using straight-edge and gage rod to determine tracing ruts.

Scheme for determining lateral evenness using present-day laser scanning systems is not significantly different (Figure 6).

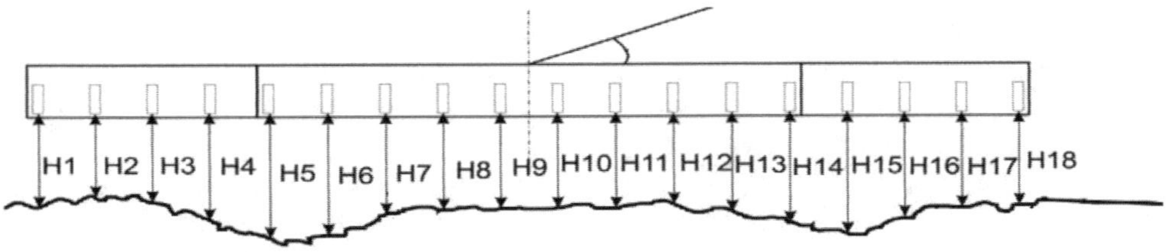

Figure 6. Scheme for determining lateral evenness using laser scanning systems

All sensors of laser scanning system are located on the same level. From this level, sensors measure the distance up to the surface of pavement. However, during measurements on stone mastic asphalt surfaces there are some problems. Structure of such pavement is shown in Figure 7.

Figure 7. Stone mastic asphalt.

As opposed to other types of pavements, surface of stone mastic asphalt is not solid. There is a significant distance between crushed agents, which very often is larger than a trace from laser sensor. Thus, laser beam may not fall on pavement surface, but fall into the gap between adjacent grains of stones. During experimental studies, it was found that this phenomenon is systematic. At the same time, it is impossible to unambiguously determine which of the sensors measured the distance up to the surface of a pavement, and which measured the distance in the gap between the grains of stones. Therefore, it is impossible to compare the obtained values with the values obtained with the help of a three-meter straight-edge. When measuring with three-meter straight-edge it actually has place an averaging of values over the width of gage rod, which is 5 cm. The lower edge of gage rod is in contact with peaks of stones and measurements are made from this level (Figure 8).

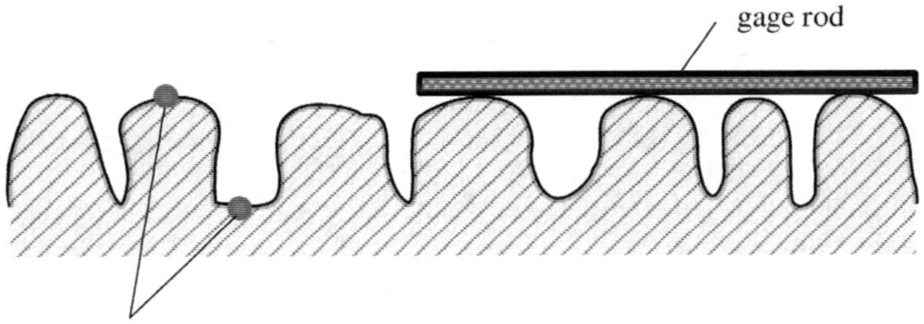

Figure 8. Comparison of measurement principle under using three-meter straight-edge and laser sensors.

It was experimentally established that the difference between measurements with using laser systems and a three-meter straight-edge can be 5-20%.

Solution to this problem may be to change operating modes of laser scanner. In places, where it is necessary to measure lateral evenness, the laser scanner has to perform at the maximum frequency and the obtained data should be processed separately. Under operating speed of measurement and maximum frequency, data can be collected every 5 mm.

As in the case, if a gage rod is used, it is necessary to collect data over 5 cm. Thus, it is possible to get about 10 points, from which it is necessary to select the minimum values, which will be the calculated ones. So, the measurement will be as identical as possible to the measurement with using a gage rod and a three-meter straight-edge

4. Conclusions

Present-day laser scanning systems have technological capabilities providing significant increase of useful data amount. In case of appropriate software upgrades, it is possible to record profile of road pavement of such objects as bridges with small step that enables identifying various specific defects. Detailed profile survey enables assessing the impact of joint condition at bridges on ride comfort as well as giving more information about their condition. Design feature of laser sensors does not allow to determine lateral evenness on roads with stone mastic asphalt pavements. This problem can also be overcome by introducing a special high-frequency mode of pavement survey in places where such measurements are necessary. Solution of this problem is also possible through software upgrades with introducing separate report for recording such data.

References

[1] Louhghalam A, Akbarian M and Ulm F-J 2015 Roughness-induced pavement–vehicle interactions: key parameters and impact on vehicle fuel consumption *J of the Transportation Research Board* **2525** pp 62–70

[2] Sayers M W and Karamihas S M 1998 *The Little Book of Profiling. Basic Information about Measuring and Interpreting Road Profiles* (Michigan: The University of Michigan Transportation Research Institute) p 306

[3] Sayers M W 1989 Two Quarter-Car Models for Defining Road Roughness: IRI and HRI *Transportation Research Record* **1215** pp 165–172

[4] Šimun M and Sršen M 2007 Roadway curtain equality on road infrastructure structures *J Građevinar* **59** pp 395–405

[5] Smolyanyuk R V, Smolyanyuk N V, Startsev V A and Zakharchenko M R 2019 Modeling the influence of technological inequalities on a bridge floor on traffic conditions *Bulletin of KhNAHU* **86(1)** pp 186–191

[6] Smolyanyuk R V, Smolyanyuk N V, Startsev V A and Zakharchenko M R 2019 Modeling the influence of technological inequalities on a bridge floor on traffic conditions *Bulletin of KhNAHU* **86(1)** pp 186–191

[7] Kiashko I V Novakovskiy D M and Smolianiuk R V 2016 Estimation of roughness of the surface of automobile roads *B Roads and road construction* **95** pp 75–86

[8] Múcka P 2017 International Roughness Index specifications around the world *J Road Materials and Pavement Design* **18(4)** pp 929–65

[9] Young Suk Park, Dong Ku Shin and Tae Ju Chung 2005 Influence of road surface roughness on dynamic impact factor of bridge by full-scale dynamic testing *Canadian Journal of Civil Engineering* **32(5)** pp 825-29

[10] K. Bogsjo, K. Podgorski, and I. Rychlik 2012 Models for road sourface roughness *Vehicle System Dynamics* **50** pp 725-747

[11] Sulaymon Eshkabilov, Abduvokhid Yunusov 2018 Measuring and Assessing Road Profile by Employing Accelerometers and IRI Assessment Tools *American Journal of Traffic and Transportation Engineering* **3(2)** pp 24-40

IOP Conf. Series: Materials Science and Engineering 1021 (2021) 012030 doi:10.1088/1757-899X/1021/1/012030

Reinforcement of a deformed structure on the pile foundation

Yu L Vynnykov[1,2], M O Kharchenko[1] and S M Manzhalii[1]

[1] Department of Oil and Gas Engineering and Technology, National University "Yuri Kondratyuk Poltava Polytechnic", Pershotravnevyi Avenue, 24, Poltava, 36011, Ukraine

[2] Email: vynnykov@ukr.net

Abstract. The results of geomonitoring of the technical state of a five-storeyed building erected on driven prismatic piles are presented. The cause of excessive deformations of the foundation base is the inability of the pile tip to reach the design depth, which led to a significant reduction in their load-bearing capacity. The constructive technological solution of strengthening of the bases from piles as a part of a tape grid by bringing under existing grids of a monolithic reinforced concrete slab is improved. A new analytical model of the system "deformed building – driven prismatic piles as a part of a tape grid – soil base with a weak underlying layer" before and after supplying the slab under the existing foundation grids. The stress-strain state (SSS) of this system is estimated using the finite element method (FEM). New data of the change in the SSS system after supplying the slab under the existing grids were obtained. The results of the project execution (without resettlement of the residents of the house) and the respective geomonitoring are presented.

1. Introduction

The causes of excessive settling of structures are errors in engineering and geological surveys, poor design, violation of the rules of construction work and operation, and for buildings on piles also the use of increasing factors on the data of laboratory tests of compressible soils; getting the tip of the piles into a layer of weak soil; immersion of piles above the design mark; overestimation of bearing capacity of piles due to non-compliance with the normative "rest"; misinterpretation of the schedule "load – pile settling"; too close placement of piles; uneven loading of them in a grid; disregard for the loading of adjacent areas under conditions of compact planning; driving in piles near buildings, development of ditches, etc. [1-3].

Reliable taking into account the influence of these factors, as well as components of absolute and relative uneven settling (compaction, uncompaction, protrusion, restructuring and due to changes in stress or strain of the foundation base during operation) on the value of strains of pile bases foundations of buildings, especially using the methods of classical soil mechanics, is either difficult or requires time-consuming and labor-intensive field and laboratory studies of soils. [4].

The most proper way to estimate the efficiency of alternatives for strengthening of pile foundation for strain building is possible by combining an analytical model of the system "strained building - pile foundation – soil base" before and after foundation strengthening and executing mathematical modeling using FEM and SSS soil this system to estimate the features of combined action of its components. [5-8].

Design and technological solutions to increase the bearing capacity of pile foundations of strained buildings and structures by pushing piles to strong soil and installation of offset piles with sufficient reliability of their performance have a very high hand labor-intensity and require a long period of work [9]. Therefore, it is advisable to improve sufficiently effective for conditions of sizable uneven strains of buildings, and at the same time less labor-intensive and more prompt solutions to increase the bearing capacity of pile foundations with their SSS changes or base soils, such as strengthening of driven prismatic piles foundations in a continuous grid by supplying a monolithic reinforced-concrete slab under the existing grids [10].

In the practice of reconstruction of buildings, including strengthening of their bases and foundations, the general system of geotechnical monitoring and scientific and technical support [11], which is even included in building norms, has arisen [12]. However, for objects with the system “strained building - driven prismatic piles as a part of tape grid - soil base with a weak subsoil” one of the most vulnerable spots is the difficulty of reliable determination the parameters of bases and foundations, in particular, the actual length of piles, and for follow-up computations – the complexity of combining the analytical model of this system before and after strengthening for reliable estimating the features of combined action of its components.

2. Purpose of the Work

The purpose of this study is to improve the structural and technological solution of foundation on prismatic cylindrical piles joint in continuous footing (pile cap) by the example of a distinctive actual object and assess the influence of supplying the monolithic reinforced concrete plate under the existing grids on the SSS of the “strained building – driven prismatic piles joint by continuous footing – soil base with loose underlying strata” system.

3. Methodology and Research

3.1. Geotechnical monitoring results

A typical full-scale object of this study, which was strengthened without the resettlement of its inhabitants during this period, is a five-storeyed residential building with a basement and crawl space, located in Horishni Plavni, Poltava region. It was built in 1977 according to a standard project. It consists of three-block sections, of which the left-handed end has undergone sizable strains (Figure 1). In terms of construction, the building is a structure with longitudinal load-bearing walls. The height of the floors is 2.8 m, and the basement is 2.1 - 2.2 m.

Figure 1. End block section of the strained building.

The walls are made of 1.5 sand-lime bricks on the cement-sand matrix. The thickness of the outer longitudinal load-bearing and end walls is 510 mm, and the thickness of the inner longitudinal load-bearing wall is 400 mm. The load-bearing walls in the basement are made of blocks and brickwork. Spatial rigidity is provided by transverse walls of a stairwell and disks of inserted floor overlapping.

The structural scheme cannot be considered rigid. The main heating system passes through the basement.

Investigating the strained block section established that during the construction and operation of the house, there were defects and damages that affected the load-bearing capacity and wearing qualities of individual elements and the building as a whole, in particular, vertical cracks with a width of up to 20 mm (Figure 2) and internal load-bearing walls mainly in the places of support of the spandrel and over (under) perforations. They are indicative of sizable strains of the base foundations of the section in the direction of partition off the main part of the house.

IOP Conf. Series: Materials Science and Engineering 1021 (2021) 012030 doi:10.1088/1757-899X/1021/1/012030

(a) (b) (c) (d)

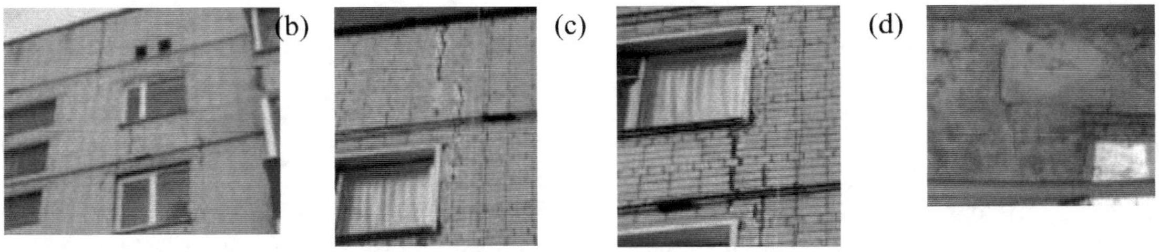

Figure 2. Vertical cracks with opening width: a – 7 mm; b – 15 mm; c – 12 mm; d – 4 mm (in the basement).

Their development significantly reduces the spatial rigidity of the section, because, for the outer and inner walls divided into parts, this rigidity is no longer provided by the transverse walls of the stairwell. There are no cracks on the outer and inner walls of the stairwell, which confirms the previous statements.

Minor cracks in the walls and between the floor slabs appeared immediately after the settlement of the house in 1977 but it underwent the largest deformations in 1993 after the heating main breakthrough.

At the same time, pre-tensioned reinforcement strands were arranged (Figure 1) from reinforcing bars with a diameter of 36 mm with tightening screws to create tension in them. Horizontal beams of the rectangular section from two welded channels No.24 established on face walls, fastening of rods to beams secured through the equilateral L-shaped member 140x10, which is installed at the corners of the building.

Since the free length of the strands between the fixing points should not exceed 15 - 20 m (in our case - 67.2 m) the cracks opened further. At the level of the third and fifth floors overlapping, there was a break in the tension bars.

At the level of the socle, marks were arranged around the perimeter of the building, which were leveled throughout the geomonitoring period. Plaster beacons are set on the external and internal load-bearing walls. Cracks in them indicate the further development of deformations. There are no cracks on the end wall of the block section. The technical condition of the load-bearing walls is condered as unsatisfactory.

The foundations are driven prismatic piles 9-35 (9 m length, 350x350 mm section), combined by a strip grid of 400 mm height (Figure 3).

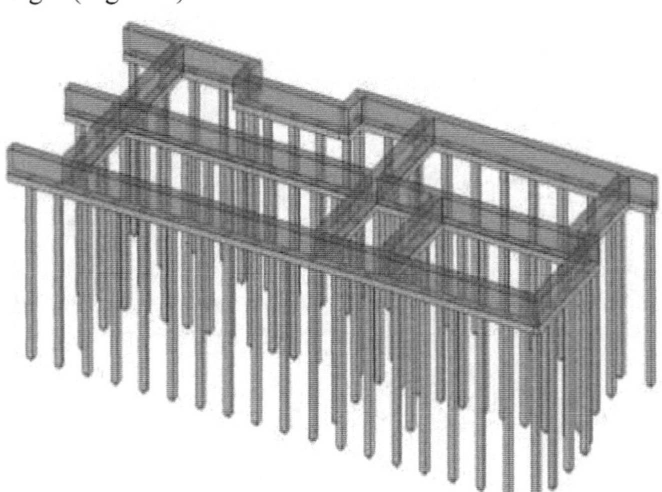

Figure 3. Spatial model of the foundation of the building section.

Under the inner load-bearing wall, the grid width is 400 mm, and under the outer wall, it is 500 mm. The increment of the piles under the inner wall is 1100 mm, under the outer wall is 1360 to 1530 mm, and under the end walls - 1590 to 1610 mm.

Within the boundary of the site under the bulk layer (Strata-1a) and fine sands (Strata-2c and Strata-2p, respectively, of medium density and dense with modulus of deformation, respectively, $E = 19.5$ and 35 MPa) with a total depth of about 7 m, there is a layer of buried soils (Strata-3) - layered sandy loams, with layers of silt and clay, fluxional ($E = 6.5$ MPa, organic matter content - 8%), which are underlined by medium-sized alluvial sands of 9-10 m deep (Strata-4, $E = 45$ MPa, and from a depth of about 18 m - by the clay.

Groundwater level (WL) at the time of the survey was 6.8 - 7.3 m from the earth surface.

Its annual and seasonal fluctuations reach 1.5 m from this level. Unfavorable engineering-geological processes have been identified within site dynamic impact on sandy soils from career explosions, which can lead to their dynamic liquefaction; mechanical suffusion during the operation of water-bearing communications; rather thick (up to 2.3 m) soil with impurities of organic matter.

The bearing capacity of the piles F_d here was estimated using the well-known three component formula

$$F_d = \gamma_c \left[\gamma_{cR} RA + \sum h_i \left(\gamma_{cf} u_i f_i + u_{oi} i_p E_i k_i \zeta_r\right)\right], \tag{1}$$

where γ_c – coefficient of pile operation's condition in soil; γ_{cR} and γ_{cf} – coefficient of soil operation under the foundation footing (bottom end) and along the lateral surface of a pile, which account for its manufacturing features; R and f_i – soil design strength under the foundation footing and along the lateral surface of a pile respectively; A – pile support area on soil; h_i – height of i-th soil strata, tangent to a pile's lateral surface; u_i – outer perimeter of i-th cross section of a pile; u_{0i} – side sizes sum of i-th cross section of a pile, which are deviated from vertical; E_i – deformation module of soil i-th strata; k_i – soil type dependent coefficient; ζ_r – rheology coefficient.

In the presented case for driven pyramidal piles the third component of the expression was equal zero. The building pile foundation settling was determined as for a reference soil massif at the pile tip level using a well-verified method of stepped summing.

$$S = \beta \sum_{1}^{n} \sigma_{zpi} h_i E_i^{-1}, \tag{2}$$

where σ_{zpi} – average value of additional stress in the i-th elementary strata; h_i, E_i – height and deformation module of the i-th soil layer respectively; n – number of elementary layers within the massif compressed under the reference foundation footing.

The project provided for penetrating the buried soil with 9-35 piles that reached Strata-4. In this case, the calculation revealed that: the load on the pile under the inner and outer bearing wall is 404.5 and 390.6 kN, respectively; pile's bearing capacity is $F_d = 1334.8$ kN; allowable design load $N = 953.4$ kN; the settling of the base of such foundation is $S = 1.44$ cm.

Considering the above and the worst possible scenario, under which accompanied by the rapture of the thermal pipeline could have the effect of "negative friction" on the lateral surface of the pile, the magnitude of which can reach 317 kN (soil layers Strata-2c and Strata-2p), allowable calculated load on the pile $N = 499.5$ kN which still exceeds the pile load from the structures 404.5 kN.

Therefore, the actual length of the piles in the foundation was checked. Control of continuity and length of piles is carried out by an acoustic method by means of Pile Integrity Tester PIT – W complex. For these tests, six trial pits were performed and the pile body was cleared by 20 cm. (Figure 4).

It was found that the actual length of the piles was 4.5 - 8.5 m. The results of the instrumental verification of the actual length of the piles and its integrity estimation are summarized in Table 1 (in particular, the pile # 119 operates on eccentric compression due to the considerable eccentricity of loading and pile's body deflection from vertical; the crack opening in the pile - 40 mm).

IOP Conf. Series: Materials Science and Engineering 1021 (2021) 012030 doi:10.1088/1757-899X/1021/1/012030

That is, part of the piles was not immersed in the design depth. The calculation was performed for pile # 119 (shaft # 1) of a 4.5 m length with a cross-section of 350x350 mm. A vertical tie of this pile to the geotechnical column is shown in Figure 5.

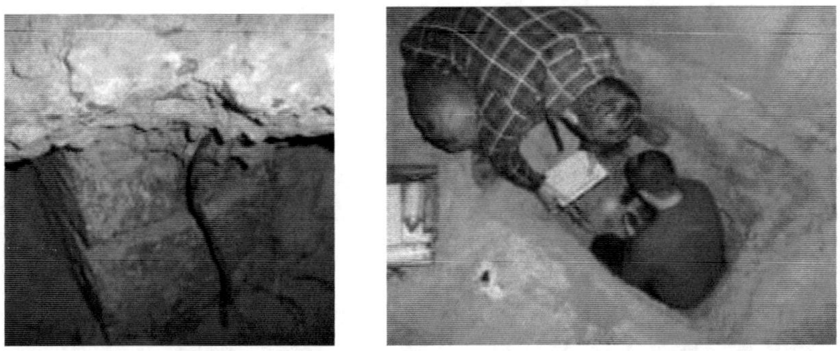

Figure 4. Pile within the grid view from the trial-pit and the pile length control.

Table 1. The results of the instrumental verification of the actual length of the piles and its integrity estimation.

Number of pitch / pile	Actual length of piles (m)
1/119	the pile is destroyed
2/105	8.5
3/9	4.5
4/77	8.5
5/171	6.0
6/33	7.5

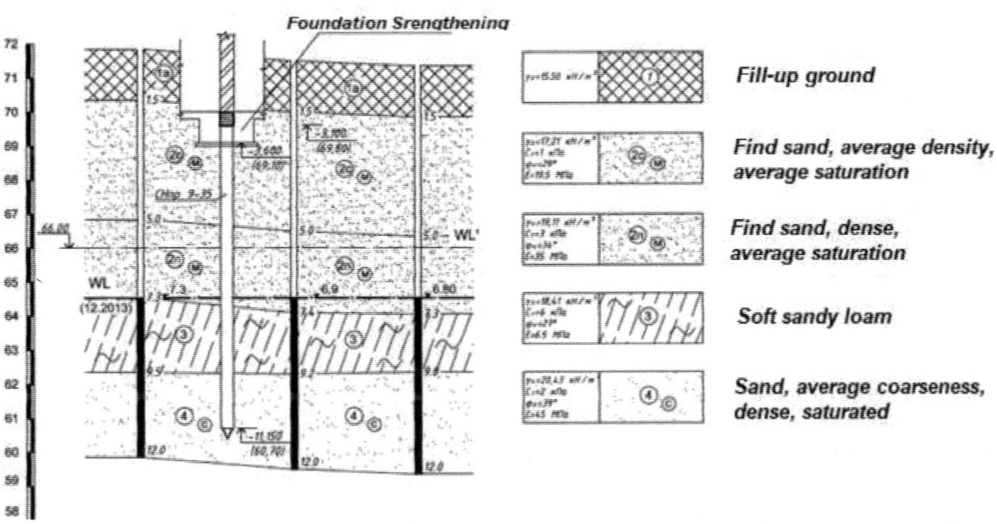

Figure 5. The vertical tie of the 4.5-meter-long pile to geotechnical column according.

The settling of the pile is S = 2.2 - 2.5 cm based on the calculation method. Thus, the difference between the settlement of piles with a length of 9 m and 4.5 m is about 1 cm, which could not cause actual over-deformation. However, given the occurrence of "negative friction" effect due to self-compacting and mechanical suffusion in the upper layers of the bulk sands after rupture of the main thermal pipeline, which was intensified by inertial forces from the explosions in the quarry, the bearing capacity of the pile decreased to F_d = 375.7 kN, it is up to N = 268.0 kN, which is less than the vertical force from the structures of the building 404.5 kN.

When the permissible load on the pile was exceeded, the pile foundation base settlement was already in the nonlinear stage, which led to the appearance and development of existing deformations of the building. Provided that if all the pile tips reached the design mark, then, of course, such deformation would not occur.

3.2. Simulation results of stress strain state of the "building – piles in continuous grid – base" system before and after supplying the slab under the caps

The technical condition of the pile foundations of the building was classified as unsatisfactory. A project of the pile foundations strengthening was developed. The strengthening meant the underlying of L-shaped monolithic reinforced concrete beams under the existing grids. The beams were joined with transverse reinforced concrete beams, and from the top – with the cast-in-situ 200 mm thick slab (Figure 6).

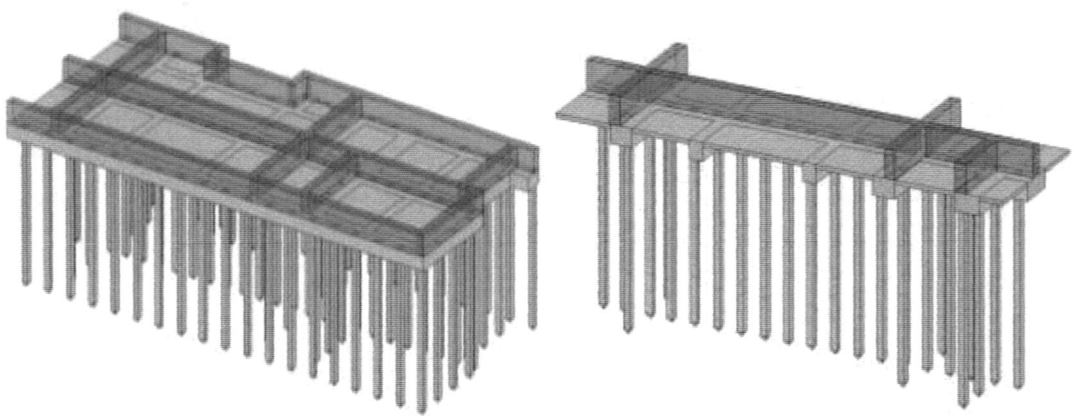

Figure 6. Spatial model of the pile foundation underpinning.

The reinforcement consisted of the supply under the grid of monolithic reinforced concrete L-shaped beams, which were joined with transverse reinforced concrete beams, and on top - with a cast-in-situ slab with a thickness of 200 mm.

To include the plate "in operation" right after it was set there was provided soil-base compaction under the plate by rubble and its ramming by vibrating plates.

When designing the reinforcement of the foundations of the building, the decisive criterion was the assessment of additional settling and its unevenness, which are realized during the works. They were taken into account depending on the category of the building according to the technical state, based on the conditions:

$$S_d \leq S_{d.u} ; \tag{3}$$

$$(\Delta\ S/L)_d \leq (\Delta S/L)_u ; \tag{4}$$

$$i_d \leq i_{d.u} , \tag{5}$$

where S_d – designed additional settling of a building in reconstruction, cm; $S_{d.u}$ – limit value of the additional settling of the building, cm; $(\Delta S/L)_d$ – calculated distortion of adjacent foundations after reconstruction; $(\Delta S/L)_u$ – the limit value of the distortion in the interval L; i_d – designed additional heeling of a building; $i_{d.u}$ – limit additional heeling of a building.

To confirm the effectiveness of the solutions made, FEM modeling was proposed. The calculation in the spatial scheme, taking into account the joint work of aboveground and underground structures, the pile foundation and soil base under it was performed. When assessing the SSS of the building, the soil base of the foundations was replaced by appropriate coefficients.

In the first stage, we could simulate the service period, taking into account the different vertical stiffness of the piles due to the reasons described above, as well as emergencies with water supply SSS to the ground part similar to that recorded during the survey.

To do this, the changed stiffness of the piles was taken into account as in Figure 7.The results of the simulation of the working period are shown in Figure 8.

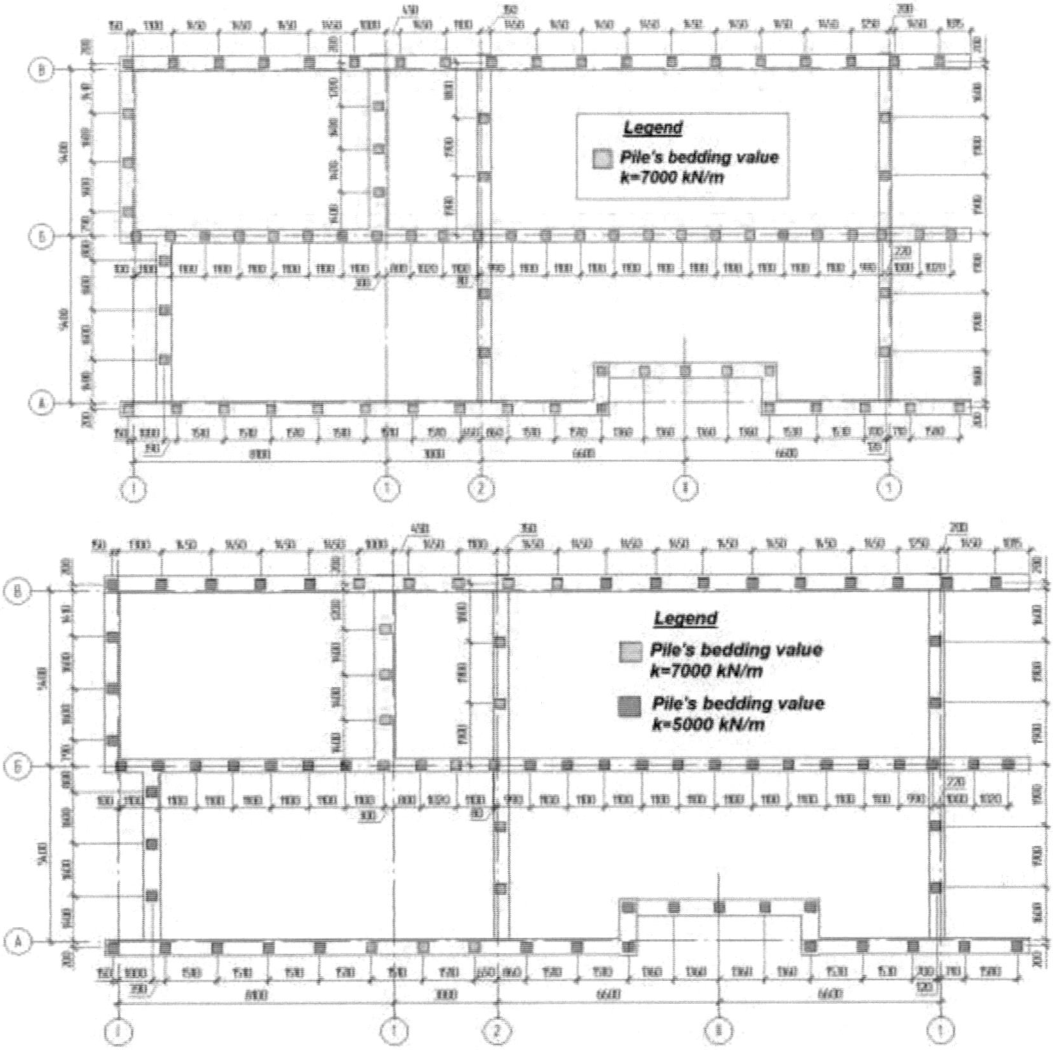

Figure 7. Initial (top) and after the soil wetting (bottom) pile stiffness accounted for in simulation.

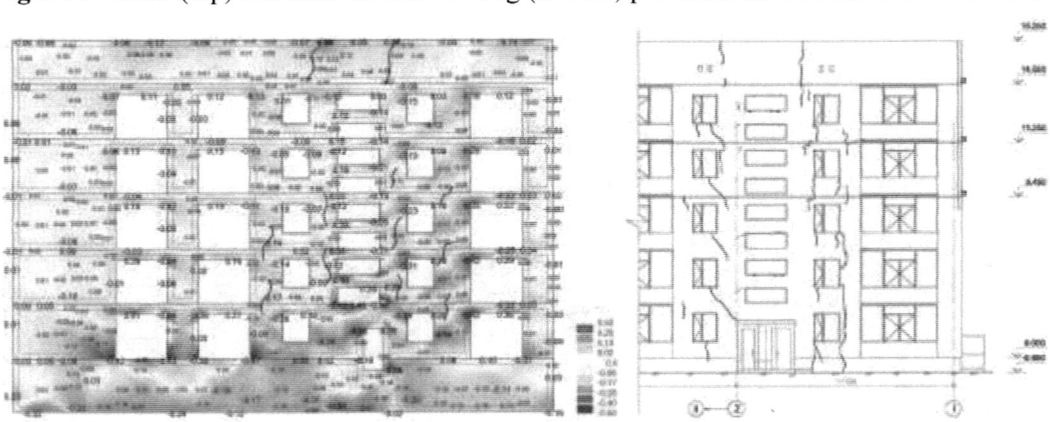

Figure 8. Map of the normal stresses distribution in the masonry of the outer walls (left) and cracks in the wall according to the results of supervision (right) after flooding the foundations.

According to the nature of the stress distribution, it can be concluded that due to flooding as a result of the accident on the pipeline, the left and right parts of the building collapsed around the stairwell, which due to the nature of the piles is more rigid. Therefore, most cracks are concentrated around the stairwell. At the third stage of the simulation the work of the elements of foundation reinforcement is taken into account (Figure 9). The stress distribution maps show that the reinforcement significantly helped to remove the uneven nature of the stress distribution and bring it closer to the initial state. The peculiarity of the distribution of tensile stresses in the masonry around the stairwell is preserved, but their values either do not exceed the tensile strength of the masonry (f_{bk2} = 0.11 MPa), or exceed it in places similar to the initial state of the building.

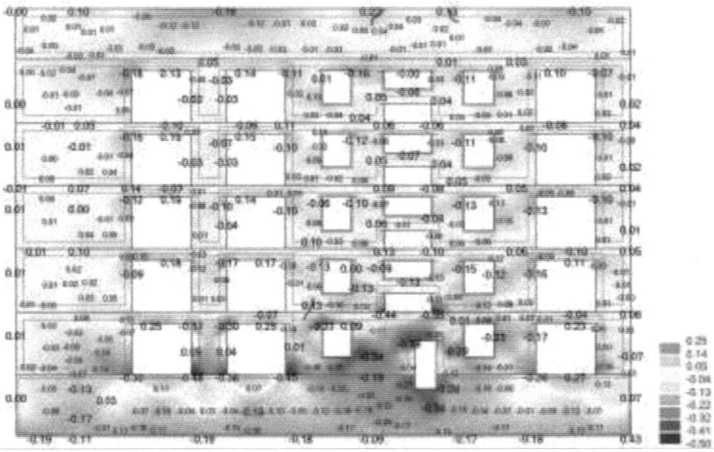

Figure 9. Map of the distribution of normal stresses in the masonry of external walls after strengthening the foundations.

3.3. Practical implementation of foundation underpinning

The underpinning operations were performed in six stages (Figure 10):

- digging out of trenches and concreting the transverse stiffening beams between the grids of the longitudinal load-bearing walls and under the grids of the transverse walls of the stairwells;
- underlying the L-shaped beams under the grids at the intersection of the longitudinal and transverse walls between the axes "2" and "1";
- the remaining L-shaped beams were underplayed under the grids in the intersection of longitudinal and transverse walls between the axes "2" and "1";
- underlying the L-shaped beams under the grids at the intersection of the longitudinal and transverse walls between the axes "I" and "2";
- underlying the L-shaped beams under the openings remaining under the grids in the intersections of longitudinal and transverse walls (between axes «I» and «2»);
- reinforcement and concreting of the upper monolithic plate, which unites all reinforcement beams into a continuous rigid spatial structure.

(a) (b) (c)

Figure 10. Building foundation underpinning: a – setting up of monolithic concrete reinforcement beams; b – reinforcing plate set up; c – basement view after the operations performance.

A ribbed reinforcement plate is obtained, which is based on Strata-2c (alluvial sand, fine, medium density). Its ribs are directed to the bottom. This design effectively redistributes stress from uneven deformations of bases and has considerable rigidity at the minimum volume of groundworks, that means that virtually slab-and-pile foundations was arranged.

As a result of geotechnical monitoring after strengthening the foundations of the object, which has lasted for more than 3.5 years, it was found that: additional settling of the foundation of the building did not exceed 1 - 2 mm and has already stabilized; after the reinforcement was completed, some of the previously destroyed gypsum beacons were restored, but no new cracks were formed in them.

4. Conclusions

Thus, the design technological solution of strengthening of the prismatic pile base as a part of a continuous grid by supplying a monolithic reinforced concrete slab under the grid is improved. Its rather high efficiency and reliability are proved. Also, the calculation scheme of the system "strained structure - prismatic piles in the continuous grid – the base with a weak underlying layer" before and after the slab supply under the grid and there was performed mathematical modeling using FEM and elastic-plastic model of soil SSS of this system to assess the features of combined action of this system's components.

References

[1] Briaud J-L 2013 *Geotechnical Engineering: Unsaturated and Saturated Soils* (Canada New Jersey: Wiley) p 1024

[2] Manjriker A and Gunarante I 2006 *Foundation Engineering* (New York: Taylor and Francis) p 608

[3] Fleming K, Weltman A, Randolph M and Elson K 2008 *Piling Engineering* (London and New York: Taylor and Francis) p 398

[4] Vynnykov Yu, Hajiyev M, Aniskin A and Miroshnychenko I 2019 Improvement of settlement calculations of building foundations by increasing the reliability of determining soil compressibility indices *Academic J. Series: Industrial Machine Building, Civil Engineering* (Poltava: PNTU) **1 (52)′** pp 115–123

[5] Chau K 2013 Numerical Methods *Proc. of the 18th Intern. Conf. on Soil Mechanics and Geotechnical Engineering* (Paris) pp 647–654

[6] Minno M, Persio R and Petrella F 2015 Finite element modeling of a piled raft for a tall building on cohesionless soil. *Proc. of the XVI ECSMGE Geotechnical Engineering for Infrastructure and Development* (Edinburg) pp 4019–4024

[7] Szerzo A and Batali L 2017 Numerical modelling of piled raft foundations. Modelling particularities and comparison with field measurements *Proc. of the 19th Intern. Conf. on Soil Mechanics and Geotechnical Engineering* (Korea, Seoul: COEX) pp 3055–3058

[8] Zotsenko N and Vinnikov Y 2016 Long-Term Settlement of Buildings Erected on Driven Cast-In-Situ Piles in Loess Soil *Soil Mechanics and Foundation Engineering* **53 (3)** (New York: Springer Science+Business Media) pp 189–195

[9] Illichov V and Mangushev R 2014 *Reference book of geotechnics. Foundations, foundations and underground structures* (Moscow: ACEU) p 728

[10] Vynnykov Yu and Manzhalii S 2019 Residential building's deformation on pile foundation *Academic J. Series: Industrial Machine Building, Civil Engineering* (Poltava: PNTU) **2 (53)′** pp 98 – 106

[11] Katzenbach R, Leppla S, Seip M and Kurze S 2015 Value Engineering as a basis for safe, optimized and sustainable design of geotechnical structures. *Proc. of the XVI European Conf. on Soil Mechanics and Geotechnical Engineering for Infrastructure and Development* (Edinburgh) pp 601-606

[12] EN 1990:2002/A1:2005/AC. 2010. *Eurocode: Basis of Structural Design*. Authority: The European Union Per Regulation 305/2011, Directive 98/34/EC, Directive 2004/18/EC p 144

Research of the behavioral of the wooden beams with fire protection lining under fire loading

M I Zmaha[1], S V Pozdieiev[2], Y V Zmaha[3,6], O V Nekora[4]and S O Sidnei[5]

[1]Science department ofCherkasy Institute of Fire Safety named after Chornobyl Heroes of National University of Civil Defense of Ukraine, 94 Chernyshevska Str. Kharkiv 61023, Ukraine
[2]Institute of Public Administration in the Sphere of Civil Protectio, 21Vuchgorodska Str. Kyiv 01011, Ukraine
[3] Department of physico-chemical bases of Cherkasy Institute of Fire Safety named after Chornobyl Heroes of National University of Civil Defense of Ukraine, 94 Chernyshevska Str. Kharkiv 61023, Ukraine
[4]Department of Civil Defense Organizations of Cherkasy Institute of Fire Safety named after Chornobyl Heroes of National University of Civil Defense of Ukraine, 94 Chernyshevska Str. Kharkiv 61023, Ukraine
[5]Department of Safety of Construction Facilities and Labor Protection of Cherkasy Institute of Fire Safety named after Chernobyl Heroes, National University of Civil Defence of Ukraine, 94 Chernyshevska Str. Kharkiv 61023, Ukraine

[6]Email: *ianazmaha1986@gmail.com*

Abstract. This article highlights the current issues of ensuring the fire resistance of wooden beams as load-bearing structures. Scientists have studied various types of fire protection of wood from the effects of fire. In the existing methods of fire protection, fire-retardant impregnating materials, fire-retardant paint, commercial fire-retardant coverings are presented. In addition, a study on the creation of fire-retardant materials was conducted. Therefore, in the field of our research there are the issues of using fire-protective plywood as a fire-retardant cladding method of protection of wooden structures. This article describes the process of manufacturing fire-retardant plywood made of 2-millimeter sheet of veneer and impregnated with a flame retardant using the method of technological baths and glued in a pressed machine to the appropriate thickness. Then in the experimental setup, fire tests were performed at standard temperature for three intervals of exposure to determine the depth of charring. These studies make it possible to draw conclusions about the effectiveness of the proposed method of fire protection of wooden structures and to determine the calculated class of fire resistance for these structures.

1.Introduction

The industrial development of construction every year dictates increased requirements for fire safety. Therefore, the conformity of building materials with the fire resistance class depending on the normative limit states is determined by DBN B 1.1-7: 2016 [1]. The substantiation of fire protection of wooden building constructions with preservation of durability, functional and aesthetic properties remains relevant.

Modern building and construction is impossible without the use of wooden building structures, because they have significant advantages, such as environmental friendliness, durability, strength, and

IOP Conf. Series: Materials Science and Engineering 1021 (2021) 012031 doi:10.1088/1757-899X/1021/1/012031

economy. The most common wooden building structures are trusses, beams, ceilings. One of the most promising ways to use wooden beams is to improve their properties.

Wooden beams must not only meet the strength and rigidity requirements, but also the fire safety requirements. Due to the high temperatures of the fire, the strength of the timber structures decreases, sometimes their complete destruction occurs, so ensuring the fire resistance level of the wooden beams is an urgent task.

In literature [2, 3] the studies of improvement it the properties of fireproofing agents for the timber structures were carried out.

Such studies consider this task incompletely, because they don't take into account the fire resistance of such structures.

Although in literature [4, 5] these issues were solved, the fireproof coating material of wooden beams in the form of fireproof plywood, was not considered. This material, however, is the most common in building nowadays.

2. Literature review and purpose of research

In the modern building industry, for the last ten years, plywood takes one of the leading places among coating materials because it has durable properties, environmentally safe, cheap, easy to use, has a wide range of applications, often used for decoration. These properties provide the maximum shelf life and cost-effectiveness. Plywood is a layered glued structure consisting of three or more sheets of wood with mutually perpendicular arrangement of fibers in adjacent layers. This gives the plywood increased strength in different directions, shape stability, which is a distinct advantage over natural wood. Plywood, like wood, is easy workable. Its surface can be arranged in different ways: laminated, covered with veneer, painted, varnished.

Different scientists have been studying this issue in different years. A.S. Belekov, M.Ya. Roitman, S.V. Pozdeeva studied fire resistance of different structures. Providing fire-retardant materials to wooden structures - T.M. Shnal, V.M. Zhartovsky, Y.V. Tsapko, S.L. Fomina. Foreign researchers who have studied the fire-resistant properties of wood products – White R.H., Forauter A., Kruppa G., Lie T.T. The available views on the issue of ensuring the fire resistance of wooden beams by the method of facing with fire-retardant plywood can be classified as not a study in full. Therefore, research on this issue is in the spotlight [2-4, 6, 10, 11].

The aim of the article is to establish the dependences of systemic signs of charring zone change on the analyzed type of fire protection of wooden beams using a standard temperature curve for hydrocarbon fire, as normatively accepted to determine the class of fire resistance of building structures. The study of this problem is a scientific basis for determining the assessment of fire resistance of wooden beams with the most effective way of fire protection.

Much attention has been paid to surface treatment issues that can be achieved using fireproofing and paints. All these measures affect the required level of fire resistance of structures. But these issues are not learned yet: increasing the level of fire resistance of plywood, additional load on the structure, large labor costs in processing, the behavior of fire-resistant plywood during fire exposure and also physical fire exposure -chemical properties of plywood and the effect of fire-retardant treatment on the fire resistance of such structures.

The topicality of our studies is to study the behavior of a wooden beam with fireproofing plywood coating during fire, as well as to loading the fire resistance class of a wood beam with flame retardant cladding and the depth of its charcoal.

Literature [4, 6] shows that in a fire condition, the failure of general structural stability usually occurs due to the destruction of the most loaded elements (columns, beams), as they are the main supporting elements in the building frame. During the destruction of beams losses reach a high level as at loss of their carrying capacity there is a destruction of technological equipment and destruction of a premise. Therefore, maintaining the load-bearing capacity of beams in a fire for a given time is an urgent task. Taking into account the mentioned above, the goal was formulated and tasks were set.

IOP Conf. Series: Materials Science and Engineering 1021 (2021) 012031 doi:10.1088/1757-899X/1021/1/012031

3. Basic theoretical provisions for the calculation

The purpose of this article is to investigate the behavior of wooden beams with fireproofing plywood coating under the action of fire when tested under standard temperature conditions. To achieve this goal, we were assigned the following tasks:

- to analyse the kind of plywood;
- to prepare test samples and conduct experimental studies of the fire resistance of wooden beams with fireproofing plywood coating at standard temperature conditions;
- to process the results of the experiment with the determination of the coefficient of charring.

Presentation of the main material with full explanation of the obtained scientific results. For the production of the experimental examples, a 2 mm birch veneer of the highest grade was used, which was impregnated with the Immersion in Technology Bathtubs method, since this method provided impregnation of the wood to 4 mm, which ensured complete impregnation of the veneer with the use of a flame retardant using the process line a 10 mm and 20 mm thick fireproof plywood was manufactured.

As the value of the level of fire resistance of wooden beams with fireproofing plywood coating is determined by testing by standard methods of fire resistance test, and the indicator of the class of fire resistance is determined by the onset of the limiting state in the loss of bearing capacity and according to [7-9, 13] it is equal to 60 minutes, so the fire tests were conducted wooden beams according to the standard temperature of the fire.

For the tests, a heating installation was used, which is a 500 × 500 × 500 mm steel casing. On the back, the chamber has a 60 mm diameter hole to install the burner nozzle. To minimize heat loss, the camera is protected inside by a 100 mm layer of non-combustible Conlit 150 insulation made by Rockwool that protects metal structures against high temperatures for up to 3 hours. Four TXA-VIII thermocouples were also used, three of which were installed in the sample and one was installed in the unit to determine the standard temperature mode complete with Digital multimeter DT 700C and Digital multimeter DT 838C (Figure 1).

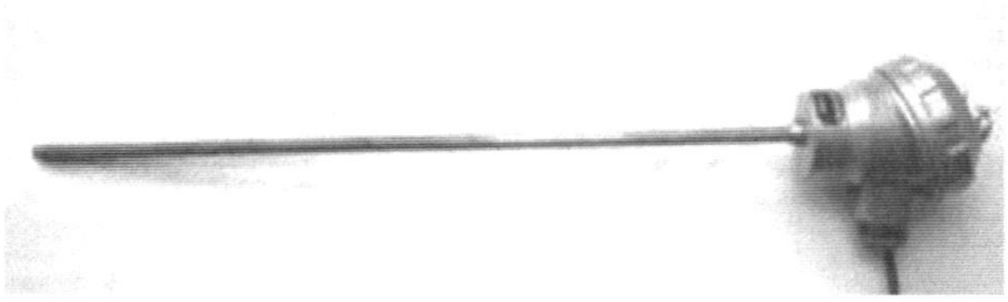

Figure1. TXA-VIII thermocouples (0 to 1100)°C.

The experimental samples were made of pine blocks measuring 250 × 70 × 350 mm and plywood 250x100 mm in the amount of three pieces for every three intervals of 15, 30 and 60 minutes for two types of thickness, i.e. in total there were twenty-four samples that are shown (Figure 2). Figure 3 shows a photo of the experimental sample with the indicated locations for thermocouples.

General view of the experimental installation for testing the fire resistance of samples - fragments of wooden beams lined with fire-retardant plywood with 10 mm and 20 mm thickness and indicating the location of thermocouples in the samples, according to the standard temperature curve is shown in Figure 4.

Thus, the installation presented in Figure 4 has high performance characteristics, a wide range of materials used for research and a small error in the results.

This experimental setup was used in the study of scientists A.V. Shvidenko, O.V. Nekora, V.M. Nuyanzina, S.O. Sydney for fire resistance of reinforced concrete structures of various configurations, as well as wooden beams and wood products – by Y.V. Zmaha, A.Yu. Novgorodchenko.

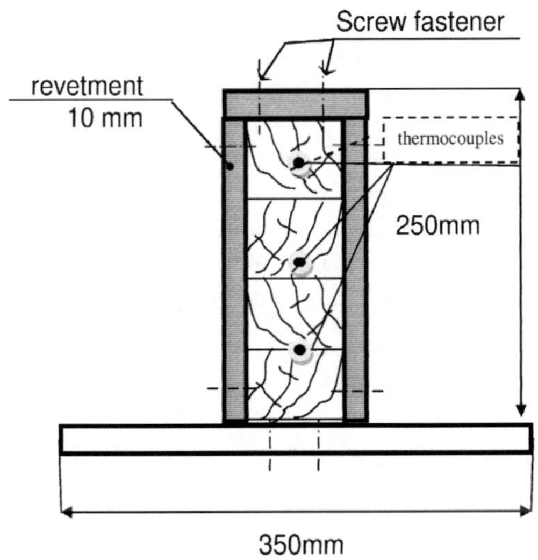

Figure 2. View of the test sample at 10 mm thick fire-protective plywood.

Figure 3. A photo of the sample with placement of thermocouples.

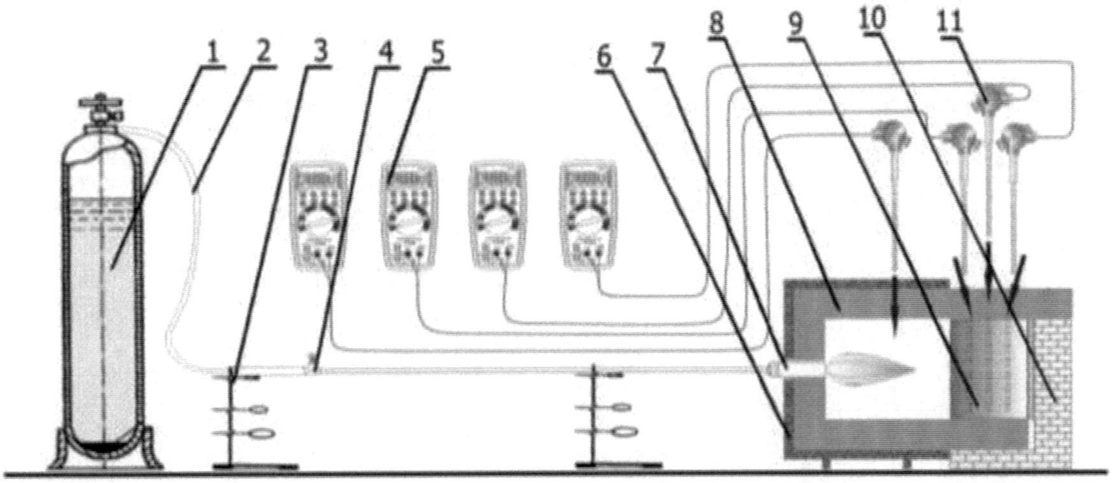

Figure4. Installation diagram: 1 - cylinder filled with gas; 2 - sleeve; 3 - tripod; 4 - valve with injector; 5 - multimeter; 6 - steel body of the furnace; 7 - gas burner; 8– thermal insulation of the chamber of thermal influence of the furnace; 9 - sample for research, installed in the body infusion chamber; 10 - brick body of the furnace; 11 – thermocouples.

The thermocouple element position allows the temperature readings to be recorded from the surface inner layers of the sample. The sample is set to the outlet of the experimental setup. When heating the specimens, a mode corresponding to the "standard" fire mode was used in the test setup chamber. The mode of heating the space of the chamber was determined by the thermocouple installed in the chamber. The thermocouple readings according to [6] were recorded from the graduated combined measuring device every 1 minute.

The temperature in the samples was recorded for 15, 30, 60 minutes, then recording stopped and the test and description of the sample was carried out.

During the complex of 15 minutes fire tests, the following behavior of the samples was observed:

- Samples with fireproof plywood lining with a single layer of 10 mm during fire tests did not emit smoke temperature rise was gradual, while the cells of self-ignition were observed.
- Samples with fireproof plywood lining with a single layer of 20 mm in a 15-minute time interval did not emit smoke, temperature rises were gradual and moderate, cases of self-ignition were not observed, which in turn can characterize the effect of the thickness of the layer on the effectiveness of protective properties.

The complex of 30 minutes fire tests gave a clearer idea of the behavior of fragments of wooden beams with fireproofing coating material during the fire, as can be seen in the graphs namely:

- Wood samples without impregnation behaved similarly to experimental studies in a 15 minute interval. The change in temperature occurred with an equally rapid increase.
- Specimens with one layer of 10 mm flame retardant were not subjected to significant fire effects during the experiment, the temperature rise was gradual and had a maximum at the first thermocouple of 160 degrees, without self-ignition, and no thick, odorless smoke was released.
- The most refractory specimens were those with a 20 mm flame retardant coating layer. They did not emit smoke, the temperature rise was slow and reached a 107-degree mark on the first thermocouple, with the specimens only charred by a few millimeters.

The following phenomena were observed in the 60 minute block of tests: specimens with lining fireproof plywood with one layer of 20 mm, characterized by high flame retardant and flame retardant properties. Collected and analyzed the actual results of measurements of the thickness of the depth of charring of samples of wooden beams allows to illustrate in (table 1), certain observations of the dependences of the geometric parameters of the thickness of fire-retardant plywood.

Table 1. Parameters of charring depth from the thickness of the facing layer and exposure time

Samples of fire-retardant plywood	Sample number	Time exp, min	B, mm	B0, mm	B1, mm	B2, mm	B3, mm	L, mm	L0, mm
10 mm	1.	15	70	0	0	0	6	250	243
	2.		70	0	0	0	8	250	244
	3.		70	0	0	0	8	250	245
20 mm	4.	15	70	0	0	0	0	250	250
	5.		70	0	0	0	0	250	250
	6.		70	0	0	0	0	250	250
10 mm	7.	30	70	14	16	20	25	250	231
	8.		70	14	15	16	26	250	236
	9.		70	14	15	17	27	250	235
20 mm	10.	30	70	3	4	5	8	250	240
	11.		70	2	4	6	8	250	243
	12.		70	1	2	3	5	250	245
10 mm	13.	60	70	19	22	26	28	250	206
	14.		70	21	24	28	27	250	208
	15.		70	18	21	25	29	250	207
20 mm	16.	60	70	15	20	24	44	250	227
	17.		70	17	19	22	26	250	215
	18.		70	17	20	23	28	250	215
Without impregnation	19.	15	70	16	18	21	27	250	187
	20.	30	70	22	24	28	31	250	165
	21.	60	70	32	41	43	53	250	132

IOP Conf. Series: Materials Science and Engineering 1021 (2021) 012031 doi:10.1088/1757-899X/1021/1/012031

4. The method of determining the carbonation depth.

The following procedures should be followed to determine the carbonation depth of wooden beams:

- Determine the performance of the sample.
- The obtained sample is subjected to one of the methods of fire protection of wooden structures.
- Conduct fire tests on a standard temperature curve for 15, 30, 60 minutes.
- After the end of the experimental part, measurements of the depth of charcoal using the caliper according to figure 2 are made.
- Conclusions are drawn as to the effectiveness of the fire retardant method for wooden structures and the depth of charcoal of wooden beams.

5. Conclusions

In this article, flame-retardant substances were analyzed and one substance was selected for further experimental study which is the most used within our country. The method of fire testing of wooden beams with fireproof plywood lining at standard temperature is also given, and the results of this experiment are published. At intervals of 45 to 60 minutes, as a rule, there is a rapid increase in temperature, and in some cases, self-ignition of the samples. This changes the structure of the wood, which in turn correlates with the change in the strength properties of the wooden beams.

References

[1] DBN V 1172016Fire safety of construction sites (Kyiv:Gosbud Ukraine)

[2] Antonov GA V Borisov OV and Orel VP 2004 Extinguishing media manual K Pozhinformtekhnika p176

[3] Nizhnik EC V Zhartovsky SV Tyshchenko OM and others 2010 Substantiation of the use of some water extinguishing agents for the fire extinguishing system for dome wooden structures of religious buildings *scientific bulletin*KUkrNIPIP**2 (22)**pp1-4

[4] Nizhnik EC V Zhartovsky SV 2005Investigation of peculiarities of the formation of combustible gases in the process of thermal decomposition of wood and their fragmentation *Scientific Bulletin of the Ukr Research Institute* **No. 1 (11)**pp14-18.

[5] Zhartovsky VM Zhartovsky SV and Grabovsky OV ***Pat.8987***2005Method of manufacturing fire-bioprotectedplywood of high water resistance

[6] Gorbachenko YA 2014Investigation of the behavior of wood beams with fireproof impregnation: scientific article Collection of scientific works. *Fire safety: theory and practice*Cherkasy ChIPB **7**pp8-13

[7] Krivtsov Yu V2008 Comparative evaluation of test methods for fire retardant compositions for wood *Fire safety***2**pp87-91

[8] Luneva NK **Pat. 1628475, MIK 08 V 5/00, V 27 K 5/00** Method of treatment of wood materials of coniferous and deciduous breeds

[9] GOST 20022.2-80Wood protection ClassificationMoscow 20

[10] GOST 30219-95Wood is fireproof. Test methods, transportation and storageMoscow20

[11] Kislov YS 2000 Physical models of combustion in a fire prevention system *A monograph.* St. Petersburgp 263

[12] Likov AV 2012Heat-mass transfer in caillary-porous bodies*in the book* Problems of heat exchange MAvtomizdatp332

[13] Ozarkiv IM Peretyatko BM 2004Fire tests of wood*Scientific Bulletin: Coll. scientific and technical wash*LvivNLTU of Ukraine**141**pp77-81

Application of energy efficient materials in reconstruction of buildings and structures of transport infrastructure

S Zolotov[1], O Pustovoitova[1], P Firsov[1, 4], Y Orel[2] and M Rezunenko[3]

[1] Building Structures Department, OM Beketov National University of Urban Economy, Marshala Bazhanova str. 17, 61002, Kharkiv, Ukraine
[2] Department of Research and Designing of Communication, Geodesy and Land Management, Ukrainian State University of Railway Transport, Feierbach Square 7, 61050, Kharkiv, Ukraine
[3] Department of Higher Mathematics, Ukrainian State University of Railway Transport, Feierbach Square 7, 61050, Kharkiv, Ukraine

[4] e-mail:pavelfirsov1991@gmail.com

Abstract. The article analyzes in detail the results of the performed reconstruction works with the application of thermosetting polymers. The method of adhesive anchoring of reinforcement rods with fixing them in the concrete body with acrylic glue was carried out on transport infrastructure facilities (railway steam shop and railway itinerary post). It has been determined that strengthening the foundation with application of the thermosetting polymers (adhesive anchor joints) gives a significant reduction in the consumption of concrete, metal and labor costs. The advantage of such acrylic adhesives is the possibility of varying the ratio of monomeric and polymeric part (powder) containing redox system, within acceptable concentrations, without significant changes in adhesion and cohesion parameters. The obtained results are very important in practice, because in the conditions of preparation of the adhesive (especially in construction conditions) it is difficult to determine the exact number of components. The experimental research data can be used for attachment joints projecting and industrial technological lines development, as well as for further polymeric adhesive materials, widely applied in industrial and civil engineering, improvement.

1. Introduction

The application of plastics based on acrylic polymers for the manufacture of load-bearing elements is quite limited. Abovementioned is due to the low values of plastics elasticity modulus and due to the high materials deformation. More widely polymers used as adhesives for fixing and connecting the joints of structural elements that have sufficient rigidity with the allowable distribution of the load [1].

Polymer adhesives are currently used in building industry mainly in three directions: during finishing works (flooring, wall decoration, etc.), during the manufacture of building structures (fencing structures, roofs, waterproofing), during the installation of sanitary equipment and pipelines [2].

The wide usage of gluing in technology is due to the fact that with the help of mineral, organic and combined adhesives strong and reliable connections of materials and structures can be obtained, and often better and more economical than with any other method. In some cases, gluing is the only possible method of bonding. Mineral and organic (natural) adhesives have long been used to join stones, bricks, wood, paper, cardboard, wall and floor tiles and other materials. To improve the adhesive properties of mortars (mineral adhesives), they have long been introduced organic impurities. With the advent of high-strength synthetic adhesives, it has become possible to combine load-bearing building structures that are exposed to significant static, dynamic, vibratory and other loads. Such adhesives provide sufficient thermal, fire and frost resistance of adhesive connection, as well as resistance to moisture, solvents, acids, alkalis, fungi and more.

Epoxy and acrylic glues are widely used as material in structural joints, as their properties meet all necessary requirements for adhesives to create reliable high-strength joints [3, 4].

2. Problem statement

Durability requirements in construction are generally lower than in mechanical engineering and other industries. However, adhesive joints in construction are used for at least 50 years, must be non-toxic during operation, have a relatively low cost [5, 6].

Both abroad and in Ukraine, extensive experience has been gained in the application of adhesive anchoring of reinforcing rods for anchor installation works during the reconstruction, repair and reinforcement of concrete and reinforced concrete structures [7]. This anchoring has a number of advantages:

- reduced terms of repair and reconstruction of buildings and structures;
- lower energy and material consumption;
- reduced cost and complexity of work;
- installation of reinforcing rods in various weather conditions is possible;
- in some cases, this method is almost the only possible one (for example: changing the thickness of the foundation by a small value with significant increases in loads);
- simplicity, reliability and high manufacturability of works;
- the ability to apply design loads to the reconstructed and reinforced structures and areas in 6-24 hours, instead of 72 hours, as for structures reinforced in the usual way;
- fastening of reinforcing rods in existing concrete and reinforced concrete structures using acrylic glue is economically justified and the most progressive;
- acrylic glue is low-component, simple and reliable in preparation. The factory supplies glue components in a set that facilitates the organization of maintenance of anchor installation works;
- acrylic glue is an environmentally friendly substance and does not affect humans and the environment.

The method of fixing the reinforcing rods with a rigid cement-sand mixture (vibrating) allows you to load the anchor device in three days. The disadvantage of vibrating is the need to use a special mechanism for compaction of the mixture, the multicomponent nature and complexity of its preparation, high complexity. Therefore, the attachment of reinforcement rods on existing concrete or reinforced concrete structures using epoxy, siloxane and especially acrylic adhesives is the most progressive, economical and less time consuming [8, 9].

3. Basis of calculation

The calculation of anchor rods was performed in accordance with [8] and Building Norms of Ukraine "Steel structures. Design standards". The magnitude, direction and character of the loads from the equipment on the rod must be specified in the design task.

According to [8], the cross-sectional area of the rod is determined from the condition of strength by the formula:

$$A_s = k_0 P \cdot (R_s)^{-1} \tag{1}$$

where $k_0 = 1.35$ for dynamic loads; $k_0 = 1.05$ for static loads.

Under the action of dynamic loads, the cross-section of the rods determined by formula (1) must be verified for endurance by formula:

$$A_s = 1.8 \chi \mu P \cdot (\alpha R_s)^{-1} \tag{2}$$

where χ is load factor (taken according to Table 1 data); μ is the coefficient taken depending on the diameter of the rod [8]; α is the coefficient that takes into account the number of load cycles [8].

During the calculation of the building structures fastenings, the pre-tightening force and the cross-sectional area of the rods are determined as for static loads, unless there are special instructions in the project.

During the group installation of rods for equipment mounting, the value of the design load P per rod should be determined for the most loaded rod by the formula:

$$P = N \cdot n^{-1} + My_1 (\sum y_i^2)^{-1} \tag{3}$$

where N is the calculated longitudinal force; M is the calculated bending moment; n is the total number of rods; y_1 is the distance from the axis of rotation to the most removed rod in the stretched joint area; y_i is the distance from the axis of rotation to the i-th rod, taking into account both stretched and compressed rods.

Table 1. Values of load factors χ and tensile strength, depending on the design of the anchor rods.

Construction of the rod	With a limb	With anchor plate		Direct	Conical (spacer)
		Deaf	Demountable		
Rod diameter, d, mm	12-48	12-140	56-125	12-48	6-48
The minimum laying depth, l_{anch}	25d	15d	30d	10d (8d)	10d (8d)
The smallest distance between the rods	6d	8d	10d	5d	8d
Load factor, χ	0.4	0.4	0.25	0.6	0.55
Pull stability coefficient, k	1.9	1.9	1.5	2.5(2)	2.3(1.8)

During the installation of rods with bends, the depth of their laying in concrete is taken to be 15d_s, for rods with anchor plates 10d_s, and for rods installed in wells - 5d_s.

During the installation of paired rods, a common anchor plate with a distance between the boreholes equal to the design distance between the axes of the rods should be provided, or single rods with a "run" in depth should be installed. The depth of laying of paired rods at a distance between their axes of 8d_s and more should be assigned 5d_s, at a distance of less than 8d_s - equal to 20d_s.

The distance from the edge of the plate to the axis of the rod should be chosen at least 2d_s, and the area of the anchor plate should be at least 32d_s.

4. Main material and results

To confirm the abovementioned advantages of the method of adhesive anchoring of reinforcement rods of ∅20 A500 and ∅25 A500 with fixing them in the body of concrete with acrylic glue, research and industrial implementations were carried out on transport infrastructure facilities (railway steam shop and railway itinerary post).

The main purpose of the steam shop reconstruction was to replace the obsolete overall boiler with a new gas boiler. Under the designed boiler it was necessary to reconstruct the existing foundation. To determine the physical condition of the foundation and the scope of the proposed work, a survey of this structural element was carried out.

The survey has found following data:

- the existing foundation has a complex configuration with the maximum geometric dimensions in plan 5.4×3.6 m. The depth of foundation laying is 3.6 m;
- the concrete body of the foundation is in good condition but needs cosmetic repairs;
- the designed gas boiler has the following parameters: length of the boiler 6,820 mm, width of the boiler 4,300 mm, height of the boiler 5050 mm, heat output - 2.9 MW.

Thus, based on the overall dimensions of the gas boiler, it is necessary to change the configuration of the existing foundation with increasing its geometric dimensions and to provide the required bearing capacity of the zones, which need to be erected.

The geometric dimensions of the reconstructed foundation need to be 7300×4800 mm (Figure 1). The depth of the foundation remained initial. The purpose of experimental and industrial implementation was to change the configuration of the foundation and its dimensions by connecting the existing concrete body with freshly laid concrete by the application of adhesive anchoring of reinforcing rods. At the same time, the volume of the existing foundation $V = 83.1$ m^3 is preserved.

Technological sequence of operations during the reconstruction of the foundation:

- digging a trench along the perimeter of the foundation to a depth of 3.6 m;

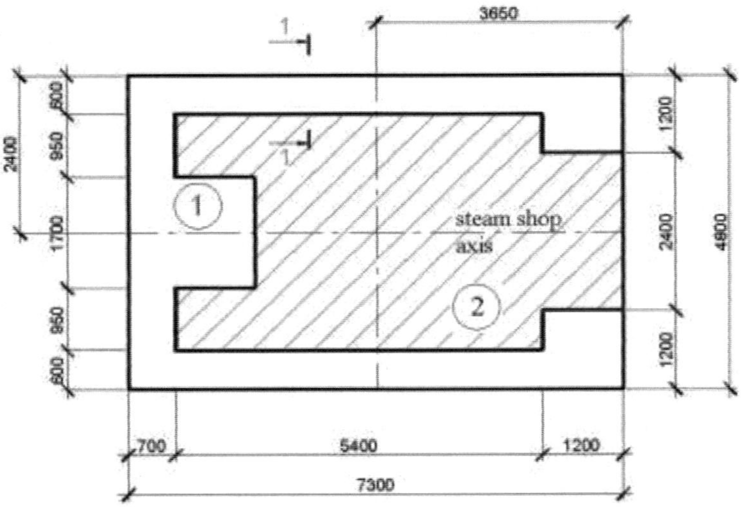

Figure 1. Reconstruction scheme of the foundation: 1 - new foundation zone, 2 - old foundation zone.

- cleaning of existing foundation surface from dust and dirt, concrete exfoliated areas reflection;
- marking of places under the holes, during which the distance of the reinforcement to the foundation boundary and between the rods must be at least $5d_s$ of the installed rods (∅20 and ∅25 mm) and is equal to 25×5 = 125 mm.

It was accepted: boreholes are located with a step of 550 mm on height of the foundation (Figure 2), and along the perimeter set the bay of the rods 300 mm;

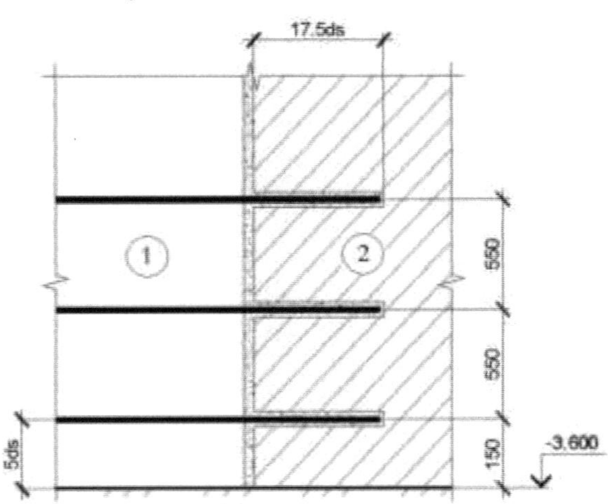

Figure 2. Cross-section of the reconstructed object. 1 - new foundation zone, 2 - old foundation zone.

- drilling of holes in the body of existing concrete. Holes were drilled to a depth of $17.5d_s$ of laying the reinforcing rod (∅20 and ∅25), which is 450 mm. The direction of the drilled holes relative to ground level is horizontal. Holes were drilled with a diameter of 30 mm;
- preparation of anchors made of crescent-shaped reinforcing rods ∅20 and ∅25 A500;
- filling holes with high-viscosity modified acrylic glue, which included special impurities that increase its adhesive and cohesive strength;
- installation of formwork from metal boards. The inner surfaces of the boards, before filling with concrete, were coated with a special mixture to prevent adhesion of metal to concrete;
- installation of reinforcing rods. The reinforcement was installed by slow immersion in a hole filled with glue. Total amount of rods, which were installed: ∅20 A500 - 311 units and ∅25 A500 – 216 units;
- to ensure the design position of the reinforcing rods, annular retainers were installed. To ensure the stability of the spatial structure, the releases of the installed anchors were connected to the vertically and horizontally located rebars with a knitting wire. The created reinforcing design provided perception of design loadings;
- application of acrylic glue without modifying impurities on the already prepared surface of the existing foundation
- laying of concrete in the formwork. At the time of laying the mobility of concrete was 3-4 cm. Concrete was laid in layers, which thickness does not exceed 1.25 of the length of the working part of the vibrator. The minimum thickness of the concrete layer was 200 mm. The total volume of stacked concrete was 61.1 m^3.

As a result of the implementation of the development and preservation of the existing foundation saved: 7.1 tons of metal, 83.1 m^3 of concrete, 290 people labor per days. In addition, the application of reinforcement anchoring with acrylic adhesives can reduce the time of commissioning of the object by 7 times.

Second object of thermosetting polymers application was reconstruction of a foundation of the railway itinerary post.

On the basis of the survey it was established: the inspected condition of the foundation is unsatisfactory; the bearing capacity is minimal.

The main purpose of experimental and industrial implementation is to restore the bearing capacity of the strip rubble foundation by increasing its cross-sectional area. Connection of the old foundation with freshly laid concrete is made using adhesive anchoring of reinforcing rods (Figure 3).

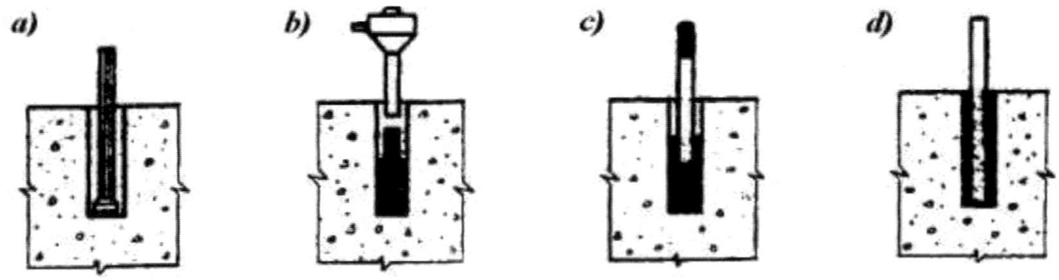

Figure 3. Technological schemes of adhesive anchoring process: a – boreholes drilling in concrete body; b – filling of acrylic adhesive into the boreholes; c - installation of a reinforcement rod in a borehole; d - installation of a reinforcement rod in design position.

The technological process of strengthening the foundation took place in the following order:

- the concrete pad and the surface of the rubble foundation were cleaned of soil, rubbish and dust. Rubble completion of the zones formed as a result of falling of a stone is performed;

- performed marking of holes for installation of reinforcement rods. According to the given scheme, holes are located along all perimeter of the foundation in 1 row with a step of 300 mm;
- vertical holes were drilled in the concrete pad to depth ≈400 mm, which is not less than $17.5d_s = 17.5\times20 = 350$ mm and a diameter of 30 mm;
- reinforcement rods were prepared in the amount of 1180 units;
- holes were filled with modified acrylic glue. This adhesive was prepared directly on the building site;
- reinforcement rods ∅20 A500 were installed vertically, by slow immersion in holes filled with acrylic adhesive;
- to ensure the design position of the reinforcement rods were installed annular retainers. After gaining the design strength of glue, the installed rods were connected to the horizontally located rods of the reinforcement with a knitting wire. The stability of the manufactured structure was ensured by anchors installed into the rubble masonry of the foundation with acrylic adhesives;
- formwork was installed from pre-made wooden boards in accordance with the requirements of technical conditions;
- the surface of the concrete pad of the foundation was covered with ordinary acrylic glue, which contains only acrylic polymer and filler in the form of quartz sand;
- laying of heavy concrete with a mobility of 3-4 cm was performed in layers, followed by compaction with a vibrator. The thickness of the compacted layer ranged from 200 to 600 mm. The total volume of concrete laid in the formwork was 47.2 m^3;
- after 72 hours or 3 days produced removal of the formwork of the structure (Figure 4).

Figure 4. Formwork removal of the reconstructed foundation.

Observations, that continue for today, on the operation of reconstructed foundations and joints of reinforced concrete elements, operating under loads, indicate the full integrity of the anchor joints.

5. Conclusions

1. As a result of the reconstruction and strengthening of the foundation with the application of thermosetting polymer materials (acrylic adhesives), a significant reduction in the consumption of concrete, metal and labor costs was achieved.

2. A new field of application of adhesive joints of metals based on thermosetting polymers are building structures, in particular the transport infrastructure facilities. The following requirements are set for adhesives and adhesive joints abovementioned structures:

- strength under static and dynamic loads;
- ability to harden without heating and at minimum pressure;
- simplicity of preparation of a surface under gluing;
- resistance to the influence of variable temperatures;
- low sensitivity of the joint strength to fluctuations in the thickness of the adhesive seam.

Some types of developed acrylic adhesive compositions meet these requirements, because they harden without heating and in the case of the filler application provide stable adhesive compounds to temperatures from -30°C to +80°C [10].

3. The application of thermosetting polymers for bonding metal during the repair of steel pipelines for gas and water supply without their dismantling is effective. Acrylic adhesives can be used to glue metal capsules that support electrical cables directly to stone or concrete walls without the use of special stands. Also, it is proposed to glue metal shapes to wood in the manufacture of large wooden beams.

The advantage of such adhesives is the possibility of varying the ratio of monomeric and polymeric part (powder) containing redox system, within acceptable concentrations, without significant changes in adhesion and cohesion parameters [11]. This is very important in practice, because in the conditions of preparation of the adhesive (especially during the construction) the exact number of components is hard to determine.

References

[1] Zolotov S M 2015 *Acrylic adhesives for building structures connection* (Kharkiv: O M Beketov National University of Urban Economy) p 184

[2] Irfan M H 1998 *Chemistry and technology of thermosetting polymers in construction application* (Heidelberg: Springer Netherlands) p 285

[3] Tutunchi A, Kamali R, Kianvash A 2015 Adhesive strength of steel-epoxy composite joints bonded with structural acrylic adhesives filled with silica nanoparticles *Journal of Adhesion Science and Technology* **Vol 29(3)** pp 195-206

[4] del Real-Romero J C, Cano M, Abenojar J and Martínez M A 2006 Adhesive bonding of aluminium with structural acrylic adhesives: durability in wet environments *Journal of Adhesion Science and Technology* **Vol 20(16)** pp 1801-1818

[5] Possart W, Brede M 2019 *Adhesive joints. Ageing and durability of epoxies and polyurethanes* (Weinheim: Wiley-VCH) p 544

[6] Amstock J S 2000 *Handbook of adhesives and sealants in construction* (New York: McGraw-Hill Professional) p 656

[7] Storozhenko L I, Davydenko Yu O, Gorb O G, Gorb O O 2015 Adhesive joints of steel and concrete *Scientific Collection of Ukrainian State University of Railway Transport* **Vol 155** pp 184-190

[8] Zolotov M S, Pustovoitova O M, Firsov P M 2019 *Cold cured thermosetting resins for restoration and reconstruction of industrial and civil engineering* (Kharkiv: O M Beketov National University of Urban Economy) p 184

[9] Zolotov S, Pustovoitova O, Firsov P, Husain Kais, Kamchatna S 2019 Stress-strained state of steel-adhesive bonding on acrylic adhesives *IOP Conference Series: Materials Science and Engineering* **Vol 708** 6 p

[10] Pocius A V 2012 *Adhesion and adhesive technology. An introduction* (Cincinnati: Hanser Publications) p 108

[11] Chamochín R, Cano M, Abenojar J, Pantoja M, Ballesteros Y and del Real-Romero J C 2010 The effect of surface treatment on the behavior of toughened acrylic adhesive/GRP (epoxy) composite joints *Journal of Adhesion Science and Technology* **Vol 24(11-12)** pp 1903-1916

The use of incineration waste to improve the properties of the bases

M L Zotsenko[1], O V Mykhailovska[1,3], I I Lartseva[1], V V Klymenko[2]

[1] National University "Yuri Kondratyuk Poltava Polytechnic", Pershotravny ave. 24, 36011, Poltava, Ukraine
[2] Central Ukrainian National Technical University, Universitetskiy ave. 8, 25000, Kropyvnytskyi, Kirovohrad region, Ukraine

[3] Email: helena_2005@ukr.net

Abstract. Innovative ways of using furnace dust to improve the strength of soil-cement are analyzed. It is proposed to solve the environmental problem of utilization of this material, when using the addition of furnace dust during the construction of soil-cement elements. The samples were kept immersed in water for 28 days in order to gain strength, until laboratory tests. The removal ash before the experiment was sieved on a 4 mm sieve. Strength of soil-cement samples with the addition of furnace dust and pellet ash (with the main content of sunflower husk with a percentage of 5% by weight of cement) was studied. It is established that the average compressive strength of the samples with the addition of Mykolaiv CHP furnace dust with a fraction of inclusions up to 4 mm in the amount of 5 % increases the compressive strength by 30%. It was found that the content of organic matter (ash from burning sunflower husks) in soil-cement reduces its strength.

1. Introduction

Construction of foundations accounts for a significant part of the total construction costs. One of the effective ways to reduce the cost of foundation construction is to improve the characteristics of the soils that form the basis for the foundation. This can be achieved by impregnating the soil with cement slurry – cementation. As a result of the process of cement hydration, a material of quite significant strength is obtained – soil-cement [1].

The use of soil-cement as a material for making foundations and strengthening foundations is an effective way to reduce the cost of construction because it uses soils that lie directly at the base of construction sites. Soil cement has recently been widely used to improve the properties of foundations, for the construction of various industrial and civil structures. The use of soil-cement structures for the main technological structures of biogas plants is promising [2].

Soil-cement structures are gaining more and more popularity in foundation construction, namely: arrangement of retaining walls of ditches; arrangement of anti-filtration curtains; strengthening of landslide slopes; strengthening the foundation of existing foundations with inclined, horizontal and vertical soil-cement elements; railway embankments; improving the construction properties of peat, loess soils, bulk, weak soils (E <5 MPa) [1,3].

From the point of view of manufacturability of production process, soil-cement piles (elements) have advantages over all other stuffed piles that their manufacturing requires no additional actions for the maintenance of walls of wells (casings, drilling muds, special shells, etc.). We know boring and mixing and jet method of their production. The latter has the advantage of its productivity and greater capabilities in depth and diameter of soil-cement piles. With the drilling method less cement and simple equipment are used, as result the estimated the cost of soil-cement is less. However, there is a problem of low strength of cement as a material. Soil-cement piles lose their bearing capacity mainly on the material and not on the soil.

To increase the strength of soil-cement, various researchers have proposed additives in the form of lime, clay, various plasticizers [5], sand and tailings (mineral waste) [4].

On the other hand, there is a problem of accumulation of waste from enterprises, namely ash, ashes and slag waste from thermal power plants (CHP), which occupy large areas.

It is known that approximately 50% of all electricity produced in Ukraine belongs to thermal power plants. According to statistics, the ash content of extracted coal is growing every year, and accordingly, thermal coal supplied to thermal power plants averages 26.3 to 36.2%, which leads to an increase in the amount of ash slag, as well as ash, which is removed with the exhaust gases. There is a constant accumulation of ash and ashes and slag waste in the ash dumps of thermal power plants, which today amount to approximately 400 million tons. As a result, the design capacity of most CHP ash dumps is being filled, and obtaining new areas is problematic. In order to solve this problem, a certain amount of ash slag after studying the content of radioactive and toxic components, is used as a raw material in the building materials industry, as well as in road construction as a backfill when laying the base of asphalt pavements [5].

Solid waste from thermal power plants – ashes and slags – are close to metallurgical slag in composition. In terms of chemical composition, 80 to 90% of these wastes consist of SiO_2, Al_2O_3, Fe_2O_3, CaO, MgO with significant fluctuations in their content. In addition, the composition of these wastes includes residues of unburned fuel particles (0.5 to 20%), compounds of titanium, vanadium, germanium, gallium, sulphur, uranium [6].

M.S. Malyovaniy (2017) proposes to use CHP waste for the purpose of making additives by mixing a dispersed mineral additive – CHP furnace dust with plasticizer. As a plasticizer using sulphate soap, and the components of the modifier are mixed in the following ratio: furnace dust of thermal power plants 97 to 99 %; sulphate soap – 1 to 3 % [7].

N.V. Blaschuk and I.V. Maevska (2017) propose to add furnace dust in order to improve the properties of soil-cement elements [8]. These authors investigated the properties of soil-cement with a content of fly ash (from 20 to 80% of the cement content, respectively from 3,4 to 15% of the soil weight). It was found that the use of ash as a mineral additive in such quantities in the production of soil-cement gives a positive effect. By adding the optimal amount of ash, the strength of soil-cement structures can be increased.

2. Purpose of the Work

The purpose of the article is to study the influence of the Mykolaiv CHP furnace dust and of ash from burning sunflower husks on the strength of soil-cement.

3. Methodology and Research

Soil-cement as a material is not just a mechanical mixture, but a system consisting of two, very complex in its content and properties, multicomponent systems – cement and soil. The main leading factor in the radical transformation of soil properties is cement – a polydisperse and polymineral system, a product of fine grinding of clinker, capable of forming a stone-like body after the addition of water [4, 10].

Soils, in turn, are considered as solid polydisperse very complex systems, many properties of which are determined by the laws of dispersion of substances. Under natural conditions, the soil in most cases is a three-phase system consisting of solid, liquid and gaseous phases. The ability of soils to show complex interactions with the substances added to it is greatly enhanced by the fact that soils are also polydisperse systems.

With the strengthening of soils with cement, the absorption of cations and anions is possible. Chemical absorption plays a significant role and is accompanied by chemisorption processes, as a result of which hydrolysis and hardening of cement in soil-cement follow slightly different patterns than in concrete or mortars. Depending on the value of metabolic capacity and composition of exchange cations, different soils, even with the same particle size distribution, have quite different

IOP Conf. Series: Materials Science and Engineering 1021 (2021) 012033 doi:10.1088/1757-899X/1021/1/012033

physical and mechanical properties that significantly affect the stability of soils in engineering structures.

In the technological process of soil-cement preparation the decisive factors are: a) the maximum degree of grinding of soil microaggregates; b) the accuracy of dosing of cement (or other binders) and the uniformity of its mixing with the soil; c) the optimal degree of moisture (water-cement ratio) and uniform distribution of moisture in the mixture; d) the maximum compaction of soil cement at the corresponding optimum humidity; e) the optimal mode of humidity and temperature during hardening of soil-cement [4].

According to the instruction, clay soils are suitable for production of soil-cement. Of these, loess and loess loams and sands meet the requirements the most.

According to experimental data, the values of physical and mechanical characteristics significantly depend on the conditions under which the soil-cement is made. In the production of soil- cement in the conditions of the construction base (stationary) from the soil of the disturbed composition with the use of mixing machines, the following compaction and even heat treatment obtain the highest characteristics. When soil-cement is produced directly in the soil massif, according to one of the above technologies, the quality of mixing the mixture decreases, its additional compaction and heat treatment are practically impossible. The presence of water in natural soils, especially below its level, significantly increases the water-cement ratio (W/C) of mixture, which reduces the density of soil-cement and respectively its mechanical characteristics.

The disadvantages of soil-cement piles include their low strength of the material. As a result, the bearing capacity of such piles by material is much less than by soil.

Recently, a significant amount of waste in Ukraine has been incineration waste (ash; furnace dust). Due to the significant rise in energy prices, the use of solid fuel boilers has recently become popular. Sunflower husk pellets are used as fuel for heating large areas of enterprises, public institutions, etc.

Pellets are used to heat residential buildings by burning in stoves, fireplaces and boilers, to provide heat and electricity to industrial facilities and small settlements (using large pellets, with a high content of bark). Pellets, for the production of which bark, sawdust, wood chips, agricultural waste (sunflower husk, straw, substandard flax), as well as organic packaging materials and cardboard packaging are used, are equivalent to coal in efficiency. Sunflower husk and straw have sufficient fuel properties, close to the properties of wood fuel. Due to this, biofuels made from sunflower can be burned in boiler equipment designed for wood biomass.

The moisture content of sawn and chipped wood intended for burning in stoves and fireplaces should be as low as possible. Made from secondary raw materials that are harmless to humans and the environment, pellets emit 10–50 times less carbon dioxide (CO_2) into the environment and 15–20 times less ash than in the case of coal combustion. Pellets are used to heat residential buildings by burning in stoves, fireplaces and boilers, to provide heat and electricity to industrial facilities and small settlements (using large pellets, with a high content of bark). In addition, pellets are cheaper than coal, liquid fuel or firewood, such biofuels are convenient to transport in packaged bags and in bulk.

Such fuel is environmentally friendly and economical, but there is a problem of waste disposal (ash).

Also, annually in Ukraine thermal power plants generate about 6.5 million tons of ash and slag waste, a significant part of this waste being furnace dust. Furnace dust is formed as a result of solid fuel combustion; after it is captured by electrostatic precipitators, in the dry state the ash is selected by means of an ash collector for production needs. Otherwise, it goes with slag and water to the ash dump. Thermal power plants are equipped with filtering devices that capture ash, preventing it from entering the air. Thus, large batches of this product are accumulated, and cement companies purchase and consume about 60,000 tons of ash emissions annually.

Furnace dust is a fine material consisting of particles up to 0.14 mm in size, formed by the combustion of solid fuels at the CHP.

The ash provides a sliding effect, so the mixture is better compacted during vibration. In the case of self-compacting concrete vibration is not even required: compaction is due to gravity. This is mainly due to the smooth surface and spherical shape of the ash particles.

Most fragments of ash have a spherical shape, smooth glassy surface texture. The size of spherical particles ranges from a few microns to 50–60 microns. The particle size of the furnace dust ranges from 3–5 to 100–150 µm. The number of large particles usually does not exceed 10 – 15%. Heavier ash particles are deposited on the firebox and fused into lump slags, which are aggregated and alloy ash particles ranging in size from 0.15 to 30 mm. [6].

The Mykolaiv CHP furnace dust before carrying out experiment was sifted on a sieve of 4 mm. The content of spherical inclusions from 1 to 4 mm was up to 40 %. The average humidity of the removal ash was 0.6 % (Figure 1).

The experiment is carried out as follows: cement and water mixed to obtain "cement milk". The amount of cement was taken as 20% by weight of dry soil for all series of tests. The W/C was taken as 1. Then to the cement mortar was added furnace dust, then obtained mixture was blended with soil (loam with a moisture content of 14%). The study is carried out for the amount of furnace dust of 5, 10 and 15% of the amount of cement. The prepared soil-cement mixture was mixed to a homogeneous mass for at least 5 minutes. After mixing, the soil-cement mixture is inserted into cylindrical molds with a diameter of 2.8 cm and a height of 3.5 – 4 cm.

A similar method of preparing the mixture was with the addition of pellets ash (waste from burning straw) (Figure 2,4). The effect of additives of 5, 10 and 15% of the amount of cement (20% by weight of dry soil) was also studied.

Figure 1. General view of the components of the mixture: 1 – furnace dust of the Mikolaiv CHP; 2 – loam

Figure 2. General view of straw ash

Prior to testing, the samples were stored immersed in water for 28 days to gain strength. The tests were carried out in accordance with DSTU B B.2.7-214:2009 as for concrete, taking into account DSTU B B.2.1-4-96. The tests were performed for 6 samples in each series.

The test results of the compression samples at the age of 28 days are shown in Tables 1 and 2.

The samples had no significant trickery up to 10% of the total sample volume (Figure 3). The cracks that preceded the final destruction of the samples were mainly vertical in nature. In this case, we observe brittle fracture due to the accumulation of mobile dislocations in front of the obstacle (subgrain boundaries), which leads to a concentration of stresses sufficient to form a crack. When the stresses reach a certain value, the crack size becomes critical and further growth is spontaneous. Fragile fracture is characterized by a sharp, often branched crack. The size of the zone of plastic deformation at the top of the crack is small. The rate of propagation of a brittle crack is significant and sudden.

As a result of the research it was found that the average compressive strength of the samples without any additive is 2,78 MPa; with the addition of furnace dust in the amount of 5% – 3.64 MPa. N.V. Blaschuk and I.V. Maevska (2017) also noted an improvement in the compressive strength of soil-cement with the addition of a small amount of Ladyzhinska CHP (3.4 – 15% of soil weight) [8].

As we can see from Table 1, the test results of soil-cement samples for compression at the amount of furnace dust additive 5, 10 and 15% are almost the same. Therefore, it is necessary to increase the range of additives and to study soil-cement samples.

Figure 3. General view of the sample of the soil-cement sample with the addition of 5% furnace dust after application of the load

Figure 4. General view of the sample of soil-cement element with the addition of 5% ash from burning straw.

Table 1. Characteristics of compressive strength of soil-cement samples with the addition of furnace dust (at the age of 28 days)

Soil-cement samples	Without additive	With the addition of furnace dust		
		5%	10%	15%
Average compressive strength, MPa	2.78	3.64	3.17	3.13
Coefficient of variation υ	0.30	0.11	0.11	0.09

The average compressive strength of soil cement samples with the addition of ash from the burning of pellets from sunflower husks decreased.

Table 2. Characteristics of compressive strength of soil-cement samples with the addition of straw ash (at the age of 28 days)

Soil-cement samples	Without additive	With the addition of ash from the combustion of straw ash		
		5%	10%	15%
Average compressive strength, MPa	2.78	1.71	1.80	1.61
Coefficient of variation υ	0.30	0.24	0.17	0.33

4. Conclusions

According to the results of the analysis, it can be concluded that the average compressive strength of soil cement samples with the addition of ash from burning pellets from straw decreased. This may be due to the inexpediency of using the method of traditional technology of preparation of materials for the manufacture of experimental samples. The ash was not previously mixed with dry cement to a homogeneous mixture. It indicates the inexpediency of using such waste for soil-cement production.

The average compressive strength of the samples with the addition of Mykolaiv CHP furnace dust with a fraction of inclusions up to 4 mm in the amount of 5 % increases by 30%.

Therefore, the use of Nikolaev CHP furnace dust as a mineral additive in the soil-cement elements production in the amount of 5% by weight of cement significantly increases the compressive strength. This expands the range of applications of soil-cement elements and increases their efficiency.

IOP Conf. Series: Materials Science and Engineering 1021 (2021) 012033 doi:10.1088/1757-899X/1021/1/012033

However, in the production of soil-cement, the rheological properties of furnace dust are important, especially the high fineness of grinding, favourable fractionation and the shape of its particles [10].

Reference

[1] Novytsky O, 2013 Methods of cement soil consolidation *Bulletin of the Donbass Academy of Civil Engineering and Architecture* Poltava **3 (101)** pp 32 – 37

[2] Karpushyn, S.O., Klymenko, V.V. & Shynder, A.V., 2018 Patent of Ukraine No.124712. Anaerobic bioreactorfor biogas and organic substrate production. Kyiv: State Patent Office of Ukraine Order 867 2016 *Rules for traction calculations for train work* (Moscow: JSC Russian Railways) p 510

[3] Nesterenko T, 2012 Development of a mounted deep vibrator for compaction of soil-cement piles Collection of scientific works (branch mechanical engineering, construction) Poltava PNTU **1 (31)** pp 257 – 262

[4] Dao Huu Do, Tuan A. Pham, 2018 Investigation of Performance of Soil-Cement Pile in Support of Foundation Systems for High-Rise Buildings **4 (2)** pp 266 – 277

[5] Burban N, Atamanyuk V Environmental aspects of the use of fly ash from thermal power plants *Environmental Protection. Energy saving. Balanced Nature Management: Proceedings of the First International Congress June 28-29* Lviv pp 16 –17

[6] Ivakhnenko A, Khlopytsky O, 2018 Formation, classification and prospects of utilization of ash-slag waste in Ukraine *Theory and practice of modern science November 23-24*, Odessa pp 111-112

[7] Painted M, Cooper M, Kanda M, 2018 A method of manufacturing a modifier for landslide structures No.126483 IPC C04B 28/14 (2006.01) Ukraine No. u201713144

[8] Maevska I, Goncharuk M, 2020 The effect of the use of fly ash for the arrangement of soil-cement piles. *Youth in science: research, problems, prospects 2019: abstracts of scientific-practical internet conference,* Vinnytsia pp 212 - 218

[9] Novitsky O, 2012 Influence of plasticizing additives on soil cement strength *Collection of scientific works (branch mechanical engineering, construction)* Poltava **4 (34)** pp 171 – 177

[10] Zotsenko M, Vinnikov Yu, Zotsenko V, 2016 *Drilling soil-cement piles, which are made by the drilling method* (Poltava: Printing House Madrid) p 94

[11] Kryvosheiev P, Farenyuk G, Tytarenko V, Boyko I, Kornienko M, Zotsenko M, Vynnykov Yu, Siedin V, Shokarev V, Krysan V 2017 Innovative projects in difficult soil conditions using artificial foundation and base, arranged without soil excavation *Proc. of the 19th Intern. Conf. on Soil Mechanics and Geotechnical Engineering Sep. 17 – 22 COEX, Seoul, Korea* pp 3007–3010

[12] Rajesh P. Shukla 2020 Pile settlement due to soil movement induced by the abrupt breakdown of retaining wall *September Acta Polytechnica* **60(4)** pp 338–348 DOI: 10.14311/AP.2020.60.0338

IOP Conf. Series: Materials Science and Engineering 1021 (2021) 012034 doi:10.1088/1757-899X/1021/1/012034

Carbon fiber plastics for vehicles manufactured by resource-saving formation technology

P Bilym[1], A Rogozin[1], P Firsov[2,3] and S Zolotov[2]

[1] Occupational and Life Safety Department, O.M. Beketov National University of Urban Economy, Marshala Bazhanova str. 17, 61002, Kharkiv, Ukraine
[2] Building Structures Department, O.M. Beketov National University of Urban Economy, Marshala Bazhanova str. 17, 61002, Kharkiv, Ukraine

[3] e-mail: pavelfirsov1991@gmail.com

Abstract The advantages of the technology of molding carbon plastics based on pre-pregs with separate application of components (SAC-technology) of epoxy binders cured by the mechanism of migration polymerization of epoxy cycles are presented. Using the methods of linear dilatometry and dynamic mechanical tests, the molecular mobility of the segmental and local types in polymer matrices of carbon fiber reinforced plastics based on traditional mixed pre-pregs and binary pre-pregs with separate application of components has been investigated. The question of the influence of the activation ability of an acid catalyst during the curing of an epoxy binder on the degree of decrease in internal stresses and the realization of an increased packing density of polymer chains in the polymer matrix of the composite is considered. The importance of choosing an acid catalyst for the curing of a polymer binder is shown, which for CFRP composites provides the necessary combination of elastic and dissipative properties. The use of the considered binders makes it possible to improve a number of important operational characteristics, especially in the case of the production of large-sized products: solidity, crack resistance, impact resistance, fracture toughness. These properties are very significant during the composite materials applying in transport engineering.

1. Introduction

The progress in the development of world mechanical engineering requires the necessity for widespread use of polymer composite materials (PCM), which have a low weight in combination with high specific strength characteristics. At the same time, the specifics of the manufacture of PCM products impose additional requirements on them in terms of manufacturing technology [1, 2]. The large-scale production of PCM products for vehicles requires a transition to a semi-finished technology - production on the basis of pre-impregnated reinforcing materials – pre-pregs. The use of pre-preg semi-finished technology in mechanical engineering makes it possible to increase its productivity, improve working conditions and enhance the production culture [3]. Intensification of the modeling process of products provides a high economic effect both due to the efficient use of autoclave equipment and due to energy-savings with a possible decrease in the molding temperature.

Its main disadvantage is the low viability of pre-pregs and limitations in the use of a number of binders, in particular binders with a high reactivity, providing the planned properties of composite materials [4].

One of the most effective technological methods for increasing the viability of pre-impregnated reinforcing materials – pre-pregs is the method of separate application of components (SAC) [5]. A characteristic feature of the processing of the resulting binary pre-pregs is that the mixing of the components occurs directly during the molding of the semi-finished product. The initial mixing of the components is achieved as a result of their joint filtration through the layers of reinforcing material under the action of the molding force throughout the volume of the semi-finished product. The completeness of the process of obtaining a composite will be determined by the completeness of the passage of the chemical interaction of the reactive groups of the binder components at the final stages of the physical structuring of the polymer matrix.

The main physicochemical feature of the process of forming a product by the method of separate application of components is a kind of frontal propagation of the curing process, in which the components of the binder-resin and hardening parts, fixed on various reinforcing elements, when the latter are joined during molding, they inter-diffuse against the background of a chemical reaction [6]. As a result, the chemical process already at the early stages is determined by the rate of interdiffusion of the components, and the diffusion process, in turn, at the later stages will increasingly depend on the intensity of the crosslink density growth. In the case of using epoxy reaction systems cured by the poly-addition mechanism, preconditions are created for the occurrence of an uneven distribution of components and, consequently, cross-link density gradient and physical parameters of the structure and realizable properties from its middle to the surface of the reinforcing filler. The combination of these factors affects the degree of homogeneity of the network structure of the composite matrix and means that when it is formed on the basis of SAC-pre-pregs, complete alignment of the components over the volume of the binder interlayer cannot be achieved and the process of formation of the structure of the polymer matrix is local in nature. Moreover, the degree of homogeneity of the structure will be the more disturbed, the greater the proportion of emerging regions of nonstoichiometric compositions.

As a result, it is of interest to conduct research on the development of composites based on SAC-pre-pregs using epoxy binders cured by the polymerization mechanism [7]. Since the polymerization of α-oxides in most cases is of a migratory nature, the completeness of mixing of the initial components, in our opinion, should not so effectively affect the degree of completion of the chemical structuring process, which is inherent in epoxy systems cured by the poly-addition mechanism (epoxyanhydride or epoxyamine).

2. Research methodology

The components of the binder were epoxy resin mark ED-20 (oligomer) with epoxy equivalents of $\approx$207 g/mol and an acid-type catalyst - a complex of boron trifluoride with an amine. The selection of the catalyst for the activation ability to epoxy resin was carried out by using amines of different reactivity. A number of amines are represented by the following compounds: aniline (I), para-toluidine (II), benzylamine (III), isophorone diamine (IV). A polymer composite based on T-2 carbon fabric of plain weave was prepared by impregnation with individual binder components: an alcohol-acetone solution of epoxy resin and an acetone solution of an acid catalyst. The latter was originally used as a 30 % solution in diethylene glycol.

After drying, the monolayers were cut, filled into a bag by alternating with a resin and an acid catalyst component, placed in a vacuum bag and molded in an autoclave according to the following temperature regime: 120 °C - 1 hour + 150 °C - 5 hours with compression parameters:

- rarefaction in the cover - 0.08 MPa;
- overpressure in the autoclave - 0.4 MPa.

Control (cast) samples of cured binders were obtained by pouring into a mold, holding for a day (until the solvents volatilized), followed by a polymerization regime similar to molding in an autoclave. The activation temperature of the curing of the compositions was determined by the method of differential thermal analysis on the Paulik-Paulik-Erdey system thermogravimetric analyzer at a heating rate of 1.3 deg/min. The magnitude of the residual stresses (σ_{res}) of the block samples was determined by the tensometric method. The linear elongation of the sample was recorded using a digital raster system at a temperature rise rate in the test chamber of 2 deg/min.

The degree of curing of the binder was controlled by the yield of the gel, the value of which corresponded to the amount of the dry residue of the crosslinked polymer after extraction with boiling acetone in a Soxhlet apparatus for 48 hours.

The tensile strength was determined on a tensile testing machine at a crosshead speed of 1 mm/min.

Dynamic mechanical tests were carried out in the mode of forced resonance oscillations of a cantilever-mounted sample in the frequency range 20-200 Hz (Figure 1).

Figure 1. Laboratory cell for dynamic-mechanical testing of polymeric materials.

3. Main material and results

Since one of the main conditions for obtaining a high-strength composite is the achievement of minimum residual stresses in it, the character of the change in σ_{res} for the indicated binders is considered in this work. It was found that a certain relationship is observed between the relaxation properties of cast (block) samples of binders and the temperature-time conditions of their gelation. For polyepoxides with a low activation temperature for curing, increased values of contact pressure were observed. With an increase in the activation temperature, which is typical for the compositions obtained in the presence of complexes of BF_3 with benzylamine and isophorone diamine, the level of σ_{res} decreases. However, after relaxation at room temperature for 60 days, the values of residual stresses for the above compositions are practically equalized (table 1).

Table 1. Influence of the substituent type in the complex on the σ_{res} of epoxy binders.

Amine substituent of the BF_3 complex	Sol-fraction substance, %	Curing activation temperature, K	Gelation time at activation temperature, min	σ_{res}	Relaxation degree of σ_{res}
(I)	9.4	309-312	2.5	10.8 / 6.5[a]	0.398
(II)	7.6	340-343	7	9.7 / 6.9[a]	0.289
(III)	7.1	383-387	9	9.2 / 7.4[a]	0.196
(IV)	6.9	392-396	12	8.9 / 7.7[a]	0.135

[a] Numerator - after 1 day, denominator - after 60 days

IOP Conf. Series: Materials Science and Engineering 1021 (2021) 012034 doi:10.1088/1757-899X/1021/1/012034

Analysis of the obtained data shows that the values of the stresses and the degree of their relaxation are uniquely determined by the type of catalyst. Thus, in the case of using catalysts with a low activation temperature, the content of the sol-fraction in the cured polymers is much higher than in the systems obtained at elevated temperatures. This unequivocally indicates that the conditions for additional packaging of structural elements of the polymerization web for polyepoxides with an increased gelation temperature deteriorate and, accordingly, the polymer material is characterized by a low relaxation ability.

The ongoing relaxation processes are reflected in a certain way on the nature of changes in the thermomechanical characteristics of composites (Figures 2, 3). Analysis of dilatometric data shows that, depending on the type of binder components used, there is a decrease in the linear expansion curve, the value of which decreases with an increase in the activation temperature of the reaction system and an increase in the glass transition temperature of the finished composite [8]. This operation of the composite is typical both for those obtained by the technology of separate application of (SAC) on the layer-preg, and for pre-pregs of the mixed type. Attention is drawn to the fact that the higher the degree of relaxation of internal stresses in block specimens (table 1), the higher is the magnitude of the decrease in the dilatometric curves of carbon fibers. It is known that epoxy polymers of the same type obtained under different temperature conditions can have different structures due to the fact that the proportion of the polymer with a non-equilibrium web formed at the stage of isothermal glass transition is not the same. Therefore, from our point of view, binders sewn at a relatively low temperature, the degree of non-equilibrium of the web should be higher. This parameter can be estimated by the indicators of relaxation spectrometry (table 2).

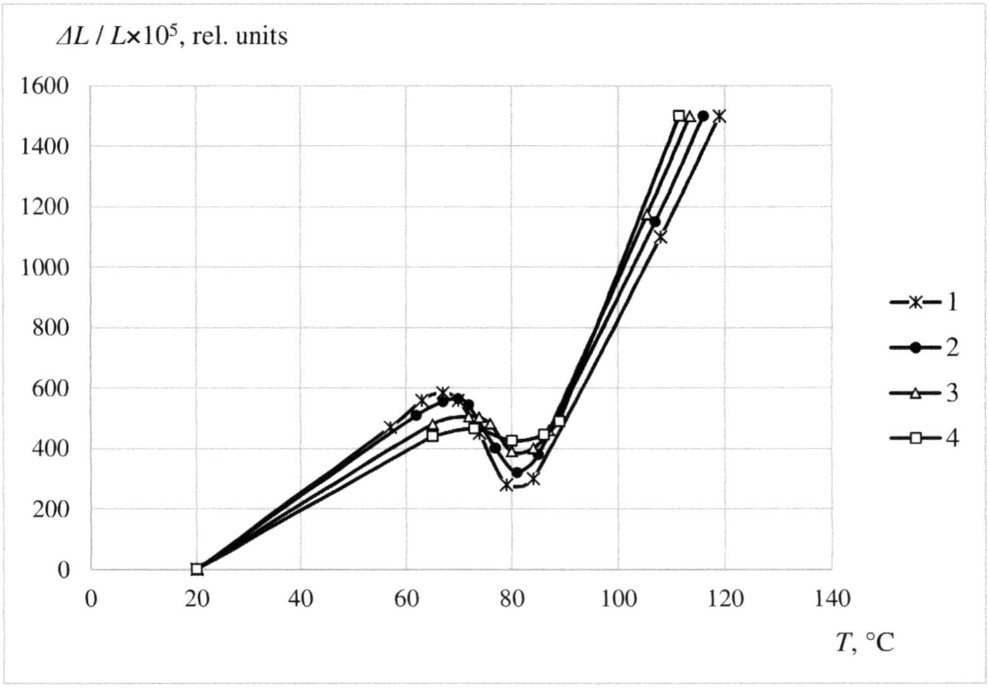

Figure 2. Dilatometric curves of carbon fiber, obtained by the mixed technology of binder applying, cured with a boron trifluoride complex with additive: 1 – aniline, 2 – para-toluidine, 3 – benzylamine, 4 – isophorone diamine.

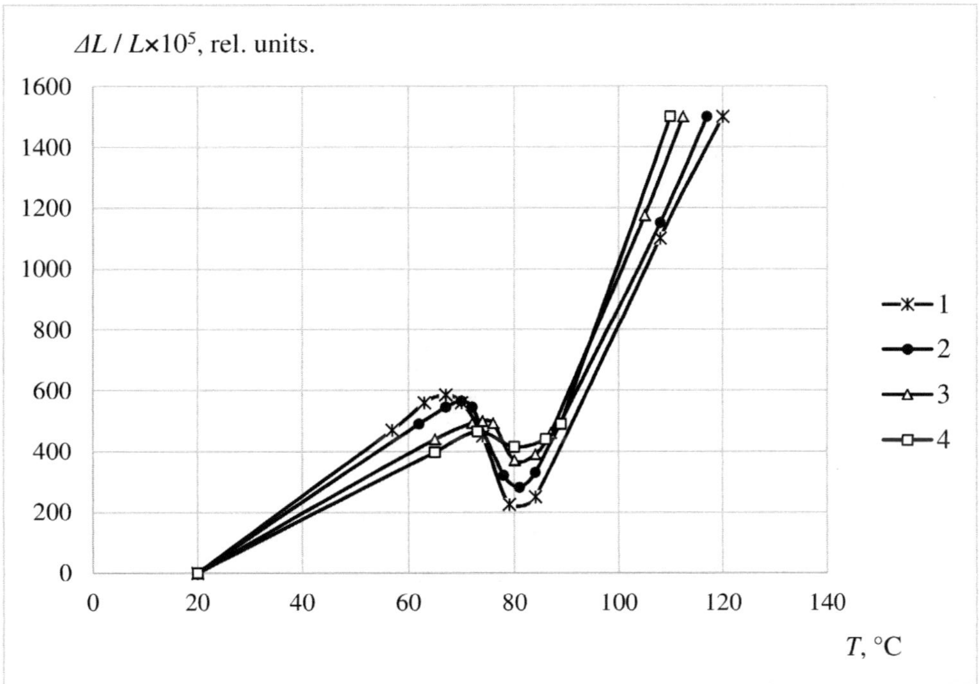

Figure 3. Dilatometric curves of carbon fiber, obtained by the SAC-technology of binder applying, cured with a boron trifluoride complex with additive: 1 – aniline, 2 – para-toluidine, 3 – benzylamine, 4 – isophorone diamine.

Table 2. Relaxation properties of carbon plastics based on epoxy binders.

Catalyst	Intensity index of the angle tangent of mechanical losses at characteristic temperatures			G'_{TC+50} - G'_{TC}, GPa
	at 293 K	β-transition	α-transition	
(I)	0.008 / 0.006[a]	0.021 / 0.018[a]	0.071 / 0.066[a]	0.12 / 0.07[a]
(II)	0.008 / 0.005[a]	0.024 / 0.021[a]	0.077 / 0.073[a]	0.13 / 0.06[a]
(III)	0.011 / 0.009[a]	0.029 / 0.024[a]	0.085 / 0.081[a]	0.07 / 0.02[a]
(IV)	0.014 / 0.009[a]	0.036 / 0.024[a]	0.088 / 0.079[a]	0.05 / 0.02[a]

[a] Numerator - CFRP obtained by the SAC-technology, denominator - CFRP obtained by the mixed technology.

So, taking into account the calculated value of the difference between the dynamic shear modulus in the highly elastic state of the composite (G'_{TC+50} - G'_{TC}) it is possible to indirectly characterize the material's ability to achieve the limiting structural parameters after molding. Judging by this indicator, the conformational mobility of macromolecules increases to the highest extent when the curing of epoxy compositions is initiated at low temperatures. Since the conditions for cooling (in an autoclave) and heating of CFRP specimens during dilatometric research were constant, the presence of anomalies (in the form of "pits") on these curves indicates a reduced degree of packing of polymer chains of the binder in the composite [9].

At the same time, it should be noted that the degree of mobility of kinetic units (fragments) of polymer chains of binders increases with an increase in the activation temperature of curing in molded composites [10]. Moreover, the effect of energy dissipation is more specific for composites obtained by the technology of separate application of components in the entire investigated temperature range. The combination of the presented indicators makes it possible to assume that the material has sufficiently high viscoelastic properties, which are responsible for its resistance to the action of loads, both in the static mode and in the mode of shock loads. Thus, to obtain carbon plastics with a complex of stable indicators from the presented series, we can recommend those, in which the binder tends to polymerize in the autoclave molding mode at elevated temperatures - in the presence of complexes of boron trifluoride with benzylamine and isophorone diamine.

As can be seen from the data (Figure 4), the strength of carbon plastics obtained by the technology of separate application of the binder components is not inferior to those obtained by the mixed technology.

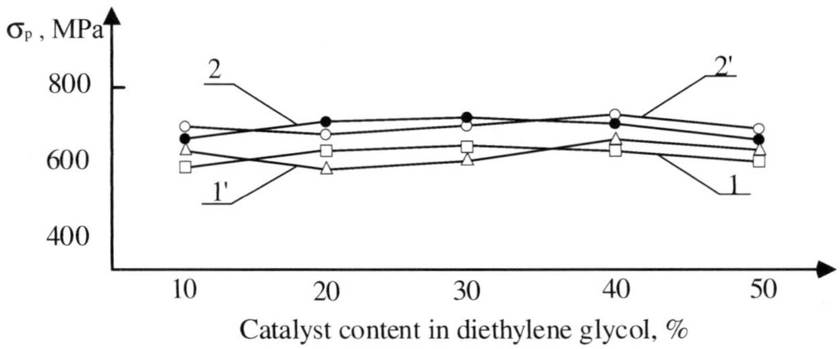

Figure 4. Dependence graphs of carbon plastics tensile strength on the amount of acid catalyst III (curves 1, 1') and acid catalyst IV (curves 2, 2') with SAC-technology (curves 1, 2) and mixed technology of binder applying (curves 1', 2').

Moreover, with an increase in the activation temperature of the curing process, which is typical for the polymerization of epoxycycles in the presence of a complex based on isophorone diamine, increased strength indicators are realized for CFRP.

Also, it should be noted that the amount of added catalyst does not significantly affect the strength of CFRP, which is provided by the weak dependence of the quantitative ratio of the components on the rate of migration polymerization of epoxycycles during the formation of the composite. This is the special appeal of this binder type and in fact guarantees reducing the likelihood of quasi-adiabatic overheating due to the use of small doses of the catalyst in the formation of large-sized objects from carbon fiber reinforced plastic.

Experimental data (table 3) shows the results of physical and mechanical tests of CFRPs based on selected epoxy binders obtained from pre-pregs using the technology of separate application of components and mixed type.

Table 3. Physical and mechanical properties of CFRPs.

Catalyst base	Density, kg/m^3	Tensile strength, MPa	Specific tensile strength, 10^3, m^2/s^2	Shear modulus, GPa	Specific shear modulus, 10^6, m^2/s^2	Impact strength, kJ/m^2
Benzylamine	1476 / 1489[a]	620 / 604[a]	42.0 / 40.5[a]	39 / 39[a]	2.6 / 2.6[a]	41.7 / 38.4[a]
Isophoron diamine	1481 / 1497[a]	718 / 697[a]	48.5 / 46.6[a]	41 / 41[a]	2.8 / 2.7[a]	48.3 / 37.9[a]

[a] Numerator - from pre-pregs using a SAC-technology, denominator - from pre-pregs of mixed type.

As can be seen from the presented data (table 3), the strength of the polymer composite based on carbon fiber is largely determined by the technological properties of the binder, its wetting ability, which, in our opinion, is realized upon relatively long heating to the effective activation temperature of the binder during formation. To a greater extent, the improvement in strength and elastic properties is clearly manifested in the example of their specific indicators. Of particular interest is the fact that composites obtained by the method of separate molding have increased impact strength. Moreover, with the transition from the curing catalyst from aromatic amine to cycloaromatic - isophorone diamine, the impact strength increases sharply, which is in full agreement with the intensity indices of the angle tangent of mechanical losses in the low temperature relaxation zone and normal conditions.

4. Conclusions

The proposed type of epoxy binder makes it possible to expand the range of available binder components within the framework of the considered technology and to carry out an almost waste-free production cycle for the manufacture of thick-walled composite sheet products for vehicles. It should be noted that the use of pre-pregs by the SAC-technology makes it possible to achieve not only significant technological advantages in comparison with mixed pre-pregs, which was the initial aim of developing a new technology. It is important to emphasize that their application can improve a number of important performance characteristics, especially in the case of the production of large-sized products: solidity, crack resistance, impact resistance, fracture toughness. These properties are very significant when using composite materials in transport engineering. The observed combination of elastic and dissipative characteristics of the obtained composites makes it possible to consider them in the future as functional elements for structural systems of various vehicles, taking into account the minimization of creep and relaxation processes in it.

References

[1] Messana A, Sisca L, Ferraris A, Airale A, Pinheiro H, Sanfilippo P, Carello M 2019 From design to manufacture of a carbon fiber monocoque for a three-wheeler vehicle prototype *Carbon fibers and their composite materials* **Vol 12** p 11

[2] Tsirogiannis E C, Stavroulakis G E, Makridis S S 2017 Design and modelling methodologies of an efficient and lightweight carbon-fiber reinforced epoxy monocoque chassis, suitable for an electric car *Materials Science and Engineering with Advanced Research* **Vol 2(1)** pp 5-12

[3] Weizhao Zhang, Xuan Ma, Jie Lu, Zixuan Zhang, Jane Wang, Xuming Su, Danielle Zeng, Mansour Mirdamadi, Jian Cao 2018 Experimental characterization and numerical modeling of the integration between carbon fiber composite prepregs during a preforming process *Journal of Manufacturing Science and Engineering Transactions of ASME* **Vol 140(8)** p 8

[4] Haydachuk A V, Wang Bo, Bychkov A S, Andreev A V 2019 Development of a complex criterion of the rational selection of polymer composite materials *Physico-Chemical Mechanics of Materials* **Vol 55(6)** pp 110-117

[5] Smirnov Yu N, Shatskaya T Ye, Natrusov V N 2004 Fiberglass formation based on binary pre-pregs in a gradient temperature field *Scientific and Technical Journal "Plastics"* **Vol 7** pp 19-23

[6] Clyne T W, Hull D 2019 *An introduction to composite materials: third edition* (Cambridge: Cambridge University Press) p 360

[7] Mihaylyuk A P, Oleynik V V, Afanasenko K A, Bilym P A 2004 Sandwich-package for the production of reinforced plastics *Scientific Collection of National Technical University "Kharkiv Polytechnic Institute"* **Vol 14** pp 121-126

[8] Smirnov Yu N, Magomedov G M, Dzhamayeva N M 2002 Structural and kinetic features of the formation of elastic, relaxation, dissipative and strength properties of epoxyphenolic CFPRs based on binary pre-pregs *Journal of Applied Chemistry* **Vol 75(1)** pp 113-119

[9] Smirnov Yu N, Glavina T A, Efremova A I 2011 Influence of molecular mobility of epoxyamine web polymers on relaxation and physico-mechanical properties *Polymer Science (Series: Polymer Physics)* **Vol 53** pp 32-38

[10] Trofimov A N, Simonov-Emel'yanov I D, Pleshkov L V, Zarubina A Y, Prokhorova Y S, Trofimov D A 2017 Controlling the softening temperature of epoxy oligomers and their mixtures *Polymer Science (Series: Polymer Chemistry)* **Vol 10(2)** pp 119-122

Justification of reference table data of dependence of wind effect on critical surface density heat flux for corrugated cardboard

A Borysova [1,3], V Nizhnyk[1], Ya Ballo[1] and O Kyrychenko[2]

[1] Institute of Public Administration and Research in Civil Protection 18, Rybalska St., Kyiv, 01011, Ukraine
[2] Cherkasy Institute of Fire Safety named after Chornobyl Heroes of National University of Civil Defense of Ukraine, 8, Onoprienko St.,Cherkasy, 18034, Ukraine

[3] E-mail: borysova_anna@ukr.net

Abstract. To date, there is no statistical base of critical values of surface heat flux density for various substances and materials, in particular those that can be used in the decoration of buildings and technological installations. This article presents conclusions from the analysis of recent studies on this issue, substantiates the reference tabular data of the dependence of wind on the critical surface heat flux density for corrugated cardboard using a full-factor experiment. The authors determined the largest and smallest indicators of significant factors influencing the value of the critical surface heat flux density, namely wind speed (v, m/s) and the set surface heat flux density (q, kW/m²). In the course of the research, a mathematical model was constructed which describes the dependence of wind influence on the critical surface heat flux density for corrugated cardboard. These tables can be used in the future to develop an improved simplified method for predicting the thermal impact of fire on adjacent construction sites, taking into account the impact of wind.

1. Introduction

In the initial stage of fire development in buildings, heat from the burning room to the adjacent is transferred by thermal conductivity through building structures, metal pipes and other utilities. In open fires, heat is transferred to the surrounding objects mainly by radiation. The speed of their propagation (ignition of objects adjacent to the burning building) depends on the conditions of heat exchange of radiation, that is. the area of thermo physical properties of the irradiating and irradiated surfaces, as well as their mutual location. Significantly affects the situation in open fires, the height of the flame zone, as it depends on the surface area of the radiation and the intensity of heat flow towards non-burning objects. The parameters of the fire interact with each other, but can't affect the development and spread of fire without interaction with the environment. The connection of environmental parameters with the processes of heat exchange between objects has an unconditional impact on human activities. The study of these factors and the ability to reduce their negative impact is an important issue today.

During the development of an open fire, the radiant heat flux from the flame falls on the surrounding buildings or combustible materials. Necessary and sufficient condition of ignition of any surface of combustible material is expressed by the ratio $q_f \geq q_{cr}$ (q_f - incident flow on the surface of the irradiated material W/m^2; q_{cr} - critical heat flux, which causes ignition of the surface of this type of combustible material, W/m²).

To date, the impact of environmental phenomena on heat transfer processes has not been thoroughly studied. The influence of wind and its speed on heat transfer processes is not scientifically substantiated. In addition, there is no reference table of values of the critical surface heat flux density for substances and materials taking into account wind exposure. Therefore, the development and substantiation of reference data on the dependence of wind influence on the critical surface heat flux density for corrugated cardboard is an urgent scientific task. Many scientists, both domestic and foreign, deal with the study of heat transfer conditions in a fire. In this case, the work of researchers such as: Roytman M [1], Hrushevsky BV [2], Koshmarov Yu [3], Pozdeev S [4], Basmanov A, [5-7], Semerak M [8], Nizhnyk V [9], Gudym V [10], from foreign authors E. Carlsson [11], and others [12-16] was analyzed.

In the course of the analysis of literature sources, normative documents and other sources of information, the authors came to the conclusion that to assess the value of the critical surface heat flux density, it is advisable to use the sign of flame combustion of substances and materials. However, to date there is no statistical base of critical values of surface heat flux density for various substances and materials, in particular those that can be used in the decoration of buildings and technological installations.

In view of the above, the processes of changing the heat flux density for different substances and materials depending on the influence of the amount of air flow introduced into the study space need further study.

The absence of data on the change in the value of the heat flux density on the adjacent surface, which is subject to thermal influence depending on the wind influence, is determined.

For a simplified calculation, it is proposed to use the tabular method described below.

The purpose and objectives of the study. The aim of the work is to improve the information and reference data of the dependence of wind influence on the critical surface heat flux density for corrugated cardboard by conducting a complete factorial experiment.

To achieve this goal:

- the choice of the most significant factors of influence and their range is carried out; affecting the processes of thermal radiation, namely the value of the critical surface density of heat flux;
- using the results of experimental studies to build a regression model;
- on the basis of experimental studies to determine the coefficients of regression;
- create an information and reference table on the values of the critical surface heat flux density for substances and materials depending on wind exposure.

2. Presenting main material

The wind speed (v, m/s) and the given surface heat flux density (q, kW/m²) are considered as a criterion that can significantly affect the critical surface heat flux density. Therefore, to create a plan of a complete factorial experiment, we take these parameters, which have a significant impact on the critical surface heat flux density.

The ranges of these parameters are substantiated in [17] and are given in table 2.1. The wind speed range for the study was selected on the basis of statistical data of the Ukrainian Hydrometeorological Center [18] and climatic data for the period from 1899 to 2019, according to which the average wind speed in Ukraine depending on the region varies between 2.5÷4.5 m/s. But in many cases on the territory of Ukraine the wind speed of 20÷30 m / s was fixed. Based on the analyzed statistical data of wind speed recorded on the territory of Ukraine, we decided to choose the maximum wind speed of 30 m/s. The defined intervals show the smallest and largest values of the parameters in the experiment, which are selected as factors.

Table 2.1. Parameter intervals in experiments that determine the quality of factors.

Wind speed (m/s)		The set surface heat flux density (kW/m²)	
The smallest value, *v*	The largest value, *v*	The smallest value, *q*	The largest value, *q*+
0	30	5	50

According to the results of experimental studies of the duration of ignition of five-layer corrugated cardboard depending on the surface heat flux density and wind conditions, the dependence of the time of occurrence of preconditions for fire propagation taking into account the heat flux and ignition duration on wind exposure parameters was determined. appropriate coefficients.

Analyzing the results of experimental studies, the mathematical model describing the dependence of wind influence on the critical surface heat flux density for corrugated cardboard can be represented as a linear dependence of ignition time and factors, which has the form of equation 1 [19]

$$t = b_0 + b_1 v_v + b_2 q_s + b_3 v_v q_s, \tag{1}$$

where v_v, q_s are parameters that take into account the selected factors, which are listed in Table 2.1; b_o, b_1, b_2, b_3, are the constants of the numerical regression equation.

To determine the constants of the numerical regression equation in accordance with formula (1), it is necessary to conduct 4 numerical experiments [18] according to the compiled planning matrix, which is recorded in the form of table 2.2.

Table 2.2. General matrix for planning a complete factorial experiment to build a mathematical model.

Experimental number	x_1	x_2	x_1 x_2
1	+	+	+
2	-	+	-
3	+	-	-
4	-	-	-

The constants of the numerical regression equation (1) can be determined using formulas (2,3,4,5) given in [4,20]:

$$b_0 = \frac{1}{N}\sum_{i=1}^{N} y_i; \tag{2}$$

$$b_1 = \frac{1}{N}\sum_{i=1}^{N} x_1 y_i; \tag{3}$$

$$b_2 = \frac{1}{N}\sum_{i=1}^{N} x_2 y_i; \tag{4}$$

$$b_3 = \frac{1}{N}\sum_{i=1}^{N} x_3 y_i, \tag{5}$$

where $N = 4$ is the number of experimental situations according to the experimental plan; x_i is the value of the parameter according to the plan matrix, y_i - the value of the specified heat flux
According to the results of the calculations, the obtained constants of the numerical regression equation are given in Table 2.3.

Table 2.3. Constants of the numerical regression equation.

Coefficient	b_0,	b_1,	b_2,	b_3,
Value	281.4	25.36	5.43	0.43

IOP Conf. Series: Materials Science and Engineering 1021 (2021) 012035 doi:10.1088/1757-899X/1021/1/012035

Using the obtained regression dependence, the corresponding surface of the dependence between the wind speed, the critical surface heat flux density and the time during which the ignition of the material is constructed is shown, which is shown in Figure 2.1.

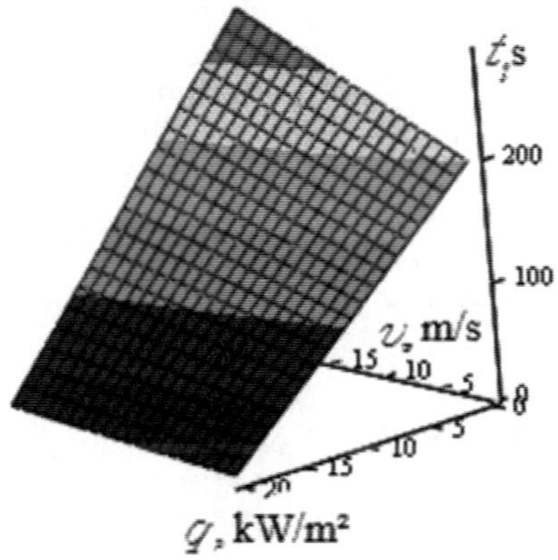

Figure 2.1. Dependence between wind speed, critical surface heat flux density and the time during which the material ignites

Using the regression dependence and with certain constants, it is possible to construct information-reference tabular data on the dependence of wind influence on the critical surface heat flux density for corrugated cardboard. The results obtained are presented in Table 2.4.

Table 2.4. Determination of the time of reaching the critical surface heat flux density for corrugated cardboard depending on the power of the heat source and the influence of wind

The value of wind speed *v*, (m/s)	The estimated value of time to reach the critical surface heat flux density depending on the power of the heat radiation source q_s kW/m² and wind exposure q_s (kW/m²)			
	5	15	35	50
	t, s			
0	254.2	199.9	91.4	9.986
from 0 to 5	370.2	294.4	142.6	28.8
from 5 to 10	486.3	388.8	193.9	47.7
from 10 to 15	602.5	483.3	245.2	66.6
from 15 to 20	718.3	577.7	296.4	85.5
from 20 to 25	834.4	672.1	347.7	104
from 25 to 30	950.4	766.6	399	123

3. Conclusions.

According to the results of the complete factorial experiment, scientifically substantiated information and reference data on the values of the critical surface heat flux density for substances and materials depending on the wind influence. The following conclusions were made:

- it is established that the most significant factors influencing the time of achieving the preconditions for fire are the wind influence and the value of the surface heat flux density, and substantiated their ranges, which are for wind exposure from 0÷30 m/s, for heat flux density 5÷50 kW/m²).
- a regression model is constructed, which describes the dependence of wind influence on the critical surface heat flux density for corrugated cardboard in the form of a linear dependence and its coefficients are determined;
- the results show that with a change in air flow rate, the critical surface heat flux density decreases, resulting in longer ignition;

- the table of the definition of time of achievement of the critical surface density of a heat stream for corrugated cardboard depending on the power of a source of thermal radiation and the influence of wind is developed.

Taking into account the identified patterns of influence of air flow, as well as the distance at which the heat radiation material is installed on the heat flux density, reference tables are substantiated to establish the critical surface heat density of corrugated cardboard depending on air flow and distance between corrugated cardboard and heat source.

The conducted researches and the generated data tables can be further used to develop an improved simplified method for forecasting the thermal impact of fire on adjacent construction sites, taking into account the wind impact.

The obtained data will also be used for further studies of flammability of substances and materials.

In the future, the development and use of statistical tabular data of values of the critical surface heat flux density for substances and materials under wind conditions will take into account the properties of substances and materials during finishing of adjacent structures, as well as their mutual location. At the same time, it is planned to achieve an economic effect, as it is known that 1 m² of undeveloped area is about 20% of the value of the same area of built-up area.

References

[1] Roytman M Ya, Komissarov Ye P, Pchelintsev V A 1977 Thermodynamics and heat transfer in firefighting 297-298

[2] Hrushevskiy B V, Yakovklev A I, Krivosheyev I A. and others Edited by Kuralenkin V F 1985 Fire Prevention in Construction M., HEFES, 4521

[3] Romanenko PN, Koshmarov YA, Bashkirtsev MP 1978. Fire prevention in construction M Stroyizdat, 363

[4] Pozdeev S V, Nekora O V, Pozdeev A V 2006 Methods for determining the modes of heating of concrete samples that simulate the state of the elements of building structures in case of fire *Problems of fire safety* 111 - 116

[5] Abramov Yu.A., Basmanov A E 2004 Modeling of tank heating under the action of fire radiation *Bulletin of the International Slavic University* 7-9

[6] Abramov Yu.A., Basmanov A E 2005 Estimation of fire danger of the tank with oil product at its heating from a flame of the next burning tank Radioelektronika i informatika №2 110-112

[7] Abramov Yu.A, Basmanov A E 2006 Thermal processes in a heated tank Municipal economy of cities. Scientific and technical collection, Issue **67** 357-362

[8] Semerak M M, Mykhailyshyn M R 2017 Influence of wind speed on heat exchange processes between vertical steel tanks (on the example of a fire at the oil depot "BRSM Nafta") Fire safety **30** 137-147

[9] Nizhnyk V V, Shchipets S, Tarasenko O, Kropyvnytsky V and Medvid B 2018 A Method of Experimental Studies of Heat Transfer Processes between Adjacent Facilities International Journal of Engineering & Technology 7 (**4.3**) 288 – 292

[10] Gudym V I, Labay V Y, Gavrilyuk A F 2013 Research and ways to reduce thermal radiation during fires Bulletin of the National University "Lviv Polytechnic" Theory and practice of construction **755** 221-226

[11] Emil Carlsson, 1999 Report 5051 - External fire spread to adjoining buildings – A review of fire safety design guidance and related research Department of Fire Safety Engineering Lund University

[12] Collier Peter 1996 Method for determining safe separation distances between building in the event of fire issn:0111-8161

[13] Incropera F.P. and De Witt D.P. 1996 Fundamentals of Heat and Mass Transfer John Wiley and Sons, New York, 4th edition

[14] Baum H.R., McGrattan K.B., and Rehm R.G. 1997 Three Dimensional Simulations of Fire Plume Dynamics *Journal of the Heat Transfer Society of Japan*, 35(139):45 – 52

[15] Zhang, D., Xue, W. 2010 Effect of Heat Radiation on Combustion Heat Release Rate of Larch. *Journal of West China Forestry Science* **39** 148
[16] Pullins C A 2011 High temperature heat flux measurement: sensor design, calibration, and applications Virginia Polytechnic Institute and State University
[17] Borysova A S, Nizhnyk V V 2020 Methods of experimental studies of the dependence of the critical surface heat flux density on wind exposure Scientific and technical collection "Municipal utilities". Series: technical sciences and architecture **2** 200–203
[18] Website of the Ukrainian Hydrometeorological Center [Electronic resource] - Access mode: https://meteo.ua/ua/vocabulary/shkala-boforta-539
[19] Nizhnyk V V, Pozdeev S V, Feshuk Yu.L 2018 Methodology for forecasting the limits of fire resistance of wooden columns with fire protection *International Scientific Journal Internauka* **14** 74 - 81
[20] Milovanov A F 1998 Stability of reinforced concrete structures in case of fire M Stroyizdat 304

Prediction of the life time of cylindrical tribosystems of a vehicle

A Dykha[1,3], J Padgurskas [2,4] and O Babak[1]

[1]Department of Tribology, Automobiles and Materials Science, Khmelnytsky national university, 11, Institutska str., 29016, Ukraine
[2]Department of Machinery and transport, Vytautas Magnus University, K. Donelaicio g. 58, LT-44248 Kaunas, Lithuania

[3]E-mail: tribosenator@gmail.com
[4]E-mail: juozas.padgurskas@vdu.lt

Abstract. Development and research of methods for calculating the life of friction parts of a vehicle is an actual problem. Ball joints of a car suspension are one of the most responsible for reliability and safety friction unit of transport vehicles. In this paper the approaches to the computational and experimental assessment of the wear and reliability of the ball joint of a vehicle are proposed. The calculation of wear will be carried out based on the solution of the wear contact problem. The equilibrium condition, the wear law, and the geometric condition in the contact are taken as the basic equations. As a result, design formulas for wear and contact pressures have been proposed. A method for calculating the reliability function with determining the coefficients of variation of the wear parameters of ball joints is proposed. To assess the effectiveness of lubricants, a method for calculating wear parameters based on solving the inverse wear-contact problem is considered.

1. Introduction

The suspension is one of the most important parts of a car in terms of the comfort of transportation of goods and passengers and, most importantly, in terms of traffic safety. Wear of friction units is the main reason for the decrease in suspension reliability. The suspension structure of the front wheels of a car contains many important friction units. Among these supports, the most important are ball bearings. These bearings have different friction pairs: steel on steel, steel on polymers, steel on rubber, etc. The study of the support structure is the first mandatory stage of calculations and tests for wear and reliability of friction units. It is also necessary to understand how the construction of the assembly works, as well as to understand the entire subsequent calculation methodology. Analysis of the design of ball joints of different cars allows us to trace the tendencies of their improvement. Suspension ball joints are applied between control arms and wheel carriers or knuckles of the suspension that allow relative articulation and rotation between the mating components. Depending on its application, the ball joint can transmit longitudinal and lateral forces with only a small proportion of forces as well as the complete vehicle weight in vertical direction. Much attention is currently paid to the study of the tribological aspects of extending the service life of the movable joints of a car [1-3]. At the same time, a comprehensive study of analytical and experimental methods for predicting the life time of vehicle friction units is an actual problem.

IOP Conf. Series: Materials Science and Engineering 1021 (2021) 012036 doi:10.1088/1757-899X/1021/1/012036

2. Wear and contact pressures in ball bearings of a vehicle suspension

To determine the effectiveness of construction and lubricants used in transport, a calculation method is proposed for determining the wear of vehicle parts. The technique takes into account the properties of materials and the operating conditions of the friction unit and makes it possible to determine the durability of the friction unit by the wear criterion. The basis is the solution of the wear contact problem for a given contact pattern of parts [4-5].

In this case, the design of the ball joint of the car suspension is considered. In the ball joint, there is a contact interaction of a rigid spherical liner and a rigid cage according to the scheme in figure 1. A ball bearing of radius R rotates around its axis along the body and wears out. It is necessary to obtain the dependence of the wear of the mating of the axes of the main factors. The angle of the liner ranges from $\varphi = \varphi_1$ to $\varphi = \varphi_2$.

The mathematical formulation of the problem consists of an equilibrium condition, a geometric relationship, and a wear law. The equilibrium condition for a spherical joint is:

$$Q = 2\pi R^2 \int_{\varphi_1}^{\varphi_2} \sigma(\varphi) \cos\varphi d\varphi , \tag{1}$$

where $\sigma(\varphi)$ is the required pressure function.

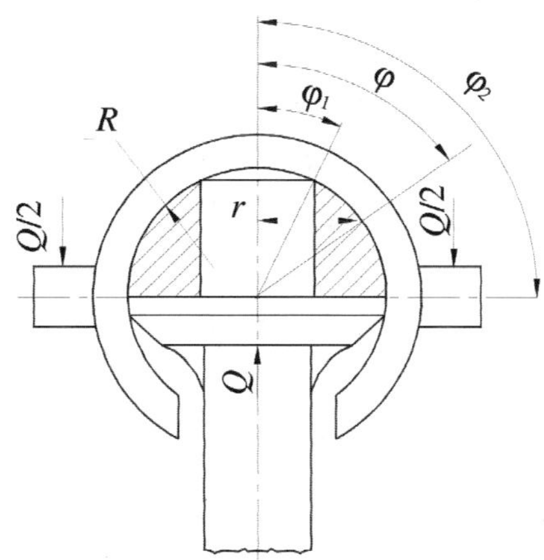

Figure 1. Design scheme of the hinge.

The geometric dependence of displacements u_0 on wear $u_w(\varphi)$ is as follows:

$$u_w(\varphi) = u_0 \cos\varphi . \tag{2}$$

The wear law with two parameters k_w and m is taken as:

$$\frac{du_w}{dt} = k_w \sigma^m 2\pi nr , \tag{3}$$

where n is the number of vibrations of the hinge per unit time t; $r = R\cos\varphi$.

As a result of differentiating condition (1) and equating (3), we obtain:

$$k_w \sigma^m 2\pi nR^2 \cos\varphi = \frac{du_0}{dt} \cos\varphi , \tag{4}$$

from here we have:

$$\sigma=\left(\frac{du_w}{dt}\right)^{\frac{1}{m}}\left(\frac{1}{k_w 2\pi nR^2}\right)^{\frac{1}{m}}. \tag{5}$$

As a result of mutual substitutions, we obtain the resolving integral equation of the problem:

$$Q=2R\left(\frac{du_w}{dt}\right)^{\frac{1}{m}}\left(\frac{1}{k_w 2\pi nR^2}\right)^{\frac{1}{m}}\int_{\varphi_1}^{\varphi_2}\cos\varphi\sin\varphi d\varphi. \tag{6}$$

After integration, we obtain the dependence for the maximum wear:

$$u_w(t)=\left(\frac{Q}{\pi R^2}\right)^m\frac{k_w 2\pi Rn}{(\sin^2\varphi_2-\sin^2\varphi_1)^m}t. \tag{7}$$

If we take the friction path for one cycle s_1, then for the time t the total friction path is: $s=s_1nt$ or $nt=\frac{s}{s_1}$. As a result, for the friction path we get:

$$u_w(s)=\left(\frac{Q}{\pi R^2}\right)^m\frac{k_w 2\pi R^2 s}{\sin^2\varphi_2-\sin^2\varphi_1}. \tag{8}$$

The formula for determining the contact pressure is obtained by substituting (8) in (5):

$$\sigma=\left(\frac{Q}{\pi R^2}\right)\frac{\left(k_w 2\pi R^2 n\right)^{\frac{1}{m}}}{\sin^2\varphi_2-\sin^2\varphi_1}\left(\frac{1}{k_w 2\pi nR^2}\right)^{\frac{1}{m}}. \tag{9}$$

After the transformations, we finally get:

$$\sigma=\frac{Q}{\pi R^2\left(\sin^2\varphi_2-\sin^2\varphi_1\right)}. \tag{10}$$

An example of calculating the wear of the lower support in the contact of a hardened steel shaft and a polyurethane polymer liner.

Initial data. Hinge load 2,000 *N*. The total path of friction in the joint for 10,000 km of run. Ball pin radius 15 mm. The parameters of the wear model for polyurethane are: m = 2.04; $k_w=0.5\cdot10^{-8}$ (MPa) $^{-m}$. Liner contact angles: $\varphi_1=35^\circ$; $\varphi_2=75^\circ$.

The calculation of the wear of the liner is performed according to dependence (7) with the selected initial data per 10,000 km of run:

$$u_w=\left(\frac{Q}{\pi R^2}\right)^m\frac{k_w 2\pi Rs}{\sin^2\varphi_2-\sin^2\varphi_1}=\left(\frac{2000}{\pi 15^2}\right)^{2{,}04}\frac{0.5\cdot10^{-8}\cdot2\pi\cdot15^2\cdot3\cdot10^4}{\sin^2 75^0-\sin^2 35^0}=0.266\text{ mm.}$$

The obtained dependencies for calculations of wear are valid for all ball bearings of the front suspension. The difference lies in the different values of the contact angles φ_1 and φ_2. The formula for calculating contact pressures is also common to all mates in ball joints. This calculates the wear due to the frictional path caused by the axial rotation of the ball pin during the rotation of the steering

knuckle when turning from the steering. The friction path from lowering and raising the wheels while driving must be added to the total friction path in accordance with the suspension kinematics.
The maximum initial contact pressure in the hinge is determined by the dependence (10):

$$\sigma = \frac{Q}{\pi R^2\left(\sin^2\varphi_2 - \sin^2\varphi_1\right)} = \frac{2{,}000}{\pi 15^2\left(\sin^2 75^0 - \sin^2 35^0\right)} = 4.68 .$$

It has been established that this value of contact pressure does not change with wear of the liner. Contact pressure is the main, basic characteristic of the friction unit. To determine pressures, it is necessary to have a solution corresponding to the contact problem. The upper and lower supports each have two types of spherical couplings, which differ both in the sign of curvature and in the properties of the contacting bodies. However, all mates can be represented as one generalized mate: the convex spherical surface of the finger and the concave spherical surface of the support body (or vice versa). In all cases, contact starts at some angle $\varphi_1 > 0$ and ends at an angle $\varphi_1 \leq 90^\circ$.

3. Reliability calculations of ball bearings

Let us assume that the current wear of the suspension ball bearings is distributed according to the normal law. In this case, the quantile up of the reliability function is determined [6, 7] by the dependence:

$$u_p = -\frac{n-1}{\left(n^2 V_{w*}^2 + V_w^2\right)^{\frac{1}{2}}}(\varphi), \tag{11}$$

where V_{w*}, V_w is variation coefficients of limit and current wear; n is safety factor for wear:

$$n = \frac{u_w^*}{u_w}. \tag{12}$$

The limit wear value can be taken constant $u_w^* = 0$, then the variation coefficient $V_{w*} = 0$. In this case, from (12) we have:

$$u_p = -\frac{n-1}{V_w}. \tag{13}$$

By quantile, the reliability function is determined by the Laplace function: $\phi\left(u_p\right)$:

$$P\left(u_w < u_w^*\right) = 0.5 + \phi\left(u_p\right). \tag{14}$$

The determination of the general variation coefficient is performed through the partial coefficients of variation, taking into account the form of the main dependence (7). The determination of the general variation coefficient is performed through the partial variation coefficients, taking into account the form of the main dependence (7). In this case, the load, the wear rate factor and the friction path are random values. The variation coefficients for these quantities V_Q, V_{k_w}, V_s are assumed to be known.

In accordance with the general methodology [5] determining the coefficient of variation of a function by the variation coefficients of the arguments, we have:

$$V_w = \left(m^2 V_Q^2 + V_{k_w}^2 + V_s^2\right)^{\frac{1}{2}}. \tag{15}$$

Let us determine the probability of failure-free operation of the ball joint of the car during 10,000 km of run. Let's use the initial data from the previous example. Let us take the coefficients of variation: $V_Q = 0.4$, $V_S = 0.4$, $V_{K_w} = 0.3$. The general coefficient of variation of wear is determined by the dependence (15):

$$V_{u_w} = \left(m^2 V_Q^2 + V_{k_w}^2 + V_s^2\right)^{\frac{1}{2}} = \left(2.04^2 0.4^2 + 0.3^2 + 0.4^2\right)^{\frac{1}{2}} = 0.957.$$

Average wear is determined by the formula (8). According to the accepted data, the wear of the ball joint was obtained for 10,000 km. $u_w = 0.266$ mm.

According to [3], if, when checking the wear in the upper ball joint, the total indicators of the indicator exceed 0.8 mm, then the hinge must be repaired. Thus, dividing this value into two joints (lower and upper) we obtain the allowable wear value in the ball joint of the front suspension: $u_w^* = 0.4\,мм$ .Safety factor: $n = \frac{u_w^*}{u_w} = \frac{0.4}{0.266} = 1.5$.

The quantile of the probability of no-failure operation is determined by the dependence (11):

$$u_p = -\frac{n-1}{V_w} = -\frac{1.5-1}{0.957} = -0.522.$$

According to the tables of the Laplace function: $\phi(-0.52) = 0.3$ .Finally, reliability function according to (14): $P = 0,5 + \phi\left(u_p\right) = 0.5 + 0.3 = 0.8$.

The given example is methodological in nature. For a more reliable assessment of reliability, it is necessary to perform an experimental determination of both the average values and their coefficients of variation.

4. Determination of wear parameters of ball bearings

In the calculation formulas, the wear parameters m and kw were taken as given. The problem of determining the parameters of the friction pair wear model is one of the stages of the computational and experimental evaluation of the wear of the ball joints. The problem of increasing the wear resistance of this unit can be solved in two directions: improving the wear-resistant properties of the liners and choosing new types of lubricants. Currently, great progress has been made in the development of lubricants with additives [9-12]. The task is to assess the effectiveness of lubricants for use in ball joints. The wear characteristics of the materials were determined from the results of wear tests. The test scheme is shown in figure 2. An automobile ball joint polyurethane liner was used as a sample. The outer surface of the insert was tested with loading with a sphere diameter *R* = 28 mm (Figure 2).

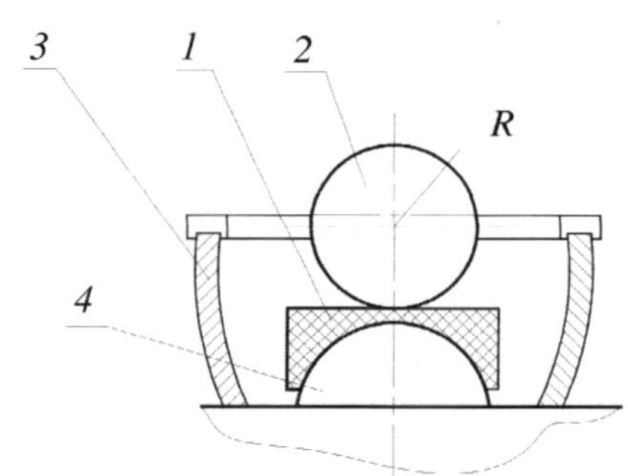

Figure 2. Scheme of testing the polyurethane liner for wear: 1– polyurethane hinge liner; 2– steel counter sample HRC 50; 3– container with grease; 4 – loading arm.

The tests were carried out at a rotational speed $n = 830\ rev^{-1}$ and a load of 10 *N*. For lubrication, we used ShRB-4 grease, Mobil grease, and Shell grease. Three samples of each variant were tested. The contact patch was measured in two directions. The average test results are presented in Table 1.

Table 1. Test results of polyurethane with different types of lubricants.

Friction path, s 10^6, mm	a, mm			
	Dry contact	ShRB-4	Mobil	Shell
1	1.414	1.311	1.231	1.247
3	1.55	1.49	1.35	1.34
5	1.685	1.525	1.452	1.471
7.5	1,74	1.65	1.52	1.51
10	1.812	1.734	1.592	1.614

The test results were approximated by a power function of the dependence of wear on the friction path in the form: $a = cs^{\beta}$. The power-law approximation parameters were determined using the Excel program (Table 2).

Table 2. Results of determining wear parameters for various lubricants

Parameters	Dry contact	ShRB-4	Mobil	Shell
β	0.107	0.121	0.111	0.112
c	0.319	0.245	0.263	0.265
m	3.642	3.118	3.4763	3.129
k_W	$14 \cdot 10^{-6}$	$0.41 \cdot 10^{-6}$	$0.39 \cdot 10^{-6}$	$0.43 \cdot 10^{-6}$
$I(6MPa) \cdot 10^{-6}$	2.177	0.0833	0.066	0.0733
$I(2MPa) \cdot 10^{-6}$	0.0398	0.0027	0.00145	0.00163

The formulas for determining the wear parameters were determined by solving the inverse wear-contact problem [4, 8]. Relationships (1-3) were used as the basic equations. As a result, we got the formulas:

$$m = \frac{1-2\beta}{2\beta}. \tag{16}$$

$$k_w = \frac{c^{2m+2}}{(2m+2)(Q/\pi)^m R}. \tag{17}$$

The results of calculating the wear parameters are shown in table. 2. Comparison of the effectiveness of lubricants is carried out in terms of the wear rate at two pressures of 6MPa and 2MPa. The calculation of the intensity is carried out according to the main dependence:

$$I = \frac{du_w}{ds} = k_w \sigma^m. \tag{18}$$

The results of calculating the wear rate for various lubricants are shown in Table 2. From the results of comparison with the base grease ShRB-4 it follows: the use of Mobil grease reduces wear at 6 MPa by 1.26 times; at 2 MPa by 1.85 times; the use of Shell lubricant reduces wear at 6 MPa by 1.14 times; at 2 MPa by 1.67 times.

5. Conclusion

1. To estimate the wear of the ball joints of the car suspension, the wear-contact problem was solved. The solution is based on: the equilibrium equation of the tribosystem, the law of wear, the geometric equation of continuity in contact. As a result, closed formulas were obtained for determining the amount of wear and contact pressure.
2. A method is proposed for calculating the reliability of ball joints of a car half-suspension by the criterion of wear. A method of sequential calculation of the coefficient of variation of wear is considered by the coefficients of variation of random parameters that are included in the formula for calculating wear.
3. Based on the solution of the inverse wear-contact problem, formulas are proposed for calculating the parameters of the wear law. A comparative assessment of the tribological properties of lubricants for ball joints of the car suspension is carried out.

References

[1] Aulin V, Derkach O., Makarenko D, Hrynkiv A & etc. 2019. Analysis of tribological efficiency of movable junctions “polymeric-composite materials - steel.” *Eastern-European Journal of Enterprise Technologies*. *4*(12–100), 6. https://doi.org/10.15587/1729-4061.2019.176845

[2] Chang-Fu Han, Hsiao-Yeh Chu, Ruei-Yang Luo & etc. 2018 Effects of groove factor and surface roughness of raceway in ball-bearing-like specimens on tribological behavior and the onsets of two instabilities of dry contacts. *Wear*. 406, 126. https://www.sciencedirect.com/science/article/abs/pii/S004316481731551X

[3] Aulin V, Hrynkiv A, Lysenko S & etc. 2019. Exploring a possibility to control the stressed-strained state of cylinder liners in diesel engines by the tribotechnology of alignment. *Eastern-European Journal of Enterprise Technologies*, *3*(12–99), 6. https://doi.org/10.15587/1729-4061.2019.171619

[4] Shuhui Cui, Le Gu, Liqin Wang, Bo Xu, Chuanwei Zhang. 2018 Numerical analysis on the dynamic contact behavior of hydrodynamic journal bearings during start-up. *Tribology International*.121 (260-268)

[5] Ali Rezaeia, WimVan Paepegema, Patrick De Baets, Wouter Ost. 2012 Adaptive finite element simulation of wear evolution in radial sliding bearings. *Wear*. 296,1–2 (660-671)

[6] Florian König, Achraf Ouald Chaib, Georg Jacobs 2019. Christopher Sous. A multiscale-approach for wear prediction in journal bearing systems – from wearing-in towards steady-state wear. *Wear*. 426–427 (1203-1211)

[7] Oksanen V, Lehtovaara A, Kallio M. 2017 Load capacity of lubricated bismuth bronze bimetal bearing under elliptical sliding motion. *Wear*. 388–389 (72-80)

[8] Sorokatyi R V, Dykha A V 2015 Analysis of Processes of Tribodamages under the Conditions of High-Speed Friction. *J. Friction and Wear*, 36, 422. doi: 10.3103/S106836661505013X

[9] Sorokatyi R, Chernets M, Dykha A, Mikosyanchyk O 2019 Phenomenological Model of Accumulation of Fatigue Tribological Damage in the Surface Layer of Materials *Mechanisms and Machine Science*, 73, 3761. https://link.springer.com/chapter/10.1007%2F978-3-030-20131-9_371

[10] Dykha A, Marchenko D, Artyukh V, Zubiekhina-Khaiiat O, & Kurepin V 2018. Study and development of the technology for hardening rope blocks by reeling. *Eastern-European Journal of Enterprise Technologies*. 2(1–92), 22. https://doi.org/10.15587/1729-4061.2018.126196

[11] Aulin V, Hrynkiv A, Lysenko S, Zamota T, Pankov A, & Tykhyi A. 2019. Determining the rational composition of tribologically active additive to oil to improve characteristics of tribosystems. *Eastern-European Journal of Enterprise Technologies*. *6*(12–102), 52. https://doi.org/10.15587/1729-4061.2019.184496

[12] Guo Xiang, Yanfeng Han, Jiaxu Wang, Jiefu Wang. 2019 Coupling transient mixed lubrication and wear for journal bearing modeling. *Tribology International*. 138 (1–15)

Influence of the flow area of distribution systems on changing the operating parameters of planetary hydraulic motors

S Kiurchev [1, 6], P Luzan[2], A Zasiadko[5], H Radionov[4] and N Boltianska[3]

[1]Department of structural materials technology, Dmytro Motornyi Tavria State Agrotechnological University, B. Khmelnitsky ave. 18, Melitopol, Ukraine
[2]Institute of Vocational Education National Academy of Pedagogical Sciences of Ukraine, Vito-Litovsky lane, 98-a, Kyiv, Ukraine
[3]Department Berdyansk college of the Tavria State Agrotechnological University, Eastern ave. 23, Berdyansk, Ukraine
[4]Department of Missile and Artillery Weapons, National Academy of the National Guard of Ukraine Zakhysnykiv, Ukrainy sq., 3, Kharkiv, Ukraine
[5]Department of technical systems and technology in livestock, Dmytro Motornyi Tavria State Agrotechnological University, B. Khmelnitsky ave. 18, Melitopol, Ukraine

[6]Email: voloshinaa2012@gmail.com

Abstract. Predicting changes in the operating parameters of a planetary hydraulic motor is an urgent direction for improving its design. The physical and mathematical models of the planetary hydraulic motor have been developed, which make it possible to simulate the processes occurring under operating conditions. A design scheme and a mathematical apparatus have been developed for determining the pressure and leaks of the pressure fluid in the gap formed by the end surfaces of the distributors, which make it possible to study the change in the operating parameters of the hydraulic motor. By studying the features of the distribution system of a serial hydraulic motor, it was found that insignificant fluctuations in the flow area (up to 3%) cause rather large (up to 71%) pressure fluctuations. That causes the loss of fluid passing through the hydraulic motor up to 10% and fluctuations in the angular velocity of the output shaft – up to 11%. The absence of pressure fluctuations in the discharge line of the distribution system of the modernized hydraulic motor ensures the stability of its operating parameters. The results of the studies carried out make it possible to predict changes in the operating parameters of planetary hydraulic motors operating as part of a hydraulic drive at the design stage.

1. Introduction

At present, the increase in the efficiency of using self-propelled equipment is largely determined by the hydrofication of its active working bodies and propelled systems by using hydraulic drives with planetary hydraulic motors [1, 2]. An analysis of the existing designs of planetary hydraulic machines found that the improvement of the operating parameters of a planetary hydraulic motor depends on the relationship between the geometric and functional parameters of its distribution system, which determine its throughput [3, 4].

By the throughput of the distribution system we mean the amount of liquid that has passed at a certain point in time through the distribution windows into the operating chambers of the hydraulic motor. The throughput is determined by the current value of the flow area of the distribution windows

[3, 4].

To date, there is no clear methodology for predicting changes in the output characteristics of hydraulic drives with planetary hydraulic motors under operating conditions. Therefore, the study of the influence of the throughput of the distribution system on the change in the operating parameters of a planetary hydraulic motor is an urgent direction. The results of the research carried out will make it possible to predict changes in the operating parameters of planetary hydraulic motors operating as part of a hydraulic drive, even at the design stage, which is very important in the development and modernization of these hydraulic machines.

2. Analysis of recent studies and publications

The increasing requirements for the performance of mechatronic systems with a hydraulic drive of active working bodies of self-propelled machines require the use of new approaches in the process of their development and design. A universal model has been proposed that describes the dynamic and static processes occurring in the elements of a mechatronic system [5], a methodology for designing hydraulic mechatronic systems with elements of multi-criteria optimization has been developed, which makes it possible to design a mechatronic system with given output characteristics [6], transient processes in the control system of a hydraulic drive have been analyzed [7], the functional characteristics of the hydraulically driven mechatronic system on the chassis were investigated [8]. Analytical and finite element models of a machine with a trochoidal rotor working surface for quasi-static conditions [9], a mathematical model of axial piston hydraulic machines for compressible flows [10] are presented. Much attention is paid to dynamic machines. A one-dimensional numerical model of a turbine of centrifugal pumps [11] was proposed, an analysis of the flow rate in the pump and turbine operating modes [12], as well as an analysis of the dynamic response of the turbine [13], a numerical calculation and experimental analysis of pressure fluctuations [14] and instability phenomena in the pump turbine [15], the internal flow in bidirectional channels of the axial flow system was investigated by the method of numerical simulation and experiment [16], the operating characteristics and characteristics of the internal flow were investigated [17, 18]. The processes occurring in hydraulic machines with cycloidal rotors have not been studied.

For the drive of active working bodies of self-propelled equipment, hydraulic machines with cycloidal displacers — orbital and planetary — are increasingly used [5–7]. With many different designs, these hydraulic machines can be combined into three main units that determine their operational efficiency: power connection [19, 20], with a special cycloidal profile of the rotors; a mechanism that compensates for the planetary (orbital) movement of the rotor; a distribution mechanism [21, 22], which creates a hydraulic field necessary for the operation of the rotors. The forces acting in the gearing were considered [23], a mathematical model was developed that describes the change in loads in the power connection formed by the rotors [24], a model of the operation of the rotors [25], as well as a program for designing hypocycloidal surfaces of the rotors [26], were proposed. The wear of the working surfaces of the rotors was investigated [27], the limiting deviations of the error in the form of manufacturing the toothed surface of the rotors [19] were substantiated, a method for determining the reliability of the hydraulic motor by modeling the change in the technical state of its rotors [20] was implemented. The influence of the shape of the windows of the movable and fixed distributors on the output characteristics of the planetary hydraulic motor [3, 21] was investigated, the angular arrangement of the distributing windows was substantiated [22], and an algorithm for designing the flow parts of the distribution system of the planetary hydraulic motor [4] was developed. The parameters of variation are substantiated, which determine the change in the output characteristics of the planetary hydraulic motor, depending on the design features of its rotor system and the distribution system [28]. It should be noted that during the operation of hydraulic drives with planetary hydraulic machines, functional failures are observed due to pressure deviations (drops, pulsations and overshoots) caused by imperfect design of the distribution system of these hydraulic machines [29]. Therefore, improving the operating parameters of a planetary hydraulic motor by changing the throughput of its distribution system is an urgent direction.

IOP Conf. Series: Materials Science and Engineering 1021 (2021) 012037 doi:10.1088/1757-899X/1021/1/012037

3. Statement of the objective and tasks of the study

To study the effect of the flow area of the distribution system of a planetary hydraulic motor on the change in its operating parameters, it is necessary:

- to develop physical and mathematical models of a planetary hydraulic motor as part of a hydraulic drive;
- to develop a design scheme and mathematical apparatus for determining the pressure and leaks of the pressure fluid in the gap formed by the end surfaces of the movable and stationary distributors;
- to investigate the influence of the flow area of the distribution system of the planetary hydraulic motor on the change in its operating parameters.

4. The basic part of the study

The physical model of a planetary hydraulic motor [5] as part of a hydraulic drive (figure 1), as a result of the performed refinements and transformations, makes it possible to study the processes occurring in the distribution systems of planetary hydraulic motors operating under operating conditions. Hydraulic pump 1 transfers the required hydraulic energy (fluid supply Q_p at pressure p_p) to hydraulic motor 2, the shaft of which is connected to load 4. To protect the hydraulic system from possible overloads arising during operation, a safety valve 3 is installed in the hydraulic drive circuit, which operates when the pressure rises in the discharge line.

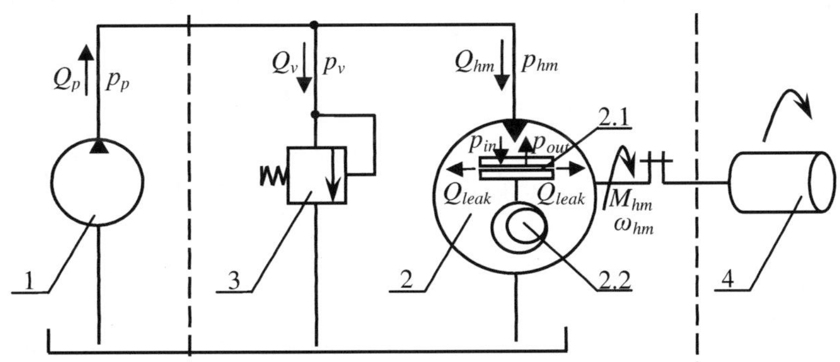

Figure 1. Physical model of a planetary hydraulic motor as part of a hydraulic drive: 1 – pump; 2 – hydraulic motor; 2.1 – pressure fluid distribution system; 2.2 – rotor system; 3 - a safety valve; 4 – load; Q_p – pump flow; p_p – the pressure at the pump outlet; Q_{hm} – the amount of fluid entering the hydraulic motor; p_{hm} – the pressure at the inlet to the hydraulic motor; Q_v – flow through the safety valve; p_v – pressure at the inlet to the safety valve; p_{in} and p_{out} are the pressure in the discharge and drain lines of the end distribution system, respectively; Q_{leak1} and Q_{leak2} – leaks of pressure fluid in the discharge and drain lines of the distribution system, respectively; ω_{hm} and M_{hm} – angular velocity and torque of the hydraulic motor shaft.

The functional parameters of the hydraulic drive under consideration can be estimated by the stability of the operating parameters of its hydraulic motor 2 (angular velocity ω_{hm} and torque M_{load}), due to the corresponding load parameters 4 (ω_{load} and M_{load}). To determine the deviations of the operating parameters of the hydraulic motor 2, we assume that the hydraulic power of the pump 1 (Q_p and p_p) and the load parameters 4 (ω_{load} and M_{load}) are constant. Then, the deviations of the operating parameters of the hydraulic motor 2 will be caused only by the pressure pulsation p_{in} in the discharge line of its distribution system 2.1 (figure 1). Deviations of the operating parameters of the planetary hydraulic motor 2 are associated with the imperfection of the design of the main elements of its distribution system 2.1.

In order to determine the relationship of the main geometric and functional parameters of the distribution system of hydraulic motors of the PRG series, its operation is considered (Figure 2). The pressure fluid under pressure p_{hm} is supplied to channel A of the fixed distributor and enters the drain through channel B. Through the distribution windows located on the contact surfaces of the movable and fixed distributors, the pressure fluid first enters the channels of the movable distributor, and then into the operating chambers of the hydraulic motor. The principle of operation of distribution systems of serial and modernized hydraulic motors is described in detail in the works [4, 21, 29]. Distribution systems are structurally distinguished by the number of operating and discharge windows located on the contact surfaces of the movable and fixed distributors. The throughput of the distribution system is determined by the number of operating and additional windows simultaneously participating in its operation [21, 29].

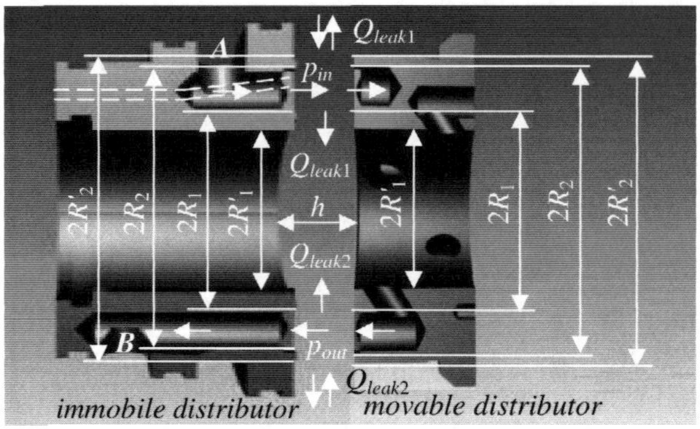

Figure 2. Calculation scheme for determining leaks in the gap formed by the "opening" of the contact surfaces of the movable and stationary distributors: A – channel for supplying the pressure fluid under pressure; B – channel for draining the pressure fluid (drain); p_{in} and p_{out} – pressure in the pressure and drain lines of the distribution system contact, respectively; Q_{leak1} and Q_{leak2} – leaks of pressure fluid in the pressure and drain lines of the distribution system contact, respectively; h – the height of the gap between the contact surfaces of the movable and immovable distributors; R_1 and R_2 – inner and outer radii of the contact surfaces of the movable and fixed distributors, respectively; R'_1 and R'_2 – are the inner and outer radii of the distribution windows, respectively.

For trouble-free operation of the hydraulic motor, the design of the elements of its distribution system provides for a constant preload of the movable and fixed distributors. The pressing force is determined by the relationship between the pressure p_{hm} at the inlet to the hydraulic motor and the current pressure value $p_{in\ i}$ in the delivery line of the distribution system, while the condition $p_{in\ i} \leq p_{hm}$ must be met. At the moments when the specified condition is not met, there is a "opening" of the contacting surfaces of the movable and fixed distributors. Through the gap formed by the contact surfaces of the distributors, the Q_{leak1} pressure fluid leaks, which leads to a change in the angular velocity of the output shaft of the hydraulic motor.

During operation of the hydraulic motor, the current value of the pressure $p_{in\ i}$ changes depending on the fluctuations in the flow area, expressed through the pulsation coefficient P_A [29] in the pressure line of the distribution system. Analysis of the change in the pulsation coefficient P_A shows that for a serial hydraulic motor there are three cases of the relationship between the current value of the flow area A_{fl} and the average flow area A_{aver}: $A_{fl\ i} > A_{aver}$, $A_{fl\ i} = A_{aver}$ and $A_{fl\ i} < A_{aver}$.

If the current value of the flow area $A_{fl\ i}$ is greater than or equal to the average value A_{aver}, that is $A_{fl\ i} \geq A_{aver}$, then the current value of the pressure $p_{in\ i}$ at the inlet to the distribution system is lower or equal to the pressure p_{hm} at the inlet to the hydraulic motor $p_{in\ i} \leq p_{hm}$. Consequently, the condition for preloading the distributors is met, which means that the leakage rate Q_{leak1} in the distribution system of the hydraulic motor is zero, $Q_{leak1} = 0$.

When the current value of the flow area $A_{fl\ i}$ is lower than the average value A_{aver}, that is $A_{fl\ i} < A_{aver}$, then the current value of the pressure $p_{in\ i}$ at the inlet to the distribution system is higher than the nominal pressure p_{nom} in the line, that is, $p_{in\ i} > p_{nom}$. In this case, the condition for preloading the distributors is not met, which means that there is a leakage rate Q_{leak1} in the distribution system of the hydraulic motor, $Q_{leak1} > 0$. Then, the amount of liquid entering the distribution system and passing through the hydraulic motor is determined by the difference in pump flow Q_p, flow Q_v through the safety valve and flow Q_{leak1} leaks in the distribution system of the hydraulic motor, $Q_{hm} = Q_p - Q_v - Q_{leak1}$.

In all the cases considered, the safety valve 3 of the hydraulic drive does not work (Figure 1), since the pulsations of the current pressure $p_{in\ i}$ at the inlet to the distribution system are not transmitted to the pressure line of the hydraulic drive. Therefore, the flow through the valve is zero, $Q_v = 0$, and the flow through the hydraulic motor is equal to the pump flow, $Q_{hm} = Q_p$.

Thus, the trouble-free operation of the distribution system depends on two operating parameters – the p_{in} pressure and Q_{leak1} leaks of the pressure fluid in the delivery line of the distribution system. Therefore, to determine the specified operating parameters - p_{in} pressure and Q_{leak1} leaks of the pressure fluid, we have developed a mathematical apparatus.

The pressure p_{in} and leakage Q_{leak1} of the pressure fluid in the end clearance of the distribution system of the planetary hydraulic motor are determined by expressions (1) and (2), respectively:

$$p_{in} = \frac{12\mu'}{h^3}\cdot\left(\frac{Q_{leak_1}}{\sum\left(\frac{\pi}{Z_2}-\Delta+\left|\alpha_i(t)-\beta_{1i}(t)\right|\right)}\cdot ln\frac{R_2'}{R_2}+\frac{Q_{leak_2}}{\sum\left(\frac{\pi}{Z_2}-\Delta+\left|\alpha_i(t)-\beta_{2i}(t)\right|\right)}\cdot ln\frac{R_1}{R_1'}\right)+ \tag{1}$$

$$+\frac{3\rho\cdot\omega_{гм}^2}{40}\cdot\left(R_2^2-R_1^2+R_1'^2-R_2'^2\right),$$

$$Q_{leak_1} = \frac{h^3}{12\mu'}\cdot\sum\left(\frac{\pi}{Z_2}-\Delta+\left|\alpha_i(t)-\beta_{1i}(t)\right|\right)\cdot\left[\frac{p_{in}}{ln\frac{R_2'}{R_2}}-\frac{3\rho\cdot\omega_{гм}^2}{40}\cdot\left(\frac{\left(R_1'^2-R_1^2\right)}{ln\frac{R_1}{R_1'}}+\frac{\left(R_2^2-R_2'^2\right)}{ln\frac{R_2'}{R_2}}\right)\right]; \tag{2}$$

where μ' is the coefficient of dynamic viscosity of the pressure fluid; α_i is the angles of the operating windows of the movable distributor [4, 29]; β_{1i}, β_{2i} is the angles of the discharge and discharge windows of the stationary distributor [4, 29]; ρ is the density of the pressure fluid; ω_{hm} is angular speed of the hydraulic motor shaft.

When developing a mathematical model of the processes taking place in a hydraulic drive with a planetary hydraulic motor (figure 1), the following assumptions were made. The hydraulic power of pump 1 and load parameters 4 are constant. The pressure p_p at the outlet of the pump is equal to the pressure p_{hm} at the inlet to the hydraulic motor and the pressure p_v at the inlet to the safety valve and is equal to the nominal pressure p_{nom}. The pressure at the outlet of the hydraulic motor and safety valve is zero. The pump flow rate is constant Q_p = const. The pressure pulsation p_{in} at the inlet to the distribution system of the hydraulic motor is caused by a change in the flow area A_{fl} of its channels.

To study the processes occurring during the operation of a hydraulic drive with a planetary hydraulic motor, a mathematical model has been developed (3) that describes these processes.

$$\begin{cases} P_p = f\left(P_A\right) \\ p_p = p_{hm} = p_{in},\ p_{in} = f\left(P_p, M_{tur}\right),\ p_{out} = 0 \\ Q_v = f\left(p_{in}\right),\ p_{in} > p_{nom}, \\ Q_{leak1} = f(\ p_{in}, Q_{in}\ ),\ Q_{leak2} = f(\ p_{out}\ ) = 0 \\ Q_{hm} = Q_{in} = Q_p - Q_v - Q_{leak1} \\ M_{tur} = M_{load} + M_j,\ P_p = 0,\ M_j = 0 \\ \omega_{hm} = f\left(Q_{hm}, p_{in}\right),\ \omega_{hm} = \dfrac{2\pi\cdot Q_{hm}\cdot\eta_{vol}}{V} \end{cases} \tag{3}$$

where P_p is coefficient of pressure pulsation of the pressure fluid; P_A is coefficient of pulsation of the flow area; M_{load} is load moment; M_j is moment of inertia; V is the operating volume of the hydraulic motor; η_{vol} is volumetric efficiency of the hydraulic motor.

IOP Conf. Series: Materials Science and Engineering 1021 (2021) 012037 doi:10.1088/1757-899X/1021/1/012037

To simulate the processes occurring in the distribution system of the planetary hydraulic motor of the PRG-22 series, the following initial data and initial conditions are taken. Serial and modernized hydraulic motors differ only in distribution systems [4, 29] with a different number of distribution and discharge windows. The pressure at the pump outlet is equal to the pressure at the inlet to the hydraulic motor and the safety valve and is equal to the nominal pressure, $p_{p(out)} = p_{hm(in)} = p_{v(in)} = p_{nom}$ = 16 MPa. The pressure at the outlet of the hydraulic motor and the safety valve is $p_{hm(out)} = p_{v(out)} = 0$ MPa. The pump flow is constant and equal to Q_p = 89 l/min = const. The load is non-inertial, constant and equal to M_{load} = 365 N·m. The operating volume of the hydraulic motor is V = 160 cm^3.

Analysis of the results of modeling the processes (figure 3) occurring in a hydraulic drive with a planetary hydraulic motor, allowed to confirm theoretical studies and establish the influence of the area flow area of distribution systems on changes in the operating parameters of planetary hydraulic motors.

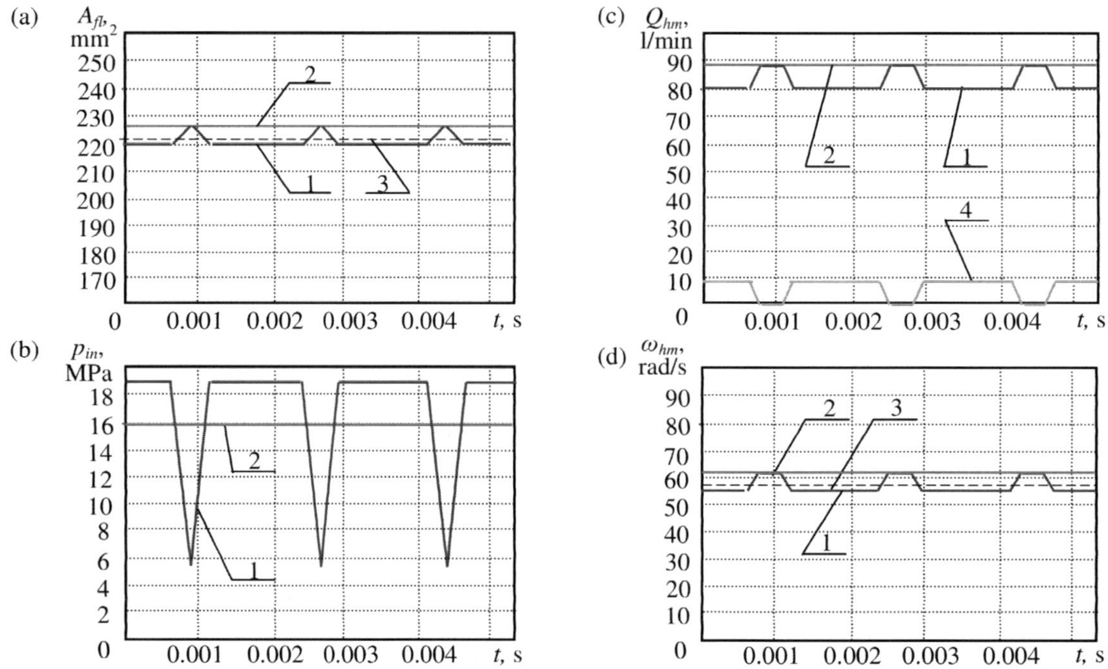

Figure 3. Changing the operating parameters of the planetary hydraulic motor of the PRG series with the end-end distribution system: a – the flow area of the distribution system [29]; b – pressure fluid pressure; c – the flow rate of the pressure fluid; d – angular speed of rotation of the hydraulic motor shaft; 1 – operating parameters of the serial distribution system; 2 – operating parameters of the modernized distribution system; 3 – average value of the operating parameters of the serial distribution system; 4 – flow rate of leaks in the end gap of the serial distribution system.

Figure 3 a shows the changes in the area flow area of the distribution system [29]. The authors found that fluctuations in the area flow area (curve 1) of the distribution system of the serial PRG-22 hydraulic motor are up to 3% and are in the range of 220 ÷ 226 mm^2, with the average value being 222mm^2 (curve 3).

Changes in the flow area of the distribution system of the modernized hydraulic motor (curve 2) show that there are no oscillations, and the flow area is 226 mm^2. This is due to the rational design of its distribution system [4, 21, 29].

By modeling the processes occurring in the distribution systems of planetary hydraulic motors, it was found that during the operation of a planetary hydraulic motor, there are changes in operating parameters caused by the irrationality of the design of the serial distribution system, occurring with a frequency that is numerically equal to the oscillation period of the flow area.

Fluctuations in the flow area of the distribution system of a serial hydraulic motor cause corresponding cyclic fluctuations in the pressure p_{in} of the pressure fluid in the pressure line of its distribution system (figure 3 b). Analysis of the pressure change p_{in} in the pressure line of the distribution system of a serial hydraulic motor (curve 1) shows that at the time when the current value of the flow area is less than the average, i.e. $A_{fl\ i} < A_{aver}$, the current pressure $p_{in\ i}$ exceeds the nominal p_{nom} and is equal to $p_{in\ i}$ = 18.7 MPa. With an increase in the current value of the area $A_{fl\ i}$ from 220 to 226 mm^2, there is a sharp drop in the current value of pressure to $p_{in\ i}$ = 5.4 MPa. With a subsequent decrease in the current value of the area flow area $A_{fl\ i}$ to its minimum value, the current value of the pressure $p_{in\ i}$ of the pressure fluid in the pressure line of the distribution system of the planetary hydraulic motor sharply increases to the value $p_{in\ i}$ = 18.7 MPa. Thus, fluctuations in the pressure p_{in} of the pressure fluid in the pressure line of the distribution system of the serial hydraulic motor reach 71%, varying in the range of 18.7 ÷ 5.4 MPa.

Analysis of the change in pressure p_{in} in the pressure line of the distribution system of the modernized hydraulic motor (figure 3 b, curve 2) shows that there are no fluctuations in the current value of pressure $p_{in\ i}$, since there are no fluctuations in the current value of the area $A_{fl\ i}$ of the flow area.

A change in the current value of the pressure $p_{in\ i}$ of the pressure fluid in the pressure line of the distribution system of a serial hydraulic motor causes short-term disruptions in its operation as a result of the "opening" of the contacting surfaces of the movable and fixed distributors (figure 3 c). In this case, the condition for preloading the distributors is not met, and there is a leakage rate Q_{leak1} in the distribution system of the hydraulic motor (curve 4), appearing synchronously with the increase in pressure $p_{in\ i}$ in the pressure line of the distribution system. Depending on the fluctuations in the current value of the pressure $p_{in\ i}$ in the pressure line, the leakage rate Q_{leak1} in the distribution system (curve 4) also fluctuates in the range from 8.9 to 0 l/min. Consequently, the amount of liquid passing through a serial hydraulic motor (curve 1) varies in the range of 80.1 ÷ 89 l/min, and the fluctuations reach 10%.

Since there are no fluctuations in the current value of the pressure $p_{in\ i}$ in the discharge line of the distribution system of the modernized hydraulic motor (figure 3 b, curve 2), the condition of pressing the valves is met, which means that the leakage rate Q_{leak1} is zero, Q_{leak1} = 0, The amount of fluid passing through the modernized the hydraulic motor (figure 3 c, curve 2) is constant (without fluctuations) and is Qhm = 89 l/min.

Analysis of the change in the values of the angular velocity of the serial hydraulic motor (figure 3 d, curve 1) shows that they are similar and correspond to the change in the flow rate of the pressure fluid in the serial hydraulic motor (figure 3 c, curve 1). Oscillations of the angular velocity of a serial hydraulic motor (curve 1) are in the range 55 ÷ 62 rad/s, reaching 11%, while the average value of the angular velocity (curve 3) is ω_{aver} = 57 rad/s. It should be noted that all changes in the operating parameters of a serial hydraulic motor occur synchronously with a change in the flow area of its distribution system.

The absence of fluctuations in the angular velocity (figure 3 d, curve 2) of the output shaft of the modernized hydraulic motor is explained by the absence of fluctuations in the flow rate of the pressure fluid (figure 3 c, curve 2).

The results of the studies carried out make it possible to predict changes in the operating parameters of planetary hydraulic motors operating as part of a hydraulic drive even at the design stage, which is very important in the development and modernization of these hydraulic machines.

5. Conclusions

Physical and mathematical models of a planetary hydraulic motor as part of a hydraulic drive have been developed, which allow simulating the processes occurring in planetary hydraulic motors and their elements under operating conditions.

A design scheme and a mathematical apparatus have been developed for determining the pressure and leaks of the pressure fluid in the gap formed by the end surfaces of the movable and stationary distributors, which make it possible to study the change in the operating parameters of the elements of the distribution system of a planetary hydraulic motor.

By studying the features of the distribution system of a serial hydraulic motor, it was found that insignificant fluctuations in the flow area (up to 3%) cause rather large (up to 71%) pressure fluctuations. That causes the loss of fluid passing through the hydraulic motor up to 10% and fluctuations in the angular velocity of the output shaft – up to 11%.

The absence of pressure fluctuations in the discharge line of the distribution system of the modernized hydraulic motor ensures the preload condition of the distributors, which means that the leakage rate is zero, which ensures the constancy of the angular displacement of the output shaft of the modernized hydraulic motor.

The developed models and mathematical apparatus will make it possible to conduct further studies to determine the influence of the distribution system capacity on the change in the output characteristics of the planetary hydraulic motor. The results of the studies carried out make it possible to predict changes in the operating parameters of planetary hydraulic motors operating as part of a hydraulic drive, even at the design stage, which is very important in the development and modernization of these hydraulic machines.

References

[1] Gamez-Montero P, Codina E and Castilla R 2019 A Review of Gerotor Technology in Hydraulic Machines *Energies* **12 (12)** 2423

[2] Panchenko A, Voloshina A, Titova O, Panchenko I and Zasiadko A 2020 The Study of Dynamic Processes of Mechatronic Systems with Planetary Hydraulic Motors *Advanced Manufacturing Processes II. Lecture Notes in Mechanical Engineering* 56–64

[3] Voloshina A, Panchenko A, Boltyansky O and Titova O 2020 Improvement of Manufacture Workability for Distribution Systems of Planetary Hydraulic Machines *Lecture Notes in Mechanical Engineering, Springer* 732-741

[4] Panchenko A, Voloshina A, Boltyansky O and others 2018 Designing the flow-through parts of distribution systems for the PRG series planetary hydraulic motors *Eastern-European Journal of Enterprise Technologies* **3 1 (93)** 67–77

[5] Panchenko A, Voloshina A, Kiurchev S and others 2018 Development of the universal model of mechatronic system with a hydraulic drive *Eastern-European Journal of Enterprise Technologies* 4 7(94) 51–60

[6] Voloshina A, Panchenko A, Panchenko I, Titova O and Caldare A 2020 Design of Hydraulic Mechatronic Systems with Specified Output Characteristics *Advances in Design, Simulation and Manufacturing III DSMIE 2020 Lecture Notes in Mechanical Engineering* 42-51

[7] Kozlov L, Polishchuk L, Piontkevych O and others 2019 Experimental research characteristics of counterbalance valve for hydraulic drive control system of mobile machine Przeglad elektrotechniczny **4(95)** 104-109

[8] Panchenko A, Voloshina A, Milaeva I and Luzan P 2019 Operating Conditions' Influence on the Change of Functional Characteristics for Mechatronic Systems with Orbital Hydraulic Motors *Modern Development Paths of Agricultural Production. Springer* 169-176

[9] Gamez-Montero P, Castilla R, Khamashta M and Codina E 2006 Contact problems of a trochoidal-gear pump *International Journal of Mechanical Sciences* **12(48)** 1471-1480

[10] Vacca A, Klop R and Ivantysynova M 2010 A numerical approach for the evaluation of the effects of air release and vapour cavitation on effective flow rate of axial piston machines *International Journal of Fluid Power* **1(11)** 33-45

[11] Amelio M and Barbarelli S 2004 A one-dimensional numerical model for calculating the efficiency of pumps as turbines for implementation in micro-hydro power plants *Proceedings of the 7th Biennial Conference on Engineering Systems Design and Analysis* **1** 65-72

[12] González J, Oro J and Argüelles-Díaz K 2009 Flow analysis for a double suction centrifugal

machine in the pump and turbine operation modes *International Journal for Numerical Methods in Fluids* **2(61)** 220-236

[13] Egusquiza E, Valero C, Presas A and others 2016 Analysis of the dynamic response of pump-turbine impellers. Influence of the rotor *Mechanical Systems and Signal Processing* **68-69** 330-341

[14] Yao J, Shi W, Wu S and others 2013 Numerical calculation and experiment on pressure fluctuation in axial flow pump *Transactions of the Chinese Society for Agricultural Machinery* **1(44)** 119-124

[15] Gentner Ch, Sallaberger M, Widmer Ch, Braun O and Staubli T 2012 Numerical and experimental analysis of instability phenomena in pump turbines *IOP Conference Series: Earth and Environmental Science* **3(15)** 0320422

[16] Liu C, Jin Y, Zhou J and others 2011 Study of internal flow in cube-type bidirection passages of axial-flow pump system by numerical simulation and experiment *Journal of Hydroelectric Engineering* **5(30)** 192-198

[17] Barrio R, Fernández J, Blanco E, Parrondo J and Marcos A 2012 Performance characteristics and internal flow patterns in a reverse-running pump-turbine *Journal of Mechanical Engineering Science* **3(226)** 695-708

[18] Mao X, Dal Monte A, Benini E and Zheng Y 2017 Numerical study on the internal flow field of a reversible turbine during continuous guide vane closing Energies **7(10)** 988

[19] Panchenko A, Voloshina A, Milaeva I, Panchenko I and Titova O 2018 The Influence of the form Error after Rotor Manufacturing on the Output Characteristics of an Orbital Hydraulic Motor *International Journal of Engineering and Technology* **7(4.3)** 1–5

[20] Panchenko A, Voloshina A, Panchenko I, Titova O and Pastushenko A 2019 Reliability design of rotors for orbital hydraulic motors *IOP Conference Series: Materials Science and Engineering* **708 (1)** 012017

[21] Voloshina A, Panchenko A, Panchenko I and Zasiadko A 2019 Geometrical Parameters for Distribution Systems of Hydraulic Machines *Modern Development Paths of Agricultural Production. Springer* 323-336

[22] Voloshina A, Panchenko A, Panchenko I, Titova O and Zasiadko A 2019 Improving the output characteristics of planetary hydraulic machines *IOP Conference Series: Materials Science and Engineering* **708 (1)**, 012038

[23] Chang Y, Kim J, Jeon C, Chul K and Jung S 2006 Development of an Integrated System for the Automated Design of a Gerotor Oil Pump *Journal of Mechanical Design* **10 (129)** 1099-1105

[24] Stryczek J, Bednarczyk S and Biernacki K 2014 Strength analysis of the polyoxymethylene cycloidal gears of the gerotor pump *Archives of Civil and Mechanical Engineering* **4 (14)** 647–660

[25] Gamez-Montero J, Garcia-Vilchez M, Raush G, Freire J and Codina E 2012 Teeth Clearance and Relief Grooves Effects in a Trochoidal-Gear Pump Using New Modules of GeroLAB *Journal of Mechanical Design* **5 (134)** No: 054502-7

[26] Stryczek J, Bednarczyk S and Biernacki K 2014 Gerotor pump with POM gears: Design, production technology, research *Archives of Civil and Mechanical Engineerin* **3 (14)** 391–397

[27] Choi T, Kim M, Lee G, Jung S, Bae J and Kim C 2012 Design of Rotor for Internal Gear Pump Using Cycloid and Circular-Arc Curves *Journal of Mechanical Design* **1 (134)** 011005-12

[28] Voloshina A, Panchenko A, Titova O, Milaeva I and Pastushenko A 2020 Prediction of Changes in the Output Characteristics for the Planetary Hydraulic Motor *Advanced Manufacturing Processes II. Lecture Notes in Mechanical Engineering* 84–93

[29] Voloshina A, Panchenko A, Boltyansky O, Panchenko I and Titova O 2018 Justification of the Kinematic Diagrams for the Distribution System of a Planetary Hydraulic Motor *International Journal of Engineering and Technology* **7 (4.3)** 6–11

Conceptual Resource-Saving Design of Real-Time Socio-Technical Systems

Anatolii Kosolapov[1,2] and Pavlo Ivin[1,3]

[1] Computers Department, National University of Railway Transport named acad. V. Lazaryan, Lazaryan Street 2, 49010 Dnipro, Ukraine

[2] Email: kosolapof@i.ua
[3] Email: pavelvi1980@gmail.com

Abstract. The article discusses the methodological resource-saving approach to the System design of socio-technical systems for real-time managing (STS RTM). The definition of STS RTM from the point of view of computer systems architecture is introduced. This approach made it possible to classify the main paradigms of using computer systems, including the modern stage, corresponding to STS RTM. For such complex systems, it is important to ensure their energy efficiency, which is ensured by the use of resource-saving models and methods for optimizing structural solutions. This is ensured by minimizing the total length of communications at the enterprise, saving computing and financial resources during system migration from a centralized structure to a functionally distributed hierarchical multi-microcontroller system. STS is a human a real-time machine system, therefore, the system's response time to the processing of requests related to the issuance of messages to the system's operating personnel is determined. The complex of formulated tasks and their solution on a common information and analytical base is extremely difficult without a unified system design methodology and automation tools for solving problems. Such a Framework is described in this article.

1. Introduction

For more than 40 years we have been developing and implementing of automated control systems for railway transport [1]: by tracing the development trends of such systems, we observe not only changes in the architecture of automation systems, but also the emergence of new users' needs based on expanding information resources for decision-making.

1.1. Architecture CS as a multistructure for the description of STS

In the work, developing the concept of CS architecture [5], a new definition of the architecture as an object-oriented multi-structure, which can be described STS in the form of the following expression, is introduced:

$$ASTS=Gs\cap Os\cap(Hs\cup Ss\cup Ms\cup Is\cup Ls\cup Ps\cup Ds\cup MLs), \qquad (1)$$

where *Gs* is the business goals of creating a system; *Os* - the set of objects or the environment to which the system operates, raising the functions of control and management to achieve business goals; *Hs* is hardware, technical support (a set of technical means of the system); *Ss* is software (general and special); *Ms* is mathematical support (complex of mathematical models, methods and algorithms); *Is* is

IOP Conf. Series: Materials Science and Engineering 1021 (2021) 012038 doi:10.1088/1757-899X/1021/1/012038

information support (description of signals, principles of classification and coding of information, description of arrays, forms, reference information and information on regulations and other types of information); *Ls* is linguistic support (a set of language means of communication with the system for personnel); *Ps* is people support, organizational support (organizational structure and instructions to operational people).

All these types of support are characterized by a set of interrelated static and dynamic structures that are formed in the process of system design and are combined by a common conceptual scheme to achieve the creation goals while minimizing total costs [4, 5].

1.2. The development of computerization paradigms

The development of paradigms or trends in computerization in the world took place in the form of the following stages. The first stage is the stage of computing systems (CS). These were hardware-centric systems, in which the main goal was to get maximum performance; the dominant features in the ASTS are *Hs*, *Ss* (1950-80s). After this comes the network-centric stage of creating information systems that provide access to information "here and now", the dominants are *Is*, *Hs* (net) (1990s - 2010). The third stage is human-centric, associated with the creation of systems with artificial intelligence, the dominants are integrated $Hs \cup Ss \cup Ms \cup Is \cup Ls \cup Ds$. The current stage is being formed in the context of the globalization of socio-economic and production processes. It will be called the sociocentric paradigm associated with the construction of STS with HYBRID TNTELECT (STS GI). The dominant in this case are *Gs, Ps, Os, Is, Hs*.

2. STS conceptual design framework CoDeCS-2020

The variety and complexity of systems architectures makes it very important to create a methodology and develop frameworks and tools based on it.

The architectural scheme of the conceptual design is related to the preliminary research, optimization and selection of rational structures and characteristics: topology of technical facilities at the company, the technical structure, the structure of information flows and the databases organization, processor performance requirements for the modes of their operation in real time and priority processing of applications, indicators of reliability, power consumption and cost of the projected STS. All these studies and calculations are carried out, as a rule, in the conditions of incompleteness of the initial data and their uncertainty. In addition, many modern STS go into the category of complex (large) systems, described by a large number of parameters. For example, the control system of a modern nuclear power plant is connected with the input and processing of up to 10.000 signals. For complex systems, an essential characteristic that determines its effectiveness is the structure [7]. In this paper we are talking about a multistructure [8].

Effective implementation of a large volume of analytical work, the use of various models and methods of research, modeling and design, including artificial intelligence (fuzzy sets, genetic algorithms, neural networks, ontologies) is impossible without the methodology of their integration into a single conceptual framework (CF) with software tools for automation of design and research works [4, 6].

The authors developed a methodology for the conceptual (system) design of complex STS based on the scientific-methodical framework for conceptual design (CF), the general algorithm of which is shown in figure 1.

The methodology is based on the following statements.

1. The process of research and design of complex systems is presented in the form of a sequential iterative scheme for the step-by-step search for rational design solutions using heuristic optimization methods that can be tuned to the sequence of early stages, phases and tasks of STS information technology design in accordance with [3]. But the structure of CF can be adapted to any methodology of conceptual design.

2. The knowledge base of the methodology is presented in the form of a set of ontologies describing the automatized company and the complex of software and hardware of industrial computers

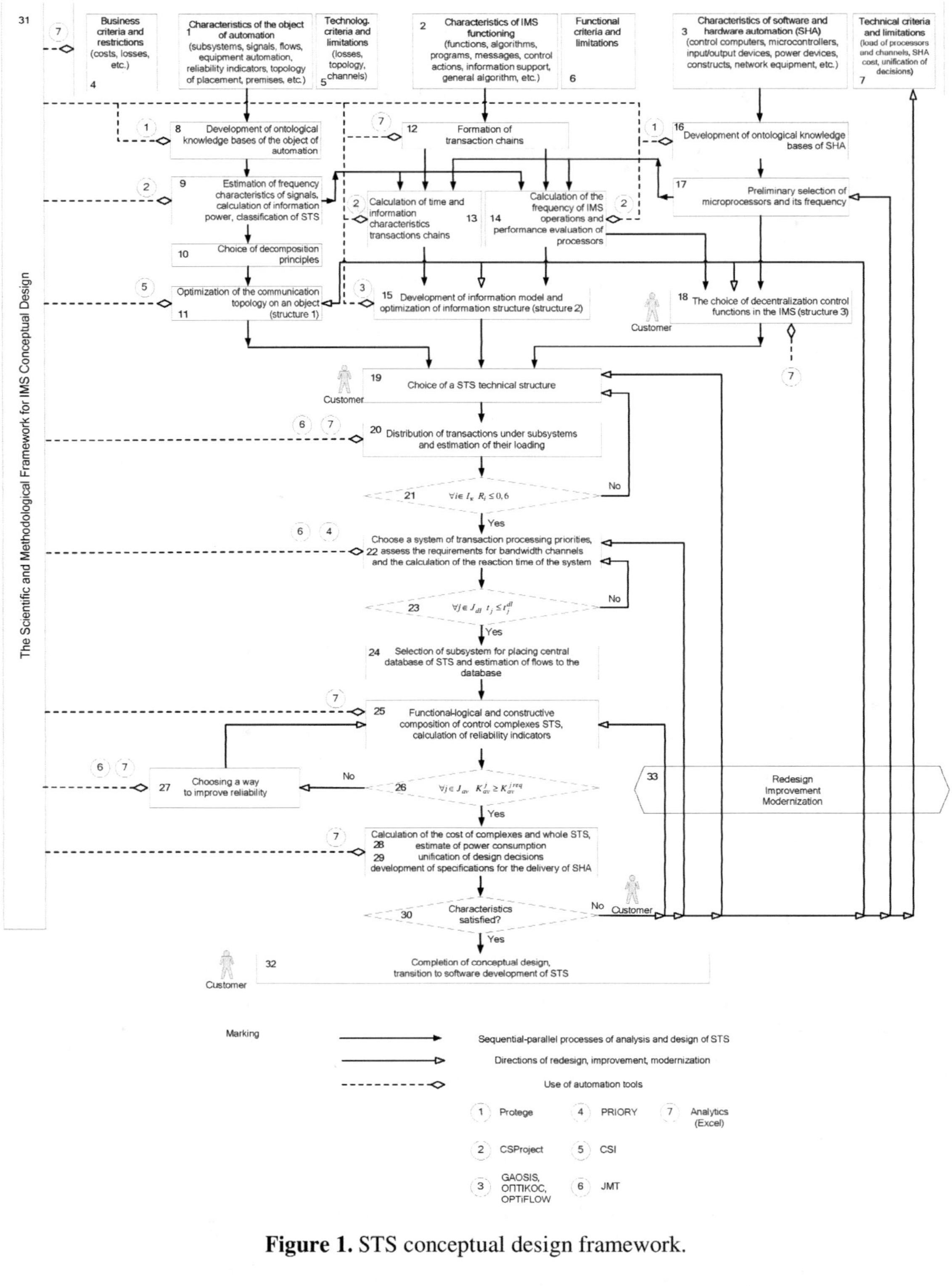

Figure 1. STS conceptual design framework.

(HSA). For example, HSA Advantech [9] is described to which development of automated systems in Germany, Russia and Ukraine is oriented.

3. The methodology is adaptive to the set of initial data available to the developer when analyzing and designing a particular system. This means that in the knowledge base, the sequence of design stages and procedures for which there is all the necessary data (either in the knowledge base or entered by the designer in the dialogue mode) is activated.

4. The methodology includes software tools to automate the work performed. The program CSI (complex system integrator), the program CSProject for the construction and calculation of information-time characteristics of real-time computer systems' transactions, the program OPTiFLOW for distribution of information flows in real-time systems, the program OPTICOS for optimization of information control systems, the program GAOSIS (based on the genetic algorithm) to optimize the structures of information systems and the program PRIORY to select priorities of application flows in STS. In addition, tools have been developed to automate the selection of design solutions under uncertainty. To create and edit ontologies in the form of OWL databases, a special program has been developed that makes it possible to use Cyrillic.

5. The initial data and results of system analysis and design are given in table-graphical forms. The initial data for creating STS are (Figure 1):

block 1 - the characteristic of the object of automation: a description of the organizational and technological characteristics of the system that is designed, including the technological subsystems, their structures, sections, flows and characteristics of signals in the sections, mechanization tools and equipment of the bottom automation, the layout of their location, buildings and premises in the company, reliability of devices, the sequence of technological operations, which is performed by the operational staff of the system (flows, location, time); peripheral equipment, which should be serviced by STS;

block 2 - the characteristic of STS functioning (in terms of the customer): functions, tasks, algorithms or programs that the projected system should perform; for new systems, this description will be at the level of functions and tasks; for systems that are being improved or modernized are developed algorithms or programs for estimating the frequency composition of operations (commands); for all levels with different layers of completeness and accuracy, the components of information support (variables, arrays) are specified; For each signal (application), the control actions or messages transmitted to personnel are indicated; the description is made in the form of transactions in the dialogue with the CSI program;

block 3 - description of software and hardware automation tools SHA: control computers, microcontrollers, input-output, normalization and switching of signals, power supplies, network equipment, software, reliability and cost indicators of devices and modules;

blocks 4, 5, 6, 7 - describe the criteria and limitations in business: resource saving, costs, losses, system costs, reliability indicators that must be provided, technological constraints (topology elements, channels), functional and technical indicators (workload of processors and channels, cost of software and hardware, unification of solutions);

The design of the system begins with three interrelated directions. The first - blocks 8, 9, 10, 11: optimization of the communication structure at the automation object by the minimization criterion for the total length of communications with fixed channels (the construction of a minimum spanning tree). The second - blocks 12, 13, 14, 15: optimization of the STS information structure by the minimization criterion for the growth of the total information flow with decreasing the number of information links in the structure). The third direction - blocks 16, 17, 18: the choice of variants for the decentralization of functions in a hierarchical structure by the minimization criteria for the total losses and the cost of STS.

Within the framework of the first and third directions, the respective ontologies of the company are formed (the ontology for the automation object and the ontology for complex of SHA software and hardware).

The first ontology is used in blocks 9, 10, 11, 19, 24, 25, 29. The second ontology is used in blocks 17, 18, 24, 25, 29.

On the basis of the formed three variants of the structure, a variant of the technical structure (block 19) is chosen with the participation of the customer, which represents the current business interests (priorities) for the preselected type of microprocessor (block 17).

In the second direction, transactions (block 12) are generated on the basis of the initial data (block 2) and their time and information characteristics (block 13) are calculated to optimize the information structure (block 15). In addition, the frequency composition of the operations solved in the system of tasks is calculated, and the performance of the processors in the MIPS and in transactions/s for finding optimal hierarchical structures (block 18).

For the chosen variant of the technical structure, the transactions are distributed among the subsystems and their load is estimated (blocks 20, 21). If this restriction is performed for a distributed STS, then the system response time is calculated for each transaction, the optimal priority is selected and channel bandwidth requirements are calculated (block 22). Further, when the time constraints are satisfied (block 23), based on the data of block 13, a subsystem is selected to place the central database in the decentralized STS (block 24).

At the next phase (block 25) for each subsystem, the functional-logical and constructive composition of the corresponding computer complex is done using the SHA ontology (block 16). Besides, technology and communication facilities are selected in the system in the same ontology. These means must match the requirements for the channel capacity (block 22). Estimation of availability factors is carried out for the obtained technical structure. If availability factors are not provided (block 26), a weak element reservation scheme is selected (block 27).

If the required reliability is achieved, the unification of the designed hardware-software decisions is carried out for the STS and the specifications for the SHA purchase are prepared, the cost of the complexes and the whole system, the power consumption and system performance are calculated (block 29). The design results are considered by the customer (block 30) and, if they suit it, the system design is completed and the development of the software for STS begins (block 32). If "No", the direction of redesign is selected (block 32). improvements or modernization of the system begins from the same place. Figure 1 shows the names of the tools used for the research and conceptual design of STS.

3. Conclusion

The conceptual design of real-time information management systems (IMRTS, now STS RT) in the conditions of increasing their dimensionality and multi-structural complexity is a necessary stage of any development. In the process of system (conceptual) design, it is important to represent the STS architecture as a set of interrelated types of support that are oriented towards the achievement of specific business goals. The proposed CoDeCS-2020 methodology is complemented by a set of tool programs for solving practical engineering tasks that can be used autonomously and can be adapted to other design standards.

References

[1] Zhukovytskyy I V 2014 The structural diagram of a digital control system of cut of cars brake control by means of retarder position, *Information and Control Systems on Railway Transport* **4** 78-81 (in Russian)

[2] INTEND. D2.1Transport projects & future technologies synopses handbook, (Version 0.4, 25/04/2018) (2018)

[3] GOST 34.003-90 1990 *Information technology. Set of standards for automated systems. Automated systems. Terms and Definitions* (in Russian)

[4] Bobrovskiy V I, Kozachenko D M, Vernigora R V and Malashkin V V, 2010 *Models, methods and algorithms for automated design of railway stations* (Dnepropetrovsk: Publishing house Makovetskiy) (in Russian)

[5] Trutnev D R *2012 Information systems architecture. Design Basics: Textbook* (St. Petersburg: NRU ITMO) 66 (in Russian)

[6] Eckert C I, Stacey and Kelly M 1999 *Artificial Intelligence for Engineering Design, Analysis*

and Manufacturing **13** 303-320

[7] Gelfand I M 1962 On some methods of controlling complex systems, *Uspekhi Matematicheskikh Nauk* **Vol. 17 No. 1 (103)** 3-25 (in Russian)

[8] Kosolapov A A 2017 Multifunctional Intelligent Management *Modern Information and Communication Technologies in Transport, Industry and Education* Abstracts of the XI International Scientific and Practical Conference (Dnipro, December 13-14, 2017) p 62 (in Russian)

[9] Advantech Industrial Automation 2018 Application Stories. Vertical Market Solutions. http://www2.advantech.com/industrial-automation

[10] Kosolapov A A 2019 Resource-Saving Method of Forming Information Infrastructure of Sorting Stations. Published online: DOI: https://doi.org/10.1051/matecconf/201929401011

IOP Conf. Series: Materials Science and Engineering 1021 (2021) 012039 doi:10.1088/1757-899X/1021/1/012039

Study on the prospects for the use of lubricants produced by Fuchs Lubritech GmbH on the Ukrainian railways in "wheel of rolling stock-rail" tribocoupling

Andrii Kravets[1,4], **Vitaliy Vlasovets**[2], **Andrii Yevtushenko**[1], **Yevgen Romanovych**[1], **Anna Kravets**[3]

[1] Construction, Track and Handling Machines Department, Ukrainian State University of Railway Transport, Feuerbach sq. 7, 61050 Kharkiv, Ukraine
[2] Educational and Scientific Institute of Mechanotronics and Management Systems, Kharkiv Petro Vasylenko National Technical University of Agriculture, Moscow ave. 45, 61050 Kharkiv, Ukraine
[3] Cargo and Commercial Work Management Department, Ukrainian State University of Railway Transport, Feuerbach sq. 7, 61050 Kharkiv, Ukraine

[4] E-mail: kravets_am@ukr.net

Abstract. Controlling friction and wear of rolling stock wheels and rails is a very urgent task for railway workers and scientists in many countries of the world, including Ukraine. The use of lubrication of the wheel-rail friction pair is rather a promising way to solve this problem, and the success of the solution depends significantly on the choice of an effective lubricant. The article considers the results of study of Locolub ECO and Tramlub F 234 MOD 2 greases, taking into account the prospects for their use in locomotive wheel flange lubricators that are operated in the Ukrainian railway. The results of laboratory and operational tests of these greases in comparison with the currently used Relsol lubricant are presented and analyzed. Quite high tribological characteristics of Locolub and Tramlub lubricants and their compliance with the operating conditions of rolling stock, wheel flange lubricators and track lubricators are noted. Based on the results of operational tests, recommendations are given on the prospects for the use of the studied lubricants in the Ukrainian railway transport.

1. Introduction

The problem of wear of the wheel flanges of rolling stock and the gauge face of rails is very relevant for railway transport in many countries [1, 2], since it affects the dynamics of rolling stock movement, its service life, noise characteristics of movement, and cargo safety [3, 4, 5]. In Ukraine, this problem has recently been becoming ever more urgent, primarily due to the increased severity of the operating conditions of rolling stock [18], and, accordingly, the track, as well as the fact that this problem has been neglected for a long time in the national railway.

This problem can be controlled in the traditional way – by applying grease in the "rolling stock wheel – rail" friction pair (lubrication) [4, 5, 6]. This method has demonstrated positive results on rail transport in many countries [1, 2, 7]. To ensure the best effectiveness of the use of lubricants on the Ukrainian railway, it should be taken into account that the factors that have the most significant impact include:

- rational choice of lubricant;

- rational organization of the lubrication process;
- correct operation of the lubricant feed system.

The second factor, which is purely organizational and technical, is the correct choice of the lubricant feed system (lubricator), taking into account the local operating conditions of the track and rolling stock. Currently, there is a certain parity between the use of locomotive wheel flange lubricators and stationary track rail lubricators at the Ukrainian railway. Although it should be noted that locomotive wheel flange lubricators are more commonly used on trunk routes, while rail lubricators are utilized on station tracks and in mountain areas, where there are many track curves of small radius.

Concerning the third factor, it is purely technological and depends primarily on the “culture of utilization” of the lubricants already in use.

It is due to the first factor that can be used to control the efficiency of the entire lubrication process without making changes to existing designs of lubricant feed systems and operating modes of rolling stock and track. For this purpose, an effective lubricant must be selected taking into account the following requirements [6]:

- good performance within a wide temperature range;
- high tribological (anti-wear) properties;
- high adhesive capacity;
- easy to pump through the lubrication system;
- no impact on the performance of components and parts of the lubrication system;
- good environmental properties (biodegradability).

Lubricants for the “rolling stock wheel-rail” friction pair are now manufactured by almost every well-known manufacturer of lubricant oils. There is a wide variety of lubricants, and the essence of the process of rational selection is a deep analysis of the local operating conditions of a specific friction pair; analysis of the design features of the lubrication systems that are already available; conducting necessary preliminary studies to justify the feasibility of using this particular lubricant.

2. The purpose and objectives of the study

Relsol greases, which are currently used on the Ukrainian railway in locomotive wheel flange lubricators, have much poorer performance than that achieved on other railways in the world where other more modern lubricants are used.

It is obvious that in order to reduce the operating costs of the locomotive facilities associated with the operation of the wheel-rail friction pair, it is now reasonable to consider applying other, more up-to-date lubricants to protect parts from wear. One of the world's recognized leaders in this industry is the manufacturer of lubricants Fuchs Lubritech GmbH. Confirmation of the effectiveness of utilization of this manufacturer's greases in the wheel-rail friction pair is their sustained use on the railway in Austria, Switzerland, Germany and other countries of the world.

The purpose of the study which will be discussed in this article was to justify the feasibility of using Locolub ECO and Tramlub F 234 MOD 2 greases by Fuchs Lubritech GmbH in locomotive wheel flange lubricators, which are installed in locomotives of the Ukrainian railway.

To achieve this goal, comparative tests of these greases and Relsol lubricants were carried out both in laboratory and in real operating conditions of locomotive wheel flange lubricators.

3. The results of laboratory studies

Tests of lubricants in laboratory conditions consist in the study of their chemmotological and tribological properties. In this case, the performance features of promising lubricants Locolub and Tramlub were compared with those of the currently used Relsol lubricant. During the study, the authors focused on the already established ideas about the quality indicators of lubricants for the wheel-rail friction pair [8] and took into account the experience of other researchers in this field [9]

As for the comparison of chemmotological indicators, the complete results of which have been already published [10], it should be noted that the properties of Locolub and Tramlub greases quite suit the operating conditions of locomotive wheel flange lubricators and can provide good pumpability in

IOP Conf. Series: Materials Science and Engineering 1021 (2021) 012039 doi:10.1088/1757-899X/1021/1/012039

the system, even at quite low temperatures, as well as adhesion on working surfaces of tribocoupling, corrosion protection of friction surfaces and parts and components of the system wheel flange lubricator and other performance properties.

To evaluate the ability to protect the wheel–rail friction pair from wear and reduce the energy cost of overcoming the friction forces in it, tests using special friction machines are more remarkable. Such tests were carried out using quite well-known and widely used methods, namely, "four balls in a pyramid" and "block-roller" [11, 12, 13].

Tests on the four-ball friction machine were carried out according to the standardized methods of ASTM D2596-15 [14] and ASTM D2266-01 (2015) [15]. The scheme of the tribocoupling during testing, which is shown in figure 1, implements the contact of friction surfaces at a point and simulates the operation of higher kinematic pairs. Several tribological parameters were determined on the four-ball machine, among them: welding load and critical load, which determine the limits of the maximum performance of the lubricant; load wear index, which characterizes the properties of the lubricant to protect surfaces under boundary friction; the diameter of the wear scar on the surface of fixed balls, which determines the anti-wear properties of greases.

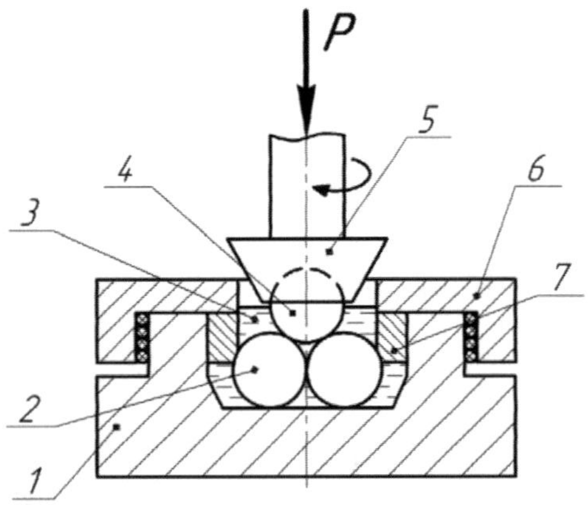

Figure 1. Test scheme on the four ball friction machine: 1 – cup; 2 – bottom fixed ball; 3 – tested lubricant; 4 – upper moving ball; 5 – collet clamp (cartridge); 6 – nut; 7 - clamping washer.

Based on the results of study on the 4 ball friction machine (table 1), the following was established:

- the diameter of the wear scar when loading on the upper ball was P=196 N is almost the same for Tramlub and Relsol greases, but it is almost 1.6 times larger for Locolub grease;
- the diameter of the wear scar when loading on the upper ball is P=392 N is 10% larger with greasing with Relsol than when greasing Locolub oil and almost 34% larger than with Tramlub
- the load wear index with Tramlub grease is almost 15% better than with Locolub, and 40% better than with Relsol;
- in terms of critical load, the lubricant exceeds Locolub by 1.33 times, and Relsol by almost 1.6 times;
- in terms of welding load, Tramlub has 1.58 times better performance than Locolub and 2.35 times better than Relsol;

Table 1. Test results on the four-ball friction machine.

Quality indicator	Locolub	Tramlub	Relsol
diameter of the wear scar at a load of 196 N, mm	0.76	0.47	0.48
diameter of the wear scar at a load of 392 N, mm	0.88	0.65	0.98
load wear index, N	429	502	306
critical load, N	980	1303	823
welding load, N	2323	3685	1568

In general, the analysis of the test results on the four-ball friction machine showed that Tramlub is superior in all respects to both Locolub oil and Relsol greases. In addition, Locolub was inferior to Relsol only in terms of diameter of the wear scar at a load of 196 N. The results also show that Tramlub and Locolub greases demonstrate their anti-wear and anti-friction properties better at higher loads, while Relsol practically does not work at high loads and, accordingly, hardly protects the tribocoupling from friction and wear.

Tests according to the "block-roller" scheme (figure 2) were carried out on the SMC-2 friction machine according to the generally accepted method [16]. This scheme implements the contact of the tribocoupling in the plane and simulates the operation of lower kinematic pairs. It provides a simulation of the operation of the tribocoupling under sliding friction, which takes place between the wheel flange of the rolling stock and the gauge face of the rail.

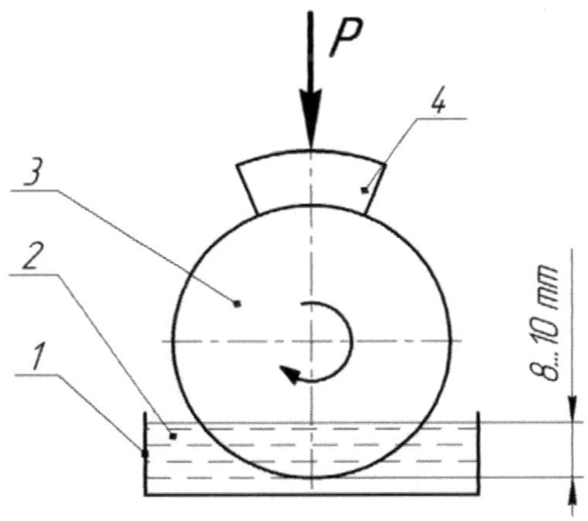

Figure 2. Schematic diagram of the "roller-pad" test on the SMC-2 friction machine: 1 – lubricant tray; 2 – lubricant level; 3 – roller; 4 – pad;

Since the force at the contact point of wheel and rail depends on many factors (such as curve radius, speed of movement, load of the rolling stock, the wheel profile, the degree of wear of both contacting elements [17, 18], the simulation of power characteristics of work of this friction pair is a very complex process. Therefore, to study the behavior of the grease in different operating conditions of the tribocoupling, tests were conducted in four programs (table 2) with different load modes. In each load mode, tests were performed four times for each grease, each time with a fresh portion of grease using new friction pair components, as recommended [16]. The rotation speed of the roller was 5 s^{-1} (300 min^{-1}).

Table 2. Test program on the SMC-2 friction machine.

Test time, min.	Load, N			
	mode I	mode II	mode III	mode IV
5	0	0	0	0
10	50	100	100	200
10	200	300	400	500
15	400	500	800	1000
20	800	1000	1500	2000
360	1600	2000	2500	3000
Σ=420				

The block and roller were made of steel and processed to a hardness that is approximately similar to the hardness of the surfaces of the wheel flanges of the rolling stock and the gauge face of the rails, which are used on the Ukrainian railway (hardness of the block is 360 to 370 HB; hardness of the roller is 280 to 300 HB).

IOP Conf. Series: Materials Science and Engineering 1021 (2021) 012039 doi:10.1088/1757-899X/1021/1/012039

The anti-wear properties of the tested greases were determined by the mass loss of the friction pair parts during the tests (the block and the roller in total). The anti-friction properties were evaluated by periodical monitoring the surface temperature of the roller during the entire time of each test.

Studies on the SMC-2 friction machine showed that the total wear of both samples (blocks and roller) (figure 3) lubricated with Tramlub grease is less than of those lubricated with Locolub and Relsol. In addition, the performance of Locolub was worse than that of Relsol only in the first test mode - at fairly insignificant contact loads. However, at the lowest test loads, the difference between the anti-wear properties of all lubricants is only minor - 1.15-1.38%, which is not at all representative and is within the permissible error for such measurements. At the same time, it is remarkable that, as in the case of tests on the 4 ball friction machine, the anti-wear properties of Tramlub and Locolub greases are more intensively implemented at greater contact loads, which is clearly evidenced by the graphs in figure 4. In load mode IV, the wear of parts greased with Relsol is 28% greater than of those greased with Locolub and 37% greater than of those greased with Tramlub.

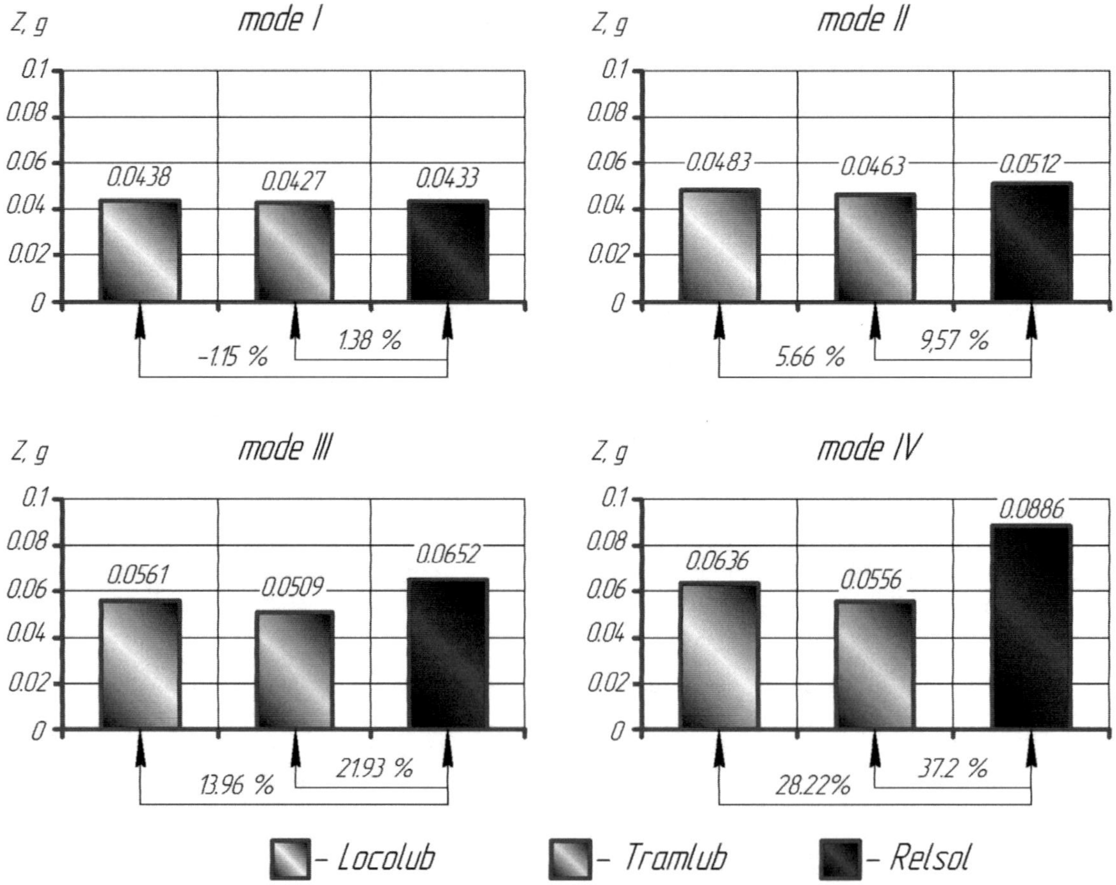

Figure 3. The average total wear Z of blocks and rollers on the SMC-2 friction machine at different load modes (Table 2.)

The best anti-wear properties of Tramlub oil are also proved by the appearance of the friction surfaces after testing (figure 5). A digital microscope photograph of the surface of a roller that was lubricated with Relsol (figure 5, b) shows numerous signs of adhesive wear, which are not observed at all on the surface of a roller that was lubricated with Tramlub (figure 5, a). The wear marks on Figure 5a are more uniform with fewer deep furrows, which indicates a more stable friction mode and, accordingly, better anti-friction and anti-wear properties of Tramlub grease.

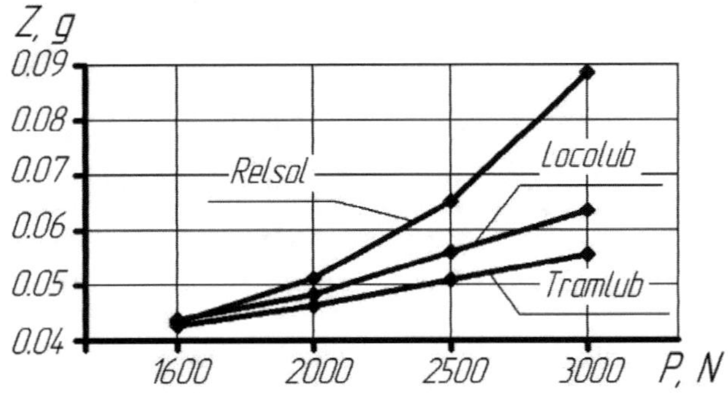

Figure 4. Change in the average wear of samples depending on the load mode

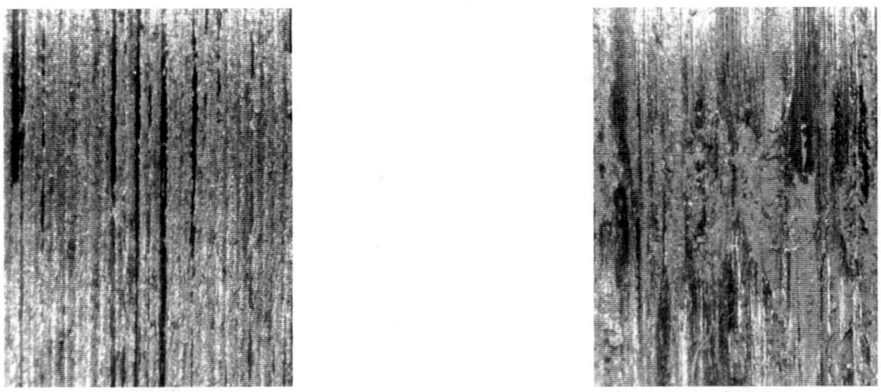

a) b)

Figure 5. The surface of the rollers which were lubricated with Tramlub grease (a) and Relsol (b) when testing on SMC-2 in loading mode IV

Control of the surface temperature of the rollers showed that in all tests, the highest temperature was on the rollers that were lubricated with the Relsol lubricant. The graph of temperature changes in the roller surface in load mode IV (averaged results of four tests), shown in figure 6, demonstrates that when the temperature of the roller surfaces lubricated with Locolub grease was 10-13°C lower, and of those lubricated with Tramlub - 15.5-19°C than when lubricated with Relsol. A similar trend was observed in other tests, only that the temperature values were slightly lower (in mode III by 7-11%, in mode II by 18-24% , and in mode I by 29-37%), which is quite natural. At the same time, with an increase in the maximum load in the friction pair, the difference in surface temperatures became more significant, i.e., the anti-friction properties of Locolub and Tramlub greases, as well as their anti-wear properties, manifested themselves with greater intensity. This suggests that even better results can be expected when these lubricants are used in a real wheel-rail friction pair, where the contact forces can be significantly greater than those experienced on the SMC-2 friction machine.

In addition, it should be noted that during the tests on friction machines, the worn particles of the tribosurface are removed from the friction zone together with the lubricant. In real operating conditions, the worn particles in the "wheel-rail" friction pair will in fact stay on the surfaces together with the lubricant and be involved in the further work of this tribocoupling. In view of this, the best anti-friction and anti-wear properties of the lubricant (for example, Tramlub) will lead to the prevention of adhesive wear and reduce the destruction of surfaces during abrasive and other types of wear, i.e. the formation of finer wear particles. As it is known, when such fine particles fall into the clearance of the friction pair, they create a positive effect on its operation, separating the surfaces, taking on the load and reducing the wear rate [19].

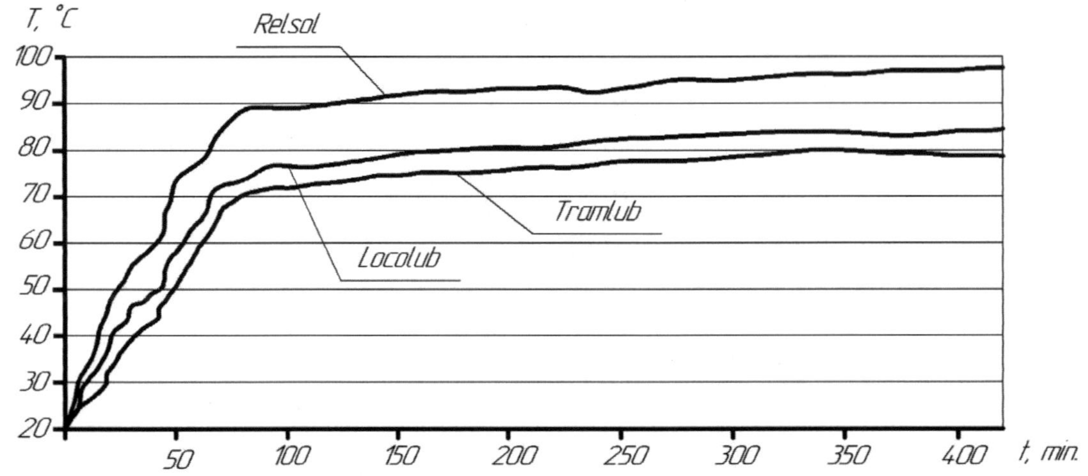

Figure 6. Change in the roller surface temperature during testing on SMC-2 in load mode IV

4. Operational tests

To confirm the results of laboratory tests and justify the feasibility of using Tramlub and Locolub greases in locomotive lubricators on the Ukrainian railway, these greases were tested in real operating conditions of wheel flange lubricators and rolling stock.

Operational tests were carried out on six two-unit locomotives (4 VL-11 locomotives and 2 VL-10 locomotives). The required number of study objects was determined using well-known methods [20]. Also, the number of locomotive wheels that are exposed to the studied lubricants, and, accordingly, their wear control, as well as the number of wheel flange lubricator systems on each locomotive were taken into account. On three locomotives, the wheel flange lubrication systems were filled with Tramlub grease and on another three – with Locolub. The mileage of the locomotives during the tests was about 150,000 km. The test lasted for 16-20 months, so that all seasons were covered and the influence of various natural factors on the performance of the tested oils were studied. During the entire time of testing, there were no issues with the operation of wheel flange lubricators with the studied lubricants.

During operational tests, the geometric parameters of the locomotive wheel flanges were controled, such as the height, thickness and dangerous shape of the flange, the thickness of the tyre, and the total amount of grease consumed was recorded. Based on the results of measurements of locomotive wheel pairs, the specific change in the average thickness of the flange (wear rate) and the specific consumption of lubricant (in relation to the mileage of the locomotive) were determined. In general, processing of the data complex obtained allowed us to determine a number of parameters related to the effectiveness of the studied lubricants in the wheel-rail friction pair and the operation of the locomotive wheel flange lubricators (table 3). For comparison, the results of the application of Relsol grease were taken from publicly available information about the use of locomotive wheel flange lubricators at the Ukrainian railway transport.

The wear rate of the tyre flanges lubricated with Tramlub oil vs. Relsol decreases by 27.12% on VL-10 electric locomotives, and by 20.65% on VL-11 electric locomotives. However, Locolub ECO grease showed poorer results and the wear of the flanges with it was greater than with Relsol grease, by 7.6% for Vl11m electric locomotives and by 28.81% for VL10 electric locomotives. The trend in the ratio of average thickness loss rates of tyres is preserved, but in somewhat other proportions.

Knowing the average wear rate, the average predicted life of locomotive tyres was estimated, provided that the average thickness of the new tyre is 95 mm (figure 7). Estimations show that when Tramlub grease is used, an increase of the service life of tyres of VL-11 electric locomotives by approximately 16.37%, and VL10 - by 33.76% can be expected; at the same time, the use of Locolub oil can lead to a decrease in the service life of tyres of Vl11m by 1.75 %, and VL10 – by 17.04 %.

Table 3. Comparison of the results of using various lubricants in the wheel flange lubrication systems of locomotives

Operational indicator	Relsol		Tramlub			Locolub		
	VL11m	VL10	VL11m	VL11m	VL10	VL11m	VL11m	VL10
1. Mileage of the locomotive during the test, km	-	-	153,878	150,224	50,235	159,895	158,863	97,996
2. Average mileage between tyre returnings, km	15,608	20,130	17,097.6	18,778	36,391	15,989	14,442.1	19,852.8
			17,938			15,215.5		
3. Wear rate of wheel flange, mm/10 thousand km	1.84	1.18	1.414	1.506	0.86	1.923	2.037	1.52
			1.46			1.98		
4. Thickness loss rate of the tyre, mm/thousand km	0.319	0.174	0.267	0.285	0.132	0.318	0.338	0.213
			0.276			0.328		
5. Service life of the tire, thousand km	171	311	206 [a]	193[a]	416 [a]	173 [a]	163 [a]	258 [a]
			199 [a]			168 [a]		
6. Specific lubricant consumption rate, l/thousand km	0.343[b]		0.377	0.346	0.338	0.294	0.264	0.245
			0.361			0.279		

[a] - estimated value, with an average thickness of a full tyre of 95 mm;
[b] - estimated indicator for the average locomotive speed of 35 km/h and the grease feed rate of the wheel flange lubrication system twice a minute.

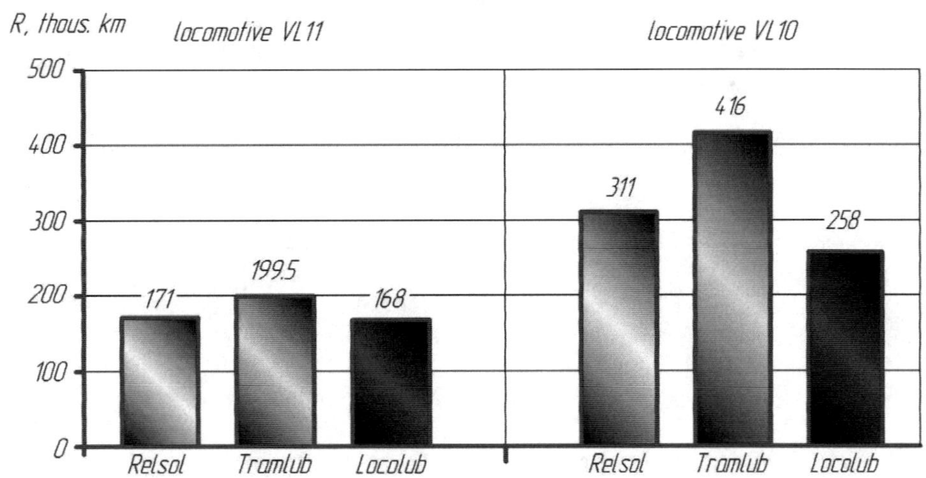

Figure 7. The estimated service life of wheel tyres of locomotives

5 Conclusion

The conducted study allows us to state the following:

1) In general, when the Relsol grease is used in locomotive wheel flange lubricators, it performs the functions of protecting the wheel-rail tribocoupling from friction and wear, but its effectiveness is insufficient, especially in view of the current operating conditions of rolling stock and track;

2) Locolub ECO and Tramlub F 234 MOD 2 greases manufactured by Fuchs Lubritech GmbH are proposed for use and have chemmotological characteristics that match the operating conditions of the wheel-rail friction pair and locomotive wheel flange lubricators, and meet the requirements for this type of grease.

3) The tribological properties of Locolub and Tramlub lubricants defined on friction machines are better than those of Relsol lubricants. In addition, these advantages are more vivid at higher loads on the friction pair.

4) Performance testing showed definite advantages of Tramlub grease vs. Relsol. However, the Locolub lubricant performed somewhat worse and lost to Relsol in the ability to protect the wheel-rail friction pair from wear in real operating conditions.

5) Quite an important factor in modern conditions is 70-90% biodegradability of Locolub and Tramlub greases declared by the manufacturer, which positively affects the environmental aspects of the use of such greases and is consistent with the modern global trends [12, 13].

6) In general, the obtained results allow us to recommend and state the prospects of using Tramlub F 234 MOD 2 oil in locomotive wheel flange lubricators in existing railway transport in Ukraine in the conditions of track and rolling stock operation.

6. References

[1] Lukyanenko V 2016 Use of modern technologies for rail lubrication *Railway transport* **8** pp 54-55

[2] Extend the life of wheels and rails by 10 times 2016 *Car Fleet, Railway publishing house «Rolling stock»* **11-12** (116-117) pp 4-5

[3] Csortos G and Augusztinovicz F 2018 Noise Mitigation with Rail Lubrication Device on Tramline *Road and Rail Infrastructure* **5** pp 1173-1179

[4] Fryza J and Omasta M 2016 The Experimental Determination of the Grease Amount to Effective Wear Reduction in the Wheel-Rail Contact *Latest Methods of Construction Design* pp 127-132

[5] Siedlowsky F 2015 Automatische Schienenschmierung gegen Lärm und Verschleiß *Eisenbahntechnische Rundschau* **3**, pp 31-34

[6] Stock R, Stanlake L, Hardwick C, Yu M, Eadie D and Lewis R 2016 Material Concepts for Top of Rail Friction Management - Classification, Characterisation and Application *Wear* **366** pp 225-232

[7] Oldknow K 2012 Controlling Friction Delivers Longer Rail Life *Railway Gazette International*, **1** pp 53-55

[8] EN 16028:2012. *Railway Applications. Wheel/Rail Friction Management. Lubricants for Trainborne and Trackside Applications. Started Acting With* (European Committee for Standardization)

[9] Razak IHA, Ahmad MA and Puasa SW 2019 Tribological and Physiochemical Properties of Greases for Rail Lubrication *Tribology Online* **14**, **5** pp 293-300

[10] Kravets A, Yevtushenko A and Pogrebnyak A 2018 New Materials for Lubrication of a Friction Pair "a Wheelset – a Rail". Results of Laboratory Studies *Railway Transport of Ukraine* **2** pp 22-30

[11] Kombalov V S 2008 *Methods and Means of tEsting for Friction and Wear of Structural and Lubricants: a handbook*, ed K.V. Frolov and E.A. Marchenko (Moscow: Engineering) p 384

[12] Ermakov C, Chmykhova T and Timoshenko A 2019 Tribological Features of Environmentally Safe Lubricant Compositions Based on Rapeseed Oil *Journal of Friction and Wear* **40**, **2** pp 194-199

[13] Fabio Perez-de Brito A, Ponce S and Francisco Perez-Robles J 2020 Linseed and Complex Rosin Ester Oils Additivated with MWCNTs and Nanopearls for Gears/Wheel-Rail Systems European Journal of Lipid Science and Technology **122**, **2** Art. number 1900331

[14] ASTM D 2596-15 2015 *Standard Test Method for Measurement of Extreme-Pressure Properties of Lubricating Grease (Four-Ball Method)* (ASTM International)

[15] ASTM D 2266-01 2015 *Standard Test Method for Wear Preventive Characteristics of Lubricating Grease (Four-Ball Method)* (ASTM International)

[16] Kravets A, Kravets V and Afanasov H 2011 *Investigation of Antiwear Properties of Lubricants*

Using a Friction Machine SMTs-2: Methodical Instructions (Kharkiv : UkrSART) p 26

[17] Demchenko V 2015 Theoretical Studies and Calculations of Interaction Forces Between the Way the Wheels Rolling of Different Degrees of Wear and Tear *Reporter of the Priazovskyi State Technical University*. Section: Technical sciences **30**, **2** pp 51-63

[18] Ushkalov V, Mokriy T and Malysheva I 2015 Mathematical model of interactions between railway and track considering distributions of contact forces throughout contact spots *Technical Mechanics* **2** pp 79-89

[19] Ventsel Ye, Orel O, Shchukin O, Saienko N and Kravets' A 2018 Dependence of Wear Intensity on Parameters of Tribo Units *Tribology in Industry* **40**, **2** pp 195-202

[20] Vlasov K P 2013 *Methods of research and organization of* experiments (Harkov: Publishing House Humanitarian Center) p 412

IOP Conf. Series: Materials Science and Engineering 1021 (2021) 012040 doi:10.1088/1757-899X/1021/1/012040

Simulation of the oil film thickness on a friction surface in the presence of fullerene compositions in the lubricant

A Kravtcov[1], Y Gradiskiy[2], B Tsymbal[3, 5], K Borak[4]

[1]Department of Transport Technologies and Logistics, Kharkiv Petro Vasylenko National Technical University of Agriculture, Kharkiv, 61002
[2]Department of Woodworking Technologies and Systems Engineering of Forest Complex, Kharkiv Petro Vasylenko National Technical University of Agriculture, Alchevsky str. 44, Kharkiv, 61002
[3]Department of occupational, technogenic and environmental safety is a leading subdivision, National University of Civil Defence of Ukraine, Chernyshevskaya str. 94, Kharkov, 61023
[4]Deputy Director for Academic Affairs, Zhytomyr Agrarian and Technical College, Pokrovska str. 96, Zhytomyr, 10031

[5]E-mail: tsembalbogdan@ukr.net

Abstract. The paper contains theoretical studies of the oil film formation on a friction surface in the presence of fullerene solutions in a lubricant. The mathematical model is developed on the basis of the interaction between electrically active heterogeneous finely dispersed systems at the interface between the friction surface and the lubricating medium and is described by the Poisson differential equation. The role of the friction surface on the formation of clusters and micelles in an adsorbed film of the lubricant near the friction surface is established. It is shown that in the stress-strain state of triboelement material surface layers, the friction surface acts as an "electrostatic field generator". Expressions are obtained for the simulation of the oil film thickness in the presence of fullerenes in the base lubricant. Based on the simulation results, the most significant factors in the formation of the oil film thickness have been established: these are the electric field strengths of the friction surface and the base lubricating medium; the load on the tribosystem; the temperature, and dynamic viscosity of the lubricating medium.

1. Introduction

Today, the use of nanocarbon additives and supplements (C_{60} fullerenes) is of great interest [1]. The work [2-4] presents the results of studies devoted to the effect of fullerene additives to lubricants on the processes of friction and wear of metals under boundary lubrication conditions; a conclusion is made here about the prospects of using such additives.

The complexity of this problem is that it is impossible to study the mechanisms of interaction between active elements of a lubricating medium, for example, fullerenes in the process of friction and their interaction with the friction surface. At the same time, during the tribosystem operation under the action of plastic and elastic deformation of a surface layer roughness and material, the friction surface acts as an "electrostatic field generator", which affects the ordering of the thin oil film structure and determines the film thickness.

Electrically active dispersed systems with a well-developed specific surface, which include clusters and micelles, particularly fullerene molecules, are one of the promising classes of modern materials. These materials will soon be used in the development of new lubricant compositions. The complexity of the structure of dispersed nanomaterial-based systems, as well as the presence of local inhomogeneities there determine the lack of an established understanding of the mechanism for reducing wear and friction losses in tribosystems where fullerenes are used in lubricants. In the presence of numerous interfaces in these complex systems, the total effect of interphase interactions becomes especially strong and is decisive for the generation of intrinsic electric fields. This interaction leads to the appearance of electrical forces, and, consequently, to the electrostatic field gradient at the friction surface.

2. Analysis of recent studies and publications

The analysis of the works devoted to the use of fullerenes as additives to lubricants made it possible to conclude that fullerenes are not dispersed in all technical oils, both mineral and synthetic [1, 2]. However, fullerenes are well dispersed by volume in vegetable oils, especially high oleic oils [6]. Therefore, using this phenomenon, it is possible to preliminarily disperse fullerenes in a high oleic vegetable oil, for example, rapeseed one, and then introduce this composition into a lubricant. Thus, it is possible to realize the mechanism of micelle formation near the electrostatically charged friction surface, which will allow the creation of strong surface layers.

In [5, 6], theoretical studies of the formation of an oil film on the friction surface when fullerenes are introduced into vegetable oil followed by their introduction into a lubricant have been carried out. The mathematical model is developed on the basis of the interaction between electrically active heterogeneous fine-dispersed systems at the interface between the friction surface and the lubricating medium and is described by the Poisson differential equation. The relationship between the electrostatic field on the friction surface and the electric field in the volume of the liquid is shown, as well as the role of the friction surface on the formation of clusters and micelles in a lubricant film at the friction surface. It was found that under the stress-strain state in surface layers, the friction surface acts as an "electrostatic field generator", which affects the formation of an electric field in the volume of an oil film with a micellar structure [9]. Expressions for calculating the strength of the total electric field in the system "friction surface + lubricant" are obtained.

3. Statement of the objective and tasks of the study

The purpose of this work is to carry out theoretical studies of the oil film thickness formation on the friction surface in the presence of fullerene compositions in the lubricant.

The structure of a mathematical model for the thickness formation of a lubricating film containing fullerenes on the friction surface of a tribosystem is described in [7, 8] and contains:

- Simulation of the electrostatic field strength of the friction surface during the tribosystem operation, which can be expressed as surface energy with dimension in J / m ^{2}, or as surface charge density with dimension in C / m ^{2};
- Simulation of the electric field strength within the lubricant volume, which is under the influence of the force field of the friction surface, and which contains fullerenes and other molecules of surfactants;
- Simulation of the oil film thickness formation on the friction surface of the tribosystem, which is formed due to the "stitching" of spatial aggregates, micelles and clusters, into a volumetric structure resembling a honeycomb structure.

The formation of "stitched" structures on the friction surface leads to a change in the arithmetic mean deviation of the profile points Ra , μm, and the average roughness width along the midline of the profile Sm , mm, for newly formed friction surfaces.

Determination of the dependences of the Ra decrease with a simultaneous increase Sm with a change in the oil film thickness allows reducing the wear rate and lowering friction losses during the operation of tribosystems.

4. Development of mathematical model

The most convenient method for determining the electric field strength is to solve the differential equation for the potential, which is presented in [7]:

$$\frac{\partial^2 \varphi}{\partial z^2} = -\frac{\rho}{\varepsilon \varepsilon_0} = -\left(E_n + E_k + E_m\right) \quad (1),$$

where φ is the electric potential, V; z is the coordinate perpendicular to the friction surface; ρ is Volumetric charge density; σ is surface charge density, C / m^2; ε is relative dielectric constant of the lubricating medium; ε_0 is a dielectric constant equal to 8.85 · 10^{-12}, C/V · m or F/m; E_n is the strength of the electrostatic field created by the friction surface during operation, V / m; E_n is the strength of the electrostatic field within the lubricating film volume due to the formation of clusters from fullerenes, V/m; E_n is the intensity of the electrostatic field within the lubricating film volume due to the formation of micelles from fullerenes, V/m.

The first term on the right-hand side of equation (1) determines the intensity of the electrostatic field created by the friction surface during operation:

$$E_n = \frac{\sigma}{2\varepsilon\varepsilon_0}, \text{ V/m,} \tag{2}$$

where σ is the surface charge density, C / m^2, which is determined by the formulas given in [7].

The second term on the right-hand side of equation (1) determines the electric field strength in the lubricating film volume due to the formation of clusters from fullerenes:

$$E_k = \frac{n_k p_k}{4\pi\varepsilon\varepsilon_0 d_k^3}, \text{ V/m,} \tag{3}$$

where n_k is the number of clusters per unit volume of the base lubricant, pcs.; p_k is the dipole moment of clusters, with dimension in Kl · m; d_k is the average distance between clusters and the friction surface equal to the average size (diameter) of the cluster, with dimension in m (metres).

The work [7] presents formulas for determining the value depending on the change in the concentration of fullerenes, the temperature of the lubricating medium, and the electrostatic field strength magnitude on the friction surface.

The third term on the right-hand side of equation (1) determines the electric field strength within the lubricating film volume due to the formation of micelles, where a fullerene molecule or a cluster of fullerenes acts as a nucleus, which surrounds attached molecules of a high-molecular acid (oleic, stearic, etc.) ... Such acids act as a "solvent" for fullerenes. The expression for determining the electric field magnitude due to the formation of micelles is written in the form:

$$E_m = \frac{n_m p_m}{4\pi\varepsilon\varepsilon_0 d_m^3}, \text{ V/m,} \tag{4}$$

where n_m is the number of micelles per unit volume of the base lubricant, pcs.; p_m is the dipole moment of the micelle, C · m; d_m is the average distance between the micelle and the friction surface, equal to the average size (diameter) of the micelle, with dimension in m.

The work [7] presents formulas for determining the value E_m depending on the change in the concentration of micelles, the temperature of the lubricating medium, and the electrostatic field strength magnitude on the friction surface.

During the functioning of the tribosystem, the process of cluster and micelle formation, as well as their destruction due to the influence of temperature, as well as the load and sliding speed, can occur simultaneously; therefore, the total electric field of the lubricating medium is determined as the sum:

$$E_g = E_k + E_m, \text{ V/m,} \tag{5}$$

Applying the dimensional analysis method, which is widely used in the theory of similarity and modelling, as well as experimental data, it is possible to obtain an expression for determining the thickness of the lubricating film, which will be formed on the friction surface under the influence of the electric force fields on the friction surface and in the liquid:

$$h = \frac{\mu(T) \cdot A_{тр} \cdot \upsilon_{скл}}{N} \left(\frac{E_n + E_{ж}}{\sqrt{E_n^2 + E_g^2}} \right), \text{ m,} \tag{6}$$

where h is the lubricating film thickness, m; $\mu(T)$ is the function of changing the dynamic viscosity of the base lubricant medium depending on temperature, Pa · s; A_{mp} is the actual friction area, which is the sum of the friction areas of all patches of actual contact, with dimension in m^2, is determined by the formulas given in the work [10]; $\upsilon_{скл}$ is sliding speed, m / s; N is the tribosystem load, N.

The operating temperature in the volume of the oil film T is determined by the expression:

$$T = T_{oc} \cdot \exp(W_{mp} \cdot f)\text{, °C,} \tag{7}$$

where T_{oc} is the ambient temperature, °C; W_{mp} is the rate of work of dissipation in the tribosystem, J/s; it is determined by the method described in the work [10]; f is the coefficient of friction; a dimensionless quantity, which is determined by the method described in the work [10].

The function of changing the dynamic viscosity of the lubricating medium with a change in temperature is determined by the expression:

$$\mu = \mu_0 \cdot \exp(-\alpha(T - T_0))\text{ , Pa s,} \tag{8}$$

where μ_0 is the dynamic viscosity of the lubricant at 100 ° C, Pa · s; α is the piezo coefficient equal to 0.013; T is the working temperature determined by the formula (7), °C; T_0 is the temperature at which μ_0 is determined, i.e. equal to 100 °C.

The result of the $\mu(T)$ change obtained using formulas (7) and (8) is substituted into expression (6) to simulate the change in the thickness of the lubricating film h.

5. The results of the calculation and their discussion

The results of simulation of the change in the oil film thickness on the friction surface depending on the strength of the electric fields of a friction surface and a liquid are shown in figure. 1. When analysing the dependences, it can be concluded that the magnitude of the electrostatic field strength on the surface E_n is a more significant factor than the magnitude of the electric field strength of the liquid E_g. As follows from figure. 1, at small values of E_n = 2,5 × 10^6 V / m, the film is of one oleic acid molecule thickness h = (1 - 7) × 10^{-11} m. With an increase in E_n = 7,5 × 10^6 V / m, the film thickness increases to h = 1 · 10^{-6} m. The obtained simulation results confirm that the friction surface, being a "generator of an electrostatic field", is a more significant factor in the formation of an oil film on the friction surface.

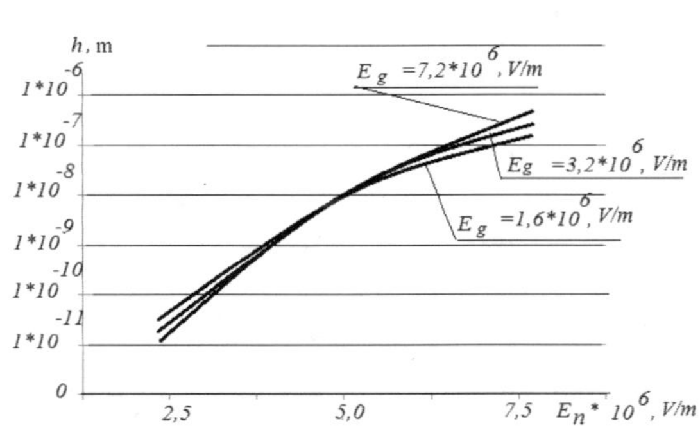

Figure 1. Dependences of the change in the oil film thickness on the electric field strengths on a friction surface and in a liquid.

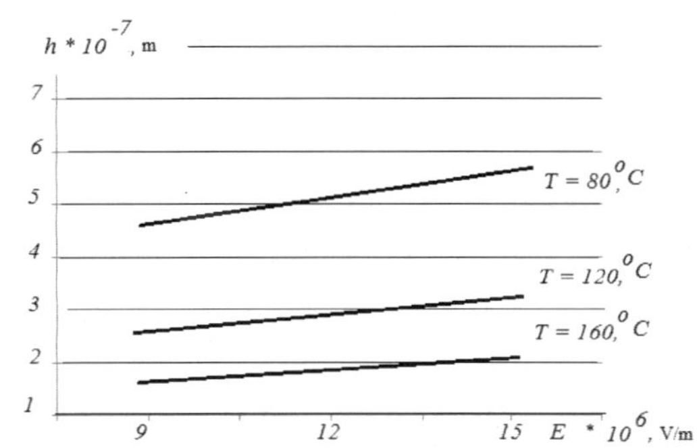

Figure 2. Dependences of the change in the lubricating film thickness on the magnitude of the total strength of electric fields and on operating temperature.

The degree of influence of the operating temperature T °C in the oil film volume, as well as the dynamic viscosity of the base lubricant μ, Pa·s on the lubricating film thickness depending on the total strength of the electric fields E=E_p+E_g is shown in Figure 2 and Figure 3.

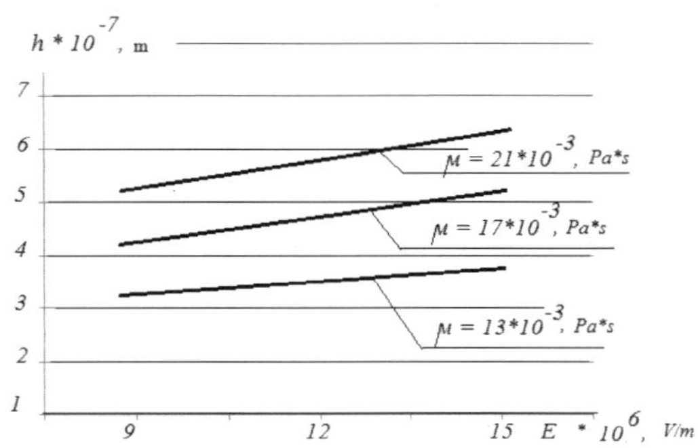

Figure 3. Dependences of the change in the lubricating film thickness on the total value of the electric field strengths and the dynamic viscosity of the base oil.

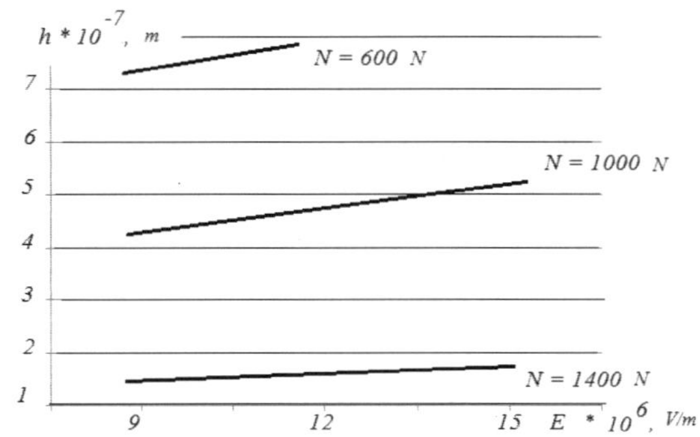

Figure 4. Dependences of the change in the lubricating film thickness on the total value of the electric field strengths and the load.

When analysing the dependences, it can be concluded that a change in temperature from $T = 80$ ° C to $T = 160$ ° C leads to a decrease in the film thickness from $h = 4.5 \cdot 10^{-7}$ m to $h = 2.5 \cdot 10^{-7}$ m. Similar dependences were obtained when changing the dynamic viscosity of the base oil, figure. 3.
The degree of tribosystem load influence on the lubricating film thickness is shown in figure. 4. It follows from the dependences that an increase in the load N, significantly reduces the oil film thickness.

As follows from the obtained theoretical dependencies, the most significant factors in the formation of the oil film thickness are: the electric field strengths on a friction surface and in a liquid; load on the tribosystem; temperature and dynamic viscosity of the lubricant.

6. Conclusions

It is shown that the electrical interaction in a lubricating medium of clusters and micelles, which include fullerenes, is described by the Poisson differential equation and its solution contains three components: the electrostatic field strength on the friction surface; the electric field strength in the volume of the liquid due to the formation of clusters; electric field strength in the volume of the liquid due to the formation of micelles.

The friction surface role in the formation of clusters and micelles in an adsorbed film of a lubricant near the friction surface is established. It is shown that in the stress-strain state of the surface layers of triboelement materials, the friction surface acts as a "generator of an electrostatic force field". Expressions are obtained for simulation of the oil film thickness in the presence of fullerenes in the base lubricant.

Based on the simulating results, the most significant factors in the formation of the oil film thickness were established: these are electric field strengths on the friction surface and in the base lubricating medium; the tribosystem load, the temperature and dynamic viscosity of the lubricating medium.

References

[1] Bezmelnitsyn V N, Yeletskiy A V, Okun M V 1998 Fullerenes in solutions, *Advancements of physical sciences* **11(168)** 1195-1220

[2] Ginzburg B M, Baydakova M V, Kireenko O F [et al.] 2000 Influence of fullerenes C_{60}, fullerene soot, and other carbon materials on the boundary sliding friction of metals, *Technical physics journal* **12(70)** 87-97

[3] Ginzburg, B M, Shibaev L A, Kireenko O F et al. 2002 Antiwear Effect of Fullerene C_{60} Additives to Lubricating Oils, *Russian Journal of Applied Chemistry* **75** 1330–1335

[4] Tuktarov A R, Khuzin A A, Dzhemilev U M 2020 Fullerene-Containing Lubricants: Achievements and Prospects, *Pet. Chem.* **60,** 113–133

[5] Yakhyaev N Ya, Begov Zh B, Batyrmurzaev Sh D A 2009 New lubricant composition for modification of surfaces of tribo-couplings in a small-sized marine diesel engine, *ASTU Bulletin. Series: Marine Engineering and Technology* **1** 47-52
[6] Semenov K N, Charykov N A, Arapov O V [et al.] 2010 Solubility of light fullerenes in some essential and vegetable oils, *Chemistry of vegetable raw materials* **2** 147-152
[7] Kravtcov A G 2017 Development of a mathematical model for the interaction between electrically active heterogeneous fine-dispersed systems at the interface “friction surface - lubricating medium”, *Problems of tribology* **1(85)** 89-99
[8] Kravtcov A G 2018 Simulation of the oil film formation on the friction surface in the presence of fullerene additives in a lubricant and the film influence on the wear rate of tribosystems, *Problems of tribology* **1(87)** 69-77
[9] Frolenkova L Yu, Shorkin V S 2013 Method of calculating the surface energy and the adhesion energy of elastic bodies, *Bulletin of PNRPU (Perm National Research Polytechnic University): Mechanics* **1** 235-259
[10] Voitov V A, Zakharchenko M B 2015 Simulation of the friction and wear processes in tribosystems under boundary lubrication conditions. Part 1. Calculation of the dissipation work speed in tribosystems, *Problems of tribology* **1** 49-57.

IOP Conf. Series: Materials Science and Engineering 1021 (2021) 012041 doi:10.1088/1757-899X/1021/1/012041

Use of modern logistics technologies in terms of saving resources

D V Lomotko[1], H O Prymachenko[1,4], O V Kovalova[2], O V Shkurenko[3] and Y I Hryhorova[1]

[1] Transport Systems and Logistics Department, Ukrainian State University of Railway Transport, Feuerbach sq. 7, 61050 Kharkiv, Ukraine
[2] Freight Commercial Railway Operation Management Department, Ukrainian State University of Railway Transport, Feuerbach sq. 7, 61050 Kharkiv, Ukraine
[3] Business Logistics and Transport Technologies Department, State University of Infrastructure and Technology, Kyrylivska str. 9, 04071 Kyiv, Ukraine

[4] Email: gannaprymachenko@kart.edu.ua

Abstract. The necessity of research of modern logistic technologies is proved. The definition of “logistics technology” is given. The connection of technological and logistical operations with the Pareto principle is shown. The logistic technology "just in time" is considered, which once appeared in connection with the changing situation in the Japanese market, its basic principles are investigated. It is shown how this technology “just in time” became the foundation for the development and practical implementation of new technology “lean production”. The historical periods of introduction of "lean production" in various branches and evolution of the technology are considered. Also, the technology of Lean transformation system is offered which is based on 5 questions to the company, which seeks to increase its efficiency and ensures understanding how to implement effectively the principles, purpose and concepts of Lean manufacturing to make continual improvements to the operations of an organization. Methods, methodology and tools of Lean were reviewed on practical examples.

1. Introduction

Today, the process of research of logistics technologies (LT) by both specialists and researchers is becoming increasingly important. LT is the main part of the process associated with the “birth” of the material flow, its movement (transportation) and storage, distribution and marketing [1]. LT must be understood as a complex of logistical operations or actions similar in direction, with the material flow in a specific time interval and space and associated with change and movement to achieve the logistical goal.

LT is a regulated sequence of logistics processes. LT combines technology and logistics operations [2]. There are operations related to the direct transformation of raw materials, semi-finished products and other into a finished product or service. These operations are associated with obtaining a material product and called technological. And there are operations that are part of the technological processes of obtaining a product or service related to the movement, transportation, warehousing, storage, forwarding of material flow on the way to its passage to the final consumer. They are called logistics [3].

2. Methodology for using logistics technologies

2.1. Problem statement

The specific complex of technological and logistical operations accords to the Pareto principle (20/80) [4]. According to this principle, only one-fifth (20%) of the total number of operations, which usually compiles the technological process, is aimed at obtaining (manufacturing) a material product or service, and the rest is about 80% of operations are logistics, transportation operations, moving, storing and so on [5].

Today, one of the most common LT in Ukraine is the technology "just in time" (JIT) [6].

JIT is the technology of the logistics system or organization of the logistics process in a particular functional area, which ensures the delivery of material resources, work in progress, final products in the required quantity, in due time and to the right place.

JIT appeared in connection with the changing situation in the Japanese market. It led to the fact that increase output stopped giving the former effect. This led to the intensification of the struggle for markets and the search for new methods of production management aimed at increasing profits. The result of this situation was a pragmatic approach, known as “just in time”.

2.2. Optimization technology "just in time"

JIT emerged after the First World War in the Japanese automotive industry. The founder of the system was the Toyota motor corporation. There is an element of chance in the emergence of this system. Automotive in that period formed from the giant US automaker - Detroit. However, the Japanese could not afford large investments to accumulate stocks unlike the Americans. Because inventories were seen as unnecessary losses due to shortage of funds. Obviously, this was one of the ideological lines of production by reducing costs of production. To reduce the cost of production must be quickly, accurately and flexibly adapting to fluctuations in consumer demand in the market. Therefore, the JIT system ensures the production of the required product in the required quantity and at the specified time. Obviously, there are three main principles for the system.

The first is that the delivery of resources is carried out to the places of processing exactly at the required time (semi-finished products, components, etc.).

The second consists of the continuity of the process improvement methods to increase productivity and reduction of production losses [7].

The third is production philosophy based on continuous improvement of production (includes technical and organizational spheres).

The purpose of JIT is a continuous production or serial services with “zero” losses. The basic philosophy is continuous development and improvement. In general, the use of JIT implementation process for the delivery of goods to customers just in time, allows an average halve the execution of orders, about 50% lower inventory levels and reduce by 50-70% the duration of the contracts the company that manufactures the product. This is important for consumers because they believe that it is better to complete an order in 10 days than it will range from 3 to 30 days.

The implementation of LT JIT on average allows:

- 60% reduction inventories of materials and components in the enterprise;
- 40% reduction in transportation costs;
- 40% reduction in the cost of materials;
- 28% improvement delivery of materials.

2.3. Optimization of "Lean Production" technology

JIT became the foundation for the development and implementation of such LT as LP (Lean Production). This technology combines [8]:

1) maintaining high quality production;
2) production of small quantities parties that are not desirable on market conditions;
3) support for low reserves during production;

4) high level of staff qualification in different categories;
5) use of “flexible manufacturing”.
The main ideas of Lean Production are the following:
- elimination of problematic operations, such as warehousing, reloading, etc.;
- organization of low-cost production in batches of small volumes;
- minimization of using resources in production.

The effectiveness of the practical application of LP is determined by the level of production capacity, which is formed by the quantitative and qualitative flexibility of the production system, strong partnerships with suppliers and consumers, the introduction of constant control and regulation of the production process.

Henry Ford's idea of "flow" arose in the 1900s. In the postwar period, small markets had to be flexible at presence deficiency of resources. In the 1950s, the Toyota production system was developed and introduced in Japan, and widely used in the automotive industry. The first examples of the use of LP outside Japan and then outside production were discovered around 1996. Today, LP tools are used in the field of service and IT [9].

There are some companies in Ukraine which use Lean transformation system, for example, Nova Poshta, Alfa Bank and others. Lean transformation system is based on 5 questions to the company:
1) What is the purpose of the company? What is necessary to solve the actual problem?
2) How to improve performance?
3) How to develop business?
4) What is the behavior of the leader and which type of management system do companies need?
5) What are the basic opinions of the company's employees?

The first question shows the future answer to what transformation is needed. Reducing staff or costs is popular now, but it is no longer relevant. Today, it is important to find something that will help to become competitive, flexible and as viable as possible on the market. For example, to reach the level of a competitor during a certain period, to create a flow in accordance with the growing demand, to offer the customer a wider range, to create a flow while expanding the range, etc. Lean has a tool for “determining the true north.” Companies use it when they want to produce as much as possible with small stocks, despite changes in demand, rapid production. The acceleration of all processes is always right transformation.

The second question helps to determine the work that needs to be done, improved, and in which ways to do it, namely, as short and fast as possible. In terms of quarantine company «Toyota» developed materials for staff. The main goal is to prove the ability of one person to work with six units of equipment in a pandemic, and when the situation improves to re-staff up to 6 people.

The third issue concerns staff development, starting with the right skills of people and ending with ways to develop these skills. People often do not want to change and learn, but Lean's approach involves having staff in the company who are constantly learn and share knowledge [10].

The fourth question helps identify management system and behavior of leaders to support the new way of working in accordance with the company's mission, to support the objectives, processes and personnel changes. Leaders improve themselves and then motivate their staff. The success of «Toyota» is the right leadership.

The fifth question is based on understanding the current basic thinking (mental models, attitudes and values) and the degree of their impact on the organization and its culture.

Thus, LT can be used for modern companies and transport companies for the purpose of right and efficient transformation. Consider the resources of Lean for railway transport in Ukraine, as the transport is a manufacturer, but produces a service for the transportation of goods, mail, luggage and passengers.

Using simple formulas, the reduction of the volume of batches can be linked to the increase of efficiency of the process cycle. The first law of “lean production” to accelerate the supply chain (used with the permission of ProfiSight Technologies, protected by US patents 5195041 and 5351195) says that the level of consumer demand (P) amount to

$$P = {Q_{\min}}/{t}, \tag{1}$$

where $Q_{\min}$ is the minimum volume of the consignment of goods; t - workplace turnover time;

or

$$T = {t}/{2} = {Q_{\min}}/{2 \cdot P}, \tag{2}$$

where T is the delay time in the supply chain.

Ultimately, the volume of work in progress decreases, and, accordingly, the area under warehouses of work in progress.

Efficient implementation of LP technology for resource conservation was carried out at PJSC "Mironivska Poultry Farm" (represented in Ukraine by "Nasha Ryaba") - a closed-loop enterprise from the production of young animals to the production of broiler meat. The commissioning of the poultry farm took place in two stages: the launch of the first phase began in 2006 and was successfully completed in 2007; the launch of the second phase began in 2008 and ended in 2009. By 2020, it is a major global player, beating 150 million chickens a year. It is not easy to transform it. The purpose of the introduction of LP technology was to increase productivity, staff moved to foreign companies. The company set a goal to increase the productivity of each person two years ago, now the situation is better (Tables 1-2).

Table 1. Dependence of marriages on the number of staff at the poultry farm.

The staff of the conical line of the poultry farm, pers.	16	14	13
Production capacity of one conical line per hour, kg	529	508	439
Defect, %	2.63	4.9	5.24

The process value creation flow map solves all the necessary processes in which the least people are involved. If there is a conveyor with chicken, there should be no pass, as the system and the person will not process this pass. This is the basis of energy efficiency (Figure 1).

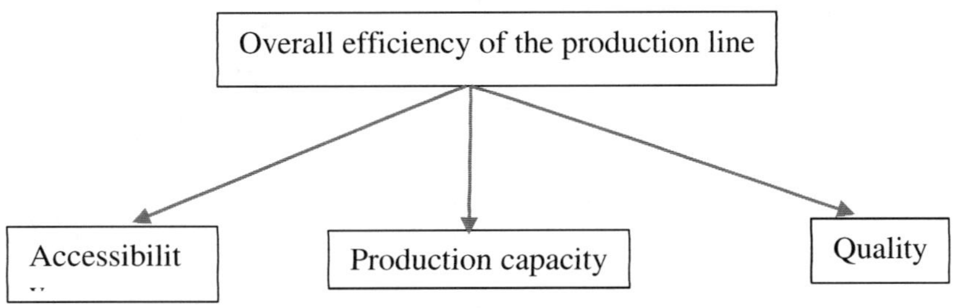

Figure 1. Components of efficiency in the production enterprise.

The Pareto curve (Figure 2) is drawn on the downtime of the equipment, the person came and worked 8 hours on the equipment without waiting until it is set up.

IOP Conf. Series: Materials Science and Engineering 1021 (2021) 012041 doi:10.1088/1757-899X/1021/1/012041

Table 2. Statistical indicators of PJSC "Mironov Poultry Farm".

Month, 2020	Total change time, min.	Just outage, min.	Total outage, %	Heads were scored, units	Actual production capacity, units / hour	Plan, units	Regulatory production capacity, units / hour	Equipment availability, %	Production capacity (staff efficiency), %	Total line efficiency (per shift), %
May	27,968	900	3.2	6,012,209	12,932	6,250,725	13,500	96.8	98.7	92.5
June	28,338	824	2.9	6,113,859	12,996	6,376,050	13,500	97.1	98.8	93.99
July	29,201	1,401	4.8	6,189,490	12,815	6,570,225	13,500	95.2	99	94.2
August	27,287	1,065	2.9	5,858,880	12,956	6,139,575	13,500	96.1	99.3	95.4
September	30,343	1,111	3.2	6,567,593	13,064	6,827,175	13,500	96.3	99.9	96.2

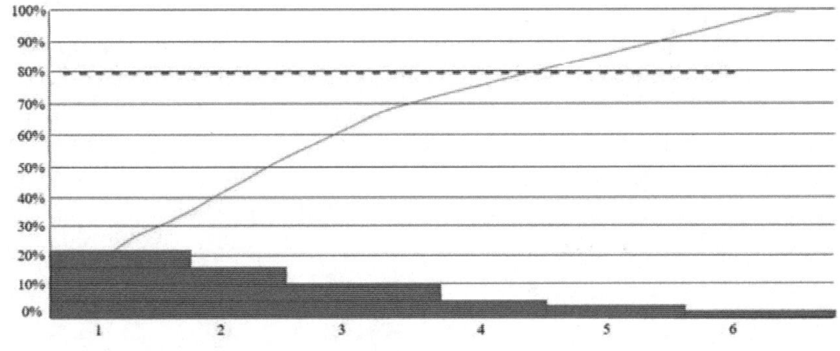

Figure 2. Stork line overall flow line efficiency (1 – tipper, 2 - the absence of Moffett, 3 - linear motor, 4 – container, 5 - no car with a bird, 6 – turntable).

3. Technology for "Ukrainian Railways" Joint Stock Company

Lean Transformation Technology for Joint Stock Company "Ukrainian Railways" (JSC "UR"):

1. Lean culture among employees: to change the thinking of employees so that problems are perceived positively, without processes loss and to help people do not afraid and do not resist changes.

2. Continuous analysis of processes to eliminate or minimize 8 types of losses: waiting (at all stages of work with cargo, train traffic, etc.), "overproduction" of services (resource planning more than the demand for rail transport services, for example, the number of wagons in passenger trains), excessive processing (for example, insufficient automation of technological processes of transportation), unrealized potential, displacement, defects (errors and delays in the train schedule), excess stocks (components, fuel, etc.) [11].

3. Processes are analyzed using a value stream mapping (VSM), i.e. visual systematization of value flow. VSM is a tool that visualizes the process of transforming various technological operations and actions into finished products of transport to consumers.

The objects of it are material, financial and information flows of resources and also time. Toyota Corporation is a pioneer in systematizing of value flows. And the term VSM was first used in 1990. VSM is a way to visualize the steps needed to turn the needs of rail service consumers into a product or service. VSM is often referred to as kaizen flow, means the continuous improvement of value flows. It is the best tool for identifying and planning opportunities to improve the production of rail transport services. Therefore, value flow maps are compiled in a way to clearly depict the current state, to plan and implement the future state with measurable goals.

Systematization of the flow of value is performed on various scales - from simple administrative procedures to large-scale production of services in international traffic. The method helps to identify steps that do not add value and need to be solved, as well as weaknesses where the process can be improved – speed up, reduce costs or provide safe working conditions [12].

Creation VSM is divided into blocks: the production or process-flow - the traditional block diagram, which is fixed from left to right way to create value from procurement of raw materials to shipment of products. If in addition to the main process there are additional or auxiliary, they are applied under the main. Thus, the main task is separated from the secondary. The information or communication flow is depicted by "arrows" at the top of the flow map and occur in parallel with the production. Both formal and informal data exchange are taken into account. Information flows are plotted on the map in free form, as they flow in reality.

Timeline and distance are the lines drawn at the bottom of the map. The timeline is divided into upper and lower parts. The lead time is displayed at the top. The duration of the cycle is plotted below. Below the timeline may be another line, at the very bottom, showing the distances that the product or personnel move within the process.

It is recommended to use this method for a service or department and to compare it with the goals of the whole unit for certain regional branches.

4. All employees have a Lean approach to solving problems in such a way as to find the root causes and make it never happen again in the future: using the A3 tool (A3 letters near the offices of managers to reflect the tasks and trends), Ishikawa diagrams (graphical method of research and determination of the most significant cause-and-effect relationships between factors and consequences in the studied situation or problem), technology "5 Why" (interactive technology questions used to identify causal relationships that underlie a particular problem [13].

The main purpose of this technique is to determine the key cause of the defect or problem by repeating the question "Why?").

5. Employees of UZ JSC in their daily activities use the following lean-tools:

- 5S - workplace rationalization system. It was developed in the post-war Japan by the Toyota Corporation. It is used mainly in the manufacturing industry and in the service sector. 5S is an idea from the Japanese concept of production, according to which discipline and cleanliness are the main requirements for improving the work of the processes running on them. Cleanliness and maintaining order are seen as the basis of quality work.

5S system is used in all areas of the organization, both in production and in services and management [14]. The direct participation of all employees is encouraged. 5S is five Japanese words: “Sorting” - a clear division of things into necessary and unnecessary and getting rid of the latter; “Keeping order” (accuracy) - the organization of storage of necessary things, which allows you to quickly and easily find and use them; "Keeping clean" (cleaning) - keeping the workplace clean and tidy; "Standardization" (maintenance of order) - a necessary condition for the implementation of the first three rules; "Improvement" (habit formation) - education of the habit of exact execution of the established rules, procedures and technological operations;

- Kanban system (routing system for economical and "on time") (JIT) production, Kanban is a system for controlling the logistics chain in terms of production, but not an inventory system.

6. UZ JSC constantly introducing Kaizen, that operates on the principle of continuous improvement.

7. UZ JSC regularly holds 3-4-days’ Kaizen events to implement rapid changes.

8. All employees strive to find true value for the customer (passenger, cargo owner or consignee) by building a value creation stream.

9. Strategic planning is carried out according to the system HoshinKanri (strategic planning technique, introduced in 1954 by Peter Drucker, Hoshin planning - a logical continuation of management by objectives (MBO), Hoshin strengthens the power of MBO and avoids its weaknesses).

10. The company regularly collects and analyzes the “Voice of the Client” to assess the effectiveness of its processes.

11. The company uses personal development plans for employees who clearly and transparently describe career opportunities.

12. The company has a system of proposals that allows employees to put forward their ideas about process optimization.

13. Decisions inside the company are made on the basis of data and facts, not assumptions.

14.The data on the efficiency of machines analyze on a regular basis (automation) - performance, availability and quality. Employees use offline tools every day.

15. Visual management (instructions, tips, visualizations) helps to improve the quality and productivity of processes.

16. In case of stopping the process (downtime, breakdowns, etc.) there is a clear scheme of escalation of the problem in the organizational structure.

17. Managers, process owners and employees regularly go to Gemba (the place where the process takes place) to understand the real situation about the processes.

Artificial intelligence is taking up the pace when it comes to global logistics and supply chain management. The on-going evolution in the areas of technologies like artificial intelligence, machine learning, and similar new technologies is said to possess the potential to bring in disruption and lead innovation within these industries.

Today, transport logistics solves a set of tasks related to the organization of the movement of goods by public transport. One of the tasks of transport logistics is the creation of transport systems, corridors and chains, in which the use of artificial intelligence approaches is of particular importance and is, even, necessary to support decision-making by the operational staff of the transportation industry [15].

The relationship between material and information flows is obvious, but the responsibility of one stream to another is conditional, since the content of the material flow, as a rule, reflects the data of the information stream, but they may not coincide in terms of time.

Material and information streams can be both unidirectional and multidirectional, and this feature allows artificial intelligence to control transport processes [16]. The path in which the information flow moves may generally not coincide with the flow path of the material flow, which significantly complicates the flow control of vehicles.

Talking about artificial intelligence means using robotics. They are used to track, locate and move inventory within the warehouses. Apart from robots, artificial intelligence is also about big data. When the insights of Big Data are used along with artificial intelligence, it helps to improve different areas of supply chain like supply chain transparency and route optimization [17].

Advantages of artificial intelligence for Supply Chain Management [18]:

1. Predictive Analytics. Forecasting of requests helps to upgrade supply chain forms. Ideal inventory levels and decreased holding costs are key advantages of exact interest forecasting.

2. Improving Inventory Management. With an effective inventory management system in place, you can help reduce costs, keep business profitable, analyze sales patterns and predict future sales, and prepare the system for the unexpected.

3. Automated Quality Inspections. The use of artificial intelligence to control automated quality inspections lessens the odds of conveying broken merchandise to customers.

4. Quick High-Yield Shipping. These technologies save time of workers, enabling them to give more an incentive to their more important duties or solve other problems in management or technical process.

The introduction of artificial intelligence into supply chain operations can propel logistic business into the future – harnessing automation, optimizing supply chain planning, and evaluating multiple scenario outcomes processes in decision-making. That's a powerful future for transport.

4. Conclusion

The introduction of the Lean Production practice in Ukraine is all the more important because the quality of high-tech domestic products produced for the mass consumer, as a rule, is rather low not only by Japanese, but also by the average European standards. At the same time, in many cases, domestic enterprises do not have access to international sales markets, are forced to use not the cheapest and not the most qualified labor force and constantly invest in the renewal of fixed assets. If in this situation we continue to adhere to the ideology of large-scale production of the beginning of the last century, then the result will be only one – bankruptcy. This bankruptcy can only bring the growing interest of international corporations closer to the Ukrainian economy. Fortunately, today a number of Ukrainian companies are already implementing their programs for introducing Lean Production elements. Thus, lean manufacturing is becoming a kind of sign of industry leadership, which makes it possible to predict a further surge of interest in this production system on the part of the most dynamically developing large and medium-sized companies operating in a competitive environment and in dire need of new breakthrough technologies for organizing business processes. In general, the introduction of lean production technologies in railway transport enterprises will consistently reduce non-production costs and improve the quality of repair and operation of railway facilities. To do this, it is very important to create working groups at each of the structural units of the railway. Their task is to identify non-production costs and losses as a result of the production process, and to determine the area where the introduction of advanced methods would give the greatest effect. This will increase productivity, reduce the time of implementation of the main technological processes in railway transport and ultimately significantly reduce costs.

References

[1] Womack J P and Jonis D T 2003 LEAN THINKING Banish waste and create wealth in your corporation (*FREE PRESS New York London Toronto Sydney Singapore*) 396

[2] Womack J P, Jonis D T and Roos D 2007 The Machine That Changed the World: The Story of Lean Production - Toyota's Secret Weapon in the Global Car Wars That Is Now Revolutionizing World Industry Paperback (*FREE PRESS New York London Toronto Sydney*) 336

[3] Gaa P 2015 Creating and Using a Shingo style Process Map: A quicker path for finding waste and eliminating process problems (*Texas: Rikai Publishing*)

[4] Liker J and Ross K 2016 The Toyota Way to Service Excellence: Lean Transformation in Service Organizations Hardcover (*McGraw-Hill Education*) 419

[5] Drucker P F 2006 The Practice of Management Paperback (*HarperBusiness A Division of HarperCollins Publishers*) 404

[6] Lomotko D, Kovalov A and Kovalova O 2015 Formation of fuzzy support system for decision-making on merchantability of rolling stock in its allocation *Eastern-European Journal of Enterprise Technologies* **6** **(3)** 11–17

[7] Ward A, Sobek D and Shook J 2014 Lean Product and Process Development (*Lean Interprise Institute Cambridge, MA, USA*) 349

[8] Shook J and Womack J 2010 Managing to Learn: Using the A3 Management Process to Solve Problems, Gain Agreement, Mentor and Lead Pap/Chrt Edition (*Lean Interprise Institute, Inc. One Cambridge Center, Cambridge, MA 02142, USA*) 144

[9] Blanckenberg N 2019 Extensive List of Fulfillment Companies and Services for eCommerce *https://blog.storeya.com/2018/11/list-of-fulfillment-companies/ (date of application: 13.10.2020)*

[10] Gilbert N 2020 Best Order Fulfillment Services & Companies of 2020 *https://financesonline.com/top-20-order-fulfillment-services/ (date of application: 13.10.2020)*

[11] Gronkvist F 2019 Top 8 Fulfillment Centers in China and Hong Kong *https://www.chinaimportal.com/blog/fulfillment-centers-china-hong-kong/ (date of application: 13.10.2020)*

[12] Heidegger J 2019 Warehouse & Distribution Center Best Practices for Your Supply Chain *https://legacyscs.com/warehouse-and-distribution-center-11-best-practices/ (date of application: 13.10.2020)*

[13] Krasnoff B 2020 The Best E-Commerce Fulfillment Services *https://www.pcmag.com/picks/the-best-e-commerce-fulfillment-services (date of application: 13.10.2020)*

[14] Pndey K 2019 How AI is Revolutionizing Global Logistics and Supply Chain Management *https://readwrite.com/2019/04/15/how-ai-is-revolutionizing-global-logistics-and-supply-chain-management/ (date of application: 13.10.2020)*

[15] Sino S 2019 Rail freight from China to Europe – Rates & Transit Time. Free Quote *https://www.sino-shipping.com/rail-freight-china/ (date of application: 13.10.2020)*

[16] Transmetrics Blog. Technologies Disrupting Logistics Order Fulfillment *https://transmetrics.eu/blog/logistics-order-fulfillment-technologies/ (date of application: 13.10.2020)*

[17] Open4Business 2019 Ukraine's foreign trade in goods – import *https://open4business.com.ua/ukraines-foreign-trade-in-goods-import-jan-march-2019/ (date of application: 13.10.2020)*

[18] ZorziniC 2020 Ecommerce Fulfillment Services (Jul 2020): The Best Order Fulfillment

Companies *https://www.loadingcenter.co.uk/e-commerce-fulfilment/ (date of application: 13.10.2020)*

Features of determining engagement line of traction gears with different degrees of wear of gear teeth

V I Moroz[1], V I Hromov[1], O V Bratchenko[1] and O A Logvinenko[1,2]

[1] Department of Mechanics and Machine Design, Ukrainian State University of Railway Transport, 61050 Feierbakh Sq. 7, Kharkiv, Ukraine
[2] Email: logvinenko@kart.edu.ua

Abstract. Calculations of strength, forecasting the residual operation life for repairing traction gears (TG) is carried out on the results of modeling their motion characteristic, which is based on the determined coordinates of the contact points of the gear and wheel profile for the period of engagement (engagement line). This article presents results of a study aimed at developing mathematical dependencies for a refined determination of TG engagement line with the established wear of gear teeth. Adequacy verification of proposed mathematical descriptions was carried out based on the results of the study of new TG of ER-2 electric train series. Their comparison with the coordinates of the engagement line, calculated by the traditional method, showed that modelling uncertainty does not exceed 0.06%. As an example of practical use of the received mathematical descriptions, results of research of motion characteristic of new TG of ER-2 electric train series with the established degrees of wear of gear and wheel teeth are presented.

1. Introduction

Mainly traction rolling stock (TRS) and multiple-unit train (MU) with excessive service life carry out a significant amount of traffic on Ukrzaliznytsia PJSC (Ukrainian Railways) networks. Therefore, strategy for the development of railway transport for the period up to 2020 (approved by the Order of Cabinet of Ministers of Ukraine No. 51555-r on December 16, 2009) and the State Program for rolling stock renewal for 2017-2021 (approved at a meeting of Ukrzaliznytsia PJSC on November 29, 2016) provide for upgrades, as well as extension of the service life of existing rolling stock on the basis of relevant research and development work. One of the areas of such research is to ensure the reliability, technical resource of the main modules of TRS and MU [1]. These include the traction drive as a whole and the traction gear (TG) functioning in its structure.

2. Recent research and publications analysis

Experience of TG repairs has shown that due to extreme wear of gear teeth, more than 90% of gears and about 60% of wheels are rejected prematurely [2]. That is, wear features of the active surfaces of the teeth should be considered one of the main factors influencing TG technical resource, which significantly depends on the resource of the traction drive [3].

According to the results of researches performed, it is established that operational wear of active profiles of gear and wheel teeth cause essential changes in engagement characteristics of functioning TG (in comparison with new gear) [4]. Distortion of profile configuration in normal sections due to wear leads to a violation of the main property of involute gearing, which consists in conversion of rotational motion from gear to wheel with a constant gear ratio ($i_{12} = -\omega_1 / \omega_2 = const$). This causes significant additional dynamic loads, reducing TG technical resource, increasing the levels of bending stresses of the teeth and contact stresses in the gear [5,6]. Therefore, the calculations of TG strength with different degrees of teeth wear involves determining the nature of the change in the value of the gear ratio (gear ratio $u_{12} = |i_{12}|$) during the period of gear pair catching. In addition, ensuring the rational formation of TG repair [4] requires the development of sound recommendations, based on the results of modeling the characteristics of gear transmission with different degrees of teeth wear of active profiles over the entire height – that is, the actual value of the overlap coefficient, patterns of change in the coefficients of specific sliding and pressure for the period of teeth pair engagement.

The key point of such works is the specified definition of the current coordinates of the contact point's trajectory of the real profiles of the teeth for the period of their engagement (engagement lines). At this time, the problem of obtaining digital descriptions of the considered real profiles of teeth with different degrees of wear has been solved at a sufficiently high level [7, 8]. Nevertheless, modeling their engagement characteristics at best is carried out mainly using graphical methods [9]. At the same time in the special scientific and technical literature, the issue of analytical research of TG engagement characteristics with the established wear of gear and wheel teeth with the use of other methods are insufficiently covered.

3. Defining the aim and objectives of this investigation

The aim of this investigation is to develop mathematical dependences to solve the problem of determining TG engagement line with different degrees of wear of gear and wheel teeth.

To achieve this aim, the following tasks were solved:

1. Development of mathematical descriptions that provide determination of the boundary points of teeth contact (position of gear and wheel teeth axes t the beginning and end of engagement).
2. Obtaining mathematical dependences to determine the coordinates of the current points of contact in the teeth engagement.

4. The main part of the study

In order to solve the first issue to determine the position of the axes of worn gear and wheel teeth of TG at the beginning (contact point B_1 at figure 1 at the intersection of radii circles r_{a2} and r_{B_1}, angles in question are $\psi_{1\Pi}$, $\psi_{2\Pi}$) and end of engagement (contact point B_2 at figure 1 at the intersection of radii circles r_{a1} and r_{B2}, angles in question are ψ_{13}, ψ_{23}), the method of projections of closed vector contours on the coordinate axis was used (V.A. Zinoviev's method) [10].

Therefore, in relation to TG, a coordinate system $X_0O_1Y_0$ was introduced, and gear geometric parameters r_{a1}, r_{a2}, r_{B_1}, r_{B2}, a_w were replaced by the corresponding vectors $\bar{r}_{a1}$, $\bar{r}_{a2}$, $\bar{r}_{B1}$, $\bar{r}_{B2}$, $\bar{a}_w$. For the moments of the beginning and the end of engagement of gear and wheel teeth vector contours $O_1B_1O_2O_1$ and $O_1B_2O_2O_1$ were considered, for which conditions of closeness are considered

$$\bar{r}_{B_1} + \bar{r}_{a2} = \bar{a}_w; \quad \bar{r}_{B_2} + \bar{r}_{a1} = \bar{a}_w. \tag{1}$$

Formulas for determining the desired angles that define the position of the axes of gear and wheel teeth at the moments of their engagement are obtained from the transformation of equation projections (1) on the coordinate axis O_1X_0 and O_1Y_0

$$\psi_{1\Pi} = \varphi_{1\Pi} + \arcsin[\bar{s}_{B_1}/(2\cdot r_{B_1})]; \quad \psi_{2\Pi} = \varphi_{2\Pi} - \arcsin[\bar{s}_{a2}/(2\cdot r_{a2})], \tag{2}$$

where $\varphi_{1\Pi} = \arccos[(a_w^2 + r_{B_1}^2 - r_{a2}^2)/(2\cdot a_w \cdot r_{B1})]$; $\varphi_{2\Pi} = \arccos[\sqrt{1-(r_{B_1}\sin\varphi_{1\Pi}/r_{a2})^2}$,

$$\psi_{13} = \varphi_{13} - \arcsin[\bar{s}_{a_1}/(2\cdot r_{a_1})]; \quad \psi_{23} = \varphi_{23} + \arcsin[\bar{s}_{B2}/(2\cdot r_{B2})], \tag{3}$$

where $\varphi_{13} = \arccos[(a_w^2 + r_{a1}^2 - r_{B_2}^2)/(2 \cdot a_w \cdot r_{a1})]$; $\varphi_{23} = \arccos[\sqrt{1-(r_{a1}\sin\varphi_{13}/r_{B_2})^2}$.

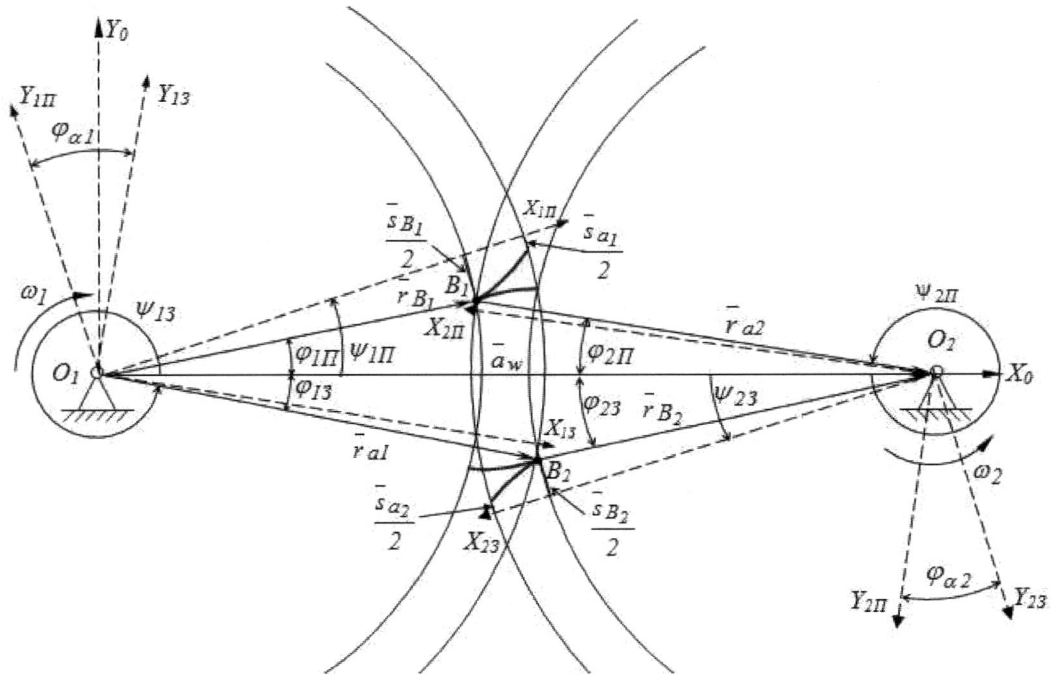

Figure 1. Before determining axes limit positions of worn gear and wheel teeth of TG.

Analytical dependences for determining the coordinates in the system $X_0O_1Y_0$ of current contact points of the teeth profiles with the established thicknesses of the circles chords of the corresponding radii of the gear and wheel (the second task of this investigation) were obtained on the basis of coordinate transformation method (G.F. Moroshkin's method) [11]. Figure 2 shows a fragment of TG motion scheme. Gear 1 and wheel 2, together with rack 0 form rotating motion pairs of the fifth class O_1, O_2.

The working profiles of the teeth form the highest motion pair of the fourth class B (contact point). Coordinates of the flow point of contact B in the coordinate system $X_1O_1Y_1$: $x_1 = O_1A_1 = l_{e1}$; $y_1 = A_1B = \overline{S}_{e1}/2$ ($\overline{S}_{e1}$ is thickness of the gear tooth, determined at a distance l_{e1} from rotation center O_1). Coordinates of the flow point of contact B in the coordinate system $X_2O_2Y_2$: $x_2 = O_2A_2 = l_{e2}$; $y_2 = A_2B = \overline{S}_{e2}/2$ ($\overline{S}_{e2}$ is thickness of the gear tooth, determined at a distance l_{e2} from rotation center O_2).

The matrix notation of equation system for the transformation of the point coordinates is the following

$$r_i = T_{ji} \cdot r_j, \tag{4}$$

where r_i is matrix-column of point coordinates in the basic coordinate system; r_j is matrix-column of point coordinates in a new coordinate system; T_{ji} is matrix of equation coefficients.

Equation matrix (4) to determine in the system $X_0O_0Y_0$ the coordinates of point B, which belongs to gear and wheel teeth, respectively

$$r_0 = \begin{Vmatrix} x_0 \\ y_0 \\ 0 \end{Vmatrix}, \quad r_1 = \begin{Vmatrix} l_{e1} \\ \bar{S}_{e1}/2 \\ 0 \end{Vmatrix}, \quad T_{10} = \begin{Vmatrix} \cos\psi_{01} & -\sin\psi_{01} & 0 \\ \sin\psi_{01} & \cos\psi_{01} & 0 \\ 0 & 0 & 1 \end{Vmatrix}; \tag{5}$$

$$r_0 = \begin{Vmatrix} x_0 \\ y_0 \\ 0 \end{Vmatrix}, \quad r_2 = \begin{Vmatrix} l_{e2} \\ \bar{S}_{e2}/2 \\ 0 \end{Vmatrix}, \quad T_{20} = \begin{Vmatrix} \cos\psi_{02} & -\sin\psi_{02} & a_w \\ \sin\psi_{02} & \cos\psi_{02} & 0 \\ 0 & 0 & 1 \end{Vmatrix}. \tag{6}$$

According to G.F. Moroshkin's method, the issue of determining the positions of TG gear and wheel for the corresponding contact point coordinates of the real profiles of the teeth is reduced to the joint solution of equations obtained from formulas (5), (6) with conditional opening of motion pair *B*.

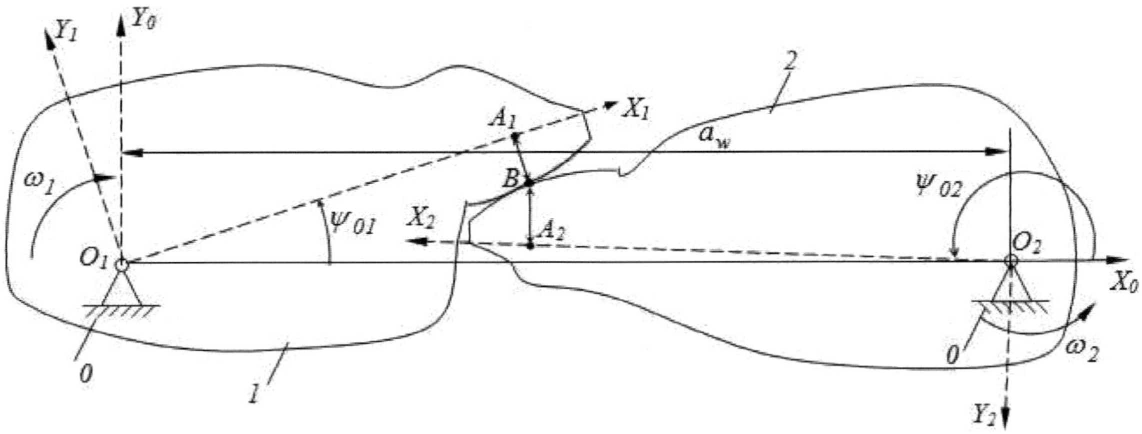

Figure 2. Before determining the coordinates of the current contact points of the profiles of the gear and wheel teeth of TG.

Then the conditions for the existence of the current contact point of the profiles of gear and wheel teeth are the equations

$$l_{e1} \cdot \cos\psi_{01} - (\bar{S}_{e1}/2) \cdot \sin\psi_{01} = l_{e2} \cdot \cos\psi_{02} - (\bar{S}_{e2}/2) \cdot \sin\psi_{02} + a_w; \tag{7}$$

$$l_{e1} \cdot \sin\psi_{01} + (\bar{S}_{e1}/2) \cdot \cos\psi_{01} = l_{e2} \cdot \sin\psi_{02} + (\bar{S}_{e2}/2) \cdot \cos\psi_{02}. \tag{8}$$

The use of the obtained mathematical dependences in TG investigation requires analytical descriptions of active profiles of gear and wheel teeth of the following type

$$\bar{S}_{e1}/2 = f(l_{e1}), \quad l_{e1} \in [r_{B_1}; r_{a1}]; \tag{9}$$

$$\bar{S}_{e2}/2 = f(l_{e2}), \quad (l_{e2} \in [r_{B_2}; r_{a2}]). \tag{10}$$

Studies have shown that in the presence of the results of experimental determination of teeth thickness according to the chords of respective circles to develop such descriptions, it is advisable to use the Lagrange interpolation formula [12].

Analytical dependences (7) to (10) are the basis for the study of the characteristics of worn TG engagement with the established thicknesses of gear and wheel teeth according to the circle chords of corresponding radii. When fixing the angle value $\psi_{01} \in [\psi_{1П}; \psi_{13}]$ and the stepwise angle setting $\psi_{02} \in [\psi_{2П}; \psi_{23}]$ (angles are calculated in accordance with figure 2), the coordinates of the corresponding contact point (X_0, Y_0) in the system $X_0O_0Y_0$, which satisfy conditions (7), (8), are determined by iterative approximations.

In order to verify the adequacy of the proposed mathematical descriptions, a calculated study of a new TG electric train EP-2 series (number of gear teeth $z_1 = 23$ and wheel teeth $z_2 = 73$; teeth modules $m = 10$ mm; intercenter distance $a_w = 485$ mm; gear displacement coefficient $x_1 = 0.478$ and wheel displacement coefficient $x_2 = 0.041$), aimed at determining the coordinates of the current contact points of gear and wheel teeth during the period of their engagement. A comparison of the obtained results with the coordinates of the engagement line B_1B_2, calculated by the traditional method (Figure 3) showed that modelling uncertainty does not exceed *0.06%*. Such results substantiate the possibility and expediency of using the proposed approach as well as the developed mathematical support in the study of the characteristics of TG engagement with different degrees of teeth wear.

Below are the results of the study of the kinematics of TG electric train EP-2 series with the established degrees of teeth wear (digital images of gear and wheel teeth are presented in Figure 4). Engagement line B_1B_2 of the studied TG, obtained using the developed mathematical descriptions, is shown in Figure 5 below. A comparison of the obtained results with experimental data [9] showed that modelling uncertainty does not exceed 1%.

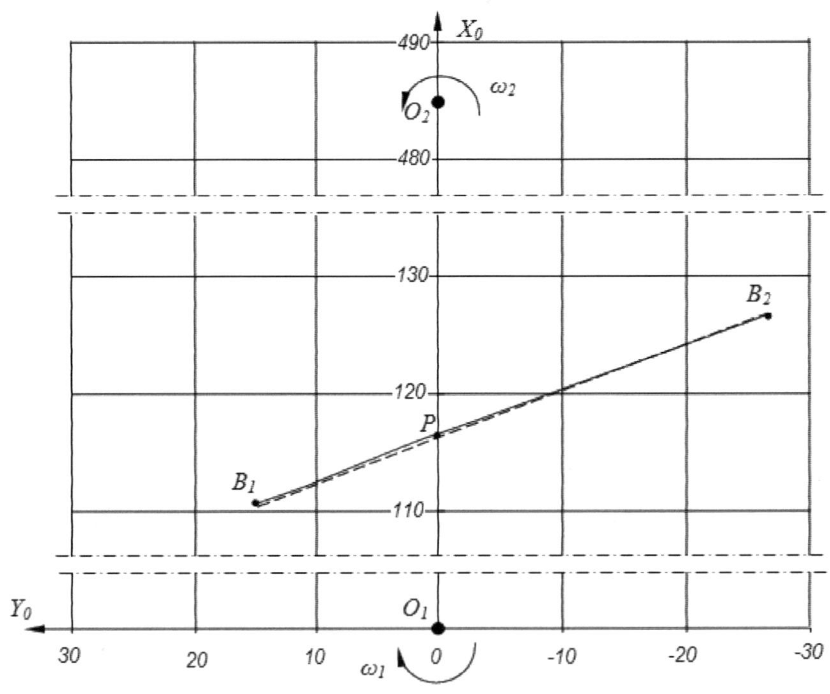

Figure 3. Results of modeling coordinates of the current contact points of the active teeth profiles for the period of engagement of the new TG electric train EP-2 series: --------- using the traditional method; using the developed ——— mathematical dependences.

During this investigation, when determining the coordinates of each *i*-th contact point within $X_0O_1Y_0$ system of active profiles of gear and wheel teeth (X_0, Y_0), the increment of the wheel rotation angle $\Delta\psi_{02_i} = \psi_{02_i} - \psi_{02_{i-1}}$ was determined for each given increment of the gear rotation angle $\Delta\psi_{01_i} = \psi_{01_i} - \psi_{01_{i-1}}$. This allowed determining the current value of the gear ratio u_{12} and the angular velocity analog ω_{q2} for the corresponding TG position

IOP Conf. Series: Materials Science and Engineering 1021 (2021) 012042 doi:10.1088/1757-899X/1021/1/012042

$$u_{12i} = \frac{\Delta\psi_{01i}}{\Delta\psi_{02i}}, \tag{11}$$

$$\omega_{q2i} = \frac{1}{u_{12i}}. \tag{12}$$

Figure 6 presents graphs of changes in gear ratio u_{12}, angular velocity analogues ω_{q2} and angular acceleration ε_{q2} depending on $\Delta\psi_{01}$ for the studied TG (graph $\varepsilon_{q2} = f(\Delta\psi_{01})$, obtained from the graph $\omega_{q2} = f(\Delta\psi_{01})$ by the method of finite differences [13]).

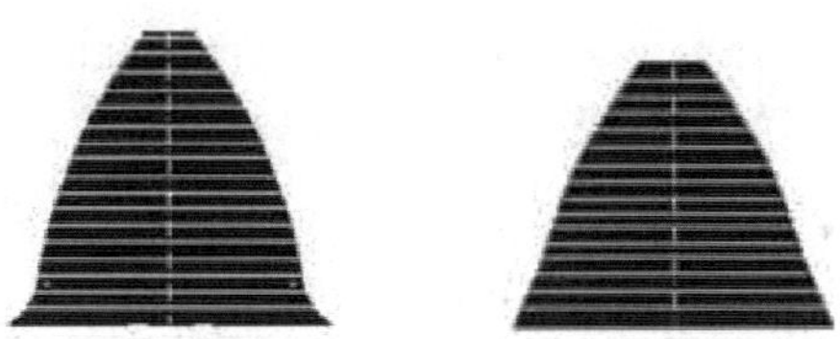

a) b)

Figure 4. Digital images of active profiles of gear teeth (*a*) and wheel teeth (*b*) of studied TG.

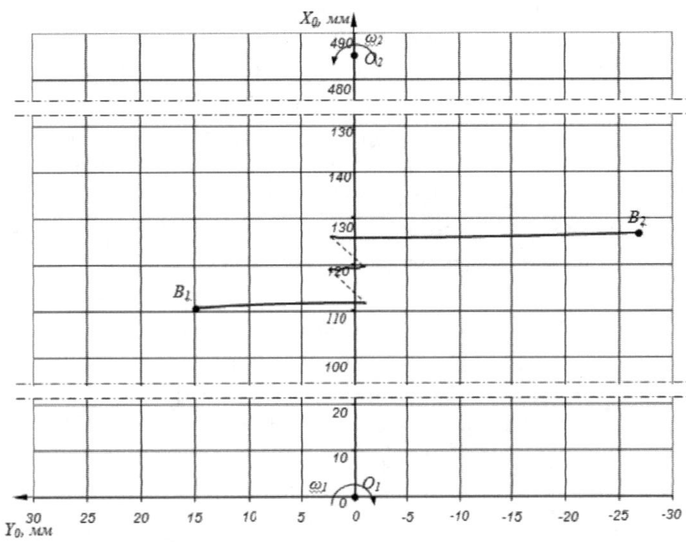

Figure 5. Engagement line of studied TG.

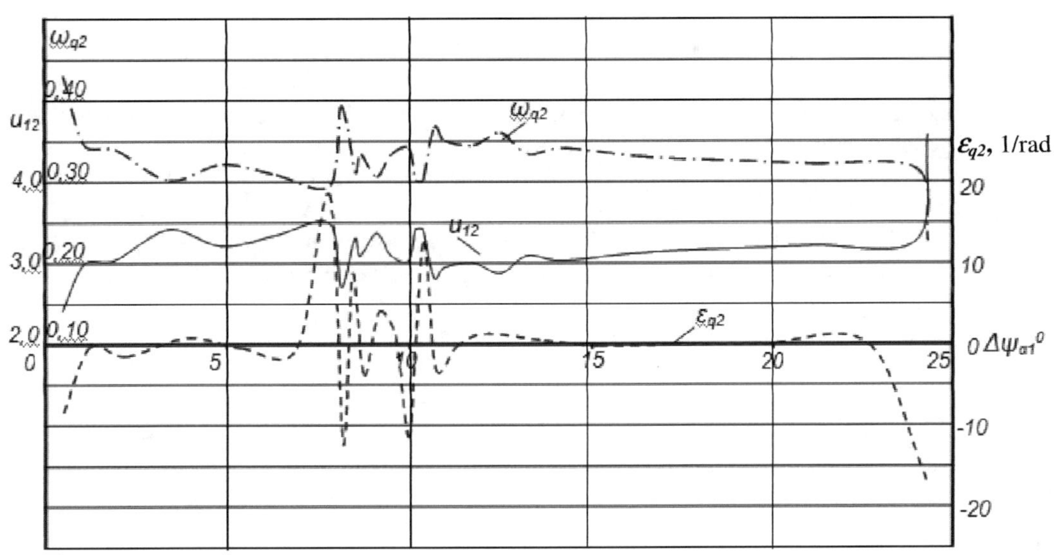

Figure 6. Results of studying of TG electric train ER-2 motion with the established degrees of gear and wheel wear

Such results are considered the basis for further investigation of TG with the established degrees of gear and wheel wear.

5. Conclusions

Analysis of the presented results proves the expediency of using the developed mathematical dependences to determine the contact points of the teeth, lines and gears, which is important for predicting TG strength and residual life.

References

[1] Gorbunov N I, Kovtanets M V and Demin R Yu 2014 Methodology of innovative development of railway transport *Bulletin of the Volodymyr Dahl East Ukrainian National University* **3(210)** 22-27

[2] Biryukov I V, Belyaev A I and Rybnikov E K 1986 Traction gears of railway electric rolling stock (Moscow: Transport) 256

[3] Kolodyazhny P V 2011 Operational damage of rolling stock traction gears and technologies of their strengthening at production *Bulletin of the Volodymyr Dahl East Ukrainian National University* Part 2 **1(155)** 88-96

[4] Moroz V I, Bratchenko O V and Bobritsky S V 2010 A new method for determining the operational wear of teeth profiles and selection of paired wheels of traction gears *Set of scientific researches of the Ukrainian State Academy of Railway Transport* **117** 12-16

[5] Wang T, Chu' F and Han Q 2017 Fault diagnosis for wing turbine planetary ring gear via a mashing resonance based filtering algorithm *ISA Transaction* **67** 173 – 182

[6] Moroz V I, Bratchenko O V, Bobritsky S V and Hromov V I 2013 Improving the technology of designing traction gears of multiple-unit train *Set of scientific researches of the Ukrainian State Academy of Railway Transport* **136** 44–49

[7] Simon Winkelbach, Sven Molkenstruck and Friedrich M. Wohl. 2006 Low-cost laser rang scanner and fast surface registration approach *DAGM* **4174** 718-728

[8] Kuznetsova A V 2011 Measuring the wear of bevel gear teeth with circular biconcave teeth by laser scanning method *Bulletin of the Volodymyr Dahl East Ukrainian National University* Part 2 **11 (165)** 77-85

[9] Bobritsky S V 2012 Improving the technology of design and repair of traction gears of multiple-unit train: author's ref. … *Candidate of Technical Sciences* 20

[10] Zinoviev V A 1975 Course in the theory of mechanisms and machines (Moscow: Science) 384

[11] Levitskaya O N and Levitsky N I 1985 Course in the theory of mechanisms and machines (Moscow: High school) 279

[12] Korn G and Korn T 1968 Mathematical handbook. For scientists and engineers (NY: McGrow-Hill Book Company) 831

[13] Rothbart G A 1960 Cam mechanisms: design, dynamics and issues of manufacturing accuracy translation from English edited by Kolchin N. I. (Leningrad: State Union publishing house of the shipbuilding industry) 335

Ways to increase the traction efficiency of modular draft device

V Nadykto[1, 3], V Kyurchev[1], A Chaplinskyi[2] and A Ayubov[1]

[1]Department of Machine Use in Agriculture, Dmytro Motornyi Tavria State Agrotechnological University, B. Khmelnitsky ave. 18, Melitopol, Ukraine
[2]Department of Technical Mechanics and Computer Engineering Dmytro Motornyi Tavria State Agrotechnological University, B. Khmelnitsky ave. 18, Melitopol, Ukraine

[3]Email: volodymyr.nadykto@tsatu.edu.ua

Abstract. The versatility of tractors can be increased by using them as a part of a modular draft device (MDD). MDD consists of energetic (EM) and technological (TM) modules. EM is a high-energy tractor with a 2WD or 4WD wheel arrangement and a traction force of 14-16 kN. TM is an additional axle with an active drive of its wheels. By connecting TM to the rear hitch linkage (RHL) of EM, the tractive effort of the entire MDD, which has a 4WD or 6WD wheel arrangement, increases to 32-36 kN. MDD can function as a road-rail vehicle. Depending on the traction resistance, MDD can be used both as part of an EM+TM or as a single EM. According to the research results, it was found that the maximum traction efficiency (TE) of MDD with a 6WD wheel arrangement is about 10% higher than that of MDD with a 4WD wheel arrangement. The TE value of MDD increases with an increase in the inclination of EM rear hitch linkage top link. The increase in MDD's tractive efficiency is facilitated by an increase in the coefficient of kinematic discrepancy in the drive of the EM and TM wheels from 1.00 to 1.05.

1. Introduction

One of the most effective ways to solve the problem associated with the nomenclature of the tractor power industry is the introduction of modular draft devices (MDD). Its high production versatility and technological adaptability are ensured by the variability of the nominal tractive effort [1]. In total, it displays the range of traction forces for tractors of 2-3 traction classes. This new property of draft devices is caused not by the use of their mechanical ballasting but by the division functions into energetic and technological within one design of the tractor.

This fundamentally new direction in the development of tractors is relevant today for almost all countries of the world. First of all, because many manufacturers, instead of the old traction concept, have begun producing tractors of a new - traction and energy concept.

The essence of the first concept consists of the approximate constancy of the energy saturation level of the tractor E_t. The latter concept represents the ratio of the tractor engine power (N_e, kW) to its operating weight (M_t, t) and is expressed in kW t^{-1}. That is, $E_t = N^e \cdot M_t^{-1} = \text{const}$.

Following the traction and energy concept, the tractor engine power is increased, while its operating weight remains practically unchanged. In this case, we have: $E_t = N^e \cdot M_t^{-1} = var$.

Our research has established that the limiting value of the parameter E_t, at which the total power of the tractor engine can be converted into the tractive effort, is approximately 15 kW t^{-1} [2]. A larger value of this parameter raises the problem of using the engine power reserve.

One of the most effective ways to solve this problem is to create energetic devices with a split flow of power from its engine. Its main part is realized by a tractor, which we propose to call the energetic module (EM). The remaining ("extra") part of the engine power is used by an additional axle with an active drive of its wheels, called a technological module (TM). The result is a modular draft device (MDD).

The MDD energetic module is essentially a tractor of a specific traction class, designed to work with a corresponding variety of machines and implements. The MDD technological module attaches to the rear hitch linkage of the energetic module. The technological module wheels are driven from a synchronous power take-off shaft of the EM, and TM is also equipped with its rear hitch linkage, a semitrailer, a brake system, etc.

When TM is attached to EM, the traction forces developed by them are summed up. As a result, the MDD, having a greater tractive effort, can be aggregated with wider and more productive agricultural machines. Due to summing the traction forces of the energetic and technological modules, MDD is a vehicle of variable traction class.

We have developed and tested in production conditions industrial models of modular draft devices of traction classes 1.4-3 (Figure 1) and 3-5 (Figure 2).

Figure 1. MDD of traction class 1.4-3

Figure 2. MDD of traction class 3-5

The study of the operation of machine and tractor units based on modular draft device is a very relevant direction in solving problems associated with an increase in the annual load of existing tractors, a reduction in the range of necessary traction equipment at enterprises, and an expansion of the MDD applicability range in various industries.

2. Analysis of recent studies and publications

The technical feasibility and economic suitability of these modular draft devices with variable traction class have been confirmed by long-term research and production tests. Particular attention was paid to the study of the engine power balance of such device. As a result, dependencies that allow determining the values of N_e, M_t, and E_t for MDD were obtained [3]. Perspectives for possible ballasting of MDD modules are considered considering the restrictions put forward by this process [4]. Problems of movement stability of machine-tractor units based on this device were investigated using a new stability criterion [5]. The issues of kinematic discrepancy influence in the drive of the MDD modules wheels on its traction indicators are detailed [6].

The analysis of the technological properties of MDD shows that it can be successfully applied not only in the agricultural sector. After a simple re-equipment, MDD is potentially suitable for rail transport, first of all, as rail vehicles of category N3 [7, 8] or category T [9, 10]. The latter is represented by the MMT-3 series motor locomotives [11-13], which are almost entirely similar to MDD [14-16]. The forecast of these studies shows that modular draft devices can be successfully used in modern bimodal transportation by rail [17, 18].

The analysis of theoretical studies shows that the design parameters [19-21] of modular draft devices almost wholly meet the criteria for the movement stability of railroad motor locomotives [22-24]. The methods for assessing the ways to reduce the rolling resistance of rail vehicles, motor locomotives, and modular draft devices are almost identical [25].

Using MDD as a motor locomotive will be more efficient, if its traction properties are higher. The universal evaluation parameter of the latter is the power delivery efficiency (PDE). The nature of its change under the influence of new requirements for the maximum permissible level of slipping of wheeled draft devices is disclosed in the study [26].

The performed studies' analysis shows that the issue of PDE components changes nature in the modular draft device has been studied very poorly, especially considering the influence of its design parameters. Therefore, this work is devoted to exploring the impact of MDD design parameters changing on its traction efficiency.

3. Statement of the objective and tasks of the study

Firstly, it is necessary to determine the design parameters that have the most significant impact on the change in traction efficiency of MDD. As shown by the preliminary analysis, such parameters include 1) wheel arrangement of the energetic module; 2) the angles of inclination of EM rear hitch top and lower links; 3) the coefficients of kinematic discrepancy in the wheel drives of MDD energetic and technological modules.

Thus, this article aims to increase the operating efficiency of the MDD by bringing it to the operating mode with the traction efficiency tending to the maximum. And also, to ensure this mode due to the optimal parameters of MDD structural elements, which affect the increase in traction efficiency. We will achieve this goal by solving the following tasks:

- study the influence of energetic module wheel arrangement on MDD traction efficiency change dynamics;
- assess of the impact of EM rear hitch links angles of inclination changes on the nature of MDD traction efficiency;
- study the influence of changing the kinematic discrepancy coefficients in the wheel drives of MDD energetic and technological modules on MDD traction efficiency.

4. The basic part of the study

We will consider the optimal interaction of the modular device traction axes, at which the maximum value of its PDE is achieved (η_o). For MDD, this parameter can be determined from the equation:

$$\eta_o = \frac{N_a(\varphi_a - f_a) + N_b(\varphi_b - f_b) + N_c(\varphi_c - f_c)}{\dfrac{N_a\varphi_a}{\eta_{ta}(1-\delta_a)} + \dfrac{N_b\varphi_b}{\eta_{tb}(1-\delta_b)} + \dfrac{N_c\varphi_c}{\eta_{tc}(1-\delta_c)}}, \quad (1)$$

where N_a, N_b, N_c – vertical loads on MDD axles (starting now, indices a and b refer to the front and rear axles of the energetic module and the index c - to the axle of the technological module); φ_a, φ_b, φ_c – coefficients of adhesion of MDD wheels to the soil; f_a, f_b, f_c – wheels rolling resistance coefficients of the appropriate MDD axles; δ_a,δ_b,δ_c – wheels slip coefficients of the appropriate MDD axles; η_{ta}, η_{tb}, η_{tc} – drive transmissions efficiency of the appropriate MDD axles.

For determining the vertical loads acting on MDD axles, a design scheme was drawn up (Figure 3).

The influence of the unit technological part on MDD is presented in the form of the main torque (M) and horizontal (P_g) and vertical (R_v) components of the main vector of forces applied to the technological module in the longitudinal-vertical plane.

In addition to these forces and torques, in the general case, the MDD is affected by (Figure 3): tangential traction forces P_{ka}, P_{kb}, P_{kc}; forces P_{fa}, P_{fb}, P_{fc} and torques M_a, M_b, M_c; rolling resistances; weight force of energetic (G_e) and technological (G_t) modules.

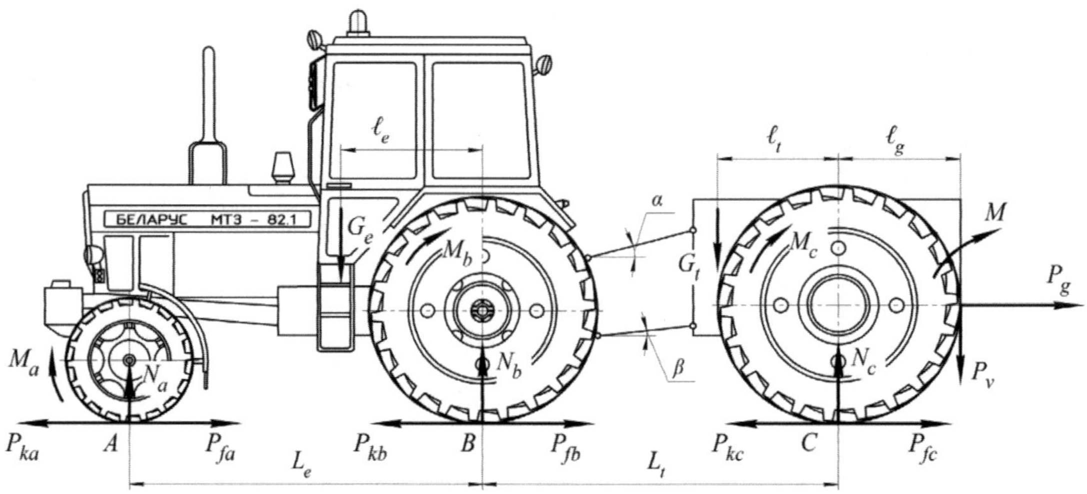

Figure 3. Diagram of forces and torques acting to MDD in longitudinal-vertical plane

Equation (1) determines the PDE value for MDD with the 6WD wheel arrangement. When using the EM with rear-wheel drive only, the MDD wheel arrangement is referred to as 4WD. The equation for determining the PDE (η_1) of such MDD is as follows:

$$\eta_1 = \frac{N_b(\varphi_b - f_b) + N_c(\varphi_c - f_c) - N_a f_a}{\dfrac{N_b \varphi_b}{\eta_{tb}(1-\delta_b)} + \dfrac{N_c \varphi_c}{\eta_{tc}(1-\delta_c)}}. \tag{2}$$

The parameters of equations (1) and (2) are described by a system of 20 additional algebraic equations, which are not given in this article but described in detail in [27, 28].

An analysis of the calculation data for equations (1) and (2) showed that the maximum PDE value for an all-wheel-drive MDD (6WD) is greater than that of a non-all-wheel drive (4WD) (Figure 4).

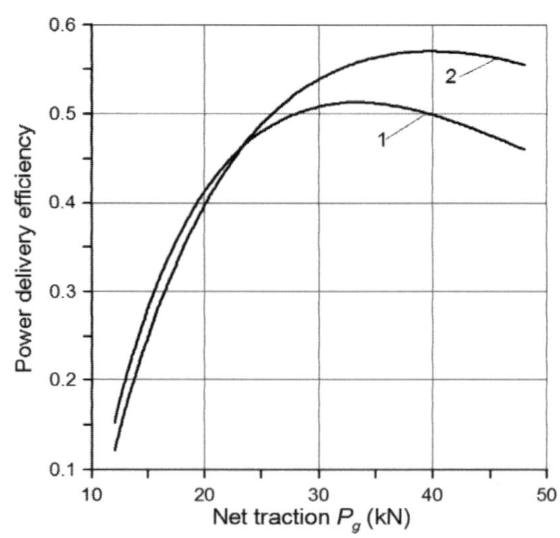

Figure 4. Dependence of MDD power delivery efficiency on its net traction: 1 – 4WD MDD; 2 – 6WD MDD

If, for the first design variant of MDD (6WD), this parameter is equal to 0.570, then for the variant of the modular draft device with the 4WD arrangement, the maximum PDE value is only 0.512 (i.e., 10% less).

Moreover, if the PDE of non-all-wheel drive MDD reaches its maximum at pulling force of 34 kN (curve 1, Figure 4), then for an all-wheel-drive device, the maximum PDE value shifts to the zone of high pulling forces. In this case, it is almost 40 kN (curve 2, figure 4).

Noteworthy is the fact that in the 12-22 kN traction range, the PDE of the non-all-wheel drive MDD is greater than that of the all-wheel-drive (curve 1, Figure 4). At high values of the MDD tractive effort, the nature of change in PDE becomes the opposite. Moreover, the greater tractive effort P_g, the greater difference between PDEs in favor of the all-wheel-drive version of the MDD.

Let us analyze what causes such a result. To do this, we should disclose the nature of those components that are included in the equation that allows us to calculate the efficiency.

According to the theory of the tractor, the traction efficiency should be determined by the equation:

$$\eta_3 = \eta_{tr} \cdot \eta_r \cdot \eta_s, \tag{3}$$

where η_{tr} – transmission efficiency; η_r – rolling efficiency; η_s– MDD wheel slip efficiency.

To clarify the reasons for the higher and shifted to the zone of large tractive forces traction efficiency of MDD with the 6WD arrangement, it is necessary to investigate the nature of the last two components.

For studying the nature of rolling efficiency and mover slipping efficiency of MDD, graphs of their dependence on tractive effort were built (Figure 5).

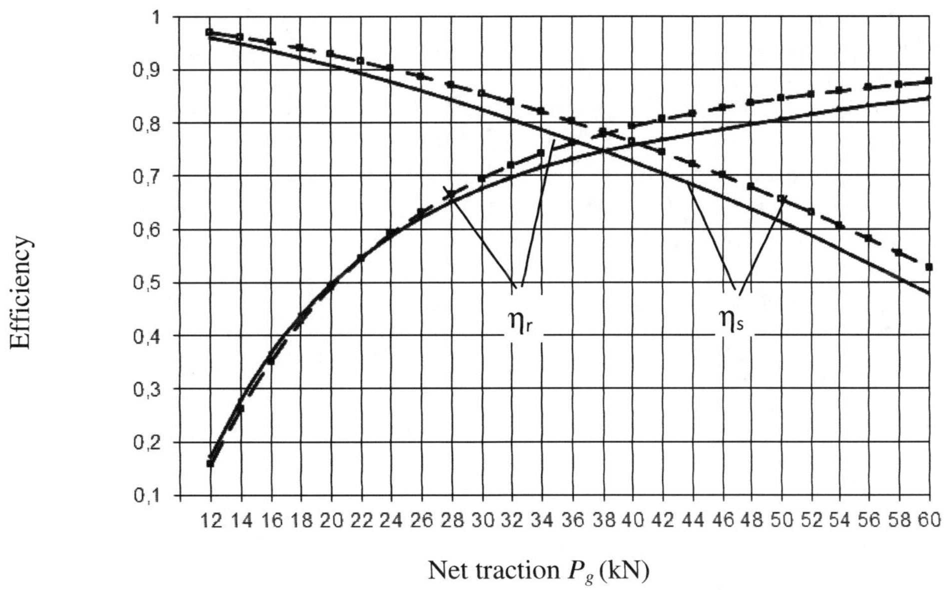

Figure 5. Rolling (η_r) and slipping (η_s) efficiency: - - - – 6WD MDD; ——— – 4WD MDD

The rolling efficiency was determined by the following equations [27-29]:

for MDD with 6WD arrangement

$$\eta_r = \frac{N_a \cdot (\varphi_a - f_a) + N_b \cdot (\varphi_b - f_b) + N_c \cdot (\varphi_c - f_c)}{N_a \cdot \varphi_a + N_b \cdot \varphi_b + N_c \cdot \varphi_c}. \tag{4}$$

for MDD with the 4WD arrangement

$$\eta_r = \frac{N_b \cdot (\varphi_b - f_b) + N_c \cdot (\varphi_c - f_c) - N_a \cdot f_a}{N_b \cdot \varphi_b + N_c \cdot \varphi_c}. \tag{5}$$

The slipping efficiency was determined from the following dependence:

$$\eta_s = 1 - \delta_s, \tag{6}$$

where δ_s is the tractor rear axle slipping.

$$\delta_s = A \cdot \frac{P_g - (\varphi_c - f_c) \cdot N_c}{N_a + N_b} + B \cdot \left[\frac{P_g - (\varphi_c - f_c) \cdot N_c}{N_a + N_b}\right]^2, \tag{7}$$

where *A, B* are slip curve approximation coefficients of the MDD.

Analyzing the nature of the slip efficiency curves (η_s), it can be seen (Figure 5) that the tractive effort of the MDD (6WD) is growing. However, it continually remains higher than the MDD (4WD) slip efficiency (Figure 5). The difference between them gradually increases; that is, the reduction intensity η_s of the second is greater than that of the first. The latter can be explained by the more advanced MDD (6WD) chassis than the MDD (4WD).

The nature of the efficiency flow (η_r), for both MDDs, in the zone of low tractive forces, is almost the same (area up to 22 kN). However, for MDD (6WD), the rolling efficiency increases more rapidly, starting from a tractive effort of 22 kN (Figure 5). Also, the difference between the values η_r is growing continuously. It is not surprising, since the MDD (6WD) slip is less in this area. This fact leads to the shift of maximum traction efficiency of MDD (6WD) into the area of high tractive forces.

Next, let us analyze how the inclination angles of the top (α) and lower (β) links of the EM rear hitch linkage (Figure 3) affect the changing nature in the power delivery efficiency of MDD.

Calculations revealed two patterns. The first of them is that an increase in the parameter α from 1 to 20° causes an increase in the PDE of the modular draft device (Figure 6). For example, with MDD pulling force 22 kN, its PDE increases from 0.420 (when $\alpha = 1°$) to 0.436 (when $\alpha = 20°$), i.e. by 3.8%.

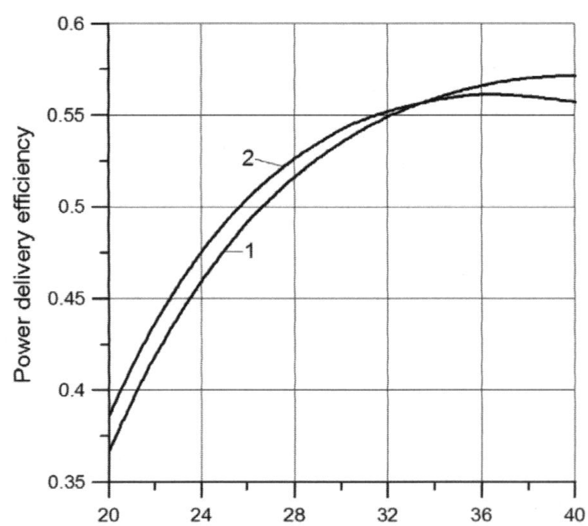

Figure 6. Dependence of MDD power delivery efficiency on its net traction: 1 – $\alpha = 1°$; 2 – $\alpha = 20°$

IOP Conf. Series: Materials Science and Engineering 1021 (2021) 012043 doi:10.1088/1757-899X/1021/1/012043

The reason for this result is the following. Research [1] established the influence of installing the MDD EM rear hitch linkage top link on the nature of vertical loads on MDD axles. Simultaneously, it has been proven that an increase in the value of the angle α leads to an increase in the vertical load on the EM rear wheels. The result is less MDD wheel slip while PDE increases.

The second regularity of the results, the graphic interpretation of which is shown in Figure 6, describes that as the tractive effort P_g increases, the influence of the inclination angle of EM rear hitch linkage top link on the change in PDE is neutralized.

In the same study [1], it is noted that the intensity of additional vertical loading growth of EM rear wheels and TM wheels gradually decreases with an increase in the value of the angle α. As a result, this determines the nature of the PDE dynamics of the modular draft device, shown in Figure 6 for P_g values over 34.5 kN.

Another design parameter that theoretically can influence the dynamics of the MDD PDE is the angle of inclination of the EM rear hitch linkage lower links β (see Figure 3).

The theoretical calculations analysis showed that changing this parameter value from 0 to 10° has little effect on the nature of the change in MDD PDE. Moreover, both for small and large values of the angle α. The reason for this result is that a change in the values of the angle β within the specified limits (0-10°) causes a slight redistribution of vertical loads on the axles of the modular draft device [1].

An increase in the kinematic discrepancy coefficient (K_{ve}) in the wheel drive of the MDD energetic module (in this case, from 1.00 to 1.05) causes its rear wheels to create a so-called "pushing" effect. As a result, this contributes to slipping reduction and, therefore, to PDE increase of the modular draft device (Figure 7).

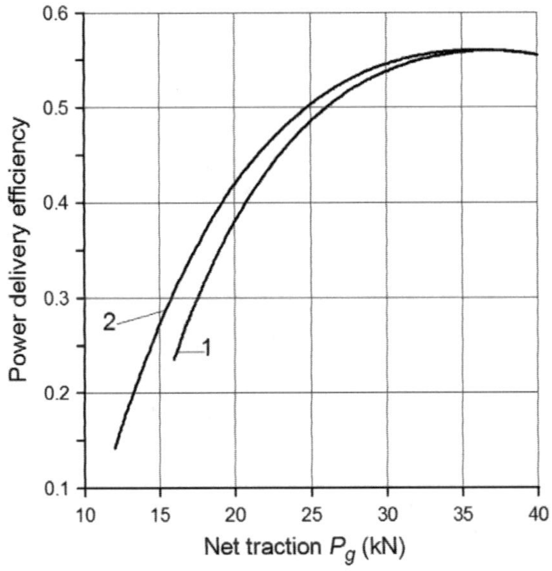

Figure 7. Dependence of MDD power delivery efficiency on its net traction: 1 – K_{ve} = 1.00; 2 – K_{ve} = 1.05

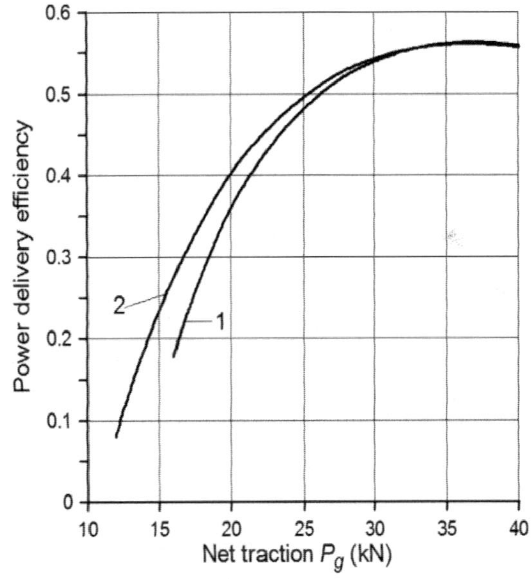

Figure 8. Dependence of MDD power delivery efficiency on its net traction: 1 – K_{vt} = 1.00; 2 – K_{vt} = 1.05

At the same time, as the P_g force increases, the MDD slip also increases. Because of this, its wheels form a deeper track. As a result, MDD's rolling resistance increases. In turn, it causes a decrease in the intensity of growth of its PDE with an increase in the value of the parameter K_{ve}.

This explains the fact that with an increase in the kinematic discrepancy coefficient in the wheel drive of the MDD energetic module, the intensity of its PDE growth, as the force P_g increases decreases. When the MDD tractive effort is more than 32 kN, the rolling resistance of its wheels increases so much that it practically neutralizes the influence of an increase in the K_{ve} coefficient on the process of changing PDE of MDD.

A similar quality result is obtained in changing the kinematic discrepancy coefficient (K_{vt}) in the wheel drive of the MDD technological module (Figure 8). The difference is only in quantitative indicators, which is entirely logical and understandable. For example, the influence of an increase in the parameter K_{vt} from 1.00 to 1.05 on MDD PDE dynamics neutralized at a slightly lower value of the tractive effort it develops. In this case, it is approximately 29 kN (see Figure 8).

5. Conclusions

Calculations have shown that the maximum power delivery efficiency of the 6WD modular draft device is approximately 10% higher than that of the 4WD one.

In the 12-22 kN traction range, the power delivery efficiency of the 4WD modular draft device is higher than the 6WD one. At high MDD tractive effort values, the nature of change in the PDE values becomes the opposite. Moreover, the greater the tractive effort of the modular draft device, the greater difference between the PDE values in favor of the 6WD arrangement version.

When choosing the installation angle of the top link of the MDD energy module's rear hitch linkage, it is desirable to give preference to large values of parameter α, since in this case, its PDE increases. Simultaneously, installing the lower rear hitch linkage links at an angle to a horizon of 0-10° does not significantly affect this indicator.

With MDD traction forces up to 29-32 kN, an increase in kinematic discrepancy coefficient in the wheel drive of MDD energetic (K_{ve}) and technological (K_{vt}) modules from 1.00 to 1.05 contributes an increase in PDE. At large MDD tractive effort values, this parameter change dynamics is practically invariant concerning the increase in coefficients K_{ve} and K_{vt}.

References

[1] Nadykto V T 2003 *Fundamentals of modular draft devices application* (Melitopol: KP "MMD")
[2] Bulgakov V, Kyurchev V, Nadykto V and Olt J 2015 Structure Development and Results of Testing a Novel Modular Power Unit *Agriculture and Agricultural Science Procedia* **7** 40-44
[3] Adamchuk V, Bulgakov V, Nadykto V, Ihnatiev Y and Olt J 2016 Theoretical research into the power and energy performance of agricultural tractors *Agronomy Research* **14(5)** 1511-18
[4] Bulgakov V, Nadykto V, Kyurchev S, Nesvidomin V, Ivanovs S and Olt J 2019 Theoretical background for increasing grip properties of wheeled tractors based on their rational ballasting *Agraarteadus* **30(2)** 78–84
[5] Bulgakov V, Adamchuk V, Nadykto V, Kyurchev V and Nozdrovicky L 2017 Theoretical consideration of the controllability indicator of machine-tractor unit movement *Acta Technological Agriculturae* **1** 11-18
[6] Bulgakov V, Pascuzzi S, Ivanovs S, Nadykto V and Nowak J 2020 Kinematic discrepancy between driving wheels evaluated for a modular traction device *Biosystem Engineering* **196** 88-96
[7] Platonov A A 2014 Classification Features Constructive-Technical Parameters of Road Rail Vehicles *Modern Problems of Science and Education* **2** 161-168
[8] Platonov A A 2016 On the prospects of application of tractors for couse combined to implement shunting operations of the wagons *Modern technics and technologies* 3(55) 78-86
[9] Popov A T and Diakonova N S 2012 Locomobile as an Alternative to Diesel Switchers *World of Transport and Transportation* **03** 66-69
[10] Tarasov D E 2017 The Development of Hybrid Road-roll Rolling Stock in Russia Switchers *World of Transport and Transportation* **15(2)** 74-80
[11] Jennings B 2009 Investigating private railroad operations and the locomotive issue *Journal of Transportation Management* **20(1)** 48-58
[12] Konarzewski M and Niezgoda T 2013 Hybrid locomotives overview of construction solutions *Journal of KONES Powertrain and Transport* **20(1)** 127-134

[13] Kozachenko D, Dovbnia M, Ochkasov O, Serdiuk V, Shepotenko A and Keršys A 2018 Rationale for Choosing the Type of Traction Rolling Stock for the Enterprise of Industrial Transport *Proceedings of 22nd International Scientific Conference. Transport Means* 991-5

[14] Kozachenko D M, Ochkasov O B, Shepotenko A P and Sannytsky N M 2017 Prospects of the private locomotives usage for goods traffic in the direction of sea ports *Science and Transport Progress, Bulletin of Dnipropetrovsk National University of Railway Transport* **6(72)** 7-19

[15] Jennings B 2009 Investigating private railroad operations and the locomotive issue *Journal of Transportation Management* **20(1)** 48-58

[16] Konarzewski M and Niezgoda T 2013 Hybrid locomotives overview of construction solutions *Journal of KONES Powertrain and Transport* **20(1)** 127-134

[17] Myamlin S, Kozachenko D and Grevtsov S 2014 Improving the competitiveness of railways through the use of bimodal technologies transportation *Ukrainian railways* **12(8)** 20-23 (in Ukrainian)

[18] Ruger B 2003 Kombinierter Verkehr als Rettung des Schienenguterverkehrs *Technische Universitat Wien (Institut fur Eisenbahnwesen, Verkehrswirtschaft und Seilbahnen)* **5** 1-27

[19] Sebesan I and Ene M-O 2015 Method of static determination of the safety against overturning of the road-rail machines *University POLITEHNICA of Bucharest Science Bulletin* **Ser. D 77(1)** 51-60

[20] Iwnicki E 2006 *Handbook of Railway Vehicle Dynamics* (CRC press Boca Raton S.U.A.)

[21] Wickens A H 2003 *Fundamentals of Rail Vehicle Dynamics: Guidance and Stability* (Lisse: Swets & Zeitlinger BV)

[22] Daniyan I A, Mpofu K, Daniyan O L and Adeodu A O 2019 Dynamic modelling and simulation of rail car suspension systems using classic controls *Cogent Engineering* **6(1)** 1602927

[23] Khromova G, Mukhamedova Z and Yutkina I 2017 Mathematical mode of oscillations of bearing body frame of emergency and repair railcars *Transport problems* **12(1)** 93-102

[24] Droździel J and Sowiński B 2008 Simulation of railway track deterioration influenced by ballast stiffness and dry friction *WIT Transactions on The Built Environment* **103** 693-702

[25] Shubin A A, Vitchuk P V and Smolovik A E 2019 Development of a backup drive in locomobile construction *IOP Conference Series: Materials Science and Engineering* **560** 1-7

[26] Nadykto V, Arak M and Olt J 2015 Theoretical research into the frictional slipping of wheel-type undercarriage taking into account the limitation of their impact on the soil *Agronomy Research* **13(1)** 148-157

[27] Chaplinskiy A 2007 Analysis of the influence of changes in the design parameters of MED of traction class 1,4-3 on its traction efficiency *Visnik Kharkiv Petro Vasylenko National Technical University of Agriculture* **67(1)** 193-201

[28] Chaplinskiy A 2012 The traction efficiency definition of the module power units of the variable drawbar category 1,4-3 *Scientific Bulletin of the Tavriya State Agrotechnological University* **3(2)** 88-94 (in Ukrainian)

IOP Conf. Series: Materials Science and Engineering 1021 (2021) 012044 doi:10.1088/1757-899X/1021/1/012044

Principles of Energy-effective Land Use in Automobile Transport in Ukraine

I O Novakovska[1], P F Zholkevskyi[1], M P Stetsiuk[1], N F Ishchenko[1,2]

[1]Department of Aerospace Geodesy and Land Management, National Aviation University (Kyiv)

[2]E-mail: nataliia.ishchenko@npp.nau.edu.ua

Abstract. The article studies contamination of the environment with motor transport and the objects of road infrastructure. It considers the experience of the European Countries in implementation of investment programs to modernize the transport and the infrastructure as a whole by their "ecologization" and "green transport" to minimize the negative impact on nature and preserve the competitive opportunities. It contemplates on types and sources of bio-fuel used in European countries determines the opportunities to grow the plants for alternative fuel production in Ukraine. The concept of "greening" in relation to the transport system of Ukraine have been considered, as in the scientific literature this concept is quite new. The authors outline the basic issues of the National transport strategy for the period till 2030 on the necessity to reduce the emissions of greenhouse gases into the atmosphere with movable sources, raising a share of public transport and electric vehicles, electrical buses, muscle-driven vehicles (bikes). The directions of providing the "greening" principle of the transport system of Ukraine have been offered. The necessity of introducing the "reservation" concept for the land of automobile transport to the following normative legal documents have been substantiated: National transport strategy till 2030, State strategy of development of highways, perspective plans of construction, reconstruction or expansion of highways and others. The article contains a plan of actions to decrease the influence of transport on the environment, namely through the implementation of power-saving technologies, the use of alternative fuels, renewal and extending possibilities to use electric transport. The comparative analysis of economical and ecological aspects of different types of transport – “green vehicles”.

1. Introduction

The motorway system, as an integral part of the united transport system of the country, is constantly developing in recent years. The flows of vehicles on the Ukrainian roads are increasing. The rapid growth in the number of vehicles, the intensity of motion, the network of international and national transport corridors, is, on the one hand, positive for economic values of the country. However, on the other hand, it harms the environment and health of people.

2. Presentation of the main research material

The principle of "greening" is based on activities related to the production, distribution and consumption of goods and services of all activities kinds and lead to an increase in human well-being over time. At the same time, no ecological damage is caused to the environment at such a level that may threaten future generations with complete or partial destruction of the biological diversity of nature. [2]

The concept of the automobile transport system of Ukraine has recently appeared in the scientific doctrine, so there is no unanimity in its interpretation. However we may admit that this is "the process of penetration of ideas, knowledge, laws of ecology, ecological thinking, penetration of ecological aspects into all spheres of activity and management of the transport system of the country.. This principle in the transport system ought to be provided in two ways.

The first principle includes measures at the stage of operation of transport complexes, the alternative fuels use, ensuring the use of environmentally friendly materials in the design and construction of transport facilities [1, 5].

The European community has understood the negative effect of an outdated and inefficient transport system. It's now on the way of a fight with its consequences. The countries of the European Union are actively implementing quite large-scale investment programs to modernize the transport and infrastructure as a whole through ecologization thereof and the "green transport" to minimize the negative effect on the environment and to preserve the competitive opportunities [10].

The European community has replied to the environmental challenge with the "green transport" – vehicles of any kind with a low negative effect on the environment. The EU countries are actively implementing the large-scale and long-term investment programs to modernize the transport and infrastructure by "ecologization" thereof to minimize the adverse effect on the environment and preservation of competitive position [3, 6].

Studying the foreign experience of protecting nature against the emissions of motor transport, one can see, that they not only make and use of electric cars. Instead, the countries of the European Union, as well as the USA, Canada, Australia, Brazil are actively implementing the programs aimed to get and use alternative kinds of fuel, namely – bio-fuel made of plants (Figure 1).

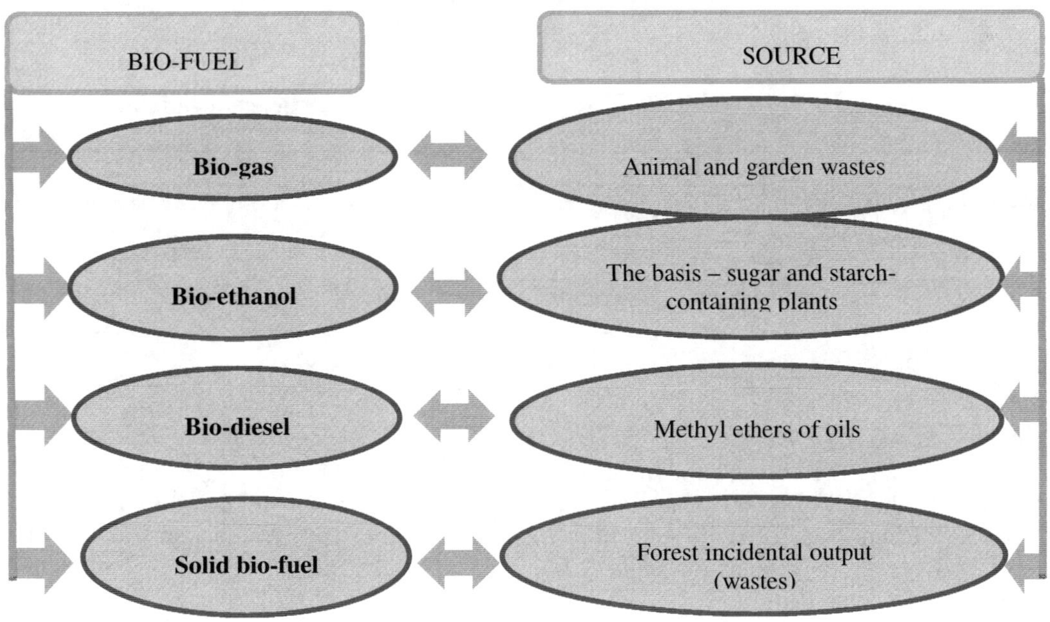

Figure 1. Types and sources of biofuel abroad.

Nowadays, the EU countries use widely just two types of bio-fuel: bio-ethanol and gasoline, and bio-diesel for diesel motors. World corporations, namely DuPont and British Petroleum, are implementing such fuel as bio-buthanol. Both bio-ethanol and bio-butanol come from the plants (corn, wheat, sugar beet, sugar cane, sorghum, and barley). Some producing countries, like Spain, Greece, Italy use sunflower oil, while the majority of European countries – canola oil. While burning, the biofuel also emits CO_2, although it is not harmful to plants and animals, is almost completely processed in water and soil. Within 28 days, the germs process almost 99% of biofuel [14].

The use of this kind of fuel could improve both environmental (by reducing emissions into the environment) and economical situation (due to the development of sales areas, increasing scopes of production, creation of additional working places in Ukraine, etc.).

Studies have shown that the use of alternative fuels and energy, which are used both in rolling stock engines and to meet the needs of stationary facilities of the transport complex have both advantages and disadvantages [1].

The development of alternative sources of fuel is topical for Ukraine, and the government has been supporting it for a long time. This support is documented in the Order by the President of Ukraine "On Measures to Develop the Production of Fuel from Biological Raw Stock" dd. September 26, 2003 [12] and the Law of Ukraine “On Changes to Some Laws of Ukraine to Stimulate Production of Motor Mix Petroleum” dd. January 01, 2011 [11]. Due to the ecological situation, the reformation of the transport policy is still in process. Thus, the EU settled the task to reduce the number of public transport means using a traditional fuel by 50% till 2030, and refuse from them completely by 2050.

According to the National Transport Strategy for the Period till 2030 [7], we have to reduce the volumes of greenhouse gas emissions from the movable sources by 60%, compared to 1990; namely – by increasing the share of public transport and electric transport, electric buses, muscle-driven vehicles (bikes). The aggregate volumes of emissions into the atmosphere from movable sources shall drop by 70% (from the level of 2015). The strategy stipulates the increase in the use of electric transport and electro mobiles, namely – the share of the first mentioned in the local transportation to 75% (by 2030); the increased use of alternative fuel from 10% to 50% (by 2030).

The necessity of changing for electric cars and their natural efficiency is substantiated in table 1.

According to the Ministry of Infrastructure of Ukraine, where there are over 9.2 bln. transport units, each emitting 0.00023 t CO_2/km and travelling approximately 100 km/day [8].

Table 1 The positive effect of electric vehicles in Ukraine

Value	Units
CO_2 emitted by a vehicle daily	0.00023*100*9.2=211.600 t
CO_2 emitted by a vehicle annually	211.600*365=77.234 t
CO_2 absorbed by one tree in a year	120 kg
Trees needed to absorb all CO_2 emissions of all Ukrainian vehicles	77.234/0.12=643.62 mln.
The areas absorbing the emissions: 50 m^2 – one tree (0.005 ha)	643.62*0.005=3218083 ha

Thus, if Ukrainian consumers replace their cars with electromobiles, significant environmental profit would be achieved.

The Ukrainian government stimulates the population to use vehicles with electric motors, providing assistance to them. For instance, according to the Law of Ukraine “On Changes to the Tax Code and some other Regulations of Ukraine to Improve Administering and Re-consider Rates of the Specific Taxes and Duties in Ukraine” of January 01, 2019 [13], the import of electric motor vehicles into Ukraine is exempted from VAT and customs duty till the end of 2022. According to the experts, this makes the electric car 17% cheaper [9].

We can determine the perspectives of electric cars compared to petroleum vehicles through a calculation of their cost recovery period. For the comparison, we took two cars, very similar in their technical characteristics, KIA Soul (Electric) – as of the moment of the purchase its cost was approximately 810 thousand UAH, and KIA Soul produced in the same year (500 thousand UAH). For calculations, let us accept the daily distance of each vehicle as 100 km. The cost recovery period of each vehicle could be determined using the equation [8]:

$$CR = \frac{Cost*S}{Year}, \tag{1}$$

where the *CR* is the cost recovery period in years; *Cost* is the cost of a vehicle, for which the cost recovery period shall be determined, UAH; *S* issavings through the nature of energy sources, UAH; *Year* – days in a year.

For visual clarity, the values of calculation of the cost recovery period for the two vehicles are provided in Table 2.

Table 2 The cost recovery of the vehicles

Value	*KIA Soul (Electric)*	*KIA Soul*
Cost of the vehicle, UAH	810.000 UAH	500.000 UAH
Fuel consumption per 100 km	27 tWt/hour	9.6 l
Cost of a power unit, UAH	1.68 UAH/kW	30 UAH/l
Cost of 100 km trip, UAH	27*1.68=45.36	9.6*30=288
Cost of 1 km, UAH	0.4536	2.88
Savings per 100 km, UAH	288-45.36=242.64	-
The cost recovery period, years	(810.000/242.64)/365=9 years	-

Hence, the cost recovery period for KIA Soul (Electric) is 17 times less, compared to a petroleum-motor vehicle. The cost of fuel for an electric vehicle is 17 times less, compared to the petroleum one. This fact is one of the most important ones for a consumer, for money-saving. Besides, these vehicles have less moving parts, and it reduces the frequency of visiting technical services.

The second direction might be provided by introducing the reservation of automobile transport land and its implementation at the legislative level, which will allow to form a new paradigm of the institutional environment for the development of this sector at the state level.

American scientists M.-K. Billy, L. Smith, J. Scott, S. Fox characterize land reservation as one of the functions of planning. That is why it is necessary to pay special attention to the perspective organization of transport lands in order to determine the social, ecological, socially necessary and economically feasible form of their use and adjacent territories [2].

It's worth to emphasize, that in the situation of the rapid technical, technological, and scientific development, the number of cars is constantly growing. It stipulates the growth of the network in international and national motor corridors. Therefore, the matter of the building and repair to the motorways and minimization of their impact on the environment is becoming more and more topical. These works demand the strict following of environmental standards.

To control these matters, Ukraine has the "Building Regulations SBR B.2.3-218-007:2012. Ecological requirements to the motorways. Designing". They settle the design requirements for building, reconstruction, and overhaul of common motorways and artificial constructions. Especially topical are those relating to environment protection [6].

These industrial regulations are aimed at the development of projects for building, reconstruction, and overhaul of common motorways, considering the need to protect the environment, ensure constant development, and overcome depression in some regions. The purpose of the regulations is to improve the state road network and transport infrastructure, to ensure the proper level of using the motorways in Ukraine.

The basic impact of motorways and artificial constructions on the environment, solutions to prevent them are described in the SBN B.2.3-218-007:2012 [4] in a form of types and solutions. (table 3).

Table 3 The impact of motorways of environment, designing* [4].

Impact type	A solution to reduce the impact
Air contamination	Noise barrier panels Design of 1st and 2nd category of roads, in circumvention of the settlements Establishing a certain schedule for motor transport
Contamination of territories, soils with building materials	Preferable use of low-evaporating materials Planting green areas
Erosions of soils due to the artificial constructions	Installing the discharge facilities Designing bridges
Low road safety	Organization of bypasses Installation of the respective road signs Arranging additional road junctions
Roadsides contamination	Arranging sites for vehicles to stop and park
Contamination of water objects with motorway sewages	Sewages purification Distribution of the sewages along the whole length of the road

It should be mentioned that the expediency of the measures stated on table 3 shall have a significant effect of the environmental situation, namely: elaboration and creation of naturally-safe constructions and building materials, improved constructions of the vehicles to reduce the harmful emissions, created an algorithm to monitor the environment and roadsides, improving the methods to manage the environment protection and transport flows to reduce the emissions of the transport.

3. Conclusion

The authors have proposed an action plan, such as using of alternative fuels, design with environmentally friendly materials, construction, reconstruction of transport infrastructure, and on the other - the inclusion of " reservation "of lands for motor transport to such normative-legal documents: National Transport Strategy until 2030, State Strategy for Road Development, long-term plans for construction, reconstruction or expansion of roads, Norms for allocating land for construction (reconstruction) of roads. The principles of the proposed plan are based on global market trends in the energy-saving technologies use in transport, the development of which, according to our calculations, in one year will save 782.414 billion UAH from funds spent today on the transport system of Ukraine. Using of "green transport" in Ukraine is increasing every year. Herewith in 2013 the number of electric cars was 50 units, in 2018 was 6740 units, which has been increasing of about 20% annually. The fuel consumption of an electric car is less than 17 times lower than that of a petrol car, and this fact is one of the biggest advantages for the consumer in terms of cost savings. On the other hand, a significant environmental effect is achieved. The processing of CO_2 emissions in the country for the year would require an area of 32.18 thousand km^2 and the use of cars on alternative fuels will reduce these areas by up to 10% and reduce harmful emissions into the environment 35%. Thus, high environmental friendliness combined with high payback creates conditions for the mass use of "green transport" as a preservation of energy efficiency of the Ukrainian transport system.

References

[1] Boichenko S, Zaporozhets O 2017 *Transport ecology manual.* (Ukraine: Center for Educational Literature) p 508

[2] Bailey M-K, Smith L., Fox S, Scott D, Grimes G. 2005 *Integrating Land Use Considerations into Transportation Projects to Address Induced Growth.* (Virginia) p 46.

[3] Chygryn O. 2016 The mechanism of the resource-saving activity at joint stock companies: the theory and implementation features (*International Journal of Ecology Development).* **31 (3)** 42–59.

[4] Ecological Demands to Motorways. Designing. GBN V.2.3-218-007:2012. Dated Aug.06, 2012 307

[5] Gobela V V (2019) Theoretical analysis of ecologization as a socio-economic phenomenon *(Ukraine: efficient economy)* ***6*** *1-7*

[6] Mayhew A (2013) *Ukraine and the European Union: financing accelerating integration* (Warsaw: Urzad Komitetu Integracji Europejskiej) p 216

[7] National transport strategy of Ukraine for the period till the year 2030: the Ordinance by the Cabinet of Ministers of Ukraine of May 30 2018 No. 430-r

[8] Novakovska I, Ishchenko N 2019 The studies of an environmental component in the economical efficiency of land management for automobile transport. (*Poland*: *International journal of innovative technologies in economy)* **4(24)** 22–28

[9] Novakovskaya I O, Ischenko N F 2017 Problems of land allocation and use for the needs of the road economy (*Ukraine: Modern issues of economics and law*) **1-2 (5.6)** p. 135-144

[10] Novakovs'ka I, Ishenko N 2017 Transport strategy of Ukraine in the context of European integration Latvia: Collective Monograph p 248

[11] On alterations to some laws of Ukraine on stimulating the production of motor mix petroleums: the Law of Ukraine of January 01 2011 No. 3502

[12] On measures to develop fuel production from biological raw stock: the Order by the President of Ukraine of September 26 2003 No. 1094

[13] On changes to the Tax Code and some other regulations of Ukraine on improvement administration and re-consideration of the tariffs of some taxes and duties: the Law of Ukraine of January 01 2019 No. 2628

[14] State Statistic Service of Ukraine (2019) Available at: http://www.ukrstat.gov.ua/

Kinematics of motion of rotors of an orbital hydraulic machine

A Panchenko [1, 3], A Voloshina[1], P Luzan[2], I Panchenko[1] and S Volkov[1]

[1]Department of Mechatronic Systems and Transport Technologies, Dmytro Motornyi Tavria State Agrotechnological University, B. Khmelnitsky ave. 18, Melitopol, Ukraine
[2]Institute of Vocational Education National Academy of Pedagogical Sciences of Ukraine, Vito-Litovsky lane, 98-a, Kyiv, Ukraine

[3]Email: tia_tgatu@ukr.net

Abstract. There is no "rigid" kinematic connection between the centers of the rotors of the orbital hydraulic motor, which leads to non-standard changes in its output characteristics. A kinematic diagram of the movement of the rotors of an orbital hydraulic motor was developed, which made it possible to substantiate the kinematics of movement of its rotors and analyze the change in output characteristics. Abnormal changes in the output characteristics of this type of hydraulic motors are changes in the overall efficiency and changes in the angular speed of rotation of its shaft. When the gap between the rotors changes in the range of 0 to 0.4 mm, the overall efficiency of the positive displacement hydraulic machine decreases 3.6 times, which is explained by the increasing pour over between its rotors. The high value (0.92) and constancy of the volumetric efficiency of the orbital hydraulic motor is explained by the property of its rotors "self-sealing" in the process of moving, eliminating overflows. A decrease in the angular speed of rotation of the shaft of an orbital hydraulic motor 35 to 27 rad/s, with an increase in the gap of 0 to 0.4 mm, is explained by additional displacements of its inner rotor.

1. Introduction

Various types of hydraulic machines [1–3] are used to drive the active working bodies of self-propelled vehicles, but special attention is paid to orbital [2, 4] hydraulic motors. Orbital motors are a new, little-studied hydraulic machine with high torque and low output shaft speed. These hydraulic machines are widely used in mechatronic systems with a gearless hydraulic drive of active working bodies and running systems of construction, railway, agricultural, drilling, municipal, logging and other self-propelled equipment.

A distinctive feature of these hydraulic motors is the presence of external and internal rotors with a special hypocycloidal tooth profile [4]. There is no "rigid" kinematic connection between the centers of the rotors of the orbital hydraulic motor, which leads to non-standard changes in its output characteristics [5]. The stability of the output characteristics of hydraulic drives of self-propelled vehicles is determined by the parameters of the hydraulic motors used in these drives.

Therefore, the study of changes in the output characteristics of an orbital hydraulic motor, by modeling the kinematics of the movement of its rotors, is an urgent direction in solving problems related to the design, manufacture and operation of orbital hydraulic motors. The solution of such problems will improve the output characteristics of hydraulic machines for mechatronic systems with a hydraulic drive of active working bodies and running systems of railway, agricultural, logging, drilling, construction, municipal and other self-propelled equipment.

2. Analysis of recent studies and publications

A semi-analytical method for analyzing the contact interaction of structural elements along leveled surfaces [6] was proposed, the coefficients of friction, wear, roughness and hardness of contacting surfaces were determined [7]. Approaches have been developed that involve a combination of the advantages of analytical models and methods for analyzing the stress-strain state taking into account contact interaction [8]. Mathematical nonlinear models of the stress-strain state and approximation methods were used to construct functions that allow evaluating the characteristics of the object under study [9]. The approaches considered make it possible to develop more adequate models for analyzing the response of thin-walled structures to the action of the load [10]. A technique is proposed that takes into account the radial clearance and contact deformations of parts [11], as well as a technique for experimental studies and construction of control characteristics [12], is investigated the wear of the flow path of vortex-chamber blowers [13]. Prediction of aeration and cavitation in gerotor pumps [14] was considered, optimization of gerotor pumps was carried out [15], the values of pressure loss, flow rate and velocity distribution over the cross section of the pipeline were determined [16]. Comparison of the SST turbulence model is carried out taking into account the curvature of streamlines and flow rotation [17], the application of the corrected SST turbulence model is considered, which makes it possible to determine all the main characteristics of the vortex flow [18]. The results of tests of tribosystems with the use of liquid crystal additives [19] are presented, methods for the production and application of carbon nanoparticles in tribology are considered [20]. Experimental studies are considered, taking into account the effect of an electrostatic field on the working fluid of a volumetric hydraulic drive [21]. Issues related to the study of mechatronic systems with a hydraulic drive of active working bodies of self-propelled equipment were not considered.

A method for designing hydraulic mechatronic systems with elements of multicriteria optimization has been developed, which makes it possible to design a mechatronic system with given output characteristics [22]. A dynamic model of the drive and methods for its simplification are proposed [23]. The dynamic processes of mechatronic systems with planetary hydraulic motors were investigated in order to predict changes in their output characteristics [24]. The functional characteristics of a hydraulically driven mechatronic system on a chassis, which consists of an axial piston variable pump and two mechatronic modules (wheel hubs) with orbital hydraulic motors, have been studied [2]. The influence of the design features of orbital hydraulic motors on their functional parameters has not been studied.

In recent years, there has been an intense search and improvement of the applied hydraulic machines in order to create the simplest and most technologically advanced designs at the lowest cost [1]. Particular attention is paid to the development of low-speed hydraulic motors, the working elements of which must have a low speed of movement and develop high torques [2, 24]. The parameters of variation that determine the change in the output characteristics of a planetary hydraulic motor depending on the design features of the system of its rotors [4, 5, 25] and the system of fluid distribution [3, 26] are substantiated. It has been established [27] that one of the main units in these hydraulic machines causing pressure pulsations is its distribution system. Geometric parameters of elements of distribution systems have been substantiated, the influence of geometric parameters and work processes within distribution systems of orbital hydraulic machines on the change in their output characteristics has been studied [3, 26, 27]. The forces acting in the cycloidal gearing [28] were considered, mathematical models were developed that describe the robot of the rotors [29], the change in the geometry of the working surfaces of the rotors [30] and the loads in the gearing [31]. Ways and methods of increasing the bearing capacity of cycloidal gears [32] are presented, an optimal tooth profile for cycloidal gears is proposed [33]. Mathematical models have been developed that describe the relationship between the design features of the rotors and the output characteristics of the orbital hydraulic motor [4]. A two-scale model is considered, which combines the effects of wear, leading to changes in the surface relief of orbital engines [34], a method for determining the reliability of a hydraulic motor by modeling changes in the technical state of rotors [35] is proposed. The kinematics of the movement of the rotors of orbital hydraulic machines was not considered.

The analysis of the performed studies shows that the question of studying the kinematics of the movement of the rotors of an orbital hydraulic motor remains open, and this to a certain extent limits the work associated with the design of new types of orbital hydraulic machines, their manufacture and operation. Therefore, this work is devoted to the study of the kinematics of movement of the rotors of an orbital hydraulic motor, in order to predict changes in its output characteristics at the stages of design, manufacture and operation.

3. Statement of the objective and tasks of the study

To study the kinematics of the movement of the rotors of an orbital hydraulic motor, in order to predict changes in its output characteristics, it is necessary:

- to develop a kinematic diagram of the movement of the rotors of the orbital hydraulic motor;
- substantiate the kinematics of motion of the rotors of an orbital hydraulic motor and a change in its output characteristics;
- investigate the change in the output characteristics of an orbital hydraulic motor by modeling the kinematics of movement of its rotors.

4. The basic part of the study

The basis of an orbital hydraulic motor is a system of its rotors (Figure 1) consisting of an external stationary rotor 1 and an internal movable rotor 3. To reduce contact loads between the corresponding teeth of the rotors, the teeth 2 of the external rotor 1 are made in the form of plug-in rollers. When the hydraulic motor is operating, under the action of the pumping pressure p_{in} of the working fluid, the inner rotor 3 rotates by an angle of movement φ relative to the outer rotor 1. When the inner rotor 3 moves, its center describes a circle with a radius e around the center of the outer rotor 1.

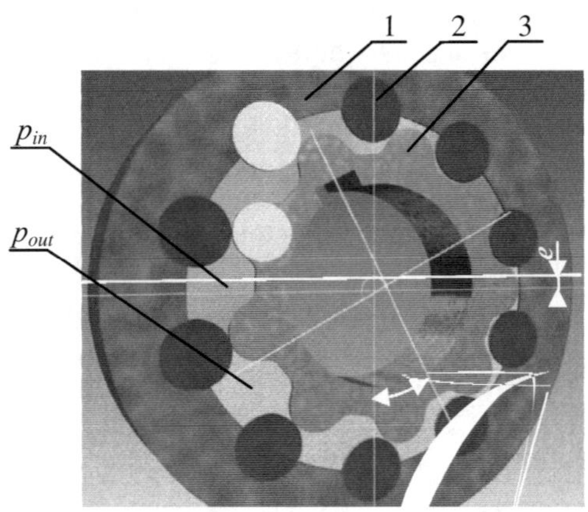

Figure 1. The layout of the rotors of the orbital hydraulic motor: 1 – outer rotor; 2 – tooth (roller) of the outer rotor; 3 – inner rotor; e – eccentricity; φ – the angle of movement of the inner rotor relative to the outer; p_{in} – pressure of the working fluid at the inlet to the hydraulic motor; p_{out} – pressure of the working fluid at the outlet of the hydraulic motor.

Errors in the approximation of the hypocycloidal contour of the profile of the teeth of the outer 1 and inner 3 rotors, the tolerances for their manufacture lead to the presence of an initial diametrical gap G in the location of the rotors (Figure 2). During the operation of an orbital hydraulic motor, as a result of wear of the teeth of its rotors, the diametral clearance G is constantly increasing [4, 5]. Therefore, when studying the kinematics of the movement of the rotors of an orbital hydraulic motor, two cases of the arrangement of the rotors are distinguished: theoretical (Figure 2, a) and real (Figure 2, b).

With the theoretical arrangement of the rotors (figure 2, a) there is no diametrical clearance G ($G = 0$). One of the teeth of the inner rotor 3 always touches two teeth 2 of the outer rotor 1 at points A and B, and the diametrically opposite teeth of the inner rotor 3 contact the teeth 2 of the outer rotor 1 at points C and D.

The points of contact of teeth A, D and B, C are always located on opposite sides of the line of centers of the considered rotors, which makes it possible to separate the high-pressure area p_{in} from the low-pressure area p_{out} without additional structural elements. It should be noted that between the remaining pairs of rotor teeth located in the range of contact points *A*, *D* and *B*, *C*, respectively, the gap $G > 0$.

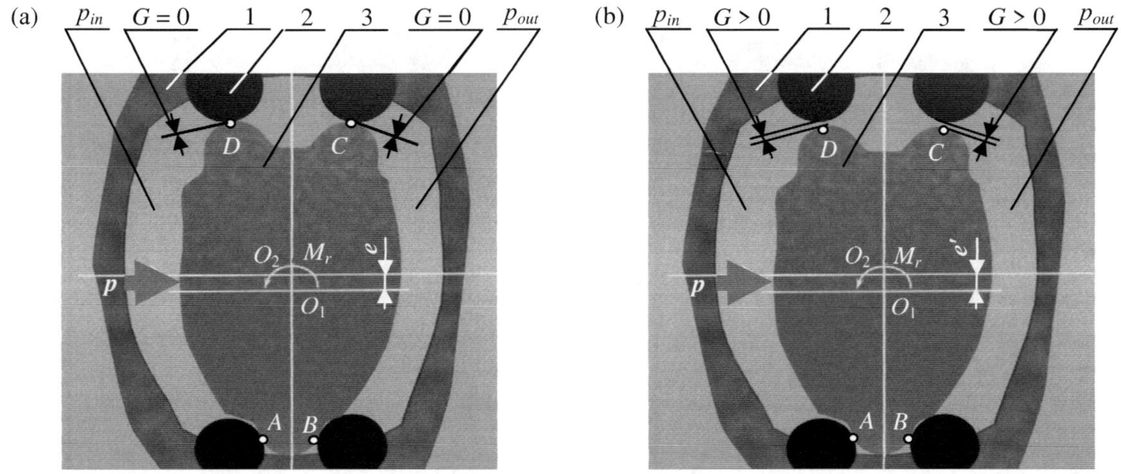

Figure 2. The initial position of the rotors for: a – theoretical position ($G = 0$); b – real position ($G > 0$); *e* and *e'*– eccentricity; O_1 and O_2 – the centers of the inner and outer rotors, respectively; M_r – moment of load resistance; 1 – outer rotor; 2 – tooth (roller); 3 – inner rotor.

Such contacting of the rotors is due only to the geometry of the mating profiles of their teeth and provides a kinematic connection between the centers of the rotors O_1 and O_2, located at a distance equal to the eccentricity *e*. The internal rotor 3 is affected by: the resulting force *P* from the discharge pressure p_{in} of the working fluid and the moment M_r of the load resistance applied at point O_1. Under the action of the resulting force *P*, the inner rotor 3 with the number of teeth z_1, moving flat-parallel, turns through an angle φ, rolling around the inner rotor 1, transmitting the torque M_{tor}.

With the actual arrangement of the rotors (Figure 2, b), there is always a diametrical gap $G > 0$, associated with the technological features of manufacturing and necessary to compensate for the thermal expansion of the rotors. In the initial position, the inner rotor 3 with one of its teeth touches two teeth 2 of the outer rotor 1 at points *A* and *B*. Similar points *C* and *D* of two diametrically opposite teeth of the rotor 3 are located at a distance *G* from the corresponding teeth 2 of the rotor 1. Since the location of the rotors there is no "rigid" kinematic connection between the centers of the rotors O_1 and O_2, the distance between them changes and is equal to $e' = e + G/2$. The high-pressure zone p_{in} with the actual arrangement of the rotors is hydraulically closed through a double slot of height *G* (at points *D* and *C*) with the low-pressure zone pout, causing overflows that reduce the volumetric efficiency of the orbital hydraulic motor. We will call this position of the rotors (Figure 2, b) position I.

In position I (Figure 2, b) the same forces act on rotor 3 as in the theoretical position of the rotors, therefore, to further study the movement of the rotors of the orbital hydraulic motor, we will use the diagram shown in Figure 3.

Under the action of the resulting force *P*, the inner rotor 3 will move from position I to position II (Figure 3, a) until the diametrically opposite tooth of the rotor 3 touches the corresponding tooth of the rotor 1 at point *C*. Moving in position II, the inner rotor 3 moves progressively upward, moving along the normal applied at point *B*, similar to the movement of the hydraulic cylinder piston. When moving from position I to position II, the inner rotor 3 cannot turn through the angle of movement φ, under the influence of the moment of load resistance M_r. Moving to position II, the rotor 3 “floats up” by the size of the diametrical clearance *G*, changing the distance between the centers O_1 and O_2 of the rotors, making it equal to $e'' = e - G/2$. This "floating" of the rotor 3 of the orbital hydraulic motor is caused

IOP Conf. Series: Materials Science and Engineering 1021 (2021) 012045 doi:10.1088/1757-899X/1021/1/012045

by the absence of a "rigid" kinematic connection between their centers O_1 and O_2, which allows the rotors to occupy different positions in the diametrical clearance range G. zones of high p_{in} and low p_{out} pressure, eliminating overflows.

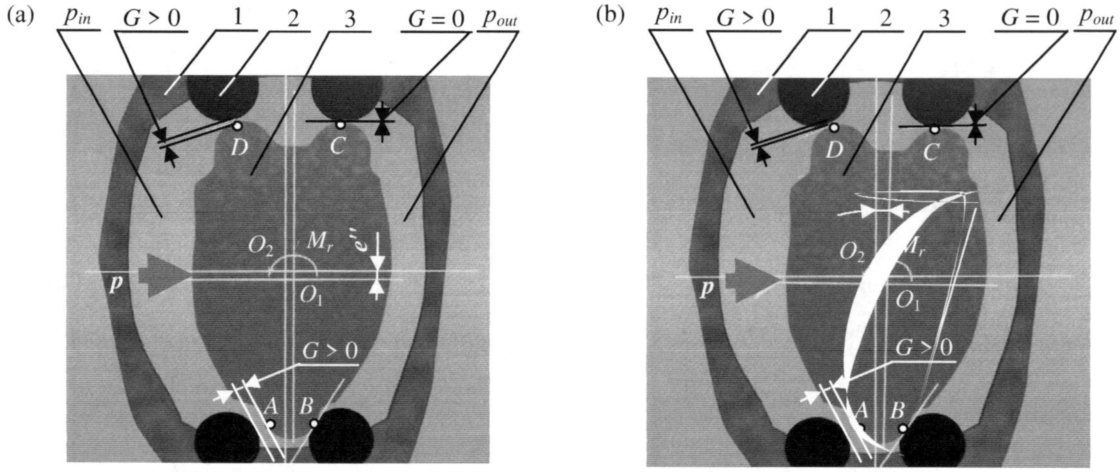

Figure 3. The scheme of movement of the inner rotor in a real position: a – "emerge" of the inner rotor; b – rotation of the inner rotor; e''– eccentricity; φ – angle of movement of the inner rotor; O_1 and O_2 – the centers of the inner and outer rotors, respectively; M_r – moment of load resistance; 1 – outer rotor; 2 – tooth (roller); 3 – inner rotor.

After "floating", under the action of the resulting force P, the inner rotor 3, transmitting the torque Mtor, rotates through the angle φ taking position III (Figure 3, b). In position III, "self-sealing" of the rotors also occurs, since their touching points B and C separate the zones of high p_{in} and low p_{out} pressures, eliminating spill over.

Consequently, the presence of a diametrical clearance G, in the actual arrangement of the rotors, does not cause leakage in the radial direction, since the absence of a "rigid" kinematic connection between the rotors allows them to occupy a position at which "self-sealing" occurs. Therefore, it can be argued that an increase in the diametrical clearance G in the process of wear of the rotors will not affect the decrease in the value of the volumetric efficiency of the orbital hydraulic motor. Attention is drawn to the "forced" movement of the inner rotor 3, in which it takes an additional intermediate position II (Figure 3, a). This movement takes additional time, which in turn affects the decrease in the angular speed of rotation of the inner rotor 3 and, as a consequence, the shaft of the orbital hydraulic motor.

Analysis of the movement kinematics of the rotors of orbital hydraulic machines allows us to conclude that the presence of a diametrical gap G and the absence of a "rigid" kinematic connection between the rotors causes additional movements of the rotors, changing the kinematics of their movement. The phenomenon of "self-sealing" of the rotors of an orbital hydraulic motor, caused by their additional displacements, causes a non-standard change in some of the output characteristics of this type of hydraulic motors. Non-standard changes in the output characteristics of an orbital hydraulic motor include: no leakage rate, which is inevitable in the presence of a gap between the rotors, and, as a consequence, stabilization of volumetric efficiency; decrease in the angular speed of rotation of the hydraulic motor shaft as a result of additional displacements of the rotor 3. Therefore, to predict the change in the rotation frequency of the orbital hydraulic motor shaft depending on the diametral clearance G, it is necessary to determine the dependence of the change in the steady-state angular speed of rotation at the theoretical and real position of its rotors.

The steady-state angular velocity ω_T of the inner rotor 3 for the theoretical arrangement of the rotors (Figure 2, a) is equal to $\omega_T = (\pi \cdot n)/30$, where n is the frequency of rotation of the shaft (inner rotor 3) of the orbital hydraulic motor [4, 5]. It is known that $n = Q/V_0$, where Q is the flow rate of the fluid passed through the hydraulic motor; V_0 is the working volume of the hydraulic motor. After transformations, we have a model of the change in the value of the steady-state angular velocity for the theoretical position of the rotors

$$\omega_T = \frac{\pi \cdot Q}{30 \cdot V_0}. \tag{1}$$

The model for determining the value of the angular velocity ω_p of the inner rotor 3 with a real arrangement of the rotors (Figure 2, b) is described in detail in [4, 5] and is represented by the dependence

$$\omega_p = \frac{\omega_T \cdot (t - 2t_1)}{t}. \tag{2}$$

where t is the time characterizing the steady rate of rotation $t_1$is the time taken to move the inner rotor to the intermediate position II.

The time characterizing the steady-state speed of rotation of the hydraulic motor shaft is determined by the dependence

$$t = \frac{2\pi}{\omega_T \cdot z_1 \cdot z_2} - 2t_1. \tag{3}$$

where z_1 and z_2 are the number of teeth of the inner and outer rotors, respectively.

The translational movement of the inner rotor from position I to position II (Figure 2, b) is similar to the movement of the hydraulic cylinder piston. Therefore, the additional time spent by the inner rotor on this movement can be determined by the known expressions, and, taking into account substitutions and transformations, can be described by the dependence

$$t_1 = \frac{G \cdot D \cdot b}{4Q \cdot \cos\delta}. \tag{4}$$

where G is the path, displacement of the inner rotor from position I to position II (Figure 2), numerically equal to the diametrical clearance G; D is the diameter of the rotor 3; b is the width of the inner rotor.

Thus, by solving successively equations (1), (3), (4) and (2), the values of the angular velocity ω_p of the inner rotor are determined for the actual arrangement of the rotors of the orbital hydraulic motor, taking into account the diametrical gap.

The study of the processes of changing the output characteristics of the hydraulic motor by modeling the kinematics of the movement of the rotors, depending on the change in the gap between them, was carried out according to the data of an orbital hydraulic motor GMP-22 with a working volume of 250 cm^3. The study of the change in the overall efficiency for a volumetric operating principle hydraulic machine was carried out according to the data of an MFS90 axial piston motor with a working volume of 89 cm^3.

When studying the processes of changing the output characteristics of the orbital hydraulic motor depending on the change in the diametral clearance G (Figure 4), the mathematical apparatus given in the works [4, 5] was used. When determining the overall efficiency of an axial piston motor, well-known mathematical relationships were used.

The analysis (Figure 4) of the dependence of the change in the functional characteristics of the orbital hydraulic motor on the diametral clearance G, taking into account the kinematics of the movement of its rotors, shows that they differ from the characteristics of hydraulic machine of volumetric of operating principle. Attention is drawn to the nature of curve 1 (Figure 4) characterizing the dependence of the change in the overall efficiency of a standard of a volumetric operating principle of a hydraulic machine on the change in the gap between the rotors. When this gap changes in the range of 0 to 0.4 mm, the overall efficiency of a volumetric operating principle of the machine decreases by 3.6 times (0.75 to 0.21), which is explained by the increasing leaks between its rotors. The overall efficiency of the orbital hydraulic motor (curve 4) with similar changes in the diametral clearance G remains practically unchanged, since the hydromechanical and volumetric efficiency of this hydraulic machine does not change their values.

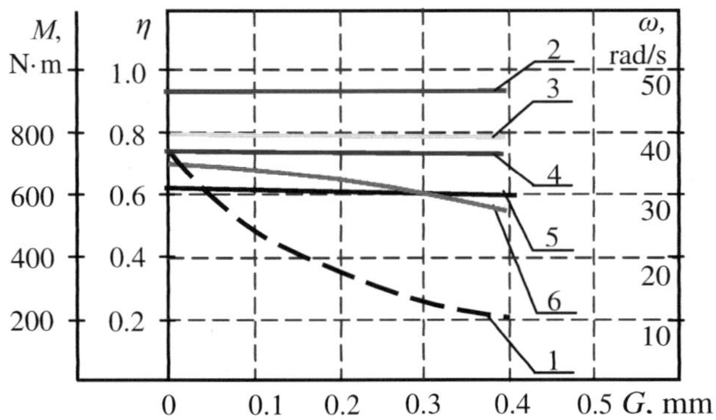

Figure 4. Dependences of changes in the functional characteristics of the orbital hydraulic motor on the diametral clearance: 1 – overall efficiency of a hydraulic machine of a volumetric operating principle; 2 – volumetric efficiency; 3 – hydromechanical efficiency; 4 – overall efficiency; 5 – torque M_{tor}; 6 – angular velocity.

The high value (0.92) and constancy of the volumetric efficiency (curve 2) of the orbital hydraulic motor is explained by the property of its rotors "self-sealing" in the process of movement, eliminating leaks in the diametrical direction.

A constant value of hydromechanical efficiency (curve 3) and torque (curve 5) with a change in the gap between the rotors of displacement hydraulic machines is a normal phenomenon for all hydraulic machines of this type.

Special attention should be paid to the change in the angular speed of rotation of the shaft of the orbital hydraulic motor (curve 6), the values of which decrease from 35 to 27 rad/s with an increase in the gap from 0 to 0.4 mm. Such a change in the angular velocity of the shaft of an orbital hydraulic motor is explained by additional displacements of its inner rotor, which arose in the presence of a diametrical gap and the absence of a “rigid” kinematic connection between the centers of its rotors.

5. Conclusions

As a result of the research carried out, a kinematic diagram of the movement of the rotors of the orbital hydraulic motor was developed, which made it possible to substantiate the kinematics of the movement of the rotors of the orbital hydraulic motor and analyze the change in its output characteristics.

It was found that in the absence of a “rigid” kinematic connection between the rotors of the orbital hydraulic motor, they occupy a position at which “self-sealing” of these rotors occurs. Additional movement of the inner rotor, in which it occupies an intermediate position, which in turn affects the decrease in the angular velocity of rotation of the shaft of the orbital hydraulic motor.

Analysis of the research on the change in the output characteristics of an orbital hydraulic motor, by modeling the kinematics of the movement of its rotors, shows that non-standard changes in the output characteristics of hydraulic motors of this type include changes in the overall efficiency and changes in the angular speed of rotation of its shaft. When the gap between the rotors changes in the range of 0 to 0.4 mm, the overall efficiency of the positive displacement hydraulic machine decreases 3.6 times (0.75 to 0.21), which is explained by the increasing spill over between its rotors. The overall efficiency of the orbital hydraulic motor with similar changes in the diametrical clearance *G* remains practically unchanged.

The high value (0.92) and the constancy of the volumetric efficiency of the orbital hydraulic motor is explained by “self-sealing” of its rotors in the process of moving, eliminating leakage in the diametrical direction.

A decrease in the angular speed of rotation of the shaft of an orbital hydraulic motor from 35 to 27 rad/s, with an increase in the gap from 0 to 0.4 mm, is explained by additional displacements of its inner rotor, which arose in the presence of a diametral gap and the absence of a “rigid” kinematic connection between the centers of its rotors.

References

[1] Gamez-Montero P, Codina E and Castilla R 2019 A Review of Gerotor Technology in Hydraulic Machines *Energies* **12 (12)** 2423

[2] Panchenko A, Voloshina A, Milaeva I and Luzan P 2019 Operating Conditions' Influence on the Change of Functional Characteristics for Mechatronic Systems with Orbital Hydraulic Motors *Modern Development Paths of Agricultural Production. Springer* 169-176

[3] Voloshina A, Panchenko A, Boltyansky O and Titova O 2019 Improvement of Manufacture Workability for Distribution Systems of Planetary Hydraulic Machines *Lecture Notes in Mechanical Engineering, Springer* 732-741

[4] Panchenko A, Voloshina A, Milaeva I, Panchenko I and Titova O 2018 The Influence of the form Error after Rotor Manufacturing on the Output Characteristics of an Orbital Hydraulic Motor *International Journal of Engineering and Technology* **7 (4.3)** 1–5

[5] Panchenko A, Voloshina A, Panchenko I, Titova O and Pastushenko A 2019 Reliability design of rotors for orbital hydraulic motors *IOP Conference Series: Materials Science and Engineering* **708 (1)** 012017

[6] Tkachuk M, Grabovskiy A, Tkachuk M, Saverska M and I. Hrechka 2020 A semi-analytical method for analys of contact interaction between structural elements along aligned surfaces *Eastern-European Journal of Enterprise Technologies* **1 7(103)** 16–25

[7] Marchenko A, Tkachuk M, Kravchenko S, Tkachuk M and Parsadanov I 2020 Experimental Tests of Discrete Strengthened Elements of Machine-Building Structures. *Advanced Manufacturing Processes Lecture Notes in Mechanical Engineering* 559–569

[8] Tkachuk M, Grabovskiy A, Tkachuk M, Hrechka I and others 2019 Investigation of multiple contact interaction of elements of shearing dies *Eastern-European Journal of Enterprise Technologies* **4 7(100)** 6–15

[9] Bondarenko M, Tkachuk M, Grabovskiy A and Hrechka I 2019 Substantiation of Thin-Walled Structures Parameters Using Nonlinear Models and Method of Response Surface Analysis *International Journal of Engineering* **44** 32–43

[10] Atroshenko O, Tkachuk M, Martynenko O, Tkachuk M and others 2019 The Study of Multicomponent Loading Effect on Thin-Walled Structures With Bolted Connections *Eastern-European Journal of Enterprise Technologies* **1 7(97)** 15–25

[11] Gaydamaka A, Kulik G, Frantsuzov V, Hrechka I and others 2019 Devising an engineering procedure for calculating the ductility of a roller bearing under a no-central radial load *Eastern-European Journal of Enterprise Technologies* **3 7(99)** 6–10

[12] Korohodskyi V, Kryshtopa,S, Migal V, Rogovyi A and others 2020 Determining the Characteristics for the Rational Adjusting of an Fuel-air Mixture Composition in a Two-stroke Engine with Internal Carburation. *Eastern-European Journal of Enterprise Technologies* **2(5)** 39–52

[13] Rogovyi A, Khovanskyy S, Grechka I and Pitel J 2019 The Wall Erosion in a Vortex Chamber Supercharger Due to Pumping Abrasive Mediums *Design, Simulation, Manufacturing: The Innovation Exchange* 682–691

[14] Smirnov V and Volkov G 2019 Computation and structural methods to expand feed channels in planetary hydraulic machines *Journal of Physics Conference Series* **1210** 012131

[15] Smirnov V and Volkov G 2018 Systematization and comparative scheme analysis of mechanisms of planetary rotary hydraulic machines *MATEC Web of Conferences* **224** 02083

[16] Chernetskaya-Beletskaya N, Rogovyi A, Shvornikova A, Baranov I and others 2018 Study on the coal-water fuel pipeline transportation taking into account the granulometric composition parameters *International Journal of Engineering & Technology* **7(4.3)** 240–245

[17] Rogovyi A, Khovanskyi S, Hrechka I and Gaydamaka A 2020 Studies of the Swirling

Submerged Flow Through a Confuser *Design, Simulation, Manufacturing: The Innovation Exchange* 85–94.

[18] Rogovyi A 2018 Energy performances of the vortex chamber supercharger *Energy* **163** 52–60

[19] Воронін С 2015 Розробка трібофізичних основ мастильної здатності рідкокристалічних присадок до базових олив *Східно-Європейський журнал передових технологій* **3 7(75)** 53–57

[20] Voronin S, Suranov A, Suranov A 2017 The effect of carbon nanoadditives on the tribological properties of industrial oils *Journal of Friction and Wear* 38(5) 359–363

[21] Voronin S, Onopreychuk D, Stefanov V, Bashkatov Ye and Panchenko V 2018 Reduction of construction duration by improving the anti-wear properties of power fluids in hydraulic drives of earth-moving machines *International Journal of Engineering & Technology* **7(4.3)** 105–109

[22] Panchenko A, Voloshina A, Titova O, Panchenko I and Zasiadko A 2020 The Study of Dynamic Processes of Mechatronic Systems with Planetary Hydraulic Motors *Advanced Manufacturing Processes II. Lecture Notes in Mechanical Engineering* 56–64

[23] Strutynskyi S 2018 Defining the dynamic accuracy of positioning of spatial drive systems through consistent analysis of processes of different range of performance, *Naukovyi Visnyk NHU* **3** 64-73

[24] Voloshina A, Panchenko A, Panchenko I, Titova O and Caldare A 2020 Design of Hydraulic Mechatronic Systems with Specified Output Characteristics *Advances in Design, Simulation and Manufacturing III DSMIE 2020 Lecture Notes in Mechanical Engineering* 42-51

[25] Voloshina A, Panchenko A, Titova O, Milaeva I and Pastushenko A 2020 Prediction of Changes in the Output Characteristics for the Planetary Hydraulic Motor *Advanced Manufacturing Processes II. Lecture Notes in Mechanical Engineering* 84–93

[26] Voloshina A, Panchenko A, Panchenko I and Zasiadko A 2019 Geometrical Parameters for Distribution Systems of Hydraulic Machines *Modern Development Paths of Agricultural Production. Springer* 323-336

[27] Voloshina A, Panchenko A, Panchenko I, Titova O and Zasiadko A 2019 Improving the output characteristics of planetary hydraulic machines *IOP Conference Series: Materials Science and Engineering* **708 (1)**, 012038

[28] Ivanovic L, Miric N, Devedzic G and Ćuković S 2010 Analysis of forces and moments in gerotor pumps *Journal of Mechanical Engineering Science* **10 (224)** 2257-2269

[29] Gamez-Montero J, Garcia-Vilchez M, Raush G, Freire J and Codina E 2012 Teeth Clearance and Relief Grooves Effects in a Trochoidal-Gear Pump Using New Modules of GeroLAB *Journal of Mechanical Design* **5 (134)** No: 054502-7

[30] Choi T, Kim M, Lee G, Jung S, Bae J and Kim C 2012 Design of Rotor for Internal Gear Pump Using Cycloid and Circular-Arc Curves *Journal of Mechanical Design* **1 (134)** 011005-12

[31] Stryczek J, Bednarczyk S and Biernacki K 2014 Strength analysis of the polyoxymethylene cycloidal gears of the gerotor pump *Archives of Civil and Mechanical Engineering* 4 (14) 647–660

[32] Biernacki K 2012 Methods of Increasing Loadability for the Plastic Cycloidal Gears *Fundamentals of machine design* **490** 156–165

[33] Biernacki K 2014 Selection of the optimum tooth profile for plastic cycloidal gears *Proceedings of the institution of mechanical engineers part c-journal of mechanical engineering science* **18 (228)** 3395–3404

[34] Furustig J, Almqvis, A, Bates C, Ennemark P and Larsson R 2015 A two scale mixed lubrication wearing-in model, applied to hydraulic motors *Tribology International* **90** 248–256

[35] Furustig J, Almqvist A and Pelcastre, L 2016 A strategy for wear analysis using numerical and experimental tools, applied to orbital type hydraulic motors *Proceedings of the institution of mechanical engineers part c-journal of mechanical engineering science* **12 (230)** 2086-2097

IOP Conf. Series: Materials Science and Engineering 1021 (2021) 012046 doi:10.1088/1757-899X/1021/1/012046

Estimation of the State of Engine of Mobile Machines in the Conditions of Operation on Basis of Onboard Diagnostics

S V Panchenko[1], M P Remarchuk[1,3], O V Kebko[1], Ya. V Chmuzh[2] and A O Zadorozhnyi[1]

[1]Department of construction, travel and cargo-handling machines, Ukrainian State University of Railway Transport, Feuerbach Square 7, Kharkov, 61050
[2]Dimerus Engineering LTD Vostochnaya Street, 8, Stariy Saltov, Kharkov, 62560

[3]E-mail: remarchyk@ukr.net

Abstract. According to the results of the analysis of literature sources, when the issue of determining the state of internal combustion engines is considered, a single basis at all stages of life has not yet been decided, because it is a complex problem. A feature that can solve this problem is the efficiency coefficient. To justify the efficiency of the engine as a parameter, at the design stage, a systematic approach was applied along with the definition of rank subordination and significance of all components with their representation in the form of functional and structural schemes. According to the schemes, there is the main system-forming component in the structure of engines, which in the operable state provides the conversion of thermal energy of the fuel into mechanical energy of the rotary movement of the engine shaft. All other components of the engine are intended to maintain the functioning of the main component of the internal combustion engine and to create comfortable operating conditions for the operator in the machine system where this engine operates. At the stage of completion of production of the internal combustion engine, bench tests are used to determine its condition on the basis of efficiency and all technical parameters of the engine. The efficiency of the internal combustion engine at the operational stage is ensured by the load of the hydraulic drive in the structure of the machines and the use of computer equipment that can remotely record diagnostic parameters.

1. Introduction

During the construction of railway structures, mobile machines are used, which include track, earthmoving and loading handling machines (TELHMs). The source of power for such machines is spark-ignition and diesel internal combustion engines, which require time-consuming and science-intensive research to determine their state under operating conditions. It is well known that if the state of the internal combustion engine during the operation of TELHM changes it results in decreased productivity and higher costs of fuel. It has been established that the technical and economic indicators of TELHMs are justified at the design stage, observed at the manufacturing stage and maintained in the operational stage, i.e., at all stages of life.

A feature that can characterize the state of the internal combustion engine at all stages of life of TELHMs must be relatively stable, sensitive to changes of the state, and the able to convert it into a signal for recording by modern measuring and computer equipment. The efficiency is accepted as such parameter [1] to determine the state of the internal combustion engine at all stages of life.

2. Analysis of recent studies and publications

According to the analysis of sources [2, 3, and 4], considerable attention is paid to determining the efficiency of TELHM engines. For instance, to determine the efficiency of engines in mass production, a variety of equipment and devices are used to measure the corresponding parameters. In particular, measurement and recording of such parameters as: – torque; – rotation speed of the

IOP Conf. Series: Materials Science and Engineering 1021 (2021) 012046 doi:10.1088/1757-899X/1021/1/012046

crankshaft; – fuel consumption for a given period of time, air; – the ignition or fuel injection advance angle; – water temperature, lubricant; – quality and chemical composition of the waste combustion products, and others.

The above parameters are determined on the test bench shown in figure 1.

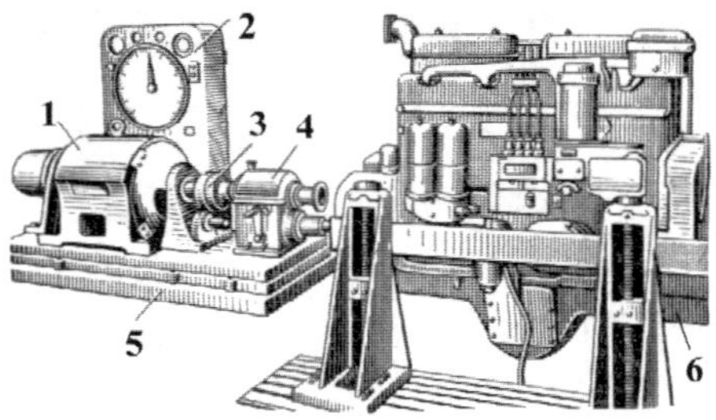

Figure 1. Test bench for diesel engines and determining of its efficiency: 1 – electric motor; 2 – control panel; 3 – elastic coupling; 4 – gearbox; 5 – foundation plate; 6 – mounting device.

So, according to [2], the value of the efficiency of the internal combustion engine η_e as an energy source for TELHMs is determined experimentally by the test bench method using the formula

$$\eta_e = (632 \cdot N_e)/(G_p \cdot Q_p), \tag{1}$$

where 632 is the heat equivalent, kcal/(h.p. hour); N_e is effective power, h.p.; G_p is fuel consumption, kg/year; Q_p is calorific value of fuel, kcal/kg.

According to the recommendation of the standard [3], the efficiency of the internal combustion engine η_e is based on the dependence

$$\eta_e = 3600/(H_u \cdot g_e), \tag{2}$$

where H_u is lower range of fuel calorific value, which is 42.7 MJ/kg; g_e is specific fuel consumption, kg/(kW year).

According to modern studies [4], the efficiency of internal combustion engine η_e can be determined by the formula

$$\eta_e = 84{,}3/g_e. \tag{3}$$

One of the necessary devices used in determining the efficiency at the stage of completion of production or after repair of the internal combustion engine is a special brake equipment, which is shown in the domestic design version figure 2 a) and in the foreign version figure 2 b).

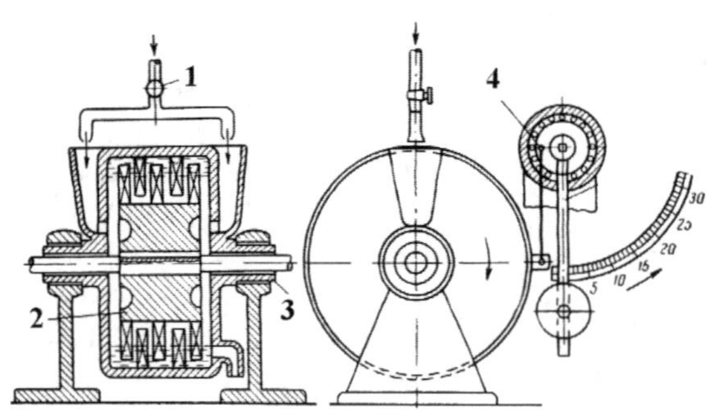

(a)

(b)

Figure 2. Hydraulic brake equipment: a) domestic brake design; b) foreign brake design; 1 – water supply pipeline; 2 – hydraulic brake; 3 – bearing support; 4 – torque weighing device.

The braking equipment shown in figure 2 (a) and (b) is a component of the test bench that is used to generate torque on the internal combustion engine shaft to maintain a given engine speed when conducting appropriate experimental tests and acceptance tests in the laboratory.

3. Statement of the objective and tasks of the study

The purpose of this article is to assess the state of the internal combustion engine, as a component of the TELHM, at the engine design, manufacture and operation stage in terms of efficiency based on the use of on-board diagnostics.

The following tasks must be completed to achieve the above goal:

– using a systematic approach to the consideration of internal combustion engines and presenting its components as a system based on the rank subordination and importance of all components to improve the quality of engine design;

- justifying the process of on-board diagnostics of internal combustion engines by the efficiency directly in the structure of hydroficated TELHMs.

4. Problem solving

At the systemic level, the internal combustion engine [5, 6, and 7] is a system that is shown as simplified as possible, in figure 3.

So, at the stage of completion of production of the internal combustion engine, according to (figure 3), its output power N_{ex} (kW) is established by conducting bench test studies. To do this, on the test bench (figure 1), when braking equipment is used (figure 2), the torque on the internal combustion engine shaft M_{kr} (kN m) is determined and the angular velocity of the engine shaft ω_v (1/s) is measured.

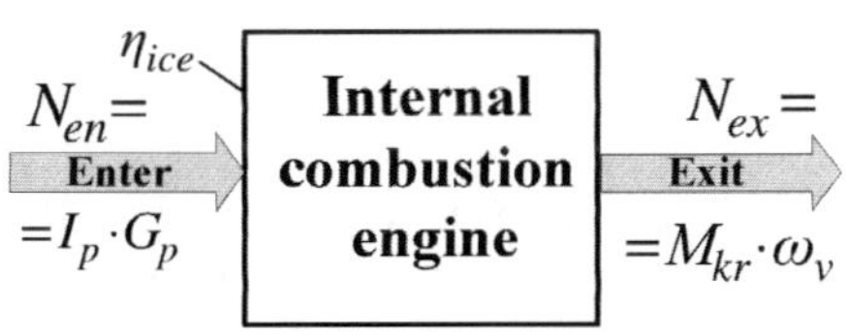

Figure 3. Schematic diagram of the internal combustion engine: N_{en}, N_{ex} is power input and output of the internal combustion engine; η_{ice} efficiency of the engine; I_p is fuel calorific value, kcal/kg; G_p is fuel consumption, kg/year; M_{kr} is torque, kN m; ω_v is angular speed of the shaft, 1/s.

Based on the results of the study, with a step-by-step increase in torque M_{kr}, thus changes in the angular speed of the shaft ω_v are recorded, which are further used to estimate the stepwise power output of the internal combustion engine, which is determined by the known dependence

$$N_{ex} = M_{kr} \cdot \omega_v . \quad (4)$$

Simultaneously with determining the output parameters of the internal combustion engine based on the use of the test bench (figure 1) and braking equipment (figure 2), the value of the change in fuel consumption G_p (kg/year) is measured in steps at the input of the internal combustion engine, according to figure 4. Moreover, the known value for this fuel is its calorific value I_p (kcal/kg).

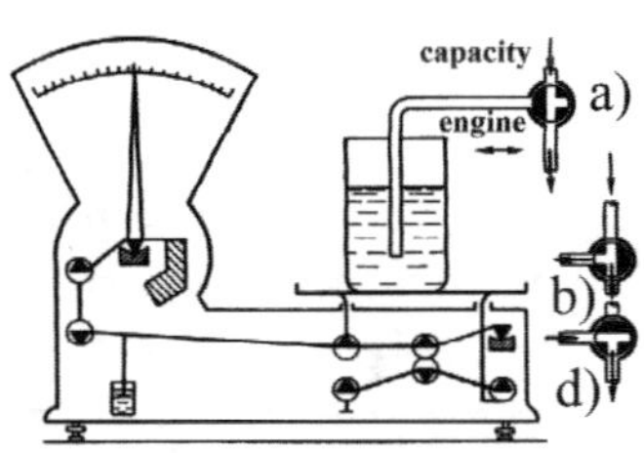

Figure 4. Measurements of the flow rate of fuel by weight in laboratory conditions: a) engine operation at fuel capacity (tank); b) the operation of the engine fuel tanks and filling of the measuring vessel; d) - operation of engine on fuel with a measuring vessel.

When the parameters G_p and I_p are determined, power [6] at the input of the internal combustion engine N_{en}, (kW), is

$$N_{en} = (G_p \cdot I_p)/3600 . \quad (5)$$

Then, the state of the internal combustion engine in terms of efficiency η_{ice} at the stage of completion of production as an integral system is determined according to the dependence

$$\eta_{ice} = (3600 \cdot M_{kr} \cdot \omega_v)/(G_p \cdot I_p). \quad (6)$$

Thus, when bench tests of the internal combustion engine are performed, the following data are recorded: N_{ex} – engine output power; ω_v – angular speed of the shaft; G_p – fuel consumption; η_{ike} – engine efficiency. These results are specification data for a specific engine at the end of production or after its repair.

The structural components of the internal combustion engine [8, 9] are: power, cooling and lubrication systems. For example, the fuel system is designed to supply fuel and air to the working chambers of the cylinders. It also includes crank and gas distribution mechanisms. The cooling system is designed to maintain a balance between the temperature of the running engine and the fan-cooled radiator. The lubrication system is designed to reduce the loss of mechanical energy caused by counteraction of the forces of friction with the use of fan cooling of the oil cooler. In addition, the internal combustion engine includes an electric generator to supply power electrical appliances.

The main system-forming component in the structure of the internal combustion engine is the fuel system, which serves to convert the thermal energy of the fuel into mechanical energy of the rotating movement of the engine shaft. In addition, the internal combustion engine design includes contains additional components that ensure the maintenance of the main component of the engine. Additional components include the cooling system and the lubrication system with fan cooling of liquids. In addition, the structure of the internal combustion engine includes an electric generator, which creates comfortable conditions for the operator to work in the machine system where this engine operates. It should be noted that in the structure of each of these components, during their operation, energy losses occur to overcome the friction forces in the fuel, power supply and lubrication systems, and during the operation of the electric generator. At this stage of the study, the share of efficiency for energy losses directly to friction cannot be identified.

When applying the system approach, rank subordination and significance of the above-mentioned structural components of the engine as an integral system, the internal combustion engine can be represented in the form of two functional and structural schemes: – with a sequential connection of structural components, figure 5 a); – with a branched connection of structural components, figure 5 b).

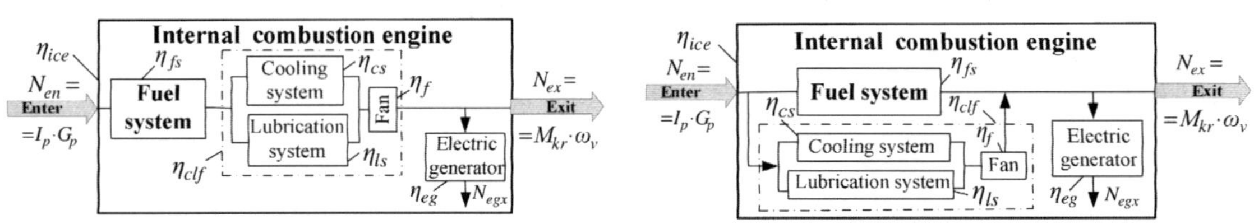

(a) (b)

Figure 5. Functional and structural diagram of the internal combustion engine: a) – sequential connection of components; b) – branched connection of components; η_{fs} – coefficient of useful action fuel system; η_{elf} – coefficient of useful action cooling system, lubrication system, fan; η_{eg} – coefficient of useful action electric generator; N_{egx} – output power of the electric generator.

The engines shown in figure 5 (a) and (b) indicate possible relations between the components and the different nature of the energy distribution in the internal combustion engine structure. As shown in figure 3, determining the efficiency of the engine in general is not particularly difficult. At the same time, determining the efficiency of each separate component, namely the power supply, cooling and lubrication systems, is a complex issue.

However, an attempt to solve this problem can be successful if the experimental simulation method is used with a real engine design that is in operable condition. Based on the use of such an engine design, the operation of all its systems is simulated without using fuel, and with maximum compliance with the operating modes according to a special program for all engine systems. At the same time, each of these systems should be connected to special energy sources to drive the crankshaft and circulation pumps of the cooling system and lubricate the movable joints of the engine. It should be noted that the engine power system can be either in the static or in the dynamic mode, in which the necessary parameters are recorded to determine the efficiency of the cooling system and the lubrication of the internal combustion engine.

At the same time, a special power source drives the crankshaft of the engine and ensures the operation of the cooling and lubrication systems. At the same time, if the specified engine operating modes are maintained, parameters are measured to determine the efficiency of the above systems. The efficiency of the η_{cs} cooling and lubrication η_{ls} system of the engine is determined by the formula

$$\eta_{cs} = (\eta_{ls}) = (N_{rs} - N_{nrs})/N_{rs}, \quad (7)$$

where N_{rs} is the amount of power that is consumed in the operation of the cooling and lubrication systems of the engine when the specific conditions are maintained, and if these systems are connected to the special energy source, and given the fact that the power system is connected to the fuel tank, and cooling and lubricating systems are filled with relevant fluids, kW; N_{nrs} is the amount of power that is consumed in the operation of the cooling and the lubrication systems of the engine under the specific conditions, and if these systems are connected to the special power source, provided that the power supply system is disconnected from the fuel tank, and the cooling and lubrication are temporarily free from working fluid, kW.

According to the same methodology, the efficiency of the η_f fan and η_{eg} electric generator is determined using formula (7). The value of the efficiency of the unit η_{clf} as part of the η_{cs} cooling system and η_{ls} the lubrication system and the η_f fan is determined taking into account the diagram of connection, figure 5. Determining the efficiency of the η_{fs} power supply system is based on determining the efficiency η_{ice} of the engine as a whole.

The most difficult stage of determining the state of the internal combustion engine in terms of efficiency is the stage of engine operation in comparison with other stages of its life. At the present level of development of internal combustion engine design, the state of individual components is determined through the use of shared electronic unit in conjunction with sensors that record, in particular: – coolant temperature; – intake air temperature; pressure lubrication; mass air flow; – crankshaft; – camshaft position, and other sensors.

However, using these features seems to be impossible to be used to determine the engine state in terms of efficiency in the conditions of operation of an internal combustion engine as part of a CRVRM. In addition, modern TELHMs are usually equipped with a hydraulic system that provides the necessary level of engine load. Then the efficiency of the internal combustion engine η_e under operating conditions is determined based on the obtained dependence

$$\eta_e = (p_n \cdot V_n \cdot n_n \cdot 60)/(G_p \cdot I_p \cdot \eta_n), \quad (8)$$

where p_n is the fluid pressure generated by the hydraulic pump, mPa; V_n is the volume of the working chamber of the pump, cm^3/rev; n_n is the rotating speed of the pump shaft, rev/min.; η_n is the total efficiency of the pump.

Since the parameters p_n and n_n can be maintained with high accuracy unchanged, and the parameters I_p and V_n are constant, the controlled parameter is the amount of fuel consumption G_p. The process of measuring fuel G_p consumption under laboratory conditions is shown in figure 4. For the operating conditions of the internal combustion engine, it is proposed to create a high-precision device for measuring fuel consumption for a given period of time and an electronic unit for determining the efficiency during the operation of modern TELHMs, based on dependence (8). The engine load to determine its efficiency by using the electronic unit on hydroficated TELHMs is provided by the use of flow regulators and other tools [10, 11 and 12].

Determining the actual value of the engine efficiency under operating conditions and comparing it with the initial value allows predicting further actions regarding the state of the internal combustion engine. It is known that if the efficiency of the internal combustion engine decreases or the engine fuel consumption increases by more than 5%, then the engine needs repairing.

5. The results of the calculation and their discussion

The presented results of the study suggest that the parameter of efficiency becomes a sign for determining the state at all stages of the life of the internal combustion engines which are part of various TELHMs. Further research in this area requires either the use of test bench equipment or the development of a new one or the use of modern measuring equipment to obtain new results with a minimum level of error.

6. Conclusions

Based on the results of the conducted studies, it was found that designing internal combustion engines according to a pre-specified efficiency value is a rather complex problem that requires further research. The most scientifically proven process for determining engine efficiency is the acceptance testing process. Currently, determining the efficiency of an internal combustion engine at the operational stage is of special scientific interest, since this is related to fuel economy in the operating conditions of modern TELHMs. It is known that if the fuel consumption increases by more than 5% or the engine efficiency decreases by more than 5% vs. the initial values, then such an engine should be sent for repair.

Timely determination of the efficiency value in the conditions of operation of the TELHMs based on the results of diagnostics of internal combustion engines will allow to repair them depending on their condition, instead of scheduled repairs and will reduce the operating cost of such machines and, accordingly, the construction cost in railway transport. In general, the economic effect can reach a significant value, which is further planned to be determined based on knowledge of all parameters.

7. References

[1] Soloviev A I 1966 *Coefficient of efficiency of mechanisms and machines* (Moscow: Mechanical Engineering) p 180

[2] Mayskiy M I and Boguslavskiy D K 1961 *Tractors and automobiles*, ed G.I. Trubnikova (Kiev: State Publishing House of the Silk Sp. Literature) p 484

[3] GOST 18509–88 (ST SEV 2560–80) 1988 *Tractor and Combine Diesels Bench Test Methods* (Moscow: Standards Publishing House)

[4] Selivanov N I 2014 *Testing of automotive engines* (Krasnoyarsk: Krasnoyarsk State Agrarian University) p 220

[5] Remarchuk N, Zadorozhnyi A and Chmuzh Y 2017 Methodology of Creating the Hydraulic Actuatorat the Stage of its Design *Eastern-European journal of enterprise technologies* No. **2/7(86)** pp 42- 50

[6] Nazarov L and Remarchuk M *2008 Cyclic efficiency as an energy indicator of the working process of construction and road machines Scientific Bulletin of Construction KhDTUBA* **45**, pp 142-153

[7] Lange O 1968 *Introduction to economic cybernetics*, ed E.Z. Mayminasa (Moscow: Progress) p 208

[8] Putintsev S V 2011 *Mechanical losses in reciprocating engines: special chapters of design, calculation and testing* (Electronic educational publication), uch. manual. by discipline "Special chapters of design and CAD" (Moscow: MSTU named after Bauman) p 288

[9] Sharoglazov B A, Farafontov M F and Klement'ev V V 2005 *Internal combustion engines: theory, modeling and calculation of processes* (Chelyabinsk: SUSU Publishing House) p 403

[10] Panchenko A, Voloshina A, Panchenko I, Titova O. and Pastushenko A. *Reliability design of rotors for orbital hydraulic motors. IOP Conference Series: Materials Science and Engineering* **708(1),** 012017

[11] Panchenko A, Voloshina A, Milaeva I and Luzan P. 2019 *Operating Conditions' Influence on the Change of Functional Characteristics for Mechatronic Systems with Orbital Hydraulic Motors.* In: Nadykto V ed Modern Development Paths of Agricultural Production. Springer, Cham, pp 169-176

[12] Zadorozhnyi A, Kovrevski A, Chovnyuk Y and Remarchuk N 2018 *Flow of a Bingham Fluid Through Circular Pipes with Variable Viscosity Coefficient Along the Pipe Length Science Publishing Corporation Publisher of International Academic Journal Special Issue* **7(43)** pp 100-104

Justification of the parameters of the life-test bench for railway wheelsets

S V Raksha[1], P G Anofriev[1], O S Kuropiatnyk[1, 2]

[1] Department of Applied Mechanics and Materials Science, Dnipro National University of Railway Transport named after Academician V. Lazarian, 2 Lazariana Street, Dnipro, Ukraine

2 E-mail: kuropyatnick@gmail.com

Abstract. One of the tasks that need to be solved in order to predict the life of wheelsets is to assess the contact strength and wear of their parts. Thus, the purpose of this work is to develop a scheme and justification of the parameters of the life-test bench for railway wheelsets in conditions close to real operating ones. This bench simulates a contact of “wheel-rail” pair as a contact of “wheel-roller” pair. To ensure compliance of the bench tests conditions with the real wheelsets operation conditions we took that the maximum contact pressure in the “wheel-roller” pair is equal to the maximum contact pressure in the “wheel-rail” pair. As results of this study we obtained the dependences of the values of the roller diameter and the working force of the bench on the characteristics of the tested wheelset and proved the mutual influence of these quantities. For railway wheelsets with an axle load of 230.5 kN and wheels with a tread diameter of 957 mm, we obtained the following values of the bench parameters: roller diameter is 75 mm; working force of the bench is 8,375 kN; wheel speed is 500 rpm; number of rollers is 8; test duration is about 220 h.

1. Introduction

Any product has an element that limits the duration of its use. For railway rolling stock, such an element is a wheelset (WS), the resource of which is an important characteristic of the vehicle.

Current regulations [1, 2] provide for a number of bench tests of wheelsets, among which are those that allow you to assess the resource of the WS – fatigue testing of WS parts. The works [3, 4, 5] deals various aspects of conducting such tests and the corresponding bench equipment.

One of the tasks that need to be solved in order to predict the life of wheelsets is to assess the contact strength and wear of WS parts, their individual surfaces. Thus, in [6] a bench was proposed for testing railway wheels for contact strength. However, the tests are simulated, i.e. the object of research is not a railway wheel, but an equivalent roller.

A number of works [7, 8, 9] are devoted to the problem of determining and increasing the life of wheelsets and their elements.

The purpose of this work is to develop a scheme and justification of the parameters of the life-test bench for railway wheelsets in conditions close to real operating conditions.

2. General description of the life-test bench

Since the life of wheelsets depends on the contact strength of the wheels, the object of the test is the railway wheel. The bench simulates the interaction of wheel with the rail under a load that corresponds to the wheelset axle load.

The general scheme of the life-test bench is given in figure 1.

Rollers 2 are installed on a frame 3 with the possibility of rotation relative to its own axes and radial movement relative to the test wheel 1. They contact with the test wheel 1 under pressure created by the loading device 4. Test wheel 1 is fixed on the axle 6 and together with it rotates by means of the bench drive 5.

Axle 6 can be both the axle for wheelsets (its fragment) and a separately manufactured part (technological axle). The bench drive 5 can be of any type and configuration. The working surface of the roller 2, in contact with the test wheel 1, corresponds to the shape of the rail head in cross section.

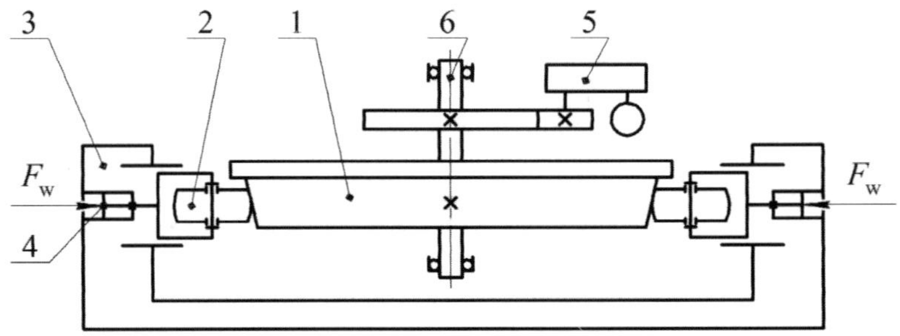

Figure 1. Scheme of the life-test bench:
1 – test wheel; 2 – roller; 3 – frame; 4 – loading device;
5 – bench drive; 6 – axle
(F_w – working force of the bench)

The test wheel rotates for a certain period of time, which will be called the test duration. It corresponds (but does not equal) to the required life of wheelsets in terms of contact strength of the wheels.

Given the design of the bench, we highlight its parameters:

- roller diameter;
- working force of the bench;
- wheel speed;
- number of rollers;
- test duration.

Other parameters of the bench within this work will not be considered.

3. Study of contact strength of wheels

Within this work, the purpose of the study of the contact strength of wheels is to establish the relationship between the individual parameters of the bench and the wheelsets characteristics. The research results are based on the use of Hertz's elastic contact theory [10].

According to the provisions of the contact mechanics [10], in the general case, each of the bodies in contact has a surface curved in two mutually perpendicular directions. Such surfaces are characterized by relative radii of curvature, which are denoted: for the first body – R_1' and R_1''; for the second body – R_2' and R_2'' (in some cases, these values can be infinite – the surface is flat in one or both directions).

The railway wheel and the roller are bodies of uncoordinated shape with external point contact. The relative radii of curvature of each of the bodies are different ($R_1' \neq R_1''$; $R_2' \neq R_2''$), therefore, due to the application of the load, an elliptical contact area will be formed. This also applies to the contact of the wheel with the rail in real operating conditions.

In further research, we believe that the values with the lower index "1" (except F_1) characterize the railway wheel, and the values with the lower index "2" characterize the part with which the wheel is in contact (rail or roller).

We note that the life-test bench must reproduce the real operating conditions of the wheelset. This can be ensured if the characteristics of the contact of the pairs "wheel-rail" and "wheel-roller" is equal. Let us consider each of the options for interaction in more detail.

3.1. Contact of the "wheel-rail" pair

The contact of the "wheel-rail" pair is schematically presented in Figure 2.

As you can see, the contact surfaces are flat in different, mutually perpendicular planes, so $R_1''=\infty$ and $R_2'=\infty$.

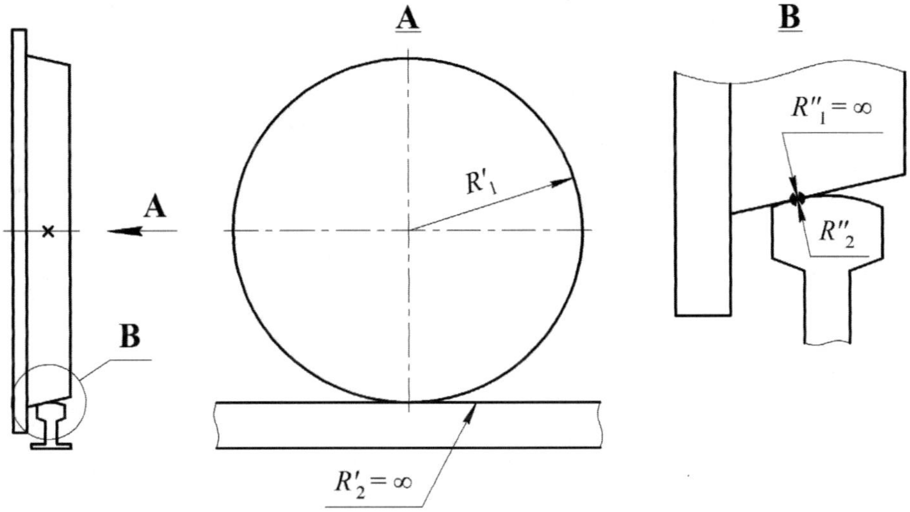

Figure 2. The scheme of contact of the "wheel-rail" pair

According to Hertz's elastic contact theory [10], we determine the following quantities:

- large relative radius of curvature

$$R' = \left(\frac{1}{R_1'} + \frac{1}{R_2'}\right)^{-1} = R_1'; \tag{1}$$

- small relative radius of curvature ($R'' < R'$)

$$R'' = \left(\frac{1}{R_1''} + \frac{1}{R_2''}\right)^{-1} = R_2''; \tag{2}$$

- equivalent radius of curvature

$$R_e = \sqrt{R'R''} = \sqrt{R_1'R_2''}; \tag{3}$$

- maximum contact pressure

$$p_0 = \left(\frac{6PE^{*2}}{\pi^3 R_e^2}\right)^{1/3} \times \left[F_1(R'/R'')\right]^{-2/3}. \tag{4}$$

In formula (4), the following designations are adopted: P is the normal force in the contact zone – a component of the load in the contact zone, which acts along the common normal to the contact surfaces at the point of contact; $F_1(R'/R'')$ is the coefficient that takes into account the eccentricity of the actual contact surface of bodies (ellipse), it is a function of the ratio of large (R') and small (R'')

relative radii of curvature, can be determined by recommendations [10]; E^* is the reduced modulus of elasticity.

Normal force in the contact zone

$$P = 0.5F\cos\beta\,, \tag{5}$$

where F is the wheelset axle load; β is the wheel tread slope (the angle $\beta = 2.86°$ corresponds to a taper of 1:10, which is specified in GOST 10791).

Reduced modulus of elasticity

$$E^* = \left(\frac{1-\nu_1^2}{E_1} + \frac{1-\nu_2^2}{E_2}\right)^{-1}, \tag{6}$$

where E_1 and ν_1 are the Young's modulus and the Poisson's ratio of the wheel material; E_2 and ν_2 are the same values for the rail material.

3.2. Contact of the "wheel-roller" pair

The contact of the "wheel-roller" pair is schematically presented in figure 3.

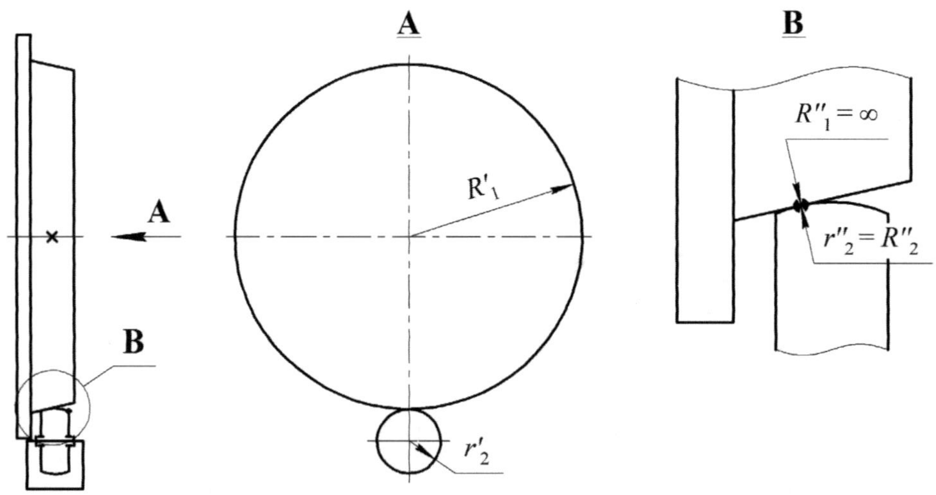

Figure 3. The scheme of contact of the "wheel-roller" pair

The roller curvature r_2'' in view B (figure 3) is equal to the rail curvature: $r_2'' = R_2''$.

The characteristics of contact are as follows:

- large relative radius of curvature

$$r' = \left(\frac{1}{R_1'} + \frac{1}{r_2'}\right)^{-1}; \tag{7}$$

- small relative radius of curvature ($r'' < r'$)

$$r'' = \left(\frac{1}{R_1''} + \frac{1}{r_2''}\right)^{-1} = R_2''; \tag{8}$$

- equivalent radius of curvature

$$r_e = \sqrt{r'r''} = (R_2'')^{1/2} \cdot \left(\frac{1}{R_1'} + \frac{1}{r_2'} \right)^{-1/2} ; \tag{9}$$

- maximum contact pressure

$$p_0 = \left(\frac{6P_b E^{*2}}{\pi^3 r_e^2} \right)^{1/3} \times [F_1(r'/r'')]^{-2/3} . \tag{10}$$

In the formula (10) the normal force in the contact zone "wheel-roller" is

$$P_b = F_w \cos\beta , \tag{11}$$

where F_w is the working force of the bench – the force applied by the loading device to the roller.

3.3. Ensuring the adequacy of the bench tests conditions

To ensure compliance of the bench tests conditions with the real operation conditions of wheelsets, it is necessary that the maximum contact pressure in the "wheel-roller" pair is equal to the maximum contact pressure in the "wheel-rail" pair. Equating the right-hand sides of equations (4) and (10), we obtain:

$$\left(\frac{6PE^{*2}}{\pi^3 R_e^2} \right)^{1/3} \times [F_1(R'/R'')]^{-2/3} = \left(\frac{6P_c E^{*2}}{\pi^3 r_e^2} \right)^{1/3} \times [F_1(r'/r'')]^{-2/3} . \tag{12}$$

After performing mathematical transformations we have:

$$\frac{P}{P_b} = \left(1 + \frac{R_1'}{r_2'} \right) \cdot \frac{F_1(R'/R'')^2}{F_1(r'/r'')^2} . \tag{13}$$

In [10] it was noted that in the first approximation the functions F_1 can be taken as equal to one. So that

$$\frac{P}{P_b} = 1 + \frac{R_1'}{r_2'} \Rightarrow P_b(r_2') = P \cdot \left(1 + \frac{R_1'}{r_2'} \right)^{-1} \tag{14}$$

The dependence analysis showed that the choice of rollers of smaller diameter (r_2' is the roller radius) allows to reduce the force P_b and, consequently, the working force of the bench.

Note that the life-test bench can be universal: the obtained dependence (14) in combination with formulas (5) and (11) allows determining the working force of the bench for different wheelset axle load, including the constant roller diameter, and the ability to move the rollers in radial direction relative to the wheel allows you to test wheels of different diameters.

It also follows from formula (14) that in the first approximation the roller profile ($r_2'' = R_2''$) does not affect the adequacy of the test conditions. Studies conducted within this work have shown that at the stage of refinement, the influence of the radius $r_2'' = R_2''$ is reflected in the coefficient F_1 and is insignificant. Therefore, there is no need to use different sets of rollers to simulate the impact of different rails, which further indicates the versatility of our life-test bench.

3.4. Determining the roller diameter and the working force of the bench

When determining the roller diameter, we will start from the roller axle diameter, which is determined from the condition of bending strength; after transformations it takes the form

$$\sigma = \frac{8F_w L}{\pi d_a^3} \le [\sigma], \tag{15}$$

where L is the distance between the axle supports; d_a is the roller axle diameter; $[\sigma]$ is the permissible axle bending stresses.

Combining formulas (5), (11), (14), counting $R_1' = 0.5D_w$ (D_w is the test wheel diameter), $r_2' = 0.5d_r$ (d_r is the roller diameter) and assuming $d_r = 3d_a$, we obtain

$$F_w = 0.5F\left(1 + \frac{D_w}{3d_a}\right)^{-1}. \tag{16}$$

Substituting (16) into (15) and performing mathematical transformations, we obtain the equation for determining the roller axle diameter:

$$ad_a^3 + bd_a^2 + d = 0, \tag{17}$$

where $a = \pi$; $b = \frac{\pi D_w}{3}$; $d = -\frac{4FL}{[\sigma]}$.

By equation (17) we determine the roller axle diameter d_a and then we determine the roller diameter as $d_r = 3d_a$.

As a special case, we determine the roller diameter and the working force of the life-test bench for railway wheelsets with wheelset axle load $F = 230.5$ kN, which contain wheels type A1 according to GOST 10791 with a tread diameter $D_w = 957$ mm.

Given the width of the rail P65 according to DSTU 4344 (75 mm) we take the distance $L = 80$ mm. Permissible bending stresses for the roller axle made of medium carbon steel $[\sigma] = 120$ MPa.

Equation (17) shows the minimum required value of the roller axle diameter is 24 mm. Given that the roller will be mounted on the axle using rolling bearings, we assume $d_a = 25$ mm. So that the roller diameter

$$d_r = 3d_a = 3 \cdot 25 = 75 \text{ mm}.$$

By formula (16) we determine the working force of the bench:

$$F_w = 0.5F\left(1 + \frac{D_w}{3d_a}\right)^{-1} = 0.5 \cdot 230.5 \cdot \left(1 + \frac{957}{3 \cdot 25}\right)^{-1} = 8.375 \text{ kN}.$$

4. Determination of test duration, wheel speed and number of rollers

The test duration (T) depends on several factors:

- required wheel life on contact strength in kilometers (L_w);
- number of rollers (k);
- wheel speed in revolutions per minute (n).

These values are related:

$$T = \frac{60L_w \cdot 10^3}{\pi n D_w k}. \tag{18}$$

When it comes to wheel life (L_w), there is somewhat conflicting data. Biserikan notes [8] that about a third of the wheels of the car fleet have a mileage between rebounds, less than 160 thousand km. The maximum overhaul mileage is 300 thousand km. The results of Vorobyov's research [6] are presented more widely with their division according to the conditions of rolling stock operation and wheeled steel grades: for freight cars – 76 to 108 thousand km, for passenger cars – 196 to 436 thousand km.

To perform approximate calculations, we take the life of wheels $L_w = 160$ thousand km. The test wheel diameter is $D_w = 957$ mm.

Let the wheel speed is $n = 500$ rpm, the number of rollers is $k = 8$. Then the test duration will be

$$T = \frac{60 \cdot 160000 \cdot 10^3}{3.14 \cdot 500 \cdot 0.957 \cdot 8} = 800 \cdot 10^3 \text{ s} \approx 220 \text{h}.$$

The obtained dependence allows setting the value of the wheel speed and the number of rollers to obtain acceptable test duration, taking into account the required life of a wheel (wheelset).

5. Conclusions

We developed the scheme of the life-test bench for railway wheelsets. This bench simulates a contact of "wheel-rail" pair as a contact of "wheel-roller" pair. To ensure compliance of the bench tests conditions with the real operation conditions of wheelsets we took that the maximum contact pressure in the "wheel-roller" pair is equal to the maximum contact pressure in the "wheel-rail" pair.

Based on the provisions of the contact mechanics, in particular, the Hertz's elastic contact theory, we obtained the dependences of the values of the roller diameter and the working force of the bench on the characteristics of the tested wheelset. The mutual influence of these quantities is proved.

A functional relationship has been established that relates the test duration, the wheel speed and the number of rollers; it also determines the influence of the wheelset characteristics on these values.

For railway wheelsets with an wheelsets axle load of 230.5 kN and wheels with a tread diameter of 957 mm, we obtained the following values of the bench parameters: roller diameter is 75 mm; working force of the bench is 8,375 kN; wheel speed is 500 rpm; number of rollers is 8; test duration is about 220 h.

References

[1] GOST 11018 2012 Traction railway stock wheelsets of 1520 mm gauge railways. General specification. Interstate standard
[2] GOST 31373 2009 Wheelsets of locomotives and motor-powered units. Strength calculations and tests. Interstate standard
[3] Kuropiatnyk O S, Raksha S V, Anofriev P G 2019 *MATEC Web of Conferences* **294** 03008
[4] Novosad M, Fajkoš R, Řeha B, Řezníček R 2010 *Procedia Engineering* **2** 2259–2268
[5] Volokhov G M, Oguenko V V 2015 *Transport of the Russian Federation* **3** 71–73
[6] Vorobyov A A 2018 Forecasting the resource and improving the technology of repairing wheels of railway rolling stock. Dissertation abstract
[7] Ushkalov V F, Bezrukavii N V 2015 *Technical mechanics* **1** 97–103
[8] Biserikan M I, Ivanov V V 2015 *Omsk Scientific Bulletin* **3** 60–65
[9] Muradian L A, Anofriev V G 2010 *Science and Transport Progress. Bulletin of DNURT* **34** 206–210
[10] Johnson K L 1985 Contact Mechanics (Cambridge University Press)

Method for determining non-uniform loading of gondola cars with bulk cargo

Yevhenii Romanovych[1,4], Andrii Kravets[1], Georgii Afanasov[1], Yevhenii Povorozhenko[2], Olga Afanasova[3]

[1]Construction, track and handling machines Department, Ukrainian State University of Railway Transport, Feuerbach sq. 7, 61050 Kharkiv, Ukraine
[2]Regionalny branch "Southern Railway" of Ukrzaliznytsya Joint-Stock Company, Yevhena Kotliara str. 7, 61052 Kharkiv, Ukraine
[3]Limited Liability Company "MPSR", Heroyiv Stalinhradu ave. 154a, 61096 Kharkiv, Ukraine

[4] E-mail: 0674274770@ukr.net

Abstract. Methods of loading gondola cars with bulk cargo are considered and possible types of non-uniform cargo distribution in the car bodies are established.
The most common methods of levelling bulk cargo in gondola cars in the Ukrainian railway network are determined. Designs of equipment for mechanization of levelling bulk cargo in gondola cars are presented, their advantages and disadvantages are considered. Based on the given analysis, it is established that the most efficient way to ensure compliance with the current regulations on the placement of cargo in cars is the timely detection of non-compliance with these requirements at the stage of loading.
The analysis of existing methods for determining the non-uniformity of cargo distribution in the car bodies was conducted. Based on this analysis, a new method for determining the non-uniform distribution of bulk cargo in the car bodies is proposed, which can be used to pre-estimate the compliance of the actual placement of bulk cargo in the car body with the requirements of current regulatory documents immediately after loading and take prompt measures to eliminate the identified inconsistencies.

1. Introduction

The volume of shipment in 2019 totaled to 312.9 million tons [1]. Among this, the share of railway transportation is more than 31.3%. In turn, bulk cargo, which is mainly transported in gondola cars, is about 65% of the total volume of railway freight transportation. One of the main indicators of the efficiency of car fleet operation is the rollover of cars, which depends, in particular, on the duration of cargo handling operations. The duration of cargo handling operations obviously depends on the number and technical condition of the means of mechanization of cargo handling points of railway stations and industrial enterprises.

The purpose of the transport industry is providing safe and high-quality railway transportation services to corporative entities and individuals [2]. Assuring the safety of transportation is also relevant when bulk cargo transported in gondola cars.

This is primarily due to the risk of rollover of gondola cars during traveling due to non-uniform distribution of bulk cargo inside the car body (Figure 1). Therefore, the location of cargo inside the car bodies should be regulated, for example [3, 4].

According to the data of Ukrzaliznytsya, about 2/3 of violations of the requirements listed in [3, 4] relate specifically to the non-uniform distribution of bulk cargo in gondola cars. If commodities are loaded in the gondola car non-uniformly, the center of mass of the "car-load" system becomes displaced relative to the geometric center of the car body along the transverse and/or longitudinal axis

IOP Conf. Series: Materials Science and Engineering 1021 (2021) 012048 doi:10.1088/1757-899X/1021/1/012048

of the car (figure 2).

Figure 1. Consequences of rollover of a gondola car with bulk cargo.

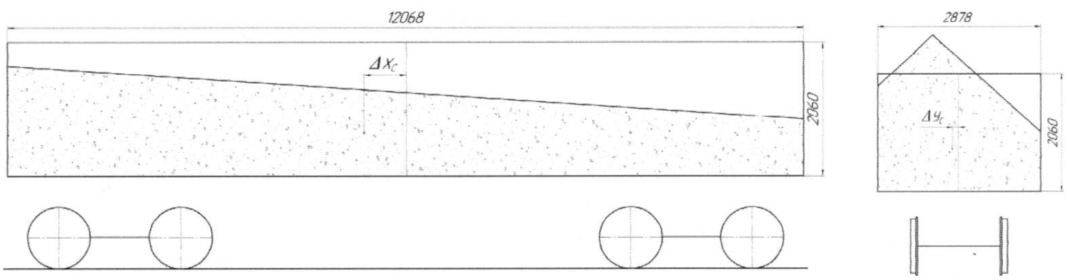

a) b)

Figure 2. Examples of non-uniform distribution of a bulk cargo in the back of a gondola car: a) longitudinally b) transversely.

2. Analysis of methods of loading bulk cargo

Non-uniform arrangement of bulk cargo in the body of the gondola car occurs actually during loading of the commodity. The following methods of loading gondola cars with bulk cargo are most commonly used at the Ukrainian railway.

Method 1. Gondola cars move slowly under the loading device, from which the commodity is continuously poured (Figure 3). Often, the cargo surface is compacted immediately after loading, using a passive compactor - a roller (Figure 4). Since the loading device is located above the longitudinal axis of the gondola car, under certain conditions (for example, if the car stops under the loading device), only longitudinal non-uniformity of loading the gondola car body with cargo is possible (Figure 2a).

Figure 3. Loading of gondola cars with bulk cargo using a continuous loading device.

IOP Conf. Series: Materials Science and Engineering 1021 (2021) 012048 doi:10.1088/1757-899X/1021/1/012048

Figure 4. Compaction of the bulk cargo surface in the gondola car body.

Continuous loading devices are used in enterprises with large volumes of loading of similar loads, such as coal mines, mining and processing plants, etc.

Method 2. Gondola cars are loaded using intermittent loading devices, such as excavators, clamshell cranes, front loaders, etc. In such devices, buckets or clamshell grabs are used as working elements (Figure 5).

Figure 5. Loading of gondola cars with bulk cargo using loading devices of intermittent operation.

This method is used in enterprises with medium or small loading volumes and/or with a wide range of commodities loaded as bulk cargo. Since the loading device operator's view of the internal part of the gondola car body is limited or impossible, a transverse non-uniformity of loading of the gondola car body with cargo may occur (Figure 2B).

3. Technical means for leveling bulk cargo

In the Ukrainian railway network, the surface of bulk cargo in the bodies of gondola cars is levelled both using a mechanized method with excavators or clamshell excavators, and a non-mechanized method with manual leveling of the surface of bulk cargo (Figure 6).

The leveler attachment for bulk cargo in gondola cars designed in the Ukrainian State University of Railway Transport is also known (Figure 7) [5, 6].

A specific feature of this device is the use of working tools in the form of rotors with attached blades. After the crane lowers the device on the top of a gondola car, power supply of rotor drives is switched on. With further lowering, the rotors immerse in the cargo and begin to level the part of the cargo located under the working tools. To level the rest of the cargo, the device is moved to the desired location.

The leveler was produced at the State Enterprise "Vinnytsiatransprilad" [7]. Its operation has shown high performance on cargo with high and medium flowability, such as crushed stone. For such cargo as chalk or iron ore concentrate, the efficiency of the rotary tools was insufficient.

Therefore, the specialists of the Ukrainian State University of Railway Transport proposed a new design of an attachment for leveling the surface of bulk cargo in gondola cars featuring the use of screw working tools (Figure 8) [8].

a)

b)

Figure 6. Common methods of levelling the bulk load:
a - manually; b – with a clamshell excavator.

Figure 7. Leveler attachment for bulk cargo in gondola cars.

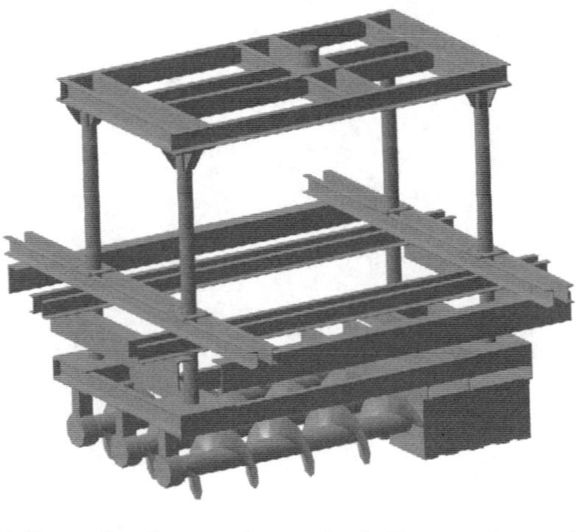

Figure 8. Screw leveler attachment for bulk cargo in gondola cars.

4. Method for determining the non-uniform placement of bulk cargo in the car bodies

It is generally known that the best way to eliminate the consequences of any emergency is to prevent the occurrence thereof. Regarding the non-uniform loading of gondola cars with bulk cargo, this may be achieved by:

- determination of loads on freight car bogies by means of car scales;
- pre-estimation of the position of the prism vertices of the cargo relative to the longitudinal and transverse axes of the car visually and/or using simple measuring devices, such as measuring tapes.

As for the first method, paragraph 4.1.8 of DSTU GOST 22235: 2015 [9] regulates the difference in the loading of car bogies, which can be determined using car scales. The obvious disadvantage of this method is the high cost of use. In addition, performing control weighing of cars and leveling the cargo in the car bodies leads to an increase in downtime of cars during cargo handling operations and, as a result, to a decrease in the efficiency of utilization of the rolling stock.

To implement the second method, the requirements of current regulatory documents can be applied, such as the Rules [3]. Figure 9 shows a graph of dependence of the maximum permissible longitudinal displacement of the center of mass of the cargo (hereinafter - CM) during loading on the weight of the cargo in the back of the car, while Figure 10 shows a graph of dependence of the maximum allowable transverse displacement of CM of the cargo during loading on the weight of cargo in the back of the car and the total height of CM of the loaded car. These graphs are based on the data provided in paragraphs 4.3 and 4.4 of Chapter 1 of the Rules [3].

This method is widely used for pre-estimation of the longitudinal displacement of the center of mass of the cargo in the body of a gondola car. However, Figure 10 shows that for a pre-estimation of the transverse displacement of the center of mass of the cargo in the body of the gondola car, the total center of mass of the "gondola-cargo" system should be determined. Performing these estimations for each car will take too much time, which consignors cannot afford. Therefore, the method of pre-estimation of the transverse displacement of the center of mass of the load in the body of a gondola car needs to be improved.

To improve the above method of pre-estimation of the transverse displacement of the center of mass of the load in the body of the gondola car, certain initial data are required which are listed in table 1.

In Ukraine, more than 31% of bulk cargo is transported in gondola cars [10], and for certain categories of cargo, the share of such cars can amount up to almost 100 %. For example, 99.96% of coal, 93.6% of coke, 55.7% of construction materials, and 90.4% of iron and manganese ores are transported in gondola cars. Also, in Ukraine, actually no other cars, except four-axle gondola cars are used. On this basis, the characteristics of a typical gondola car can be set (table 2).

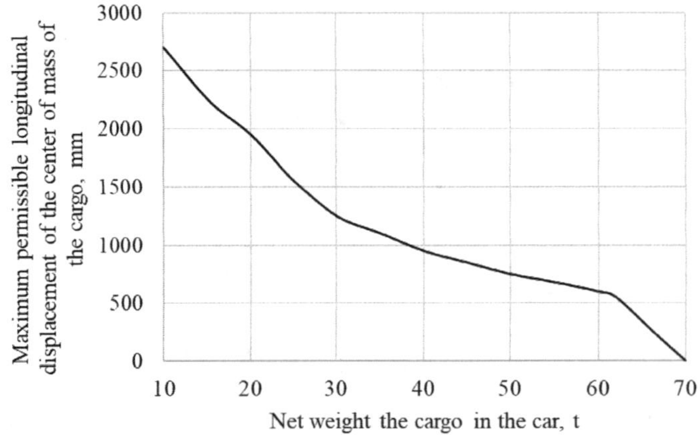

Figure 9. Dependence of the maximum permissible longitudinal displacement of the center of mass of the cargo during loading on the weight of the cargo in the car body.

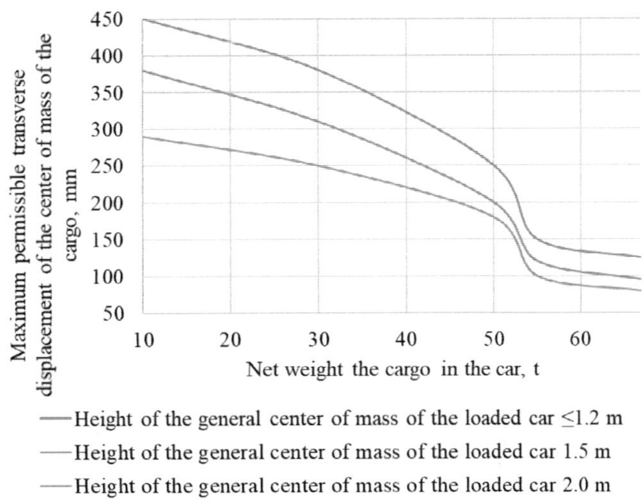

Figure 10. Dependence of the maximum permissible transverse displacement of the center of mass of the cargo during loading on the weight of the cargo in the car body and on the total height of the center of mass of the loaded car.

Table 1. Initial data for pre-estimation of the transverse displacement of the center of mass of the load in the body of the gondola car

Indicator	Unit	Note
1 Indicators that depend on the specification of the car		
1.1 Internal length of the body	m	
1.2 Internal width of the body	m	
1.3 Internal height of the body	m	
1.4 Tare weight of the car	t	
1.5 Load capacity of the car	t	
1.6 Height of the center of mass of an empty car from the top of the rail	m	paragraph 11.4.1 of the Rules [3]
2 Indicators that depend on the specification of the cargo		
2.1 Density	t/m^3	
2.2 Natural slope angle	degree	

Table 2. Specification of a typical gondola car

Indicator	Value
1 Internal length of the body, m	12.0
2 Internal width of the body, m	2.88
3 Internal height of the body, m	2.06
4 Tare weight of the car, t	22.2
5 Load capacity of the car, t	69.0
6 Height of the center of mass of an empty car from the top of the rail (according to paragraph 11.4.1 of the Rules [3])	1.13

Among the entire range of bulk cargo, three most common commodities were selected: coal, crushed granite stone, and iron ore concentrate, the specification of which are shown in table 3.

It is logical to assume that to increase the efficiency of transport, the carriers will try to maximize the use of the capacity of gondola car, and the dependence of displacement of CM of the load from the displacement of vertices of the prism of cargo in the back of the gondola car can be determined. On

IOP Conf. Series: Materials Science and Engineering 1021 (2021) 012048 doi:10.1088/1757-899X/1021/1/012048

this basis, the weight of the cargo in the gondola car was set (67 t). Figure 11 shows the estimation schemes for determining the transverse displacement of the CM of crushed stone.

Table 3. Specification of bulk cargo

Indicator	Name of cargo		
	coal	crushed granite stone	iron ore concentrate
1 Density, t/m^3	0.95	1.5	3.1
2 Natural slope angle, degrees	42	40	38

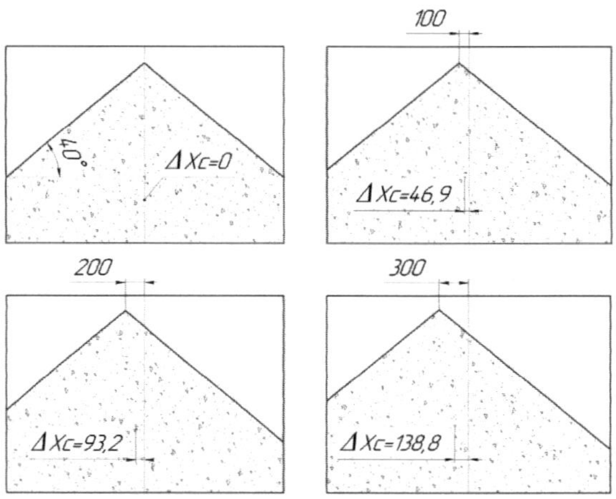

Figure 11. Estimation scheme for determining the transverse displacement of the center of mass of crushed stone from the transverse displacement of the vertex of the prism of the cargo.

The results of the estimations were the dependence of the displacement of the cargo CM on the displacement of the vertices of the prism of the cargo (Figures 12-14). In the article [11] it is proved that the longitudinal displacement of the CM of the cargo only slightly affects the stability of the car during movement. Therefore, only the transverse displacement of the CM of the cargo was considered here.

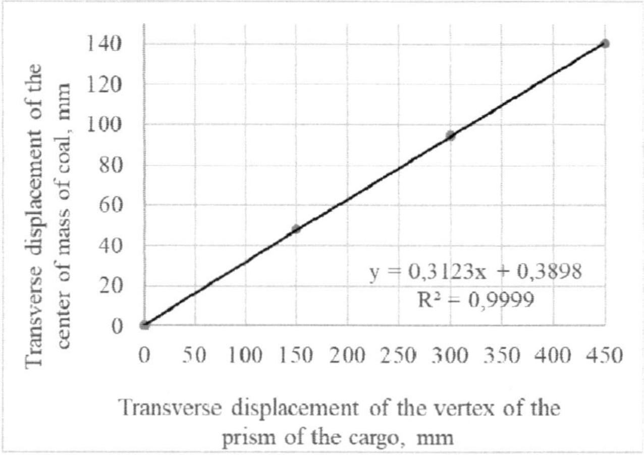

Figure 12. Transverse displacement of the center of mass of coal from the transverse displacement of the vertex of the prism of this cargo.

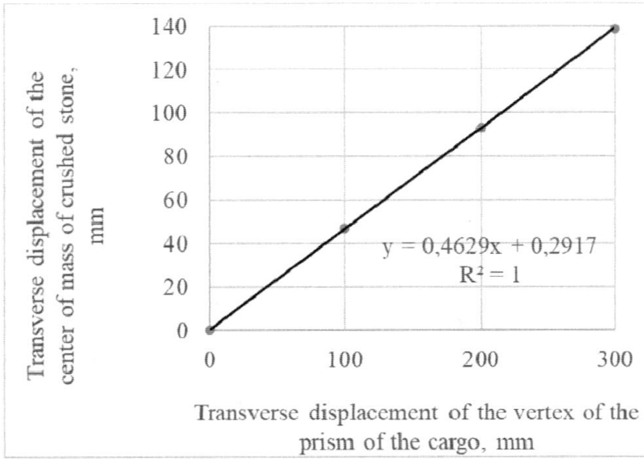

Figure 13. Transverse displacement of the center of mass of crushed stone from the transverse displacement of the vertex of the prism of this cargo.

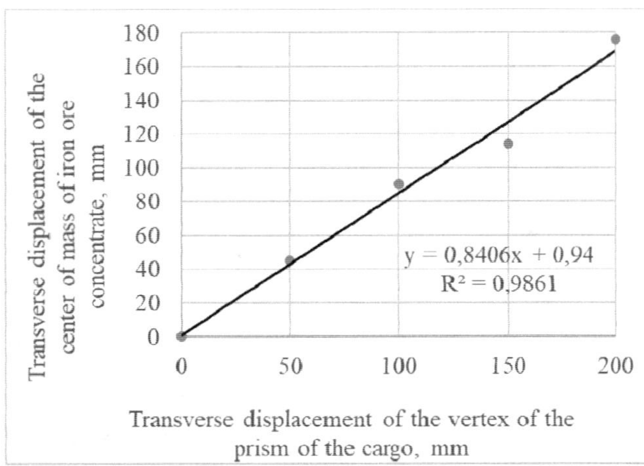

Figure 14. Transverse displacement of the center of mass of iron ore concentrate from the transverse displacement of the vertex of the prism of this cargo.

Figures 12-14 show that the regression equations based on the calculated points has the form

$$y=a{\cdot}x+b, \qquad (1)$$

where y is transverse displacement of the CM of the bulk cargo from the vertical axis of symmetry of the gondola car body; a, b are constant coefficients; x is transverse displacement of the top of the bulk cargo from the vertical axis of symmetry of the gondola car body.

In this case, the height of the CM of the "car-cargo" system increases insignificantly (Figure 15).

Thus, the procedure for pre-estimation of the transverse non-uniformity of loading a gondola car with bulk cargo is as follows.

- The total height of the CM of the "car-cargo" system is determined. In practice, the following can be assumed: if the cargo density is less than 1 t/m^3, the height of the CM of the "car-cargo" system from the top of the rails above 2 m, and if the cargo density is more than 1 t/m^3, the height of the CM "car-cargo" system from top of the rail heads level less than 2 m.
- In accordance with paragraph 4.4 of Chapter 1 of the Rules [3], the maximum permissible transverse displacement of the CM of the "car-cargo" system is assigned.
- The transverse displacement of the vertex of the prism of the cargo is measured.
- Using the graphs, examples of which are shown in Figures 12-14, the calculated value of the

transverse displacement of the CM of the "car-load" system is set and compared with the maximum allowable value, which was assigned in paragraph 2.

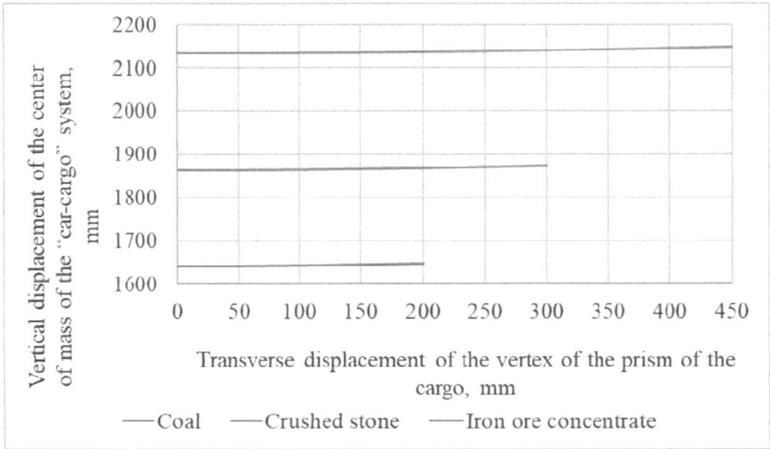

Figure 15. Vertical displacement of the center of mass of the "car-cargo" system from the transverse displacement of the vertex of the prism of the cargo.

5.Conclusions

Based on the above, the following conclusions can be drawn:

- methods of loading gondola cars with bulk cargo are considered and possible types of non-uniform placement of cargo in the car bodies are established;
- the existing methods for determining the non-uniformity of cargo placement in the car bodies are considered, their disadvantages and difficulties of application are established;
- a new method for determining the non-uniform placement of bulk cargo in the car bodies is proposed, which allows you to pre-estimate the compliance of the actual location of bulk cargo in the car body with the requirements of the applicable regulatory documents immediately after loading and promptly take measures to eliminate inconsistencies, if any;
- to use the proposed method successfully, dependencies must be available on displacement of CM of the "car-cargo" system on the displacement o vertices of the prism of cargo in the back of the gondola car for the main item of bulk cargo transported in cars.

References

[1] Transportation of goods by rail by type of cargo in 2019 (economic statistics, economic activity, transport) 2020 (Kyiv: State Statistics Service of Ukraine). Retrieved from http://www.ukrstat.gov.ua

[2] No. 735 Resolution of the Cabinet of Ministers of Ukraine. Issues of Ukrzaliznytsya joint-stock company (2015)

[3] Annex 14 to Section 4 of Article 9 of the Agreement on International Goods Transport by Rail. Rules for placing and securing cargo in cags and containers (2015)

[4] No. 796/5987 Rules for cargo transportation in open cars (2001)

[5] Romanovych, YeV, Afanasov, GM, Povorozhenko, YeV, Korobchuk, MB, Svistunov, YuB (2014). Patent Ukraine №107150. Kyiv: State Service of Intellectual Power of Ukraine

[6] Romanovych, YeV, Afanasov, GM, Povorozhenko, YeV, Svistunov, YuB (2016). Prystrii dlia rozrivniuvannia nasypnykh vantazhiv u pivvahonakh. *Visnyk Kharkivskoho natsionalnoho avtomobilno-dorozhnoho universytetu*, ***75***, 145-149

[7] State Enterprise «Vinnytsiatransprylad» (2014). *A device for leveling bulk cargo in open wagons. Consolidated price list of the company*. Retrieved from http://www.vtranspribor.com.ua

[8] Romanovych, YeV, Afanasov, GM, Kozar, LM, But, VV (2014). Zasoby dlia mekhanizatsii rozrivniuvannia nasypnykh vantazhiv u napivvahonakh. *Zbirnyk naukovykh prats Ukrainskoi derzhavnoi akademii zaliznychnoho transportu*, ***148(1)***, 116-121

[9] DSTU GOST 22235: 2015. Freight cars of mainline railways of 1520 mm gauge. General safety requirements during loading, unloading and shunting operations (GOST 22235-2010, IDT) (2015)

[10] Information on the volume, cargo turnover and payment of carriage charges from cargo transportation (by nomenclature groups of commodities, types of cars, types of transport and tariff classes) 2020 (Unified state open data web portal). Retrieved from http://www.data.gov.ua.

[11] Shvets A. O. 2018 Influence of longitudinal and transverse displacement of the center of gravity of cargo in gondola cars on their dynamic indicators *Science and Progress of Transport. Bulletin of the Dnepropetrovsk National University of Railway Transport,* **5(77)**. Retrieved from http://eadnurt.diit.edu.ua.

IOP Conf. Series: Materials Science and Engineering 1021 (2021) 012049 doi:10.1088/1757-899X/1021/1/012049

Mathematical modeling of dynamic loads on the ground robotic complex of special purpose

V B. Strutinsky[1], A. M. Gurzhii[2] and O. Y. Yurchyshyn[1,3]

[1]Machine Design Department, National Technical University of Ukraine "Igor Sikorsky Kyiv Polytechnic Institute", 37, Prosp. Peremohy, 03056, Kyiv, Ukraine
[2]National Academy of Educational Sciences of Ukraine, 52-A, Sichovykh Striltsiv Street, 04053 Kyiv, Ukraine

[3] Email: oksanayurchishin@gmail.com

Abstract. A special technique to study the impact of the discharge on the ground robotic complex equipped with small arms in this article is proposed. The development of the technique is based on mathematical modeling of the dynamic system of the robotic complex. The magnitude and nature of dynamic loads on the ground robotic complex are determined. The experimental sample of the ground robotic complex equipped with small arms has been developed by the authors. Measures to ensure a stable position of the complex and methods of compensation for uncontrolled changes in the position of the complex under dynamic loads are proposed. To study the processes of firing, their individual components are considered: the problem of internal ballistics, the process of movement of the retractable parts of the barrel, dynamic loads on the complex. The conducted study gives the chance to create highly effective samples of special equipment. These are, in particular, military ground-based robotic complexes, which will be competitive by increasing the efficiency of their operation.

1. Introduction
Ground robotic complexes equipped with small arms are effective military means. They are widely used in various military operations. Therefore, the study of robotic complexes of this type is relevant.

The issue of developing ground-based robotic systems equipped with small arms is important to ensure the national security of Ukraine.

The issue is related to the important tasks of efficiency increase of the Armed Forces of Ukraine which are of key importance for the national security and defense of Ukraine.

2. Analysis of recent research and publications
Modern ground-based robotic systems are considered and their features are identified in recent studies and publications [1]. Small size of the complexes (about 1 m) and low mass are the cause of complex dynamic processes that take place in them [2]. Ground-based robotic systems are characterized by limited rigidity of the bearing system and drives [3]. It causes possible decrease in the accuracy of positioning systems [4]. Damping systems are recommended in some cases to increase accuracy [5]. The main reason for the low accuracy of the complexes is the uncertain conditions of their support on the ground [6]. It is recommended to increase the accuracy of weapon positioning and, accordingly, the efficiency of its use by special control systems [7].

Issues of guidance accuracy and stability of the position are basic in the creation of military ground-based robotic systems [8]. Effective experimental methods are used to study the accuracy of robots [9]. It is accepted that the rational method of research is mathematical modeling [10]. Effective methods of modeling high-frequency shock processes have been developed [11].

These results do not take into account the specifics of the dynamic loads on the robotic complex during the shot. Therefore, they have limited application.

A significant number of studies have been devoted to the analysis of loads in artillery systems [12]. Experimental research and modeling are carried out [13]. Methods of experimental measurements and equipment are described in separate publications [14]. Methods of modeling shock loads during shot are given [15]. A significant number of works is devoted to solving problems of internal and intermediate ballistics [16]. Criteria for estimating impact loads on a moving object have been proposed [17]. Some works contain a description of special devices to simulate the action of the shot [18]. Methods for estimating impulse (shock) loads on an object are described [19]. They determine the displacement of an object under the action of irregular shock loads [20]. The significant amount of research does not contain practical methods and tools to ensure the accuracy of military ground-based robotic systems.

3. Defining the purpose and objectives of the study

Thus, one of the previously unresolved parts of the overall problem is to ensure the required firing accuracy of military ground robotic systems. Research in this area is carried out mainly by means of experimental methods. It requires complicated tests in a special firing range and it is not appropriate. Therefore, the aim of the research is mathematical modeling of dynamic processes during the shot of a military ground-based robotic complex and determination of dynamic loads on a ground-based robotic complex.

The tasks of research are based on solving the problem of internal ballistics to determine the kinematic and dynamic parameters of the shot on a robotic complex in the shape of mutually agreed kinematic and dynamic propellers.

4. The basic part of the study

Ground robotic complexes equipped with small arms, as a rule, have a compact design (Figure 1).

Figure 1. Experimental sample of a ground-based robotic complex equipped with small arms

Complex and multidimensional dynamic processes occur during firing. Both simplified (elementary) and special methods are used to describe them. Dynamic processes determine the nature and magnitude of dynamic loads on a mobile robotic complex. Therefore, their definition is the main and mandatory factor in the reliability of the complex. Dynamic processes during firing are manifested in the action of kinematic and dynamic force factors.

Dynamic action during firing leads to limited oscillating movements in the dynamic system of the complex. The basic ones are transverse and angular displacements corresponding to the spherical motion of the complex according to the center of mass C and are described by means of angular coordinates ψ, θ, φ. Finally, it leads to transverse and angular displacements of the barrel and equivalent displacements of the dual cut in the plane perpendicular to the axis of the barrel.

To study the processes of firing, their individual components are considered: the problem of internal ballistics; the process of movement of the recoiling parts of the barrel; dynamic loads on the complex. The task of internal ballistics is to determine the kinematic parameters of the bullet in the barrel (Figure 2).

To determine the time change of the dynamic action of a moving bullet, a dynamic calculation of the displacement of the bullet was performed. A simplified scheme of bullet movement in the barrel channel is applied. It was assumed that the force of resistance is proportional to the velocity of the bullet, and the pressure (P = Re = const) is constant in the first approximation. The equation of movement of the bullet in the barrel within the accepted assumptions:

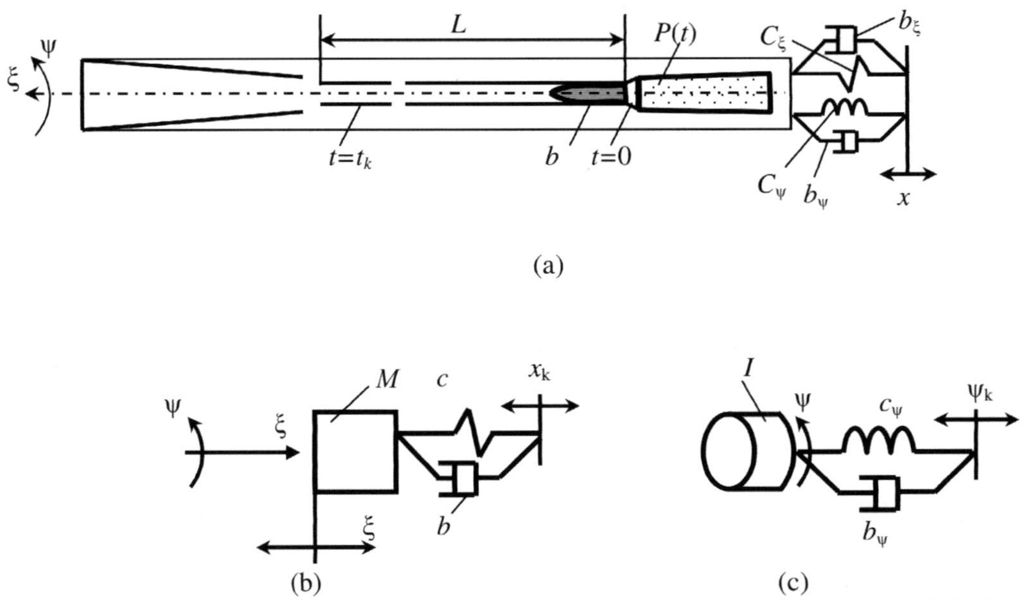

Figure 2. Scheme of bullet displacement in the barrel channel (a) and dynamic model of recoiling parts in the shape of reciprocating movements of moving parts of the barrel (b) and transverse and angular movements (c)

$$m\frac{d^2x}{dt^2} = P_e \cdot f \cdot 1(t) - b\frac{dx}{dt}, \quad (1)$$

where m - mass of the bullet; x - displacement of the bullet along the barrel; P_e - equivalent constant value of pressure in the barrel channel; f - the effective area of the end of the bullet which is affected by powder gases; $1(t)$ - single stepped function; b - equivalent coefficient of resistance of the bullet when it moves in the barrel; t - time.

Simplified equation (1) is quite approximate, but it allows you to describe the workflows that accompany the passage of the bullet through the barrel.

Equation (1) is reduced to the standard form by entering the designation:

$$\frac{dx}{dt} = V, \quad (2)$$

$$\frac{m}{b} = T_V, \quad (3)$$

$$K_p = \frac{P_e f}{b}. \quad (4)$$

Thus, the equation will take the form:

$$T_V \frac{dV}{dt} + V = K_p \cdot 1(t), \quad (5)$$

$$T_C \frac{d^2 x}{dt^2} + \frac{dx}{dt^2} = p_e \frac{f}{b} - \frac{F_0}{bV_K}. \quad (6)$$

Solution of the equation:

$$V = V_k \left(1 - e^{-\frac{t}{T_C}}\right), \quad (7)$$

$$e^{-\frac{t}{T_C}} = 1 - \frac{V}{V_k}. \quad (8)$$

Time displacement is determined by integration (Figure 3):

$$x = \int_0^t V(t)dt = V_k t + T_C V_k \left(e^{-\frac{t}{T_C}} - 1\right) = V_k \left[t + T_C \left(e^{-\frac{t}{T_C}} - 1\right)\right] \quad (9)$$

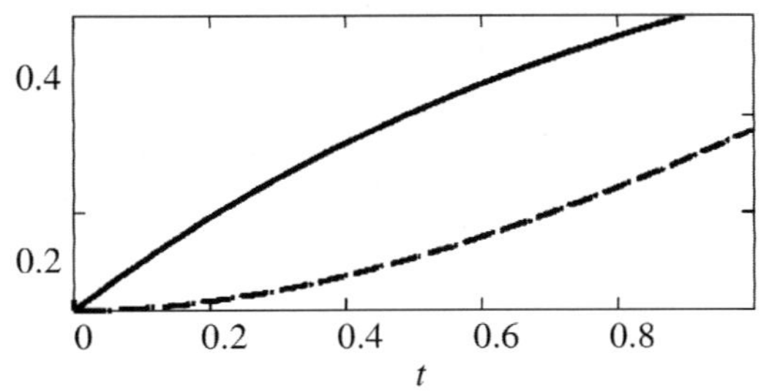

Figure 3. Time dependence of the bullet displacement in the barrel:

v(t) —— , *x(t)* ········ , *xp(t)* - - -

The parameters of the movement of the recoiling parts of the barrel are determined while considering the reciprocating and transverse and angular oscillating motion of the moving parts. We assume linear dynamic models of the minimum level of complexity in the first approximation (see Figure 2b, c).

Dynamic models correspond to the reciprocaing displacements of moving parts in the direction ξ and torsional displacements in the shape of rotation at an angle ψ.

The initial conditions through the velocity V_0 of mass M and angular velocity of rotation of the moving parts at the initial moment are determined for dynamic models. They are found from the conditions of momentum conservation and momentum of motion speed

$$m_k V_k = MV_0, \quad (10)$$

$$I_k \omega_k = I\omega_0, \quad (11)$$

where m_k is the mass of the bullet and the moment of inertia of the bullet; *t M, I* is mass and moment of inertia of moving parts.

Accordingly, the initial conditions for dynamic models are defined at t=0, ξ=0, ψ=0 as $\frac{d\xi}{dt} = V_0 = \frac{m_k V_k}{M}$, $\frac{d\psi}{dt} = \omega_0 = \frac{I_k \omega_k}{I}$.

Differential equations of displacement of moving parts at a shot are made:

$$m\frac{d^2\xi}{dt^2} + b\frac{d\xi}{dt} + c\xi = \delta(t), \quad (12)$$

$$I\frac{d^2\psi}{dt^2} + b_\psi \frac{d\psi}{dt} + c_\psi \psi = \delta(t), \quad (13)$$

where b, c are coefficient of resistance and equivalent stiffness of the moving parts of the barrel during recoil; b_ψ, c_ψ are coefficient of resistance and torsional rigidity of the moving parts of the barrel; δ(t) is delta function.

These equations are transformed to the standard form:

$$T_\xi^2 \frac{d^2\xi}{dt^2} + 2\xi_\xi T_\xi \frac{d\xi}{dt} + \xi = K_\xi \delta(t), \quad (14)$$

$$T_\psi^2 \frac{d^2\psi}{dt^2} + 2\xi_\psi T_\psi \frac{d\psi}{dt} + \xi = K_\psi \delta(t) \ , \quad (15)$$

where $T_\xi^2 = \frac{M}{C}$, $\xi_\xi = \frac{b}{2\sqrt{MC}}$, $K_\xi = \frac{1}{C}$, $T_\psi^2 = \frac{I}{C_\psi}$, $\xi_\psi = \frac{b_\psi}{2\sqrt{IC_\psi}}$, $K_\psi = \frac{1}{C_\psi}$ the dynamic load acting on the part of the moving parts on the complex will depend on the displacement of the moving parts and the dynamic displacements of the complex.

We will assume the dynamic process of firing to be infinitely fast in comparison with the oscillating process of spatial displacements of the complex as a whole in the first approximation,

The reciprocating and transverse and angular displacement of the moving parts during firing will be determined from the solution of differential equations under the introduced initial conditions. The solution of the equations is as following:

$$\zeta = \frac{K_\zeta}{T_\zeta \sqrt{1-\zeta_\zeta^2}} \cdot e^{\frac{-\zeta_\zeta t}{T_\zeta}} \cdot \sin\left(\frac{\sqrt{1-\zeta_\zeta^2}}{T_\zeta} t\right), \quad (16)$$

$$\psi = \frac{K_\psi}{T_\psi \sqrt{1-\zeta_\psi^2}} \cdot e^{\frac{-\zeta_\psi t}{T_\psi}} \cdot \sin\left(\frac{\sqrt{1-\zeta_\psi^2}}{T_\psi} t\right). \quad (17)$$

Displacement of the recoiling parts of the table cause a dynamic effect on the complex.

Dynamic loads acting on the complex have the character of a power screw "dynamo", which includes force $\vec{F}$ and torque $\vec{M}$. Dynamo parameters are determined on the basis of the dependence

$$F = C\xi + b\frac{d\xi}{dt}, \quad (18)$$

$$M = C_\psi \psi + b_\psi \frac{d\psi}{dt} . \quad (19)$$

After substitution of values of reciprocating and transverse and angular displacements and their derivatives in these formulas, we obtain similar expressions of the load vector

$$F(t) = \xi(t) + 0{,}1 \cdot V\xi(t)$$

They are oscillating processes (Figure 4).

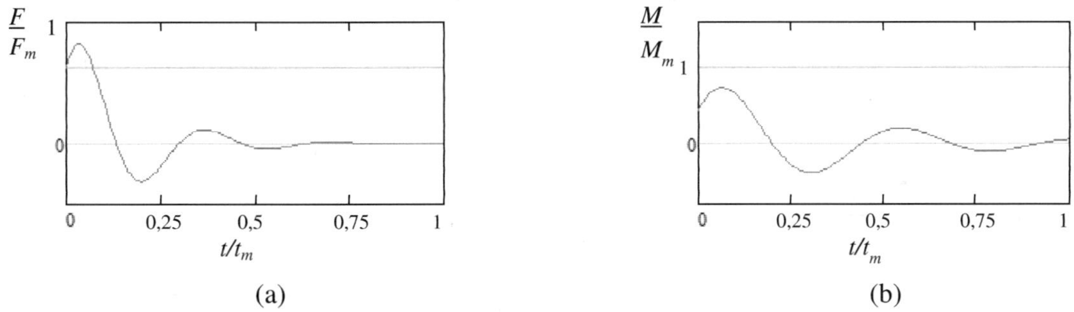

Figure 4. Relative dynamic loads on the complex caused by the shot: a - module-vector of force directed along the axis of the barrel; b - torque module acting relative to the axis of the barrel

The force factor in the shape of a dynamic screw changes during the shot according to the diagram (Figure 5).

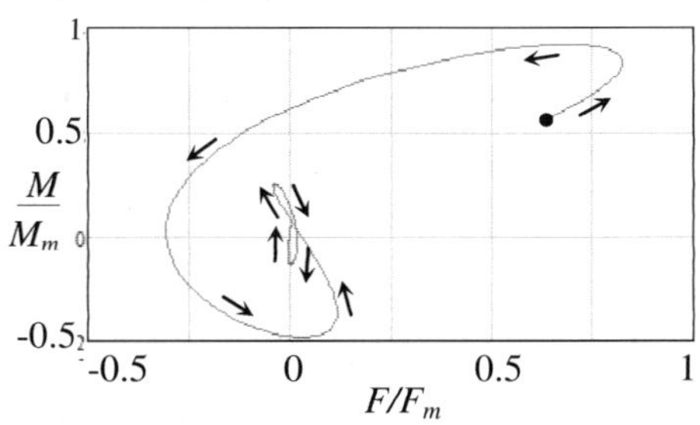

Figure 5. Hodograph of the interdependence of the components of the dynamic screw of the module-vector of force F and torque M

5. Conclusions

The nature of the spatial displacement of the ground robotic complex under the action of impulse loads due to the shot is established. Measures to ensure a stable position of the complex and methods of compensation for uncontrolled changes in the position of the complex under dynamic loads are proposed. The conducted study gives the chance to create highly effective samples of special equipment, military ground robotic complexes, in particular.

References

[1] Strutynskyi S, Kravchuk V, Semenchuk R 2018 Mathematical modelling of a specialized vehicle caterpillar mover dynamic processes under condition of the distributing the parameters of the caterpillar *International Journal of Engineering & Techology* **7 (4/3)**, 40-46

[2] Jun Qian, Bin Zi, Daoming Wang, Yangang Ma, Dan Zhang 2017 The design and development of an omni-directional mobile robot oriennnted to an intelligent manufacturing system *Sensors 2017* **17** 2073

[3] Korayem M H, Dehkordi S F 2018 Derivation of motion equation for mobile manipulator with viscoelastic links and revolute–prismatic flexible joints via recursive Gibbs–Appell formulations *Robotics and Autonomous Systems Volume* **103** 175-198

[4] Strutynsky V, Hurzhii A, Kolot O, Polunichev V 2016 Determination of development grounds and characteristics of mobile multi-coordinate robotic machines for materials machining in field conditions *Naukovyi Visnyk Natsionalnoho Hirnychoho Universytetu* **5** 43-51

[5] The U.S. Army 2017 *Robotic and Autonomus Systems Strategy* 31

[6] Baoquan Li, Yongchun Fang, Guoqiang Hu, Xuebo Zhang 2016 Model-Free Unified Tracking and Regulation Visual Servoing of Wheeled Mobile Robots *Journal Sensors and Actuators A Physical, IEEE Transactions on Control Systems Technology* **24** 1328–39

[7] Kot Tomas, Novak Petr 2018 Application of virtual reality in teleoperation of the military mobile robotic system TAROS *International journal of advanced robotic systems, January-February* 1-6

[8] Strutynskyi S 2018 Defining the dynamic accuracy of positioning of spatial drive systems through consistent analysis of processes of different range of performance *Scientific Bulletin of the National Mining University. Dnipro* **3** 64 – 73

[9] Strutynskyi S V, Hurzhii A A 2017 Definition of vibro displacements of drive systems with laser triangulation meters and setting their integral characteristics via hyper-spectral analysis methods *Scientific Bulletin of the National Mining University* **1** 43-51

[10] Panchenko A, Voloshina A, Milaeva I, Luzan P. 2019 Operating Conditions' Influence on the Change of Functional Characteristics for Mechatronic Systems with Orbital Hydraulic Motors. Modern *Development Paths of Agricultural Production.* Springe Cham 169-176

[11] Voloshina A, Panchenko ., Panchenko I, Zasiadko A 2019 Geometrical Parameters for Distribution Systems of Hydraulic Machines Modern Development Paths of Agricultural Production *Springer Cham* 323-333

[12] William G Braun, III Kim Richard Nossal & Stefanie von Hlatky 2018 Robotics and Military Operations *Kingston conference on International security series* 75

[13] Varshavskiy A Y 2017 Problemy razvitiya progressivnykh tekhnologiy: robototekhnika *Modernizatsiya Innovatsii Razvitiye* 682-697

[14] Strutynsky V, Hurzhii A, Kozlov L 2019 Determination of static equilibrium conditions of a mobile terrestrial robotic complex *Naukovyi Visnyk Natsionalnoho Hirnychoho Universytetu* **5** 79-86

[15] The U.S. Army Robotic and Autonomus Systems Strategy 2017 Maneuver, Aviation, and Soldier Division Army Capabilities Integration Center

[16] Chepkov I, Dovhopolyy A, Huslyakov O 2019 Kontseptual'ni zasady stvorennya vitchyznyanykh udarno-rozviduval'nykh nazemnykh robotyzovanykh kompleksiv vazhkoho klasu *Ozbroyennya ta viys'kova tekhnika* KievTSNDI OVT **3 (23)**16-25

[17] Analytical overview of the global robotics market 2019 *https://www.sberbank.ru/common/img/uploaded/pdf/sberbank_robotics_review_2019_17.07.2019_m.pdf* **06.10.2020** 272

[18] Ugis Romanovs, Maris Andzans 2019 Digital infantry battlefield solution. Research and Innovation DIBS project **Part Three** *Milrem Robotics*

[19] Hyun-Min Joe Jun-Ho Oh 2018 Balance recovery through model predictive control based on capture point dynamics for biped walking robot

[20] Serhii Strutynskyi, Waldemar Wójcik, Sandugash Orazalieva 2019 Grounds for mechatronic system development of the optical devices position dynamic stabilization of a mobile terrestrial robotic *system Photonics Applications in Astronomy, Communications, Industry, and High-Energy Physics Experiments* **11176** 1117610

Determination of the influence of soil irregularities on dynamic processes in caterpillar engines of a terrestrial robotic complex

S V Strutynskyi[1,2], R V Semenchuk[1]

[1] Department of Applied fluid mechanics and mechatronics, Igor Sikorsky Kyiv Polytechnic Institute, Prosp. Peremohy 37, Kyiv, 03056, Ukraine

[2] Email: strutynskyi@gmail.com
Abstract. The paper describes the change of the geometry of the caterpillar of the terrestrial robotic complex depending on external factors. It is established that as a result of the action of centrifugal forces the caterpillar elongates, on the surface of the support rollers, the caterpillar repeats their shape, the upper part of the caterpillar acquires a convex shape, and the lower part of the caterpillar repeats the shape of the curve, which determines the unevenness of the soil. The dependence of the caterpillar shape on soil irregularities is established. The peculiarity of the geometry is that on the lower part of the caterpillar there is a ripple with the number of waves between the support rollers 1-2. It is established that the presence of soil irregularities, the harmonic components of the profiles of which have a period of about 4/3 of the distance between the extreme support rollers, leads to the greatest tensile loads in the caterpillar and the corresponding changes in caterpillar length. The dependences determining the elongation of the caterpillar, which is proportional to the square of the amplitude of the harmonic component of the soil roughness of the corresponding period are obtained.

1. Introduction.

Terrestrial robotic complexesare are usually designed to perform special operations in difficult road conditions. They move at significant speeds on uneven ground. In this case, the caterpillars have intense dynamic loads. The study of dynamic processes in caterpillars is an urgent scientific problem.

The problem is associated with important theoretical and practical tasks of creating and using ground-based robotic systems for special purposes.

2. Analysis of recent research and publications.

Recent studies and publications have considered a number of designs of terrestrial robotic systems [1]. We established features of their characteristics [2] and determined the nature of dynamic loads on a mobile robotic complex [3]. In separate works, the studies of drives of complexes [4], in particular their manipulators [5] are described. Theoretical studies of robotic complexes are carried out [6] and methods of mathematical modeling are developed [7]. The results of the study, as a rule, relate to individual structures of complexes and do not allow to perform the design of robotic complexes for special purposes [8]. Special problems arise in determining the dynamic characteristics of robotic systems [9]. Dynamics analysis requires complex methods and equipment [10]. Experimental studies are performed using special equipment [11]. The studies cited in the literature refer to robotic complexes that move on smooth roads [12]. Some sources [13] include studies performed in the real road conditions. Methods for determining the dynamic characteristics of complexes are given [14].

There are studies of the characteristic modes of movement of the chassis with a caterpillar. [15] A number of publications present a study of the movement of the complex on the slope [16]. The basic requirements for the robotic complexes intended for work in field conditions are formulated [17]. In [18] the researches of perspective base of manipulators of robotic complexes are presented.

As a result of the analysis of the available information sources the conclusion is made about the absence in them of results of definition of influence of roughnesses of soil on dynamic processes in a caterpillar of a ground robotic complex. These questions relate to the previously unsolved part of the general problem of studying the characteristics of the caterpillar propulsion of a small complex of special purpose.

3. Defining the purpose and objectives of the study.

The purpose of the study is to determine the quantitative parameters of the influence of soil irregularities on the dynamic processes in the caterpillars of the ground robotic complex of special purpose. The objectives of the study include experimental determination of dynamic displacements of a caterpillar during the movement of a complex on open ground, substantiation of the scheme and mathematical model of interaction of a caterpillar with a road surface on modeling of influence of soil irregularities on dynamic processes in caterpillars of a ground robotic complex.

4. The main part of the study.

In terrestrial robotic complexes, caterpillar engines with caterpillars made of elastic polymeric materials that interact with the road surface are used. At the same time there are significant deformations of the caterpillar, which are especially significant when moving the robotic complex on the ground with an uneven surface.

Experimental determination of the shape of the caterpillar of the mobile ground-based robotic complex was performed when moving the complex on open ground and it includes video recording of the shape of the caterpillar of the mobile robotic complex with subsequent frame-by-frame video processing (Figure 1).

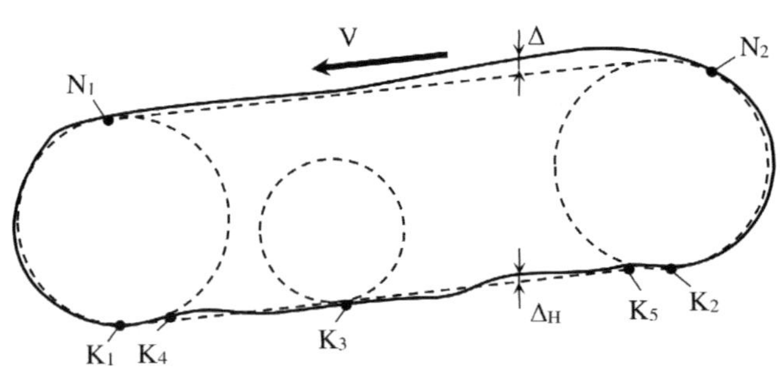

(a) (b)

Figure 1. The video frame of the mobile robotic complex when moving the complex at a speed of 3 km/h (a) and the corresponding contour of the caterpillar (b) is determined by this video frame.

A dynamic analysis of the shape of the caterpillar with the establishment of its characteristic areas was performed on the video footage. The upper part of the caterpillar between the points of its contact with the N_1N_2 drums corresponds to the scheme of a moving massive thread with minimal bending resistance, which performs transverse oscillations. The lower part of the caterpillar has an area of contact with the ground between the points K_1K_2.

In this video frame there is some sagging of the upper part of the caterpillar. There is a curved elastic line of the lower part of the caterpillar. There is a slight undulation at the bottom of the caterpillar.

The contour of the upper part of the caterpillar, depicted in the video frame, has a characteristic protrusion between the points N_1N_2. The appearance of this protrusion is due to inertial forces acting on the flexible caterpillar, and the lower part of the caterpillar is a wavy curve that repeats the unevenness of the soil. In this case, there is a deformation of the soil under the support rollers in the areas K_1, K_2 and K_3. The deformation is most intense in the area of interaction of the front roller with the soil surface in the area K_1K_4. On the contour of the caterpillar there are areas in the form of arcs of circles K_2N_2 and N_1K_1, which correspond to the interaction of the caterpillar with the peripheral surfaces of the support rollers. There are wavy areas on the lower part of the caterpillar. There are three complete waves, the maximum of which is Δ_H between the reference sections K_3K_5.

Changing the shape of the caterpillar depends on the speed of movement of the complex, the shape of the roadway and the angle of ascent and descent. To establish the features of the caterpillar shape, a series of video frames was processed (figure 2).

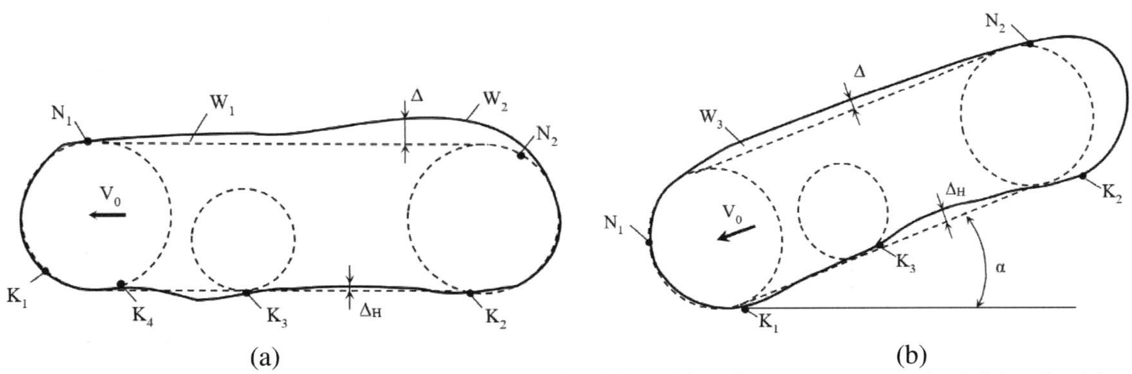

Figure 2. The shape of the caterpillar is determined by video frames at a speed of 5 km/h: (a) — movement on a horizontal surface; (b) — inclination of about 20°.

Increasing the speed of the complex does not change the qualitative picture of movements and the shape of the caterpillar. As before, the upper part of the caterpillar has some sagging. The protrusion W_2 on the contour in the area N_2 (Figure 2a) is explained by the interaction of the teeth of the drive roller with the holes of the caterpillar and the action of centrifugal forces.

The contour of the lower part of the caterpillar has a ripple, which is determined by the location of the intermediate support roller. Transverse waves can also be traced on the contour of the lower part of the caterpillar. Their maxima Δ_H are located between the support rollers K_3, K_2. The lower part of the caterpillar repeats the soil surface, taking into account its deformation (crumpling).

The movement of the complex in the areas of ascent or descent (Figure 2b) leads to an additional complication of the shape of the caterpillar. On the contour of the caterpillar, the area of convexity ΔH is traced, which repeats the protrusions on the ground between the regions K_2 and K_3.

The upper part of the caterpillar has a convex shape and a protrusion W_3 between the points N_1N_2, due to the action of centrifugal forces and forces of inertia. The lower part of the caterpillar also acquires a convex shape.

During the movement of the ground-based robotic complex, the caterpillar is subjected to significant dynamic loads. Based on the experimentally determined shape of the caterpillar of the mobile robotic complex, the scheme of caterpillar loads due to the action of forces, in particular centrifugal forces, is substantiated. Under the action of the moment M, the drive roller rotates with an angular velocity ω, moving the complex with a velocity V (Figure 3).

IOP Conf. Series: Materials Science and Engineering 1021 (2021) 012050 doi:10.1088/1757-899X/1021/1/012050

Consider the movement of the complex on a level road. In the caterpillar, which wraps around the right and left support rollers, there are centrifugal forces. To determine them, we use a moving coordinate system X_VZ, which moves with speed V_0.

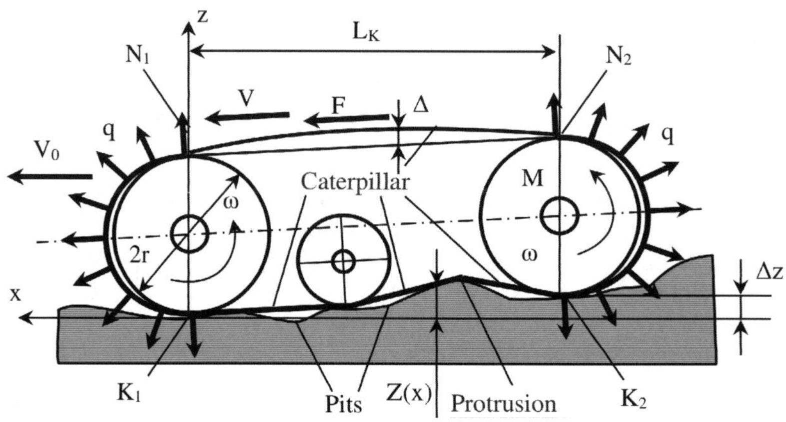

Figure 3. The scheme of interaction of the caterpillar with the road surface and the forces acting on it.

When passing a section of the caterpillar on the surface of the support roller on this section a distributed centrifugal force with intensity acts:

$$q = \frac{mV^2}{r}, \tag{1}$$

where m[kg/m] is running mass of the caterpillar; $V = 2V_0$ is circumferential speed of the caterpillar; r is the radius of the cylindrical surface of the drum.

The force that occurs at the intersection of the caterpillar is:

$$F_V = q \cdot r = mV^2. \tag{2}$$

The action of centrifugal forces, which causes the longitudinal load of the caterpillar leads to its stretching. In the first approximation, we assume a linear relationship between the force and deformation of the caterpillar. The change in caterpillar length will be:

$$\Delta L_V = (2L_K + 2\pi r)\frac{mV^2}{ES}, \tag{3}$$

where *E* is the modulus of elasticity of the caterpillar material; S is the area of its cross section.

In the initial state, the length of the caterpillar is greater than the nominal value due to the free installation (sagging) of the caterpillar. Denote the difference between the actual and nominal lengths ΔL_0. Under the action of centrifugal forces, the total length of the caterpillar will be greater than the nominal length by $\Delta L_0 + \Delta L_v$. Thus, the caterpillar will interact freely with the support rollers and contact them only in the areas K_1 and K_2. Between these points, the caterpillar rests on the ground and its shape and tension are determined by the laws of interaction of the lower part of the caterpillar with the soil.

The projections of the centrifugal forces on the x-axis have different directions in different areas of the caterpillar. Therefore, the cross-sections of the caterpillar are pushed away from the surfaces of the support rollers, and the contour of the caterpillar is deformed and stretched in the direction of the x-axis. The caterpillar is in contact with the rollers in the areas N_1 and N_2, and between these points, the position of the caterpillar corresponds to a straight line with some deflection by the value of Δ in the middle of its upper part.In sections K_1N_1 and K_2N_2, the centrifugal forces are balanced by tensile forces acting on the axial line of the caterpillar. In the upper section of the caterpillar between the points N_1N_2, the caterpillar is loaded at the edges by centrifugal forces in the sections N_1N_1 'and N_2N_2'. The caterpillar practically does not resist a bend under the influence of loadings, on edges it acquires the form of a convex curve with an arrow of a deflection Δ.

IOP Conf. Series: Materials Science and Engineering 1021 (2021) 012050 doi:10.1088/1757-899X/1021/1/012050

Assume that between the points K_1 and K_2 of contact of the rollers with the road surface (Figure 3) the caterpillar is stretched by some force, which depends on the centrifugal forces and irregularities of the road.

In the presence of depressions of the soil (pits) between points K_1 and K_2, the length of the caterpillar in this area remains unchanged and is L_K. In the presence of a protrusion (Protrusion), the caterpillar wraps around it and lengthens by some value ΔL, which is determined by the difference in the length of the curve that determines the profile of the road surface in the area between points K_1 and K_2 and the length of the straight line L_K. Assume that the road profile is described by some dependence:

$$z = z(x). \tag{4}$$

Let us define the derivative of the road profile:

$$\frac{dz}{dx} = \frac{dz(x)}{dx} = \text{tg}\alpha, \tag{5}$$

where α is the angle of inclination of the tangent profile.

The elementary section of the arc of the road profile has a length *dL* and is determined by the derivative of the contour of the caterpillar. A Fourier series was used to mathematically describe the contour of the caterpillar. In order to simplify the calculations, a symmetric problem is considered by extending the contour of the caterpillar from the point K_1 to the left coordinate half-plane. The resulting symmetrical contour of the caterpillar will have a period of $2L_K$ from point K_2' to K_2. To describe a symmetric contour, the Fourier series will have only cosine components and will be determined by the dependence:

$$z(x) = \sum_{k=0}^{\infty} a_k \cdot \cos\left(k \frac{\pi}{L_k} \cdot x\right). \tag{6}$$

When the shape of the caterpillar changes, its total length changes. Consider an infinitesimal element of the caterpillar, respectively, its length:

$$dx = dL \cdot \cos\alpha. \tag{7}$$

Using the dependence (5), we find:

$$dL = dx\sqrt{1 + (dz/dx)^2}. \tag{8}$$

Let us determine the total length of the contour of the caterpillar between points K_1 and K_2:

$$L_H = \int_0^{L_K} dL\sqrt{1 + (dz/dx)^2}dx. \tag{9}$$

The change in the length of the caterpillar due to the unevenness of the road will be:

$$\Delta L_H = L_H - L_k. \tag{10}$$

Let us determine the derivative of the dependence of the shape of the caterpillar on the longitudinal coordinate. To do this, use formula (6). After differentiation we obtain:

$$\frac{dz}{dx} = -\sum_{k=1}^{\infty} a_k \cdot \frac{k\pi}{L_k} \cdot \sin\left(\frac{k\pi}{L_x} x\right). \tag{11}$$

Dependence (11) defines the derivative as the sum of a series. For a separate sinusoidal component of the series we have:

$$\left(\frac{dz}{dx}\right)_k = \frac{a_k \cdot k\pi}{L_k} \sin\left(\frac{k\pi}{L_x} x\right). \tag{12}$$

Let us determine the change in the length of the caterpillar for one sinusoidal component of the series:

$$\Delta L_H = \int_0^{L_k} \sqrt{1 - \left(\frac{k\pi}{L_k} a_k\right)^2 \sin^2\left(\frac{k\pi x}{L_x}\right)} dx - L_k \tag{13}$$

IOP Conf. Series: Materials Science and Engineering 1021 (2021) 012050 doi:10.1088/1757-899X/1021/1/012050

To calculate the integral (13), it is reduced to the form of a normal elliptical integral of the second kind of the following type:

$$E(\phi, \gamma) = \int_0^{\phi} \sqrt{1 - \gamma^2 \sin^2 \phi} \cdot \mathrm{d}\phi. \quad (14)$$

To do this, new constants are introduced in dependence (13):

$$\phi = \frac{k\pi x}{L_k}, \quad (15)$$

$$\gamma = \frac{k\pi a_k}{L_k}. \quad (16)$$

From relations (15) and (16) it follows:

$$x = \phi \cdot L_k/(k\pi), \quad (17)$$

$$\mathrm{d}x = \mathrm{d}\phi \cdot L_K/(k\pi). \quad (18)$$

Accordingly, formula (13) takes the form:

$$\Delta L_H = \frac{L_k}{k\pi} \int_0^{k\pi} \sqrt{1 - \gamma^2 \cdot \sin^2 \phi} \cdot \mathrm{d}\phi - L_k. \quad (19)$$

To calculate the integral included in the dependences (15) and (16), it is reduced to the form of a tabular complete elliptic integral of Legendre of the second kind:

$$E_n(\gamma) = \int_0^{\pi/2} \sqrt{1 - \gamma^2 \cdot \sin^2 \phi} \cdot \mathrm{d}\phi. \quad (20)$$

Consider the contour shapes of the lower part of the caterpillar, described by cosine functions with different periods, which coincide with the distance between the support rollers of the caterpillar and correspond to the components of the series (6). There are significant components of the series with numbers k = 1.2, which correspond to the characteristic irregularities of the roadway (Figure 4).

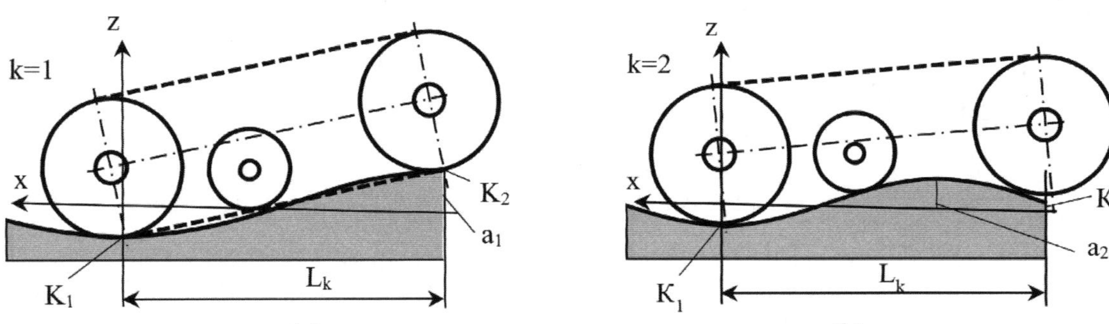

Figure 4. The scheme of interaction of the caterpillar with the inequalities described by sinusoidal dependences with the period: (a) — corresponding to the first harmonic (k = 1); (b) — corresponding to the second harmonic (k = 2).

The analysis of the deformation of the lower part of the caterpillar when moving the complex along the characteristic irregularities corresponding to the individual harmonics is performed. Road irregularities correspond to the first harmonic of the series (k = 1) and have a period of change equal to twice the distance between the extreme support rollers (Figure 4a). Changes in the transverse-angular position of the complex during its movement on the road with such irregularities compensate for the impact of road irregularities on the caterpillar. Therefore, in this case, the movements of the lower part of the caterpillar are insignificant, and the lower part of the caterpillar is close to a straight line.

As a result of the conducted researches it is established that the most significant transverse displacements of the caterpillar take place in the presence of irregularities on the road, which have harmonic components with the number k = 2 (Figure 4b). These harmonics have a wave period of 4/3 of the distance between the extreme support rollers. In this case, the intermediate roller has little effect on the deformation of the soil and does not smooth the protrusions of the roadway. Road bumps with this period are accepted as the most influential on the deformation of the caterpillar. Therefore, the characteristic changes in the length of the caterpillar are determined by the dependence (13) at k = 2. The number of quarters of cosine periods that must be taken into account when calculating the elliptical integral will be 2 (two) because the depression on the road does not increase the length of the caterpillar.

Accordingly, to determine the changes in caterpillar length, the dependence was obtained:

$$\Delta L_H = L_k\left[\frac{1}{2\pi}E_\Pi\left(\frac{2\pi a_2}{L_k}\right) - 1\right], \tag{21}$$

where $E_\Pi(\gamma)$ is the complete elliptic Legendre integral of the second kind given in the form (20).

For small values of arguments, this integral is approximately determined by the dependence:

$$E_\Pi(\gamma) = \frac{\pi}{2}\left(1 - \frac{\gamma^2}{4} - \frac{9\gamma^4}{192} - \ldots\right). \tag{22}$$

After substituting the value of the integral from (22) to (21) taking into account (15), (16) we obtain:

$$\Delta L_H = -L_H\left[\frac{\pi^2 a_2^2}{L_k^2} + \frac{144}{192}\frac{\pi^4 a_2^4}{L_k^4}\right]. \tag{23}$$

Thus, it was found that the most significant changes in caterpillar length occur when moving the complex on a road that has the form (figure 5b), and the largest changes in caterpillar length according to (23) are proportional to the square and fourth degree amplitude a_2 of road irregularities.

Based on the calculated change in the length of the caterpillar due to road irregularities according to Hooke's law there is an increase in the force of quasi-static tension of the caterpillar:

$$\Delta F = \Delta L_H \cdot ES/(2L_k + 2\pi r). \tag{24}$$

Thus, the centrifugal forces and the action of road conditions lead to a change in the length and tension of the caterpillar. The action of centrifugal forces leads to the elongation of the caterpillar, and the unevenness of the road reduces the length of the upper free part of the caterpillar. The general changes in the length of the caterpillar relative to the nominal value are:

$$\Delta L >> \Delta L_0 + \Delta L_V - \Delta L_H. \tag{25}$$

In some cases, the presence of appropriate protrusions on the road is set approximately equal to the length of the upper part of the caterpillar and its nominal value, and accordingly $\Delta L = 0$.

With significant road protrusions, when:

$$\Delta L_H > \Delta L_0 + \Delta L_V, \tag{26}$$

there is an additional tension of the caterpillar with the appearance of additional tensile force.

Determining the tension force of the caterpillar determines the reliability of its operation and durability of the ground robotic complex [19]. Therefore, the results are an important contribution to the development of special robotics [20].

5. Conclusions

As a result of the study, it was found that the shape of the caterpillar depends little on the speed of the complex, and mainly depends on the unevenness of the soil. The upper part of the caterpillar acquires a convex shape, and the lower part of the caterpillar has a ripple with the number of waves between the support rollers 1-2, and the lower part of the caterpillar repeats the shape of the curve, which determines the unevenness of the soil.

The presence of soil irregularities, the harmonic components of the profiles of which have a period of about 4/3 of the distance between the extreme support rollers, leads to the greatest tensile loads in the caterpillar and the corresponding changes in its length. The elongation of the caterpillar is proportional to the square of the amplitude of the harmonic component of the soil roughness of the specified period.

References

[1] Jun Qian, Bin Zi, Daoming Wang, Yangang Ma, Dan Zhang 2017 The design and development of an omni-directional mobile robot oriennnted to an intelligent manufacturing system *Sensors* 2017, 17, 2073.

[2] Sberbank robotics laboratory 2019 *Analiticheskiy obzor mirovogo rynka robototekhniki.* p 272.

[3] Baoquan Li ; Yongchun Fang ; Guoqiang Hu ; Xuebo Zhang 2016 Model-Free Unified Tracking and Regulation Visual Servoing of Wheeled Mobile *Robots Journal Sensors and Actuators A: Physical, IEEE Transactions on Control Systems Technology* (Volume 24, Issue: 4), pp 1328–39.

[4] Voloshina A., Panchenko A., Panchenko I., Zasiadko A. 2019 Geometrical Parameters for Distribution Systems of Hydraulic Machines *Modern Development Paths of Agricultural Production.* Springer, Cham, pp 323-336.

[5] Korayem M.H., Dehkordi S.F. 2018 Derivation of motion equation for mobile manipulator with viscoelastic links and revolute–prismatic flexible joints via recursive Gibbs–Appell formulations *Robotics and Autonomous Systems* Volume 103, pp 175-198.

[6] Chepkov I.B., Dovhopolyy A.S., Huslyakov O.M. 2019 Kontseptual'ni zasady stvorennya vitchyznyanykh udarno-rozviduval'nykh nazemnykh robotyzovanykh kompleksiv vazhkoho klasu. *Ozbroyennya ta viys'kova tekhnika.* Kyyiv: TSNDI OVT №3 (23). pp16-25.

[7] Serhii Strutynskyi, Waldemar Wójcik, Sandugash Orazalieva 2019 Grounds for mechatronic system development of the optical devices position dynamic stabilization of a mobile terrestrial robotic system *Photonics Applications in Astronomy, Communications, Industry, and High-Energy Physics Experiments* Volume 11176, 1117610.

[8] The U.S. Army 2017 *Robotic and Autonomus Systems Strategy.* 31p.

[9] Hyun-Min Joe Jun-Ho Oh 2018 Balance recovery through model predictive control based on capture point dynamics for biped walking robot *Robotics and Autonomous Systems* Volume 105, pp 1-10.

[10] Strutynskyi S.V., Hurzhii A.A. 2017 Definition of vibro displacements of drive systems with laser triangulation meters and setting their integral characteristics via hyper-spectral analysis methods *Scientific Bulletin of the National Mining University* №1. pp 43-51.

[11] Panchenko A., Voloshina A., Milaeva I., Luzan P. 2019 Operating Conditions' Influence on the Change of Functional Characteristics for Mechatronic Systems with Orbital Hydraulic Motors. *Modern Development Paths of Agricultural Production.* Springer, Cham, pp 169-176.

[12] Kot Tomas, Novak Petr 2018 Application of virtual reality in teleoperation of the military mobile robotic system TAROS *International journal of advanced robotic systems,* January-February. pp 1-6.

[13] Paul Ritzen, Erik Roebroek, Nathan van de Wouw, Zhong-Ping Jiang 2016 Trailer Steering Control of a Tractor–Trailer Robot *IEEE Transactions on Control Systems Technology* (Volume:24 , Issue: 4). pp 1240–52.

[14] Strutynskyi S., Kravchuk V., Semenchuk R. 2018 Mathematical modelling of a specialized vehicle caterpillar mover dynamic processes under condition of the distributing the parameters of the caterpillar *International Journal of Engineering & Techology,* 7 (4/3), pp 40-46 .

[15] Volosnikov S.A. 2016 The methodology for determining the critical velocity of the track of the skid platform *Mechanics and mechanical engineering.* Kharkiv: NTU "KPI" №1 pp 36-44.

[16] Strutynsky V., Hurzhii A., Kozlov L. 2019 Determination of static equilibrium conditions of a

mobile terrestrial robotic complex *Naukovyi Visnyk Natsionalnoho Hirnychoho Universytetu* №5 pp 79-86.
[17] Ugis Romanovs Maris Andzans 2019 Digital infantry battlefield solution. Research and Innovation. DIBS project. Part Three. *Milrem robotics* p 120.
[18] Strutynskyi S., Nochnichenko, I. 2019 Design of parallel link mobile robot manipulator mechanisms based on function oriented element *base Eastern-European Journal of Enterprise Technologies,* № 4/7(100). pp54-64.
[19] William G. Braun, III Kim Richard Nossal & Stefanie von Hlatky 2018 *Robotics and Military Operations. Kingston conference on International security series.* 75 p.
[20] A. Y. Varshavskiy 2017 *Problemy razvitiya progressivnykh tekhnologiy: robototekhnika* (MIR (Modernizatsiya. Innovatsii. Razvitiye.)) pp 682-697.

IOP Conf. Series: Materials Science and Engineering 1021 (2021) 012051 doi:10.1088/1757-899X/1021/1/012051

Simulation modeling of the automobile braking system performance

G Viselga[1], E B Ugnenko[2], E N Uzhviieva [2], O N Tymchenko [2] and N I Sorochuk[2,3]

[1] Department of Mechanical and Material Engineering Vilnius Gediminas Technical University, J. Basanavicius str. 28, LT-03224, Vilnius, Lithuania.
[2] Department of Surveying and Designing of Communication, Surveying and Land Management, Ukrainian State University of Railway Transport, Feierbakh Square 7, 61050, Kharkiv, Ukraine

[3] Email: nat.sorochuk50@gmail.com

Abstract. Simulation model of the automobile braking system performance taking into account not only the wheel, brake pads, brake drum dynamics but the dynamics of the hydraulic brake system: the main hydraulic cylinder, the brake gear hydraulic cylinders has been presented. Effect of changing parameters of the braking system elements on the braking distance of the automobile has been considered. Road users' safety relies heavily on effectiveness of vehicle braking system. Thus, braking system design is essential when designing a new vehicle. Modeling of the functionality of the designed system is one of the most important phases of design. Time and material expenses for imitational modeling are always less than actually producing models and conducting experiments on them. Experiments should prove theoretical abstracts stated while modeling. Thus, the imitational model should be well designed and adequate. Imitational model allows efficient operation of brake system with different system of control.

1. Introduction

Safety of travelers depends on effectiveness of performance of the automobile braking system. Thus, especial attention at designing a new automobile is paid to development of the braking system. One of its major stages is simulation of performance of the synthesized braking system and its automatic control system. Material and time expenditures on simulation modeling are always less than on making prototypes and trial models to conduct full-scale experiments. Experiments with the trial model are to prove theoretical conclusions made in the course of simulation completely. Therefore, the simulation model is to have a high degree of adequacy and high accuracy.

At simulation of the car braking system it is necessary to synthesize interconnection of different physical elements and systems – linear and rotation motion of mechanical parts, operation of the hydraulic system with regard to the braking fluid properties, electric control circuits with automatic control system based on a microcontroller in the same simulation model. Such simulation model with high accuracy and speed of the simulation experiment proceeding can be synthesized by means of MatLab. The developer can also optimize functioning of the braking system performance algorithm, optimization of parameters of the braking system elements and use the simulation model as a control system to detect disturbances at operation of automatic control system on a real object in a simulation model made in MatLab. The purpose of the work is minimization of material and time expenditures on the braking system development due to the braking system simulation modeling, conducting simulation experiments for optimization of algorithm of functioning of the braking system automatic system control and parameters of elements of the developed braking system.

In the work there have been represented a simulation model of performance of the braking system in different modes. It demonstrates dynamics of motion of all mechanical parts of the braking system – hydraulic cylinders, pads, wheels and visualizes the operation of algorithm of the braking system performance. After optimization of the control system parameters the simulation model allows to switch to the board with the mechanical part of the braking system and check the control system operation effectiveness. During the check the observing system conducts estimation of external and internal disturbances for more effective performance of the automobile braking system.

2. Publications analysis

Deceleration of a vehicle is a very complex process and of great importance for the traffic safety. This complexity originates from the occurrence of many events in different types of mechanical, electrical, thermal, pneumatic, etc. during braking. These operations occur in various points of the vehicles at varying densities and effect several parts of the mode of transport [1]. The mechanical energy of the vehicle is mostly transformed into thermal energy and spread to environment [2].

The automobile braking system consists of movable mechanical parts and the hydraulic system. Modern braking systems have a complex structure of interconnected elements [4]. The main elements of the system are: a vacuum booster, a main hydraulic cylinder, front and rear wheels brake gears (a hydraulic cylinder, pads, a brake disk, a brake drum), a pressure regulator valve, antilocking system, automatic control system [5]. At building up a simulation model it is necessary to take into account dynamics of the braking system elements and dynamics of the car movement depending on the braking system performance [6]. To ensure interconnection of elements belonging to different physical systems in one simulation model is possible in the Matlab/Simulink problem-solving environment [7]. In the suggested environment there can be realized a programmable system and obtained graphics of transients with high accuracy [10]. Matlab can also be used for controlling the braking system intelligence panel and perform follow-up control on the basis of the response discrepancy of the panel and simulation model to control actions. This allows to detect external and internal disturbances affecting the real braking system of the automobile and to develop a control algorithm for their compensation [11].

3. Purpose and statement of problem

The purpose of the work is to increase effectiveness of the process of the braking system synthesis by means of development of simulation model of performance of the braking system and its elements and conducting of simulation experiments for determining the optimal algorithm of operation of the braking system control. To achieve the purpose it is necessary to solve the following tasks:

- analysis of functioning of the braking system elements;
- development of the simulation model structure;
- synthesis of the simulation model of the braking system performance;
- optimization of the regulator parameters and of the operation algorithm.

For realization of the set tasks it is necessary to describe dynamics of the mechanical elements motion with the help of transfer functions and use specialized libraries Matlab/Simulink for building up the hydraulic part of the braking system.

4. Simulation model

A general view of the simulation model of the automobile braking system is represented in Figure 1.

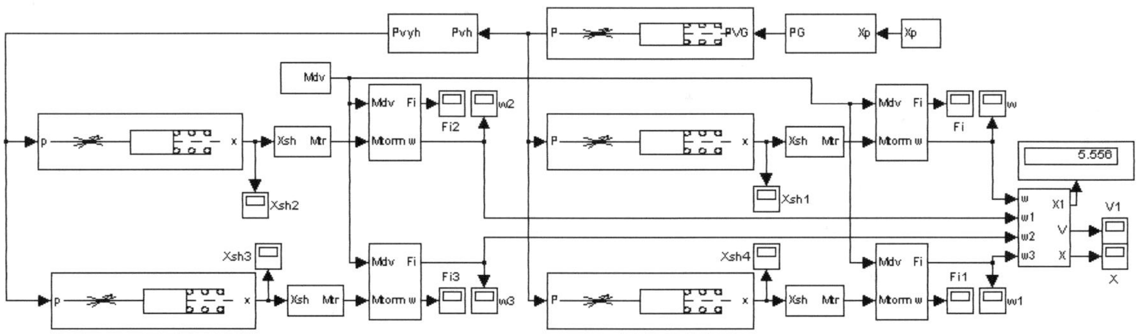

Figure 1. The simulation model of the automobile braking system

To simulate functioning of the model it is necessary to specify parameters of the braking system elements [12]. Parameters of each element of the braking system are introduced separately in a dialogue box of a correspondent subsystem. In dialogue boxes there presented short information about the required parameters as well as about the layout and dimensionality to be adhered when introducing parameters of the braking system elements. Information about geometrical parameters of each wheel and their weight is given for the subsystem simulating dynamics of the wheels rotation. For the subsystem describing the automobile dynamics it is necessary to specify the automobile weight and general resistance affecting it during the motion [12]. For the subsystem simulating the brake pads parameters denoting interrelation between a degree of a pad fitting and the moment of braking, which it causes, are to be specified. Parameters of hydraulic cylinders of the brake gears – geometry, changes of pressure, resistance of their rods are specified in specialized elements of the Simulink libraries. Parameters of conversion of the brake pedal travel into pressure of braking fluid in the braking system arising at shifting the main cylinder are specified in the subsystem of the vacuum booster. To register the load affecting the automobile rear axle in the subsystem simulating the pressure regulator of braking fluid in the rear brake gears a value of the axle load is specified, which results in pressure amplification in the rear axle braking circuit. In the subsystem simulating the internal combustion engine there specified a torsion torque and dynamics of its changing with regard to the clutch engagement/disengagement.

The simulation model demonstrates dynamics of all four automobile wheels rotation under the action of the engine torque and braking torque from the brake pads. The model allows estimating dynamics of the wheels rotation at braking operation both at the engaged and disengaged clutch. Calculation of the automobile braking distance is performed automatically, which enables estimating the braking system effectiveness at different speeds of the automobile.

Conducting simulation experiments with the help of a synthesized braking system simulation model significantly simplifies the task of the designer. He does not have to carry out full-scale experiments to determine the braking system effectiveness, which reduces material and time expenditures on its development. It also enables estimation of ineffective braking systems avoiding the car breakdown or accidents making its repairing impossible. The given simulation model allows the designer carrying out several experiments simultaneously as well as compare several braking system under development in the automated mode. Thus, the synthesized model ensures increasing effectiveness of designing the automobile braking system.

Figure 2 represents a graph of changes in the automobile speed: smooth acceleration of the car to the speed of more than forty kilometers per hour and in thirty seconds after starting acceleration the clutch being disengaged and the brake pedal depressed.

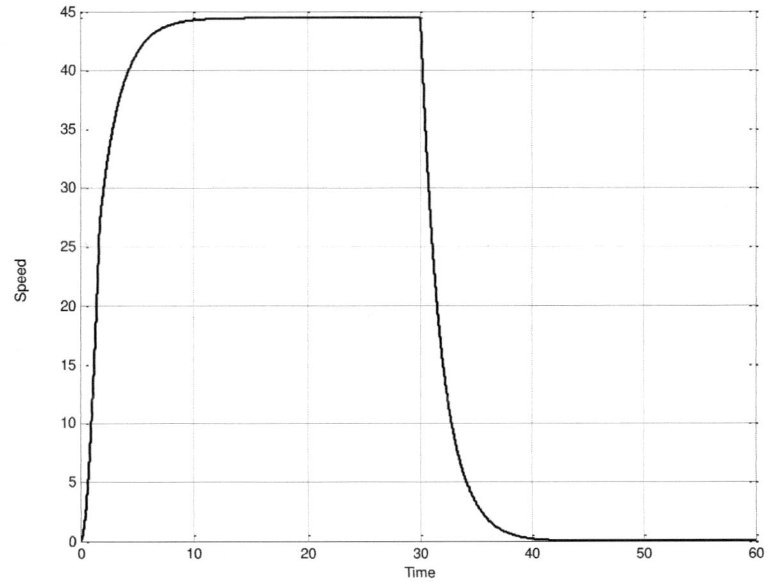

Figure 2. Changes in the automobile speed

The graph demonstrates a smooth loss of speed to a full stopping of the car. So it is possible to judge visually how hard the process of braking is and to adjust parameters of the braking system elements. At adjusting the simulation model allows to try out different braking modes: smooth braking without clutch disengaging and not up to the full stop of the automobile; smooth braking with clutch disengaging up to the full stop of the automobile; sudden emergency stop for a minimal time period.

Figure 3 represents a graph characterizing the function of the automobile movement.

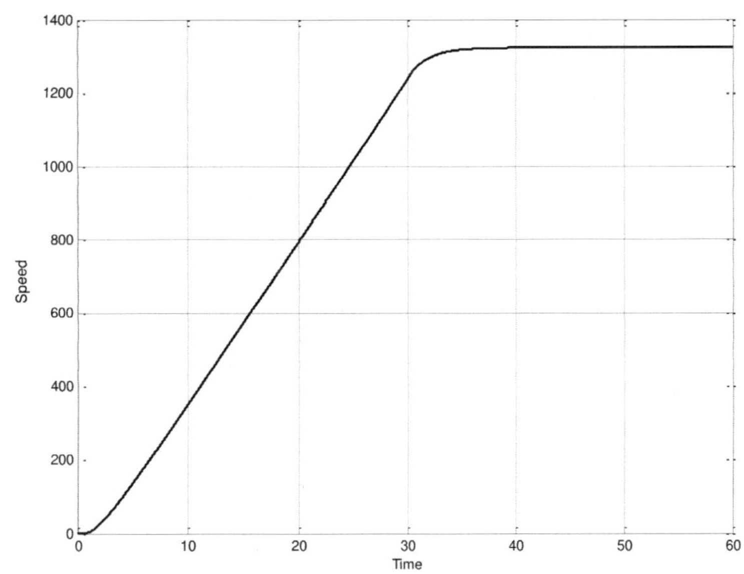

Figure 3. Movement of the automobile

IOP Conf. Series: Materials Science and Engineering 1021 (2021) 012051 doi:10.1088/1757-899X/1021/1/012051

The automobile braking distance can be determined from the graph. In the developed simulation model, the braking distance is determined automatically and is displayed in meters on a digital indicator. In the automobile subsystem at specifying the road resistance the adherence of the tires with the road surface is taken into account. Thus, estimation of effectiveness of the braking system operation in different road conditions is performed.

The simulation model allows to estimate effectiveness of the braking system operation on many parameters: the automobile braking distance, smoothness of braking depending on the mode, regulation of pressure in the circuit of the rear braking gear depending on the rear axle load, changing of pressure in the braking system, effectiveness of interaction of the braking pads with the braking disks and drums.

The simulation model allows to estimate not only dynamics of the automobile at braking but dynamics of each wheel as well, which is represented in Figure 4 – changing of the wheels rotational speed.

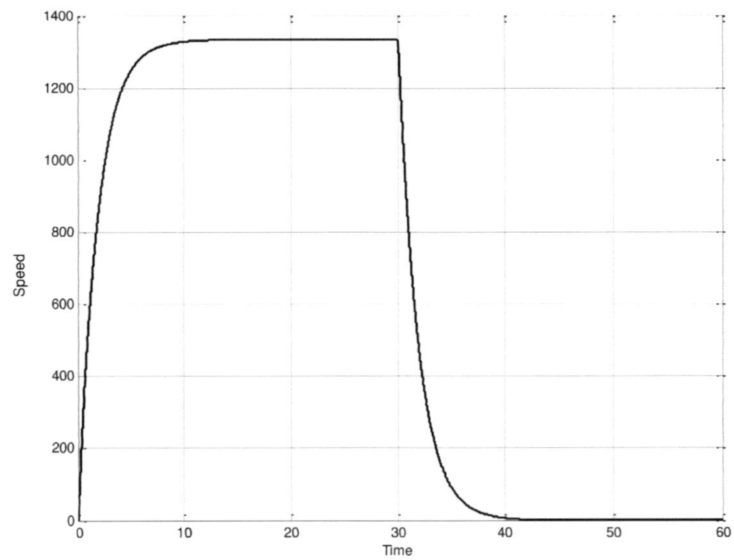

Figure 4. Changes of the wheels rotational speed

Speed of the wheels differs because of the road surface unevenness. Changing of the wheels rotational speed from the average value is realized by means of supplementing of a random variable from the fixed range.

5. Conclusion

The simulation model of the braking system has been built up. The model allows to observe dynamics of mechanical and hydraulic elements of the system being developed, as well as of the existing ones by their operational characteristics, to optimize parameters of the braking system elements as well as the algorithm of its control system operation depending on the task set. Thus, the synthesized simulation model ensures saving time and material expenditures on the braking system development due to conducting simulation experiments.

References

[1] Günay M, Korkmaz M and Özmen R 2020 An investigation on braking systems used in railway vehicles. Engineering science and technology, an iInternational journal. Vol. 23. pp 421-431.

[2] Shyrokau B, Wang D, Savitski D and Ivanov V 2013 Vehicle dynamics control with energy

recuperation based on control allocation for independent wheel motors and brake system. *International Journal of Powertrains*, Vol. **2**, Nos. **2-3**. pp 153-181. https://doi.org/10.1504/IJPT.2013.054158.

[3] Milenković P D; Jovanović S J; Janković A S; Milovanović M D; Vitošević N D.; Dordević M. V and Raicević M M 2010 The influence of brake pads thermal conductivity on passenger car brake system efficiency. *Thermal Science*. Vol. **14**, Issue suppl., pp 221-230. https://doi.org/10.2298/TSCI100505016M.

[4] Derevyanko V A 2001Passenger car braking systems Moscow: Petit, p 248.

[5] Uzhvieva E 2013 Multicriterion estimation of performance of the automobile breaking system. *Proceedings of the 8th International Scientific Conference TRANSBALTICA 2013* Vilnius. VGTU pp 251-253.

[6] Jurevičius M, Vekteris V, Turla V, Kilikevičius A and Viselga G 2019a Investigation of the dynamic efficiency of complex passive low-frequency vibration isolation systems. *Journal of low frequency noise, vibration and active control*. London: Sage publications Ltd. Vol. **38**, iss. **2**. pp 608-614.

[7] Dabney J and Harman T 2003 Simulink 4 secrets of skill. Moscow: Binom. Laboratory of knowledge. p 404.

[8] Anufriev I E 2005 MatLab 7 Saint Petersburg: BHV-peterburg. p 1104.

[9] Diakonov V P 2008 MATLAB 6.5 SP1/7 + Simulink 5/6®. Basis for the use of. Moscow: SOLON-PRESS. p 805.

[10] Jurevičius M, Vekteris V, Viselga G, Turla V, Kilikevičius A and Iljin I 2019b Dynamic research on a low-frequency vibration isolation system of quasi-zero stiffness. *Journal of low frequency noise, vibration and active control*. London: Sage publications Ltd.. Vol. **38**, iss. **2**. pp 684-691.

[11] Vekteris V, Trumpa A; Turla V, Mokšin V, Viselga G and Jurkonis E 2020 Diagnosing faults in rolling-element bearings of rotor systems equipped with vibration dampers. *Advances in mechanical engineering*. London: SAGE Publishing. Vol. **12**, iss. **4**. pp 1-5.

[12] Di Martino R 2005 Modelling and simulation of the dynamic behaviour of the automobile. Automatic. Université de Haute Alsace – Mulhouse. p 189.

IOP Conf. Series: Materials Science and Engineering 1021 (2021) 012052 doi:10.1088/1757-899X/1021/1/012052

Simulation of change in rheological properties of structure of combined materials in tribosystem

A Voitov[1, 3], K Fenenko[2] and O Fenenko[2]

[1]Department of Woodworking technologies and forestry system engineering, Kharkiv Petro Vasylenko National Technical University of Agriculture, Alchevskih street 44, Kharkiv, 61002
[2]Department of Transport technology and logistics, Kharkiv Petro Vasylenko National Technical University of Agriculture, Alchevskih street 44, Kharkiv, 61002

[3]E-mail: K1kavoitov@gmail.com

Abstract. The paper presents experimental and theoretical dependences of changes in the rheological properties of the structure of materials of the surface layers of triboelements in the tribosystem during running-in, the methodical approach to definition of such properties with use of transverse ultrasonic waves is substantiated. It has been experimentally established, that the logarithmic decrement of the attenuation of ultrasonic vibrations in the structure of the material of the surface layer during the period of running-in increases by 32 to 45 % from the initial structure. A parameter for estimating the rheological properties of the structure of conjugate materials in the tribosystem as a function of time is proposed. This is a physical quantity that characterizes the ability of the materials included in the design of the tribosystem, irreversibly convert mechanical energy into thermal energy in the process of deformation of surface layers. Determined by ultrasonic vibrations passing through the material and reduced to a distance of one meter, the dimension 1/m.

1. Introduction

For modeling and forecasting of wear resistance of tribosystems, and also for calculation speed of wear and losses on friction, in mathematical expressions it is necessary to take into account the factor which is the characteristic of materials from which triboelements are made. This factor must be a structurally sensitive characteristic and respond to changes in the surface layers of materials during the running-in process, as well as the magnitude and gradients of temperatures and stresses in the materials during running-in. As proved in the works [1, 2] such a factor may be the logarithmic decrement of the attenuation of ultrasonic vibrations in the structure of materials.

During the operation of the tribosystem, mechanical energy, due to the dynamic contact of the triboelements, is transmitted to the material by stress waves, transforming into internal friction. Relaxation of stresses on the spots of actual contact is due to the rheological properties of the structure of the materials of the surface layers, which leads to a more uniform redistribution of stresses by volume and prevents the accumulation of defects that cause stress concentration. Therefore, the greater the proportion of mechanical energy supplied by friction to the moving and stationary triboelements, dissipated in the form of heat, the higher should be the wear resistance of the tribosystem.

2. Analysis of recent studies and publications

The first work, which is set interconnection between the internal friction of the structure of materials and wear resistance, were the works of V.V. Sheveli [3-7]. In the works it is proved that relaxation processes show a higher structural sensitivity to changes in the stress-strain state of the material under dynamic loading in comparison with the physical and mechanical properties of materials. Based on the analysis of works [3-7] a conclusion can be drawn that rheological properties of structure of materials, from which the tribosystem is made, are function of wear resistance, losses on friction, and also time of running-in.

In works [8-10] the authors carried out research and presented experimental data on the effect of the structure of tribosystem materials on wear resistance. It is shown in the works that the grain size of the structure of conjugated materials are functions of wear resistance and friction loss.

In work [11] the analysis of occurrence of peaks of internal friction in copper at various temperatures, for example, at temperature is presented 175°C and at temperature 230°C. It is shown that with increasing temperature, the maxima of internal friction caused by relaxation along the grain boundaries are observed. These data allow us to conclude about the significant effect of temperature on the amount of internal friction of the structure of materials. Similar studies and conclusions are presented in work [12].

During the operation of the tribosystem there is an increase in the temperature of the surface layers of triboelements, which will cause changes in the rheological properties and thus affect the wear resistance of the tribosystem. We can assume that such processes will be active during the running-in, which is confirmed in the work [4]. In the specified work dependences of increase in a logarithmic decrement of attenuation of ultrasonic vibrations in structure of materials of triboelements during running-in are presented. It is established that the increase in internal friction during the running-in period can reach to 30%.

In the works [1, 2] experimental dependences of change of rheological properties of structure of materials of surface layers of triboelements in tribosystem during running-in are presented, the methodical approach to definition of such properties with use of cross ultrasonic waves is substantiated. It is experimentally established that the logarithmic decrement of the attenuation of ultrasonic vibrations in the structure of the surface layer material during the running-in period increases by 32…45 % depending on the initial structure of the materials. On the basis of experimental data the mathematical expression which allows to define increase in rheological properties of structure of materials of triboelements for the period of running-in is received. The use of such dependences will increase the accuracy of modeling the processes of friction and wear.

3. Statement of the objective and tasks of the study

To carry out modeling of change of rheological properties of structure of the connected materials in a tribosystem during running-in, obtain mathematical expressions for their approximation and perform experimental studies to confirm the adequacy of their approximation.

4. Development of mathematical model

In work [1] obtained an expression that characterizes the logarithmic decrement of the attenuation of ultrasonic vibrations in the structure of the material and is a structurally sensitive characteristic of this material:

$$\delta = \frac{20\lg(A_0 / A_1) - 20\lg(A_0 / A_2)}{2H} \cdot \cos a, \tag{1}$$

where δ is logarithmic decrement of attenuation of ultrasonic vibrations in the structure of the material, dB/m; A_o – the magnitude of the amplitude of the initial pulse generated by the flaw detector is not measured during the experiment; A_1 – the magnitude of the amplitude of the first, single reflected pulse of the flaw detector, mV; A_2 – the magnitude of the amplitude of the second, twice reflected pulse of the flaw detector, mV; H – material thickness, m; a – the angle of introduction of the ultrasonic wave into the material, degree.

Applying transverse oscillations excited by an inclined piezoelectric transducer (PET) and measuring with an ultrasonic flaw detector UD2-12 the magnitude of the first and second pulses in dB according to the formula (1) the value δ, dB/m, can be calculated which characterizes the rheological properties of the material structure.

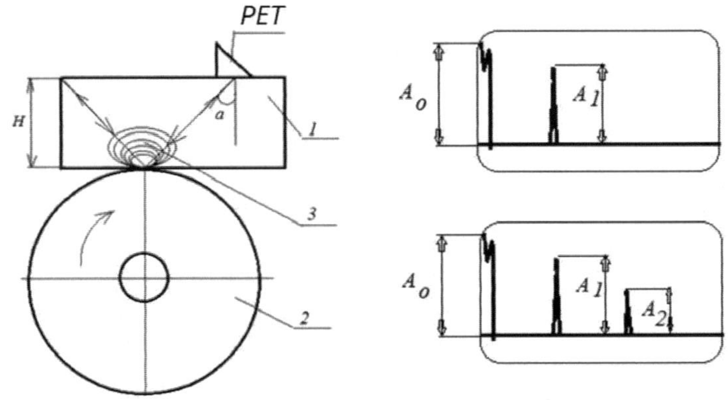

Figure 1. The scheme of application of transverse waves in determining the internal friction of the structure of materials in the tribosystem during running-in: PET is piezoelectric transducer; 1 – fixed triboelement; 2 - movable triboelement; 3 – contact area; A_o is the magnitude of the amplitude of the initial pulse of the flaw detector, during the experiment, is not measured; A_1 is single wave reflection; A_2 is double wave reflection; *a* is the angle of introduction of the ultrasonic wave into the material; *H* is thickness of the material.

The study of the dependences of the change in the logarithmic decrement of attenuation during the running-in of the tribosystem was performed according to the scheme «disk–pad» by friction machine SMT-1. A flat sample size 10*10 mm and length 35 mm was used as a "pad". PET was installed on the flat face of the sample, as shown in Figure 1.

Movable triboelement – a roller – was made of steel 40H (HRC 58). Fixed triboelements – flat samples – were made of steel 40H (HRC 58); bronze Br.AZh 9-4 (HB 110); brass LMcSKA 58-2-2-1-1 (HB 125).

As used lubricating medium transmission oil GL-4, E_u=3,6*10^{14} J./m^3; which was applied to the friction zone by drip. Sliding speed 0,5 m/s, loading 400 – 1600 N, depending on the connected materials in the tribosystem.

The initial values of the logarithmic decrement of attenuation, which is denoted by δ_0, take from work [1]:

– steel 40H (HRC 58), δ_0=443.6 dB/m;
– Br.AZh 9-4 (HB 110), δ_0=784.0 dB/m;
– LMcSKA 58-2-2-1-1 (HB 125), δ_0=840.5 dB/m.

Combinations of materials in the tribosystem that is in the works [3, 4] called compatibility of materials, will affect the tribological characteristics. The main conclusion of the work [1, 2] is that in the process of running the rheological properties of the structure of the materials of the surface layers of the mobile and fixed triboelements increase, which is reflected in the change logarithmic decrement of the attenuation of ultrasonic vibrations as they pass through the material from δ_0 to δ_{max}. This dependence is approximated by the expression presented in the paper [2]:

$$\delta_{\max} = \delta_0 + \delta_0 \cdot \left(\frac{W_{TR,f}}{f_{fr(\max)} \cdot N \cdot v_{sl}} \right) \cdot th\left(\frac{t_i}{t_{com}}\right), \tag{2}$$

where $W_{TR,f}$ is the rate of dissipation in the fixed triboelement of the tribosystem is calculated according to the work [13], dimension J/s; $f_{fr(max)}$ is the maximum value of the coefficient of friction during the running-in of the tribosystem; N is load on the tribosystem, N; v_{sl} is sliding speed, m/s; t_i is time in the process of running-in, s; t_{com} is time of completion of running-in, s.

Using this expression, it is possible to perform modeling of changes in the rheological properties of the structure of materials of movable and fixed elements of the tribosystem.

The formula is used for a fixed triboelement:

$$\delta_{f(\max)} = \delta_{f(0)} + \delta_{f(0)} \cdot \left(\frac{W_{TR,f}}{f_{fr(\max)} \cdot N \cdot v_{sl}} \right) \cdot th\left(\frac{t_i}{t_{com}} \right), \quad (3)$$

where $\delta_{f(max)}$ is the maximum value of the logarithmic decrement of the attenuation of ultrasonic vibrations in the structure of the material of the fixed triboelement after completion of running-in, the dimension dB/m; $\delta_{f(0)}$ is value of logarithmic decrement of attenuation of ultrasonic oscillations of initial structure of material of fixed triboelement, dimension dB/m.

The formula is used for the movable triboelement:

$$\delta_{mov(\max)} = \delta_{mov(0)} + \delta_{mov(0)} \cdot \left(\frac{W_{TR,mov}}{f_{fr(\max)} \cdot N \cdot v_{sl}} \right) \cdot th\left(\frac{t_i}{t_{com}} \right), \quad (4)$$

where $\delta_{mov(max)}$ is the maximum value of the logarithmic decrement of the attenuation of ultrasonic vibrations in the structure of the material of the movable triboelement after completion of running-in, dimension dB/м; $\delta_{mov(0)}$ is the value of the logarithmic decrement of the attenuation of ultrasonic vibrations in the original structure of the material of the movable triboelement, dimension dB/m; $W_{TR,mov}$ is the rate of dissipation in the movable triboelement of the tribosystem, calculated according to the work [13], dimension J./s.

After completion of running-in time of operation of a tribosystem t_i is equal to the time of completion of running-in t_{com}, i.e., the stationary process of operation of the tribosystem is considered.

5. The results of the calculation and their discussion

Dependences of change of logarithmic decrement of attenuation of various constructional materials, in combination with a mobile sample from steel 40H, as functions of time of running-in, are shown in Figure 2.

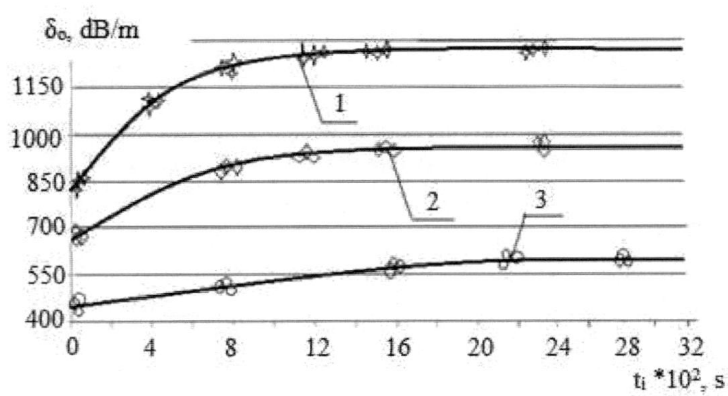

Figure 2. Dependences of change of logarithmic decrement of attenuation of ultrasonic oscillations in structure of materials in tribosystem δ_o, as a function of running time t_i for various materials of a fixed triboelement: 1- LMcSKA 58-2-2-1-1; 2 - Br.AZh 9-4; 3 - steel 40H.

The analysis of the presented experimental dependences allows to draw a conclusion that in the course of running-in rheological properties of structure of materials increase. For example, for steel 40H the logarithmic decrement of attenuation increases with values δ_0 =374 dB/m to values δ_{max} =584 dB/m, which is 31.8 %. For bronze Br.AZh 9-4 the logarithmic decrement of attenuation increases with values δ_0 =679 dB/m to values δ_{max} =984 dB/m, which is 44.9 %. For brass LMcSKA 58-2-2-1-1 the logarithmic decrement of attenuation increases with values δ_0 =840 dB/m to values δ_{max} =1218 dB/m, which is 45.0 %.

To evaluate the rheological properties of the structure of conjugate materials in a tribosystem consisting of movable and fixed triboelements, on the basis of works [14], let's use the expression:

$$RS_{TS(\max)} = \sqrt{\frac{\delta_{mov(\max)} \cdot \delta_{f(\max)}}{\pi}}, 1/m, \quad (5)$$

where $RS_{TS(max)}$ is maximum value of rheological properties of bonded materials in the tribosystem after completion of running-in, dimension 1/m.

Using formulas (3) – (4) and substituting them into a formula (5) changes in the rheological properties of the structure of the materials of the tribosystem $RS_{TS(max)}$ are simulated depending on their combination in the design. Simulation will be performed for the established mode of operation. The results of the simulation are summarized in Table 1.

Table 1. - The results of changes in the rheological properties of different structures of tribosystems

Tribosystem design	Initial value $RS_{TS(0)}$	Maximum value $RS_{TS(max)}$	% increase during running-in time
steel 45+steel 40H	229.7	318.2	38.5
steel 20H+steel ShH15	230.8	319.0	38.2
steel 40H+steel 40H	249.9	346.7	38.7
steel 40H+cast iron VCh70	309.7	431.9	39.4
steel 40H+Br.AZh 9-4	332.5	479.0	44.0
steel 40H+ LMcSKA 58-2-2-1-1	344.2	494.9	43.7
cast iron CCh15+ LMcSKA 58-2-2-1-1	406.9	586.2	44.0
cast iron VCh70+Br.AZh 9-4	412.0	593.0	43.9

Analysis of the results that are presented in the table 1, allows us to conclude that in the process of running-in, the change in the rheological properties of the structure of conjugated materials in the tribosystem reaches the value 38.2 to 44.0% in the direction of increase.

It should be noted that the results are given in the table 1, obtained for tribosystems with the following parameters: K_f= 12.5 1/m; K_{ov}= 0.5; lubricating medium GL-4, E_u=3.6*10^{14} J./m3; load N=2000 N; sliding speed v_{sl} = 0.5 m/s. Variable factors are combinations of conjugate materials in the tribosystem.

To evaluate the rheological properties of the structure of conjugated materials in the tribosystem during running-in [15], when t_i varies from zero to t_{com}, let's use the expression:

$$RS_{TS}(t_i) = \sqrt{\frac{\delta_{mov}(t_i) \cdot \delta_f(t_i)}{\pi}}, 1/m, \quad (6)$$

where $RS_{TS(ti)}$ - values of rheological properties of connected materials in the tribosystem, as a function of running-in time, dimension 1/m; $\delta_{mov(ti)}$ is the value of the logarithmic decrement of the attenuation of ultrasonic vibrations in the structure of the material of the movable triboelement, as a function of the running-in time, the dimension dB/m, calculated by expression (4); $\delta_{f(ti)}$ is the value of the logarithmic decrement of the attenuation of ultrasonic vibrations in the structure of the material of the fixed triboelement, as a function of the running-in time, the dimension dB/m, calculated by expression (3).

Modeling, according to formula (6), the theoretical dependences of the change in the rheological properties of conjugated materials in the tribosystem during running-in were obtained, which are presented in figure 3. From the analysis of the obtained dependences it follows that the rheological properties of conjugate materials in the tribosystem RS_{TS} increase from the original value, at $t_i = 0$, to values that characterize the changes in the structure of the materials of the surface layers of the mobile and fixed triboelements during the running-in $t_i = t_{np}$.

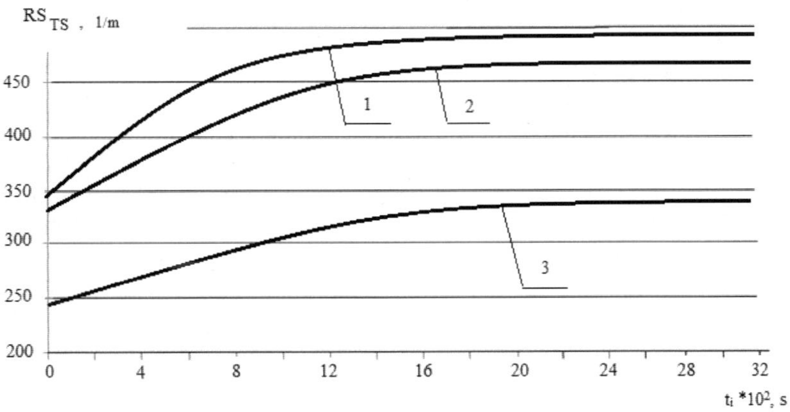

Figure 3. Dependences of change of rheological properties of structure of connected materials in tribosystem RS_{TS}, as a function of running time t_i for various bonded materials: 1 - steel 40H+ LMcSKA 58-2-2-1-1; 2 - steel 40H+ Br.AZh 9-4; 3 - steel 40H+steel 40H

Time t_i, upon reaching which the value is stabilized RS_{Ts}, corresponds to the time of completion of running-in. This fact was verified experimentally on various tribosystems with parameters: K_f= 12.5 1/m; K_{ov}= 0.5; lubricating medium GL-4, E_u=3.6*10^{14} J./m^3; load N=2000 H; sliding speed v_{sl} = 0.5 m/s. Variable factors are combinations of conjugate materials in the tribosystem.

For the tribosystem "steel 40H+steel 40H" stabilization of the moment of friction and temperature of a fixed triboelement occurred at t_i=2200 s, that can be considered as time of completion of running-in.

For the tribosystem "steel 40H+ Br.AZh 9-4" stabilization of the moment of friction and temperature of a fixed triboelement occurred at t_i=1600 s, that can be considered as time of completion of running-in. For the tribosystem "steel 40H+ LMcSKA 58-2-2-1-1" stabilization of the moment of friction and temperature of a fixed triboelement occurred at t_i=1200 s, that can be considered as time of completion of running-in.

The completion time, which was obtained theoretically, was verified experimentally. The sample size during the experimental studies was checked for compliance with the normal distribution law. Analysis of the obtained results allows us to conclude that the experimental sample size corresponds to the normal distribution law.

The results of experimental studies to determine the stabilization time of the value RS_{TS} were tested for homogeneity and reproducibility from experience to experience by the Cochran's test. The analysis of the obtained results allows concluding that the results are homogeneous and reproducible.

Rheological properties of the structure of conjugated materials in the tribosystem RS_{TS} is a physical quantity that characterizes the ability of materials included in the structure of the tribosystem to irreversibly convert mechanical energy into thermal energy in the process of deformation of surface layers. The dimension 1/m was determined by ultrasonic vibrations passing through the material and reduced to a distance of one meter.

Rheological properties of the structure of bonded materials in the tribosystem should be used in modeling and forecasting the wear resistance of tribosystems, as well as to calculate the wear rate and friction losses. Parameter RS_{TS}, which is a structurally sensitive characteristic of the materials from which the triboelements are made. As follows from the results of table 1 and dependencies, figure 3, this parameter responds to changes in the surface layers of triboelements in the process of running-in, as well as to the magnitude and gradients of temperatures and stresses in the process of running-in.

6. Conclusions

1. The paper presents experimental and theoretical dependences of changes in the rheological properties of the structure of materials of the surface layers of triboelements in the tribosystem during running-in, substantiates a methodical approach to determining such properties using transverse ultrasonic waves. It is experimentally established that the logarithmic decrement of the attenuation of ultrasonic vibrations in the structure of the surface layer material during the running-in period increases by 32 to 45 % from the initial structure of the materials.

2. A parameter for estimating the rheological properties of the structure of conjugate materials in the tribosystem as a function of time is proposed. This is a physical quantity that characterizes the ability of materials included in the structure of the tribosystem to irreversibly convert mechanical energy into thermal energy in the process of deformation of surface layers. The dimension 1/m was determined by ultrasonic vibrations passing through the material and reduced to a distance of one meter.

References

[1] Vojtov V A, Voitov A V 2020 Metodychnyy pidkhid do vyznachennya reolohichnykh vlastivostey struktury spoluchennya materialiv u tribosistemi. *Problemy tertya ta znoshuvannya* **2 (87)** 4-10

[2] Vojtov V A, Bekirov A SH 2018 Matematicheskaya model' perekhodnykh protsessov v tribosistemakh i rezul'taty modelirovaniya. *Problemi tribologíí* **1** 18-27

[3] Shevelya V V 1993 Rheology of wear resistance and friction pairs compatibility. *Trenie i Iznos* **14(1)** 48-62

[4] Shevelya V V, Vojtov V A, Sukhanov M I, Isakov D I 1995 Regularities of changing in internal friction during operation of tribosystem and its consideration when selecting compatible materials. *Trenie i Iznos* **16(4)** 734-744

[5] Shevelya V V, Oleksandrenko V P, Kalda G S 2003 O roli relaksatsionnykh yavleniy i substrukturnykh prevrashcheniy pri trenii metallov. *Problemi tribologíí* **2** 3-11

[6] Shevelya V.V., Oleksandrenko V.P. 2006 Tribokhimiya i reologiya iznosostoykosti, *Khmel'nitskiy* 278

[7] Shevelya V V, Trytek A 2010 Reologiya vyazkouprugogo friktsionnogo kontakta. *Problemi tribologíí* **4** 6-16

[8] Weiwei Z, Cancan Z, Yiwen Z, Chi T K, Junhua L, Zengbao J, Fuzeng R 2020 Achieving exceptional wear resistance in a compositionally complex alloy via tuning the interfacial structure and chemistry. *Acta Materialia* **188** 697-710

[9] Zhao X, Zhao B, Liu Y, Cai Y, Hu C 2017 Research on friction and wear behavior of gradient nano-structured 40Cr steel induced by high frequency impacting and rolling. *Engineering Failure Analysis* **83** 167-177

[10] Bhadauria N, Pandey S, Pandey P M 2020 Wear and enhancement of wear resistance – A review. *Materialstoday: Proceedings* **26(2)** 2986-2991

[11] Gryaznov M YU, Chuvil'deyev V N, Sysoyev A N, Kopylov V I 2010 Zernogranichnoye vnutrenneye treniye i sverkhplastichnost' nano- i mikrokristallicheskikh metallov i splavov. *Vestnik Nizhegorodskogo universiteta im. N.I. Lobachevskogo* **5(2)** 147-158

[12] Chengru L, Xiangtao D, Long H, Ye J, Zhaodong W 2019 Effect of temperature on microstructure, properties and sliding wear behavior of low alloy wear-resistant martensitic steel. *Wear* **442–443** 203125

[13] Vojtov V A, Zakharchenko M B 2015 Modelirovaniye protsessov treniya iznashivaniya v tribosistemakh v usloviyakh granichnoy smazki. Chast' 1. Raschet skorosti raboty dissipatsii v tribosistemakh. *Problemi tribologíí* **1** 49-57

[14] Vojtov V, Biekirov A, Voitov A 2018 The quality of the tribosystem as a factor of wear resistance. *International Journal of Engineering & Technology* **7(4.3)** 25-29
[15] Vojtov V A, Biekirov A Sh, Voitov A V and Tsymbal B M 2019 Running-in procedures and performance tests for tribosystems. *Journal of Friction and Wear* **40(5)** 376–383

IOP Conf. Series: Materials Science and Engineering 1021 (2021) 012053 doi:10.1088/1757-899X/1021/1/012053

Fractals Theory Application for Evaluation of Influence of Non Metallic inclusions on Mechanical Properties of S355J2 Steel

V M Volchuk[1,3], **O V Uzlov**[1], **O V Puchikov**[2] **and S V Ivantsov**[1]

[1] Department of Materials and Materials Processing, State Higher Education Institution "Prydniprovska State Academy of Civil Engineering and Architecture", 24-a, Chernyshevskoho Str., Dnipro, 49600, Ukraine

[2] Department of Constructional Steels, Institute of Ferrous Metallurgy of Academy of Sciences of Ukraine, 1-a, Starodubov sqr., Dnipro, 49005, Ukraine

[3]Email: volchuky@gmail.com

Abstract. Influence of non-metallic inclusions in constructional S355J2 steel on mechanical properties has been evaluated by using fractal theory. Since point evaluation of non-metallic inclusions is fault tolerant to mechanical properties change. Point evaluation of experimental steel grades has been evaluated by 1-2 points. Correlation between mechanical properties of the steel (KV^{-20}, $\sigma_{0.2}$, σ_B, δ) and fractal dimensions of non-metallic inclusions has been found. Mechanical properties of the steel have been ranking according to influence of fractal dimensions of non-metallic inclusions on the properties. It has been revealed that non-metallic inclusions, evaluated by their fractal dimensions, have strong effect on impact toughness of the steel S355J2 ($R^2 = 0.89$). It can be explained by reducing the metal matrix resistance to crack propagation due to non-metallic inclusions in the matrix. Method of mechanical properties ranging by non-metallic inclusions fractal dimensions allow to vary chemical composition of steel for main parameter obtaining. In this particular case for necessary impact toughness of constructional S355 steel narrower chemical composition of the steel should be chosen.

1. Introduction

As it known, non-metallic inclusions have strong effect on mechanical properties of the steels. Steel consumers, especially railway cars producers, increasing demands to mechanical properties of the steels almost annually. Demands to non-metallic inclusions numbers and sizes regulate by standards and regulatory documentation on the steel rolled and casted products. But the point evaluation of the non-metallic inclusions is semi-quantitate and sometime can't be used for rapid correspondence to mechanical properties of the steel. Moreover, at industrial steels non-metallic inclusions frequently grow as complexes of different non-metallic inclusions with complicated structure [1, 2]. Such complex morphology of the non-metallic inclusions is featured to non-metallic inclusions of S355J2 steel that can be used for freight railway car sheathing production. For description the objects with complicated structure can be used fractal geometry with promising results [3]. Thus, in [4] the impact of cast iron fractal dimensions to mechanical properties of the cast iron has been shown. It allow evaluate the value of cast iron structure to mechanical properties by means of ranging and use the fractal formalism for multi parametrical technologies. It should be mentioned, that fractal

dimension is a significant factor for agglomeration and dissolution of irregularly shaped non-metallic inclusions [5]. Fractal nature of metals at different size levels has been proved at papers [6-8].
Therefore, fractal approach can be used for evaluation of non-metallic inclusions content influence on mechanical properties of S355J2 steel and ranging by determining factor.

2. Experimental
Chemical composition of investigated steels is presented at Table 1.

Table 1. Chemical composition of investigated steels (weight %).

Grade	C	Mn	Si	S	P	Cr	Ni	Cu	Al	N	Ca	H
Grade 1	0.100	1.610	0.560	0.002	0.009	0.030	0.020	0.040	0.032	0.006	0.0020	0.0002
Grade 2	0.110	1.620	0.560	0.002	0.008	0.030	0.020	0.040	0.027	0.006	0.0021	0.0002
Grade 3	0.110	1.600	0.550	0.002	0.013	0.030	0.030	0.040	0.030	0.006	0.0020	0.0004
S355J2	≤0.20	≤1.65	≤0.56	≤0.025	≤0.025	-	-	≤0.55	-	-	-	-

Steel S355J2 has ferrite-pearlite structure at as rolled condition (Figure 1). Morphology of non-metallic inclusions and microstructure of the steel has been investigated by light microscopy. Neophot-2 light microscope has been applied.

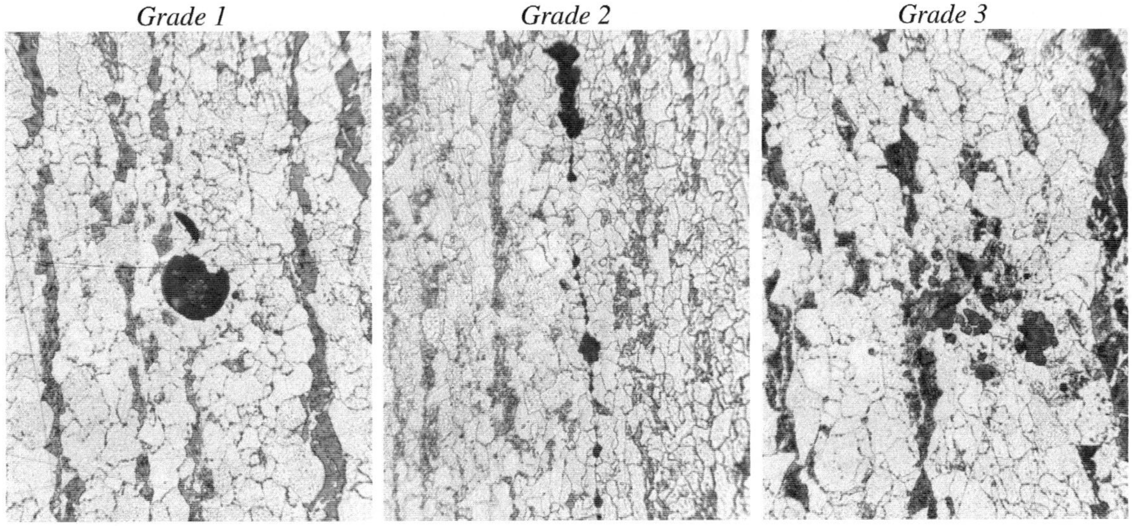

Figure 1. Microstructure (ferrite-pearlite) and non-metallic inclusions in steel S355J2, ×400.

Contamination of steel by non-metallic inclusions has been measured according to GOST 1778. Cluster of non-metallic inclusions at Grades 1 and 3 distributed non-uniformly thorough entire steel plate thickness. At Grade 2 multiple clusters of fine oxides have been revealed. Oxide clusters have been found on surface and quarter of plate thickness. The center structure of the steel plate has small quantity of oxides. Fine fragments of angular deep-grey inclusions (probably magnesium spinel) have been found at structure of the steel. Typical non-metallic inclusions of steel S355J2 are shown in Figure 2.

Non deforming silicates at all three Grades of S355J2 steel have complex morphology. The sizes of non-deforming silicates determine by point scale (semi-quantity), e.g. 2a(1), 3a(1). Brittle silicates have been determined by point scale 1b(1), 2c(1), as shown in table 2.

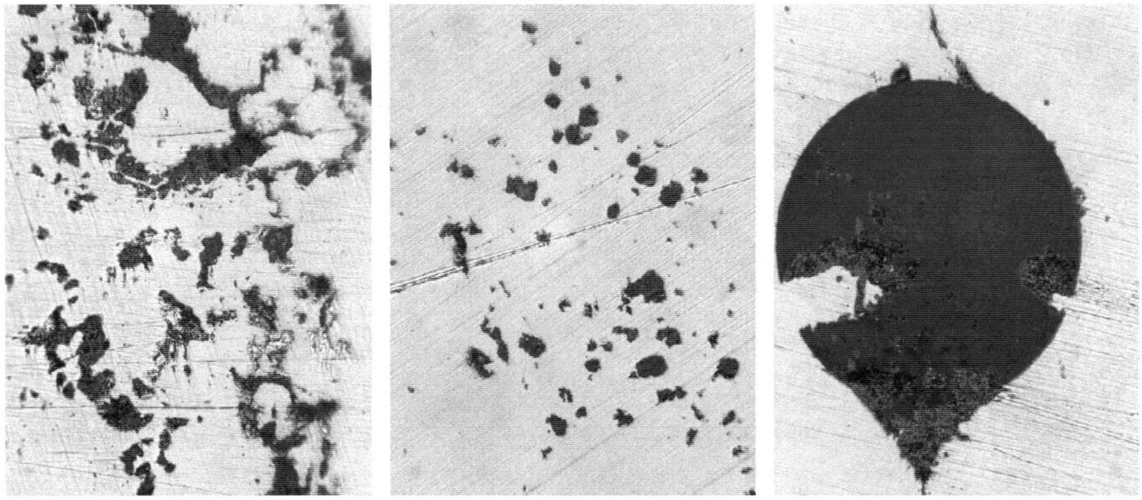

Figure 2. Morphology of non-metallic inclusions in steel S355J2, ×400.

Table 2. Point scale of non-metallic inclusions at steel S355J2 according to GOST 1778.

№ grade	Non-deforming silicates, point scale (CH)	Brittle silicates, point scale (CX)
1	2a(1), 3a(1)	1b(1), 2c1)
2	2a(1), 3a(1)	1b(1), 2c1)
3	2a(1), 3a(1)	1b(1), 2c(1)

Mechanical properties of S355J2 steel have been specified according to GOST 1497 and GOST 9454 (Table 3).

Table 3. Mechanical properties of S355J2 steel.

№ grade	KV^{20} (J/cm^2)	$\sigma_{0.2}$ (MPa)	σ_B (MPa)	δ(%)
1	190-224	540-550	670-680	25.0-30.0
2	162-293	400-425	509-543	29.5-32.0
3	203-258	400-440	500-540	30.0-31.5

For Grade 1 steel plate with dimensions 38mm×2240mm×10200mm has been investigated; for Grade 2 steel plate with dimensions 16mm×2000mm×6000mm has been analyzed, and for Grade 3 steel plate with dimensions 16mm×2000mm×6000mm.

3. Discussion

Correlation of non-metallic inclusions content to mechanical properties according standard evaluation (table 2 and table 3) is quite weak. Nevertheless, some papers [9] prove the strong impact of non-metallic inclusions to mechanical properties of constructional steels. Therefore, in the present work the impact of fractal dimension of non-metallic inclusions to mechanical properties of S355J2 steel has been investigated.

Fractal dimension of non-metallic inclusions has been calculated according. The methodology based on convergence of fractal dimensions calculated by cell and point methods. The cell method is based on F. Hausdorff equation (1) [10]:

$$D = -\lim_{\delta \to 0} \frac{\ln N(l)}{\ln l}, \quad (1)$$

where N is the number of cells with size l, that cover the object. Point dimension has been calculated as ratio (2) [11]:

$$\tilde{N}(L) = \sum_{m=1}^{K} (1/m) P(m,L) \sim L^{-D}, \quad (2)$$

where $P(m,L)$ is the probability of m points of fractal contain inside cell of L size, and $\tilde{N}(L)$ is average number of cells that cover the investigated object.

The correlation of mechanical properties and non-metallic inclusions content are shown in Figure 3.

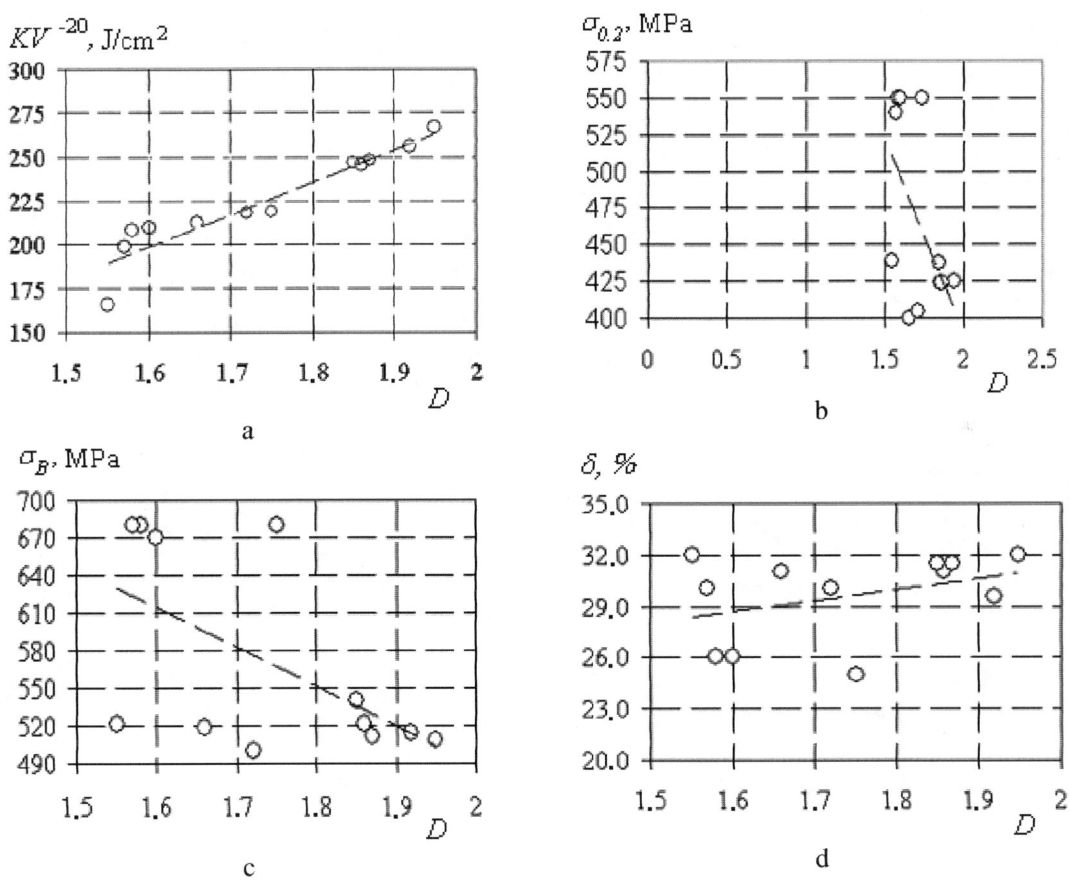

Figure 3. Correlation of mechanical properties of S355J2 steel and fractal dimension D of non-metallic inclusions at the steel: a – Impact toughness of steel at -20 ^{0}C; b – YTS; c – UTS; d – elongation.

Ratio (1) (Figure 3a) analysis shows that impact toughness of steel S355J2 can be predicted by fractal approach ($R^2 = 0.89$). All other models have relatively low correlation coefficients $R^2 = 0.14$-0.34 (3÷6) to mechanical properties. It can be explained rather low impact of non-metallic inclusions to strength and elongation in contrast to ferrite-pearlite structure. For example, at [12] high correlation of fractal dimension to mechanical properties of low carbon steel A550 at different heat treatment modes has been shown. The bias of mechanical properties prediction was 7%.

$KV^{-20} = 185.02D - 97.548,$ $R^2 = 0.89$ (figure 3 a) (3)
$\sigma_{0.2} = -261.94D + 916.61,$ $R^2 = 0.34$ (figure 3 b) (4)
$\sigma_B = -312.57D + 1114,$ $R^2 = 0.33$ (figure 3 c) (5)
$\delta = 6.3291D + 18.612.$ $R^2 = 0.14$ (figure 3 d) (6)

The analysis of equation coefficients (3÷6) gives the range of mechanical properties of steel by means of non-metallic inclusions impact to the properties. It has been revealed that non-metallic inclusions content in steel S355J2 has most effect on impact toughness. It can be explained by decreasing of crack propagation resistance of the metal matrix due to high non-metallic inclusions content in steel. The effect of non-metallic inclusions on impact toughness of steel has been shown in many papers. For example, at [1] has been noticed that uneven movement of vacancies flows, interstitial atoms and matrix atoms near inclusion surface may lead to inclusion shape change. It occurs due to different diffusion rates at surface of inclusion. Surface distortion and internal stresses in inclusion occur unless diffusional flows rates near inclusion will be equalized. Shape change of non-metallic inclusion lead to change of fractal dimension of inclusion.

All other mechanical properties have much lower range rates (coefficient $Y_{\sigma 0.2}$ = 0.47 etc. (Figure 4)).

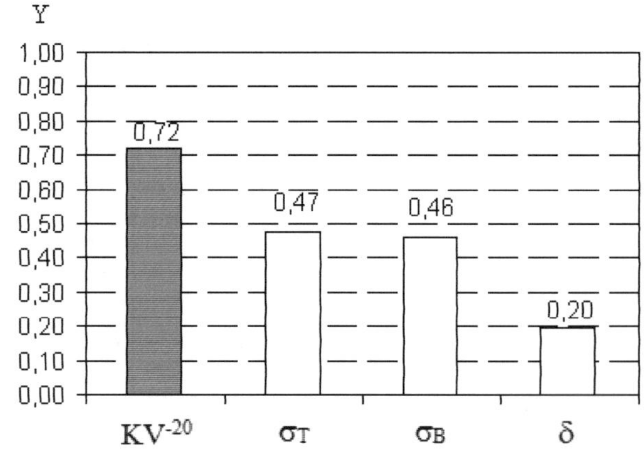

Figure 4. Ranging of different mechanical properties of steel S355J2 according to impact of fractal structure of non-metallic inclusions on the properties.

As it is shown in Figure 4, the range rate for impact toughness is highest for investigated mechanical properties and equal to 0.72. It may be the evidence of high sensitivity of impact toughness to steel contamination. Therefore, control of non-metallic inclusions in steel is vitally important for high strength rolled plates for freight railway cars production.

4. Conclusions

The possibility of fractal approach for evaluation of steel S355J2 contamination by non-metallic inclusions has been shown. It has been shown that point evaluation of non-metallic inclusions is fault tolerant to change of mechanical properties of steel. It has been proven by point evaluation of non-metallic inclusions content in range 1÷2 point insensitively to mechanical properties. Relevance between mechanical properties of steel S355J2 and fractal dimensions of non-metallic inclusions has been revealed. It helps to predict and ranging properties of steel relative to determining factor. Therefore, investigation the effect of non-metallic inclusions content on impact toughness of constructional steel by means of fractal dimension has advantages compared to point evaluation.

References

[1] Gubenko S I 1991 *Transformation of Non-Metallic Inclusions* (Moscow: Metallurgy)

[2] Uzlov O, Malchere A, Bolshakov V I and Esnouf C 2007 Investigation of acicular ferrite structure and properties of C-Mn-Al-Ti-N steels *AMR* **23** 209–312

[3] Mandelbrot B B 1982 *The Fractal Geometry of Nature* (New York–San Francisco: Freeman)

[4] Volchuk V, Klymenko I, Kroviakov S and Orešković M 2018 Method of material quality estimation with usage of multifractal formalism *Tehnički glasnik – Technical Journal* **12(2)** 93–97

[5] Guo, Luo-fang, Li, H. and Wang, Yajun 2012 Fractal dimension of nonmetallic inclusions and relationship with its morphology in steel *J. Iron Steel Res. Int.* **24** 33–38

[6] Hornbogen E 1986 Fine, Coarse and Fractal microstructures *Prakt. Metallographie* **23** 258-267

[7] Takayasu H 1990 *Fractals in Physical Sciences* (Manchester, New York: Manchester University Press)

[8] Li J M, Lu L, Lai M O and Ralph 2003 *Image-Based Fractal Description of Microstructures* (: Kluwer Academic Publishers: Springer Science & Business Media, LLC)

[9] Arreola-Herrera R, Cruz A, Rivera-Salinas, J, Romero-Serrano J and Sánchez-Alvarado R 2018 The effect of non-metallic inclusions on the mechanical properties of 32 CDV 13 steel and their mechanical stress analysis by numerical simulation *Theor. Appl. Fract. Mech.* **94** 134–146

[10] Hausdorff F 1919 Dimension und äußeres Maß *Math. Ann.* **79** 157–179

[11] Crownover R M 1995 *Introduction to Fractals and Chaos* (Boston, London: Jones and Bartlett Publishers, Inc.)

[12] Bolshakov V I and Volchuk V M 2011 Materials Science Aspects of Using of Wavelet-Multifractal Approach to an Evaluation of Structure and Properties of Low-Carbon Low-Alloyed Steels *Metallofiz. Noveishie Tekhnol.* **33(3)**, 347–360

Experimental studies of a throughput of the distribution systems of planetary hydraulic motors

A Voloshina[1, 5], A Panchenko[1], O Titova[2], V Pashchenko[3] and A Zasiadko[4]

[1]Department of Mechatronic Systems and Transport Technologies, Dmytro Motornyi Tavria State Agrotechnological University, B. Khmelnitsky ave. 18, Melitopol, Ukraine
[2]Department of Foreign Languages, Dmytro Motornyi Tavria State Agrotechnological University, B. Khmelnitsky ave. 18, Melitopol, Ukraine
[3]Department of Tactics, National Academy of the National Guard of Ukraine Zakhysnykiv, Ukrainy sq., 3, Kharkiv, Ukraine
[4]Department Berdyansk college of the Tavria State Agrotechnological University, Eastern ave. 23, Berdyansk, Ukraine

[5]Email: voloshinaa2012@gmail.com

Abstract. As a result of the study, an experimental test bench was developed for testing high-torque hydraulic motors. A methodology for experimental studies of a unified series of planetary hydraulic motors has been developed. It has been established that an increase in the throughput of a hydraulic motor with a modernized distributed system makes it possible to increase the hydromechanical efficiency of the modernized hydraulic motor by 11 to 14%, the volumetric efficiency by 2 to 4%, and the overall efficiency by 7 to 9%. The increase in the values of the hydromechanical, volumetric and overall efficiency of the modernized hydraulic motor explains the increase in the throughput of its distributed system. It has been established that the torque of the developed serial hydraulic motor is 300 to 325 min^{-1}, which is a sign of the rigid characteristics of this hydraulic motor. The stability of the torque (within 1.5%) developed by the modernized hydraulic motor 370 to 375 H·m in the frequency range of 75 to 450 min^{-1} is confirmed by its rigid characteristic.

1. Introduction

High-torque low-speed gerotor [1, 2], orbital [3] and planetary [4] hydraulic motors are most commonly used to drive active working bodies and running systems of road, construction, agricultural and other self-propelled equipment. One of the main units causing a decrease in the functional parameters of hydraulic motors is a working fluid distribution system that creates a rotating hydraulic field necessary for the movement of the inner rotor of these hydraulic machines [5, 6]. The main disadvantage of the distribution system is the fluctuations in the flow of the working fluid supplied to the hydraulic motor, caused by the unevenness of its throughput [6].

The throughput of the distribution system of the planetary hydraulic motor is understood as the total area of overlap of its working windows. Irrational design of the elements of the distribution system leads to fluctuations in the overlap area and a change in the output characteristics of the hydraulic motor as a whole, which is confirmed by parametric studies [5, 6]. Therefore, conducting experimental studies of planetary hydraulic motors in order to determine the effect of the throughput of their distribution systems on changing the functional parameters of planetary hydraulic motors is an urgent task.

2. Analysis of recent studies and publications

A mechatronic system of active control of dynamic spatial positioning of the executive body is considered [7]. A method for designing hydraulic mechatronic systems with elements of multicriteria optimization has been developed, which makes it possible to design a mechatronic system with given output characteristics [8]. The dynamic model of the drive is proposed [9]. The dynamic processes occurring in mechatronic systems with planetary hydraulic motors are investigated in order to predict changes in their output characteristics [10]. The parameters of variation are substantiated, which determine the change in the output characteristics of the hydraulic motor, depending on the design features of its system of rotors and the distribution system. Regression equations are obtained that describe the change in the output characteristics of the hydraulic motor during operation, for a given range of changes in its operating parameters [11]. A methodology for experimental research and construction of control characteristics is proposed [12]. Experimental studies of mechatronic systems with volumetric hydraulic machines were not considered.

The conditions for static equilibrium of the mobile complex were established [13]. An engineering method for determining radial compliance is proposed, taking into account the radial clearance and contact deformations of parts [14]. Using nonlinear models and the method of analyzing the response surface, the parameters of thin-walled structures have been substantiated [15], studies have been carried out to develop adequate models for analyzing the response of thin-walled structures to a load [16]. The dependences of the contact pressure on the external forces that act on them [17] are considered, the distribution of contact pressures is determined [18], the regularities of the distribution of contact pressure are established [19], the wear of the flow path of vortex-chamber blowers is investigated [20]. Issues related to research in the design and operation of hydraulic machines with cycloidal rotors were not considered.

It is known [1, 3, 8, 11] that in mechatronic systems with a hydraulic drive, gerotor [1, 2], orbital [3, 5] and planetary [4, 6] hydraulic machines with cycloidal rotors are mainly used. The proposed theoretical method for predicting the characteristics of an engine with constant difference orbital piston engines [21]. Experiments have been carried out to confirm the theoretical model [22]. Numerical modeling, modeling and forecasting of the production depth were carried out, which as a result is compared with the results of experimental studies [23]. The implemented method for determining the reliability of a hydraulic motor by modeling changes in the technical state of internal and external rotors [24] analyzes the performance of the orbital engine and conduct experimental research [25]. A comprehensive study of a hydrostatic unit with a low speed of rotation of an orbital rotor in a hydrostatic transmission system is presented. Using the actual parametric values, the general dynamic model is confirmed experimentally [26]. Issues related to experimental study of planetary hydraulic machines were not considered.

Calculation formulas are proposed and a comparative assessment is carried out according to the criterion of the channel cross-section of a hydraulic machine with a planetary rotor with a central gear wheel in the shape of a wave and floating satellites [27]. The systematization and comparative analysis of the schemes of mechanisms of planetary rotary hydraulic machines [28]. Cavitation phenomena occurring in the working fluid distribution zone have been investigated [29]. Promising methods for the production and application of carbon nanoparticles in tribology [30] were reviewed, the results of tribosystems testing with the use of liquid crystal additives [31] are given, experimental studies are considered, taking into account the effect of an electrostatic field on the working fluid of a volumetric hydraulic drive [32]. The values of pressure loss, flow rate and velocity distribution over the cross-section of the pipeline in a straight section and in a turn have been determined [33]. A comparison of the use of the SST turbulence model taking into account the curvature of streamlines and flow rotation [34] is presented, the application of the RANS approach using the corrected SST turbulence model is considered, which allows one to determine all the main characteristics of the vortex flow [35]. A 3D model has been proposed that takes into account the dynamically changing volume of the working fluid in the working chambers [36], mathematical expressions are given to assess the flow rate of the working fluid [37] in gerotor hydraulic machines. It should be noted that the operation of the rotors of a gerotor pump differs fundamentally from the operation of the rotors of a planetary (orbital) hydraulic motor. For functioning of the planetary hydraulic motor, a rotating hydraulic field of the working fluid created by the distribution system is required [6, 38]. The cyclic rotation of the hydraulic field depends on the number of working chambers formed by the external and internal rotors and is characterized by the kinematic diagram of the distribution system [6]. The kinematic diagrams of the distribution systems of planetary hydraulic motors have been substantiated [6], a technique has been developed for displacing the distribution windows of a movable distributor, which makes it possible to reduce fluctuations in the flow of the working fluid (throughput), and as a result, to improve the output characteristics of the planetary hydraulic motor [39]. The influence of the shape of distribution windows on the output characteristics of a planetary hydraulic motor, which can be made in the form of a segment [38], a circle [4] and a groove [40], has been investigated. Experimental studies of the throughput of the distribution system have not been carried out.

When carrying out parametric studies of the influence of the distribution system of the planetary hydraulic motor, a number of assumptions and limitations were made [4, 6, 38, 40]. Therefore, comparative experimental research of hydraulic motors with serial and modernized distribution systems is one of the most important and urgent tasks.

3. Statement of the objective and tasks of the study

To conduct experimental studies, in order to determine the influence of the throughput of the distribution system of the planetary hydraulic motor on the change in its functional parameters, it is necessary:

- to develop a test bench for testing high-torque low-speed planetary-type hydraulic motors;
- to develop a methodology for conducting study of the influence of the throughput of the distribution system of a planetary hydraulic motor on the change in its output characteristics experimental studies of the influence of the distribution system of a planetary hydraulic motor on the change in its output characteristics;
- to study the influence of the throughput of the distribution system of a planetary hydraulic motor on the change in its output characteristics.

4. The basic part of the study

Earlier parametric studies [4, 6, 39] determining the influence of the transmission capacity of the distribution system of a planetary hydraulic motor on the change in its functional parameters are associated with a number of assumptions and limitations. Therefore, in order to confirm and correct the previously obtained results of parametric studies, comparative bench tests of planetary hydraulic motors with serial and modernized distribution systems should be conducted.

The developed test bench for testing planetary hydraulic motors (Figure 1, a) consists of four main units: a pumping station (Figure 1, b), a loading device (Figure 1, c), a unit for measuring the flow rate and changing the direction of the working fluid flow (Figure 1, d) and the control panel (Figure 1, f). This test bench allows testing a family of unified series of planetary hydraulic motors with a power of 6.5; 11; 22 and 33 kW, respectively. The pumping station consists (Figure 1, b) of a frame with a tank for working fluid with a thermometer, an axial piston variable pump with a working volume of 89 cm^3 with a drive motor, a feed pump and three safety (overflow) valves. For the conditioning of the working fluid, the pumping station has two fine filters, two heat exchangers, two pressure gauges and shut-off valves.

The loading device (Figure 1, c) is designed to simulate the load on the hydraulic motor shaft by braking and is a powder brake placed on the frame together with the tested hydraulic motor. When measuring the output characteristics of the hydraulic motor under test, two pressure gauges are used to control the pressure and a tachometer with a pulse sensor to determine the angular velocity.

The unit for measuring the flow rate and changing the direction of the flow of the working fluid (Figure 1, d) is designed to measure the amount of working fluid passed through the hydraulic motor

IOP Conf. Series: Materials Science and Engineering 1021 (2021) 012054 doi:10.1088/1757-899X/1021/1/012054

and for reversing the activation of hydraulic motors during testing. It consists of a frame with two electrically controlled distributors, two calibrated axial piston hydraulic motors with tachometers and impulse sensors with discs, and two fine filters.

Figure 1. Experimental test bench for testing planetary hydraulic motors: a – general view; b – pumping station; c – load device; d – unit for measuring the flow rate and changing the direction of the flow of the working fluid; f – control panel.

The control panel (Figure 1, f) is designed to control the electrical elements of the test bench according to the corresponding algorithm: drive motors, flow distributors, tachometers, powder brakes, etc.

Test bench studies were carried out on a serial and modernized PRG-22 series hydraulic motors with a working volume of 160 cm^3. The modernization of the hydraulic motor consisted in changing the geometric and functional parameters of the movable and stationary distributors, other parts of the hydraulic motor remained serial. The studies were carried out for planetary hydraulic motors with a kinematic distribution system 7/6 [6]. The number of distribution windows of the movable distributor is 12 (6 working and 6 discharge windows), and the number of distribution ports of the fixed distributor is 14 (7 delivery ports and 7 drain windows). In the serial distribution system, 3 discharge windows are used as additional working ones, and the gap between the windows of the movable and fixed distributor is 0°51'. In the modernized distribution system of the hydraulic motor, 2 unloading windows are used as additional working ones, and the gap between the distribution windows is 0 [39].

The throughput of the distribution systems of the tested hydraulic motors was determined by the total flow area of the working windows of the system under consideration. For a serial hydraulic motor, the flow area of the working windows ranges from 222 to 226 mm^2, while the average area was 223 mm^2. For a modernized hydraulic motor, the flow area of the working windows is constant and equal to 226 mm^2.

The required value of the flow rate of the working fluid passing through the hydraulic motor is equal to 50, 70, 90 and 110 l/min, respectively, was set using a variable pump when the test hydraulic motor was idling (no load). The required load was set using the test bench brake in the range of 25...400 N·m with a step of 25 N·m.

The main factors that determine the change in the functional parameters of the planetary hydraulic motor in the study of the distribution system are: the rotational speed of the hydraulic motor shaft, the pressure difference and the gap between the distribution windows of the movable and fixed distributors [4, 6, 38, 40]. The study of changes in the functional parameters of the serial and modernized hydraulic motors was evaluated based on the results of comparative tests of these hydraulic motors.

The mechanical, volumetric and overall efficiency of the tested hydraulic motor was determined according to the standard method according to the results of measuring the pressure drop, torque, rotation frequency of the hydraulic motor shaft and the flow rate of the working fluid.

As a result of the research, the dependences of the efficiency of the tested hydraulic motor on the frequency of rotation of its shaft were determined (Figure 2). It has been established that the nature of the patterns of change in the efficiency of serial and modernized hydraulic motors in the entire range of change in rotational speeds is similar.

With an increase in the speed of the hydraulic motor shaft to 300 min^{-1}, the hydromechanical efficiency (Figure 2, a) of both hydraulic motors increases, taking the value 0.75 to 0.81 for the serial (curve 1) and 0.89 to 0.92 for the modernized (curve 2) hydraulic motors. The maximum value of the hydromechanical efficiency of both hydraulic motors is in the range of rotation frequencies of its shaft 150 to 450 min^{-1}. An increase in the rotation frequency of the hydraulic motor shaft to 670 min^{-1} leads to a significant decrease in its hydromechanical efficiency to values of 0.65 for the serial and 0.75 for the modernized hydraulic motors. The hydromechanical efficiency of a hydraulic motor (Figure 2, a) with a modernized distribution system (curve 2) is 11 to 14% higher than that of a hydraulic motor with a serial distribution system (curve 1).

The volumetric efficiency of a hydraulic motor (Figure 2, b) with an upgraded distribution system (curve 2) is 2 to 4% higher than that of a hydraulic motor with a serial distribution system (curve 1) and remains practically unchanged over the entire range of speed variation.

The change in the overall efficiency from the rotational speed (Figure 2, s) has the same character as the dependence of the hydromechanical efficiency, which is explained by low volumetric losses. The overall efficiency of a hydraulic motor with a modernized distribution system (curve 2) is 7 to 9% higher than that of a hydraulic motor with a serial system distribution (curve 1).

The increase in the value of the hydromechanical, volumetric and overall efficiency of the modernized hydraulic motor is explained by the throughput of its distribution system.

Studies of changes in the output characteristics of a planetary hydraulic motor (Figure 3) have established that the nature of the change in torque depending on the rotation frequency (Figure 3, a) for serial and modernized hydraulic motors are similar. In the rotation frequency range from 75 to 550

min^{-1}, the torque developed by the serial hydraulic motor is 300 to 325 N·m (curve 1) and changes insignificantly (within 8%), which is a sign of the rigid characteristics of this hydraulic motor. The stability of the torque (within 1.5%) developed by the modernized hydraulic motor 370 to 375 N·m (curve 2) in the rotation frequency range of 75 to 450 min^{-1} confirms its rigid characteristic. The stiffness of the characteristics of a hydraulic motor is understood as its ability to provide stable torque readings over a wide range of speed changes, especially at low rpm.

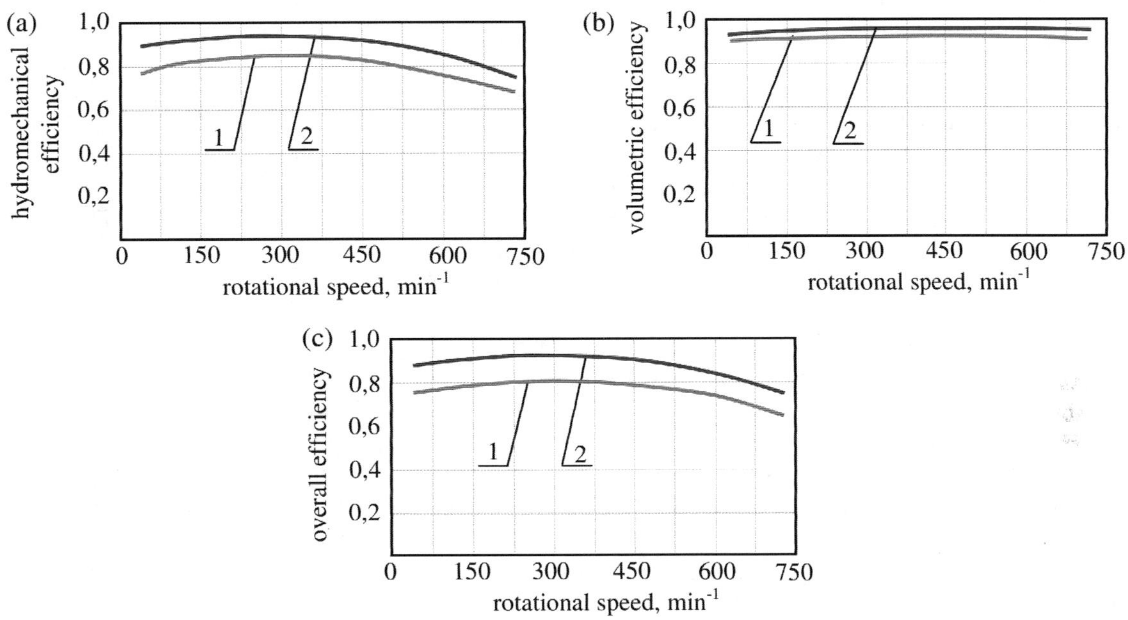

Figure 2. Dependence of efficiency on the frequency of rotation of the output shaft of the hydraulic motor at the nominal pressure: a – hydromechanical; b – volumetric; c – general; 1 – serial hydraulic motor; 2 – modernized hydraulic motor.

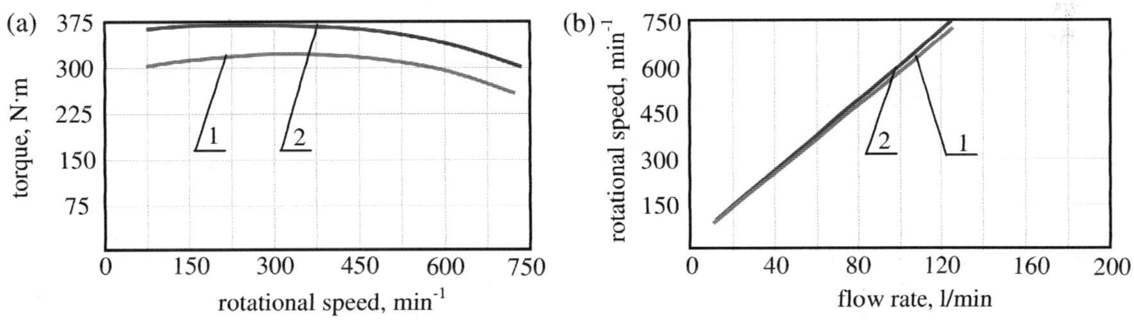

Figure 3. Change in the output characteristics of the planetary hydraulic motor: a – the dependence of the torque on the rotational speed of the output shaft of the hydraulic motor; b – dependence of the speed of rotation of the output shaft of the hydraulic motor on the flow rate of the working fluid; 1 – serial hydraulic motor; 2 – modernized hydraulic motor .

The increased stability of the torque developed by the modernized hydraulic motor and the 14% increase in the absolute value of the torque is explained by the rational design of the structure of the elements of the distribution system of this hydraulic motor.

Analysis of the dependence of the change in the speed of the hydraulic motor shaft on the flow rate of the working fluid (Figure 3, b) shows that these parameters, both for the serial (curve 1) and for the modernized (curve 2) hydraulic motors, are in a linear relationship and with an increase in the flow rate of the working fluid increase. Modernization of the distribution system has practically no effect on the change in the considered dependencies.

Thus, studies of the influence of the distribution system of a planetary hydraulic motor on the change in its output characteristics have established that elimination of the pulsation of the working fluid flow and an increase in the throughput of a hydraulic motor with a modernized distribution system makes it possible to increase the hydromechanical efficiency of a modernized hydraulic motor by 11 to 14%, volumetric efficiency by 2 to 4%, and the total – by 7 to 9%.

5. Conclusions

As a result of the study, an experimental test bench was developed for testing high-torque hydraulic motors, which allows testing a family of unified series of planetary hydraulic motors with a power of 6.5; 11; 22 and 33 kW, respectively. A methodology for experimental studies of a unified series of planetary hydraulic motors has been developed.

The maximum value of the hydromechanical efficiency of both hydraulic motors is in the range of rotation frequencies of its shaft 150 to 450 min^{-1}. The hydromechanical efficiency of a hydraulic motor with a modernized distribution system is 11 to 14% higher than that of a hydraulic motor with a serial distribution system.

The volumetric efficiency of a hydraulic motor with a modernized distribution system is 2 to 4% higher than that of a hydraulic motor with a serial distribution system and remains practically unchanged over the entire range of speed variation. The overall efficiency of a hydraulic motor with a modernized distribution system is 7 to 9% higher than that of a hydraulic motor with a serial distribution system. The increase in the values of the hydromechanical, volumetric and general efficiency of the modernized hydraulic motor is accounted for by the increase in the throughput of its distribution system.

Studies of changes in the output characteristics of a planetary hydraulic motor (Figure 3) have established that in the range of rotation frequencies of 75 to 550 min^{-1}, the torque developed by a serial hydraulic motor is 300 to 325 H·m and changes slightly (within 8%), which is a sign of a rigid characteristics of this hydraulic motor. The stability of the torque (within 1.5%) developed by the modernized hydraulic motor 370 to 375 H·m in the range of rotation speeds of 75 ÷ 450 min^{-1} confirms its tough characteristic.

References

[1] Gamez-Montero P, Codina E and Castilla R 2019 A Review of Gerotor Technology in Hydraulic Machines *Energies* **12 (12)** 2423

[2] Strmcnik E and Majdic F 2018 The Pressure and Efficiency Characteristic of Hydraulic Gerotor Motor with the Floating Outer Ring *Tehnicki vjesnik-technical gazette* **2 (25)** 609–515

[3] Panchenko A, Voloshina A, Milaeva I and Luzan P 2019 Operating Conditions' Influence on the Change of Functional Characteristics for Mechatronic Systems with Orbital Hydraulic Motors *Modern Development Paths of Agricultural Production. Springer* 169-176

[4] Voloshina A, Panchenko A, Boltyansky O and Titova O 2019 Improvement of Manufacture Workability for Distribution Systems of Planetary Hydraulic Machines *Lecture Notes in Mechanical Engineering, Springer* 732-741

[5] Panchenko A, Voloshina A, Milaeva I, Panchenko I and Titova O 2018 The Influence of the form Error after Rotor Manufacturing on the Output Characteristics of an Orbital Hydraulic Motor *International Journal of Engineering and Technology* **7 (4.3)** 1–5 - убрать

[6] Voloshina A, Panchenko A, Boltyansky O, Panchenko I and Titova O 2018 Justification of the Kinematic Diagrams for the Distribution System of a Planetary Hydraulic Motor *International Journal of Engineering and Technology* **7 (4.3)** 6–11

[7] Strutinsky V and Demyanenko A 2016 The development of mechatronic active control system of tool spatial position of parallel kinematics machine tool *Journal of Theoretical and Applied Mechanics* **54 (3)** 757–768

[8] Voloshina A, Panchenko A, Panchenko I, Titova O and Caldare A 2020 Design of Hydraulic Mechatronic Systems with Specified Output Characteristics *Advances in Design, Simulation and Manufacturing III DSMIE 2020 Lecture Notes in Mechanical Engineering* 42-51

[9] Strutynskyi S 2018 Defining the dynamic accuracy of positioning of spatial drive systems through consistent analysis of processes of different range of performance, *Naukovyi Visnyk NHU* **3** 64-73

[10] Panchenko A, Voloshina A, Titova O, Panchenko I and Zasiadko A 2020 The Study of Dynamic Processes of Mechatronic Systems with Planetary Hydraulic Motors *Advanced Manufacturing Processes II. Lecture Notes in Mechanical Engineering* 56–64

[11] Voloshina A, Panchenko A, Titova O, Milaeva I and Pastushenko A 2020 Prediction of Changes in the Output Characteristics for the Planetary Hydraulic Motor *Advanced Manufacturing Processes II. Lecture Notes in Mechanical Engineering* 84–93

[12] Korohodskyi V, Kryshtopa,S, Migal V, Rogovyi A and others 2020 Determining the Characteristics for the Rational Adjusting of an Fuel-air Mixture Composition in a Two-stroke Engine with Internal Carburation. *Eastern-European Journal of Enterprise Technologies* **2(5)** 39-52

[13] Strutynsky V, Hurzhi A and Kozlov L 2019 Determination of static equilibrium conditions of mobile terrestrial complex with lever-type manipulator *Naukovyi Visnyk Natsionalnoho Hirnychoho Universytetu* **5** 79-86

[14] Gaydamaka A, Kulik G, Frantsuzov V, Hrechka I and others 2019 Devising an engineering procedure for calculating the ductility of a roller bearing under a no-central radial load *Eastern-European Journal of Enterprise Technologies* **3 7(99)** 6-10

[15] Bondarenko M, Tkachuk M, Grabovskiy A and Hrechka I 2019 Substantiation of Thin-Walled Structures Parameters Using Nonlinear Models and Method of Response Surface Analysis *International Journal of Engineering* **44** 32–43

[16] Atroshenko O, Tkachuk M, Martynenko O, Tkachuk M and others 2019 The Study of Multicomponent Loading Effect on Thin-Walled Structures with Bolted Connections *Eastern-European Journal of Enterprise Technologies* **1 7(97)** 15–25

[17] Tkachuk M, Grabovskiy A, Tkachuk M, Saverska M and I. Hrechka 2020 A semi-analytical method for analys of contact interaction between structural elements along aligned surfaces *Eastern-European Journal of Enterprise Technologies* **1 7(103)** 16–25

[18] Marchenko A, Tkachuk M, Kravchenko S, Tkachuk M and Parsadanov I 2020 Experimental Tests of Discrete Strengthened Elements of Machine-Building Structures. *Advanced Manufacturing Processes Lecture Notes in Mechanical Engineering* 559–569

[19] Tkachuk M, Grabovskiy A, Tkachuk M, Hrechka I and others 2019 Investigation of multiple contact interaction of elements of shearing dies *Eastern-European Journal of Enterprise Technologies* **4 7(100)** 6–15

[20] Rogovyi A, Khovanskyy S, Grechka I and Pitel J 2019 The Wall Erosion in a Vortex Chamber Supercharger Due to Pumping Abrasive Mediums *Design, Simulation, Manufacturing: The Innovation Exchange* 682–691

[21] Maiti R and Nagao M 1999 Prediction of starting torque characteristics of epitrochoid generated orbital rotary piston hydraulic motors *JSME international journal series c-mechanical systems machine elements and manufacturing* **2 (42)** 416–426

[22] Dasgupta K, Mukherjee A and Maiti R 1996 Theoretical and experimental studies of the steady state performance of an orbital rotor low-speed high-torque hydraulic motor *Proceedings of the institution of mechanical engineers Part A - Journal of power and energy* **6 (210)** 423–429

[23] Furustig J, Almqvist A and Pelcastre, L 2016 A strategy for wear analysis using numerical and experimental tools, applied to orbital type hydraulic motors *Proceedings of the institution of mechanical engineers part c-journal of mechanical engineering science* **12 (230)** 2086-2097

[24] Panchenko A, Voloshina A, Panchenko I, Titova O and Pastushenko A 2019 Reliability design

IOP Conf. Series: Materials Science and Engineering 1021 (2021) 012054 doi:10.1088/1757-899X/1021/1/012054

of rotors for orbital hydraulic motors *IOP Conference Series: Materials Science and Engineering* **708 (1)** 012017

[25] Xing K, Zhang Y and Jin X 2009 Performance analysis and experimental research of the orbital motor *Proceedings of the seventh international conference on fluid power transmission and control* 131–135.

[26] Dasgupta K, Mukherjee A and Maiti R 1999 Estimation of critical system parameters that affect orbit motor performance-combining simulation and experiments *Journal of manufacturing science and engineering-transactions of the ASME* **2 (121)** 300–306

[27] Shah, Y. Vacca A and Dabiri S 2018 A fast lumped parameter approach for the prediction of both aeration and cavitation in Gerotor pumps *Meccanica* **1-2 (53)** 175–191

[28] Robison J and Vacca A 2017 Kinematic multi-objective optimization of circular-toothed gerotor pumps by genetic algorithm *Proceedings of the asme/bath symposium on fluid power and motion control* UNSP V001701A016

[29] Velev E 2016 Study Cavitation Gerotor Motors *Renewable Energies and Innovative Technologies* 64–66

[30] Voronin S, Suranov A, Suranov A 2017 The effect of carbon nanoadditives on the tribological properties of industrial oils *Journal of Friction and Wear* 38(5) 359–363

[31] Воронін С 2015 Розробка трібофізичних основ мастильної здатності рідкокристалічних присадок до базових олив *Східно-Європейський журнал передових технологій* **3 7(75)** 53–57

[32] Voronin S, Onopreychuk D, Stefanov V, Bashkatov Ye and Panchenko V 2018 Reduction of construction duration by improving the anti-wear properties of power fluids in hydraulic drives of earth-moving machines *International Journal of Engineering & Technology* **7(4.3)** 105–109

[33] Chernetskaya-Beletskaya N, Rogovyi A, Shvornikova A, Baranov I and others 2018 Study on the coal-water fuel pipeline transportation taking into account the granulometric composition parameters *International Journal of Engineering & Technology* **7(4.3)** 240–245

[34] Rogovyi A, Khovanskyi S, Hrechka I and Gaydamaka A 2020 Studies of the Swirling Submerged Flow Through a Confuser *Design, Simulation, Manufacturing: The Innovation Exchange* 85–94

[35] Rogovyi A 2018 Energy performances of the vortex chamber supercharger *Energy* **163** 52–60

[36] Ding H, Lu J and Jiang B 2012 A CFD model for orbital gerotor motor *IOP Conference Series: Earth and Environmental Science* **6 (15)** 062006

[37] Yang D, Yan J and Tong S 2010 Flowrate Formulation of Deviation Function Based Gerotor Pumps // Journal of Mechanical Design **132 (6)** 064503-5

[38] Panchenko A, Voloshina A, Boltyansky O and others 2018 Designing the flow-through parts of distribution systems for the PRG series planetary hydraulic motors *Eastern-European Journal of Enterprise Technologies* **3 1 (93)** 67–77

[39] Voloshina A, Panchenko A, Panchenko I, Titova O and Zasiadko A 2019 Improving the output characteristics of planetary hydraulic machines *IOP Conference Series: Materials Science and Engineering* **708 (1)**, 012038

[40] Voloshina A, Panchenko A, Panchenko I and Zasiadko A 2019 Geometrical Parameters for Distribution Systems of Hydraulic Machines *Modern Development Paths of Agricultural Production. Springer* 323-336

IOP Conf. Series: Materials Science and Engineering 1021 (2021) 012055 doi:10.1088/1757-899X/1021/1/012055

Influence of Temperature and Concentration of Additives on the Lubricity of Track Machine Oils

S Voronin[1, 3], V Stefanov[1], D Onopreichuk[1], O Ovchynnikov [1] and O Pashchenko[2]

[1]Department of construction, travel and cargo-handling machines, Ukrainian State University of Railway Transport, Feuerbach Square 7, Kharkov, 61050
[2]Department of Tactics, National Academy of the National Guard of Ukraine, Zakhysnykiv Ukrainy sq., 3, Kharkov, 61001

[3]E-mail: voronin.sergey@ukr.net

Abstract. The promising types of additives to hydraulic oils of track machines, which contribute to increasing the lubricity of oils, that is necessary for the introduction of resource saving in railway transport, were substantiated. On the basis of the conducted justification, two types of additives were selected for further study: pure stearic acid and a three-component additive based on carbamide A, stearic and oleic acid. These additives were used in the studies, being added to hydraulic oil I-20A in different concentrations. Methods and results of laboratory studies of the indicators of the lubricity of hydraulic oil with additives depending on the concentration of the additive and the operating temperature of the oil are given. The following indicators of oil lubricity were studied: the dynamic viscosity of the oil measured in a rotary viscometer; the thickness of the boundary lubricant film measured using the "ball stack" method; the load-carrying capacity of the boundary lubricant film measured in a special load device by establishing a load in which the electrical resistance in the point of contact between two lubricated steel balls abruptly approaches the values of resistance of the dry metal contact.

1. Introduction

One of the most promising trends for implementing resource-saving technologies for the railway transport is quality assurance for lubricants [1]. This direction is especially relevant for hydraulic oils of track machines, most of which do not contain any anti-wear and anti-friction additives, such as industrial oils. These oils are commonly used in hydraulic drives of track machines, because of their low cost, but they do not contribute to resource savings due to low lubricity.

The solution to the problem of improving the lubricity of industrial oils lies in the search and study of new and at the same time moderately priced additives which form a lubricant film with increased thickness and load-carrying capacity on the friction surfaces of hydraulic units. In addition, these additives should not significantly affect the viscosity and temperature characteristics of oils; otherwise, oil viscosity values cannot be maintained within the permissible range and the oil cannot be used in specific hydraulic systems at all. Therefore, in the article the main indicators of the lubricity of industrial oil are studied when two different types of additives are added, each meeting the requirements stated above.

2. Analysis of recent studies and publications

The paper [1] contains general ideas about the lubricity of oils, lists indicators of lubricity, test methods, and the influence of oil structure and external factors on them. According to this information, there are three main indicators of lubricity, namely: the dynamic viscosity of the oil, the thickness and load-carrying capacity of the oil film. These indicators reflect phenomena with the different mechanisms of action. For example, the dynamic viscosity of oil affects the efficiency of forming a hydrodynamic lubricating film [2], while the thickness and load-carrying capacity of the oil film are more relevant for boundary friction of parts [2, 3].

During the operation of hydraulic units of track machines, especially hydrostatic ones, all three oil indicators must be maintained at a high or set level, since during operation of these units both the hydrodynamic lubrication and the boundary lubrication exist, because of the design features of the units [4] and wide ranges of loads, relative speeds and temperature [5]. In addition, the viscosity characteristic of the oil significantly affects the efficiency of its application on the friction surface [6], which also confirms the importance of ensuring its performance at the necessary level.

Today, one of the main ways to improve the lubricity of hydraulic oils is to add a certain amount of functional additives to them during the manufacture [1, 7, 8]. However, in the actual operating conditions of track machines, the use of oils doped with modern additives is controversial, since such oils are hight-priced, while the hydraulic system can require filling with 2 tons of oil at a time or even more. Therefore, it is reasonable to search for such additives that, on the one hand, improve the lubricating ability of the oil at low concentrations, and on the other hand, are manufactured from reasonably-priced raw materials and have good solubility in hydrocarbons. In [7], 2 types of additives are proposed that meet the specified requirements, but their effect on the lubricity of hydraulic oils has been studied only theoretically. However, for further practical application of such additives, an experimental study of their effect on the lubricity at different concentrations and external conditions is required.

3. Statement of the objective and tasks of the study

The purpose of the study is to determine dependence of the main indicators of the lubricity of hydraulic oils of track machines on the concentration of two types of additives and the oil temperature. To achieve this goal, the following tasks were solved:

- determination of dependence of the dynamic viscosity of the oil on the concentration of the selected types of additives and temperature;
- determination of dependence of the maximum thickness of the lubricant film on the concentration of the selected types of additives and temperature;
- determination of dependence of the load-carrying capacity of the boundary lubricant film on the concentration of the selected types of additives and temperature.

4. Materials, equipment and research program

Hydraulic oil I-20A (GOST 20799-88), which is commonly used in most hydraulic systems of track machines in Ukraine, was selected for the study. Pure stearic acid (GOST 6484-96) and a three-component additive consisting of carbamid A (GOST 2081-92), stearic acid (GOST 6484-96) and oleic acid (GOST 7580-91) were used as additives. Such additives are well soluble in the tested oil within the studied concentration range and were selected according to the physical representations of the works [8, 9]. Stearic acid is a surfactant that, at a certain concentration, is inclined to aggregation of molecules in solution to form "reverse" micelles. The studied concentration range was selected from 0 to 0.6 % by volume, which exceeds the values of the critical micelle concentration. The three-component additive is dissolved in the oil to the level of ordered molecular "adducts", which retain their high surface activity throughout the studied concentration range. In the studies, this range was limited only by the solubility limit of the additive, so it varied from 0 to 0.2 % by volume.

The dynamic viscosity of the oil was measured using a rotary viscometer Rheomat RM100i with a relative measurement error of ± 1 %. The liquid temperature was maintained in the LNG-500 heater.

The dynamic viscosity study program included measuring the oil viscosity at a stearic acid concentration in the range of 0 to 0.6 % with an increment of 0.15 %; measuring the oil viscosity at a three-component additive concentration in the range of 0 to 0.2 % with an increment of 0.05 %. In both series of studies, oil temperature was maintained in the range of 40 to 100 °C with an increment of 20 °C.

The thickness of the boundary lubricant film was measured in a special laboratory device that uses the "ball stack" method [8] (figure 1).

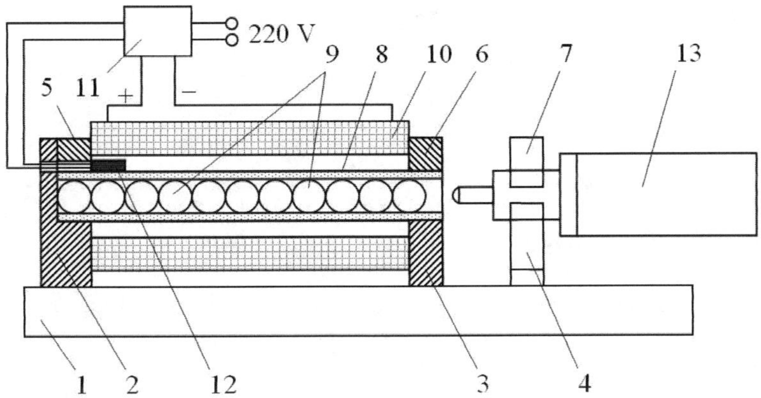

Figure 1. Schematic diagram of the boundary lubricant film thickness measurement system: 1 - plate; 2, 3, 4 - guide supports; 5, 6, 7 - covers; 8 - glass tube; 9 - steel balls; 10 – heater; 11 – temperature controller; 12 – temperature sensor; 13 – dial-type micrometer.

According to the proposed method, the total thickness of the lubricant films on the balls is

$$\Sigma h = 2\cdot(n+1)\cdot h_i, \quad (1)$$

where n is the number of balls in the stack; h_i is the thickness of the film on one surface.

The minimum number of balls in the stack is determined based on the micrometric accuracy of the device and the predicted value of the change in film thickness after an additive is added.

$$n = \frac{e}{2\cdot \Delta h} - 1, \quad (2)$$

where e is the accuracy of the measuring device; Δh is the predicted change in the thickness of the limiting film.

The program for study of the thickness of the boundary lubricant film included: measuring the thickness at a concentration of stearic acid in the range of 0 to 0.4 % with an increment of 0.05 %; measuring the thickness at a concentration of the three-component additive in the range of 0 to 0.2 % with an increment of 0.02%. In both series of tests, the oil temperature was maintained in the range of 40 to 80 °C with an increment of 20 °C.

The load-carrying capacity of the lubricant film was studied in a special laboratory device, in which the direct physical measurement of the electrical resistance in the contact point of two steel balls was used (figure 2). The load-carrying capacity is understood as the minimum load at which the electrical resistance of the lubricated contact rapidly fell to the value of the electrical resistance of the dry metal contact. I.e., it was considered that at this moment the lubricant film is destroyed, and as a result, metal-to-metal contact occurs.

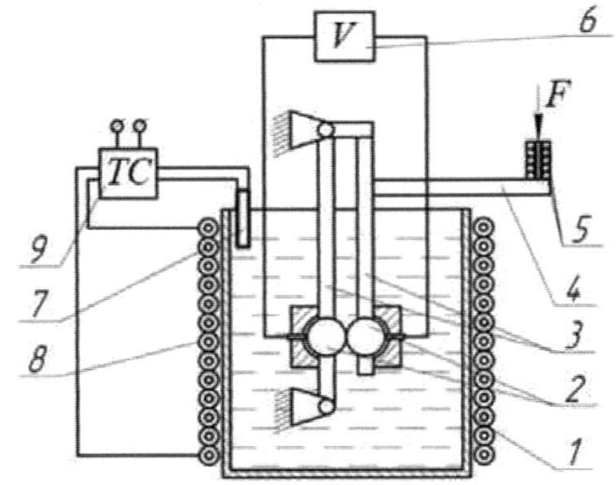

Figure 2. Schematic diagram of the device for measuring the maximum load-carrying capacity of the lubricant film: 1 - bowl; 2 - steel balls; 3 - lever system; 4 - load lever; 5 - calibrated loads; 6 - universal voltmeter; 7 - temperature sensor; 8 - heating element; 9 – temperature controller.

The program of studies of the load-carrying capacity of the lubricant film included: the measurement of the critical load when the concentration of stearic acid was in the range 0 to 0.6 % with an increment of 0.3 %; measurement of critical loads at a concentration of three-component additive in the range 0 to 0.2 % with an increment of 0.1 %. In both series of tests, the oil temperature was maintained in the range of 20 to 140 °C with an increment of 20 °C.

5. Results of the study and discussion

The results of dynamic viscosity studies are shown in figure 3.

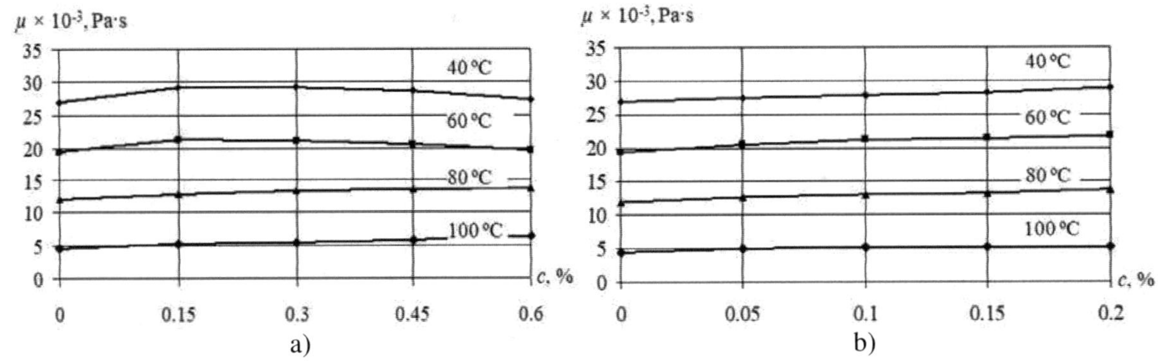

Figure 3. Results of studies of the dynamic viscosity of the oil: a) stearic acid as an additive; b) three-component additive.

According to the results shown in Figure 3, there are two dependences of oil viscosity on the concentration of additives. First, when stearic acid is added at a solution temperature of 40 °C and 60 °C, the maximum increase in viscosity occurs at the concentration of 0.3 % and 0.15 %, respectively. The increase in viscosity for these concentrations is about 7 %. With further increase in the concentration, the oil viscosity slightly decreases (figure 3 a)). The indicated values of concentrations are consistent with the data of colloid chemistry regarding the critical micelle concentrations (CMC) of stearic acid dissolved in hydrocarbons [9]. I.e., at concentrations lower than CMC, stearic acid molecules are predominantly in the monomeric state, which means that they increase the free energy of the solution. At concentrations above CMC, the additive molecules are in the micellar state. Such micelles are known to be "reverse": the surface-active parts of the molecules are in the middle, while the activity of micelles drops to almost zero [10]. As a result, the free energy of the solution falls almost to the level of the initial value. Therefore, at solution temperatures of 80 °C and 100 °C, the viscosity increases steadily over the entire concentration range, since micelle formation does not occur at such temperatures (figure 3 a)).

When a three-component additive is added to oils, the viscosity steadily increases by 5 to 10 % in all concentration and temperature ranges (figure 3 b), which indicates that there is no micelle formation of the three-component additive in these ranges.

The results of studies of the thickness of the boundary lubricant film are shown in figure 4.

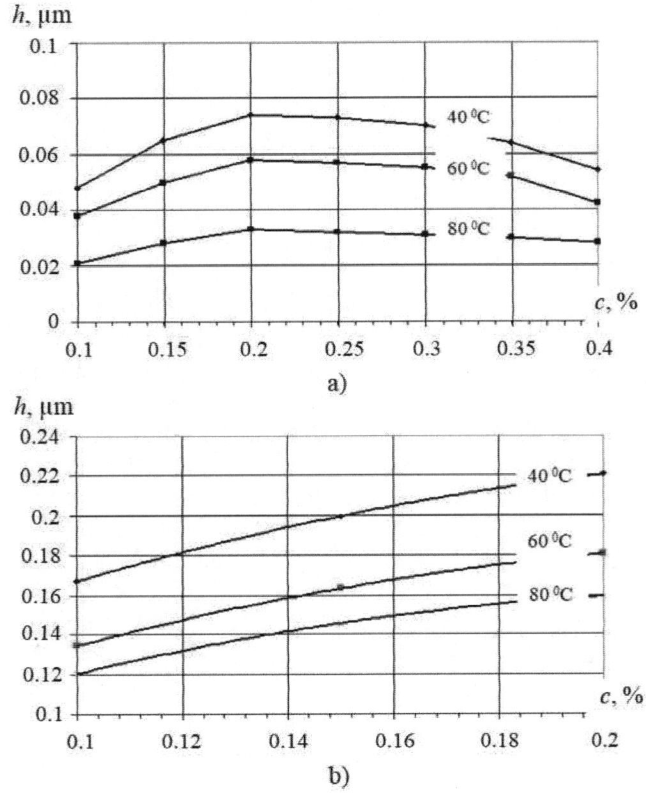

Figure 4. Results of studies of the thickness of the boundary lubricant film: a) stearic acid as an additive; b) three-component additive.

When stearic acid is added to oils, the dependence of the film thickness on the additive concentration also has a maximum (figure 4 a)). It was found that the maximum thickness increase is observed at a concentration of 0.2 % for all temperatures. A further increase in the concentration leads, as in the case of dynamic viscosity, to a decrease in the film thickness, which also indicates the influence of micelle formation. In general, when stearic acid is used, the maximum increase in the thickness of the lubricating film is achieved by 55 to 65 %.

When a three-component additive is added to the oil, the film thickness is significantly higher than when stearic acid is added, namely 2.5 to 6 times at additive concentrations of 0.1% (figure 4 a) and figure 4 b)). The reason is the high surface activity and size of the molecular "adducts" of the three-component additive. When the concentration of the three-component additive increases from 0.1 % to 0.2 %, the film thickness steadily increases by 32 to 33 %, which also indicates that there is no micelle formation.

The results of studies of the load-carrying capacity of the boundary lubricant film are shown in figure 5.

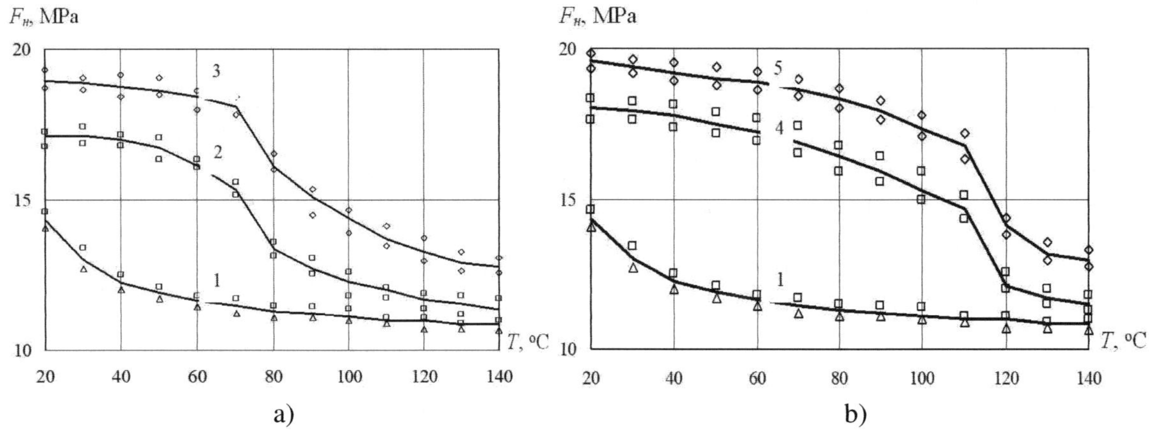

Figure 5. Results of the studies of the load-carrying capacity of the boundary lubricating film: a) stearic acid as an additive; b) three-component additive.

The results of the study shown in figure 5 are presented in the form of temperature curves, since the load-carrying capacity of the film is the main indicator of lubricity of the oil, which should ensure effective lubrication of the surfaces of the parts within a wide temperature range.

The load-carrying capacity of the oil film without additives decreases steadily with temperature and is 14 MPa for 20 °C and 11 MPa for 140 °C (curve 1). For the oil with both additives, the temperature dependence of the load-carrying capacity is different. In each case (curves 2-5), there is a gradual decrease in the load-carrying capacity to a certain critical temperature, after which the load-carrying capacity decreases abruptly, and all curves become similar to curve 1. For stearic acid, the load-carrying capacity changes abruptly at 70 °C, and for a three-component additive - at 110 °C. Interestingly, these values are close to the melting temperatures of both substances. I.e., as soon as these temperatures are reached, a phase change occurs in the boundary films of additives – from a more ordered phase to a less ordered one, as a result, the load-carrying capacity decreases abruptly. In general, when stearic acid is used, an increase in the load-carrying capacity of the oil by 16 to 30 % can be achieved. For a three-component additive, this increase varies from 17 to 37 %.

6. Conclusions

1. Adding two types of additives to industrial oil in the selected concentrations leads to a maximum increase in dynamic viscosity by 5 to 10 %, which does not change the scope of application of oil in hydraulic drives of track machines.

2. The use of the studied additives results in a significant improvement in the lubricity of the oil according to the criteria of thickness and load-carrying capacity of the lubricant film. When a particular type of additive is used, its tendency to micelle formation should be taken into account. For example, stearic acid maximizes the lubricity at concentrations of 0.15 to 0.2 % and a temperature not higher than 70°C. The maximum efficiency of the three-component additive is achieved at a concentration of 0.2 % and a temperature not exceeding 110°C.

3. In general, the use of the studied additives, especially the three-component additive, will lead to the implementation of resource saving in hydraulic drives of track machines, since increasing the thickness and load-carrying capacity of the oil film are direct factors in reducing the wear rate of friction parts of hydraulic units.

References

[1] Chichinadze A., Berliner E., Braun E. 2003 *Trenie, iznos i smazka. Tribologiya i tribotechnika*, (Moscow, Mashinostroenie), 576 p

IOP Conf. Series: Materials Science and Engineering 1021 (2021) 012055 doi:10.1088/1757-899X/1021/1/012055

[2] Mikocyanchyk O., Mnatsakanov R., Zaporozhets A., Kostynik R. 2016 Influence of the nature of boundary lubricating layers on adhesion component of friction coefficient under rolling conditions, *Eastern-European Journal of Enterprise Technologies*, 4, 1 (82), pp 24–31

[3] Kindrachuk M., Radionenko O., Kryzhanovskyi A., Marchuk V. 2014 The Friction mechanism between surfaces with regular micro under boundary lubrication, *Aviation*, 18 2 pp 64–71

[4] Voloshina A., Panchenko A., Panchenko I. and Zasiadko A. 2019 Geometrical Parameters for Distribution Systems of Hydraulic Machines. In: Nadykto V. (eds), *Modern Development Paths of Agricultural Production. Springer, Cham*, pp 323-336

[5] Panchenko A., Voloshina A., Milaeva I., Luzan P. 2019 Operating Conditions' Influence on the Change of Functional Characteristics for Mechatronic Systems with Orbital Hydraulic Motors. In: Nadykto V. (eds), *Modern Development Paths of Agricultural Production, Springer, Cham*, pp 169-176

[6] Voronin S., Hrunyk I., Stefanov V., Onopreychuk D. and Asadov B. 2018 Study of the process of lubricant application on the friction surface using aerosol lubricators, *International Journal of Engineering & Technology,* 7 (4.3) pp 20-24

[7] Voronin S. 2015 Development of tribophysical foundations of lubricity of liquid crystal additives to base oil, *Eastern-European Journal of Enterprise Technologies*, 3, 7 pp 53-57

[8] Ahmatov A. 1963 *Molekulyarnaya fizika granichnogo treniya*, (Moscow, Fizmatgiz), 472 p

[9] Rusanov A. 1992 *Mitselloobrazovanie v rastvorah poverhnostno-aktivnyih veschestv*, (St. Petersburg, Himiya), 279 p

[10] Fuks G. 1983 Adsorbtsiya i smazochnaya sposobnost masel, *Trenie i iznos*, 4 (3) pp 398-412

IOP Conf. Series: Materials Science and Engineering 1021 (2021) 012056 doi:10.1088/1757-899X/1021/1/012056

Correlation of rheological parameters with the flow of non-Newtonian fluids

A A Zadorozhnyi[1, 5], M P Remarchuk[1], A P Kovrevski[2], Y V Chovnyuk[3] and S A Buhaievskyi[4]

[1]Department of construction, travel and cargo-handling machines, Ukrainian State University of Railway Transport, Feuerbach Square 7, Kharkiv, 61050
[2]Department of Theoretical Mechanics, Kharkiv National University of Construction and Architecture, Sumska Street 40, Kharkiv, 61002
[3]Department of Machinery and Equipment Design, National University of Bioresources and Life Sciences of Ukraine, Heroiv Oborony Steet15, Kyiv, 03041
[4]Department of Bridges, Structures and Construction Mechanics, Kharkiv National Automobile and Highway University, Petrovskogo Street 25, Kharkiv, 61002

[5]E-mail: zsnj1971@ukr.net

Abstract. When new design and technological solutions are developed for equipment intended for the supply and transportation of non-Newtonian liquids through pipelines, research and design organizations face the problem of accurate understanding and behavior of such media as concrete and mortar mixtures. Currently, dozens of models of various complex fluid systems are known, most of which, however, are only empirical. The reason of this situation is the lack of a sufficiently generalized theory that would allow calculating the characteristics of the mechanical behavior of the flow of various media based on their structure. The specialist literature presents numerous purely empirical and semi-empirical rheological models of limited purpose. In most works, the well-proven and justified rheological Shvedov–Bingham, Bulkley-Herschel, Ostwald-de Waele, Briand, Shulman, and Casson models are not used to a sufficient extent due to the narrow technological specialization of the studied non-Newtonian fluids. Classical rheological models have a strong potential for improvement and rational modification to generalize rheological models of the flow for non-Newtonian fluids.

1. Introduction

Ever increasing global production facilities, such as energy engineering, machine-building, processing of synthetic and natural materials (in particular, the construction industry), oil production and petrochemical industry, pharmacy, food, paper, paint and varnish production, have been stimulating research and engineering developments to study rheological models of complex media, such as concrete and mortar mixtures. The costs associated with transporting non-Newtonian materials in the modern world have reached an immense value. The need for a reasonable choice and optimal estimation for efficient transportation of complex fluids (primarily those with high viscosity) is a relevant problem from an economic point of view. When engineers are designing pipeline transportation equipment, mortar and concrete pipelines, they are faced with the difficult challenge of finding the optimal characteristics (technological diameter of the pipeline, flow velocity of the mixture, etc.) to provide the optimal costs necessary for transporting a non-Newtonian (concrete) mixture to a technologically specified distance. In this case, determining optimal values becomes very problematic due to the need to correctly account for the non-Newtonian factor, i.e., knowledge of the true rheological "pressure–flow" curve for various fluid flow conditions.

2. Analysis of recent studies and publications

Non-Newtonian fluids, such as concrete and mortar mixtures, are commonly considered as multi-component Shvedov-Bingham media. Such fluids obey the rheological laws of Bingham fluid flow [1, 2, 3, 4, 5 and 6]. Theoretical and experimental studies of the flow processes of Bingham fluids and the construction of mathematical models were carried out by such well-Known specialists as: W L Wilkinson, A V Gnoevoy, V Prager, B M Smolsky, Z P Shulman, V V Gorislavets, S S Kutateladze, and many others. The obtained mathematical models are widely used for rigid visco-plastic media, in particular, concrete mixtures. Most of the classical equations of the rheological state of fluids are generalized [7, 8, 9, 10 and 11].

3. Statement of the objective and tasks of the study

The purpose of this work is to conduct a study based on mathematical modeling methods of the known Casson dependence for various rheological power indicators m and n, , and consequently conduct a comparative analysis; to identify inaccuracies in determining the parameter τ_y. In theoretical plotting of flow curves of Bingham fluids that affect the flow characteristics in concrete pipelines, to approach reasonably estimations in designing technological equipment for transporting mortar and concrete mixtures [12, 13, 14, and 15]. To achieve the purposes of this work, the main task is to determine the productivity Q (relative flow rate) and velocity $U(\rho)$ depending on the parameters of the mixture and dimensionless radius ρ.

4. Development of mathematical model

The rheological equation of the Casson-Shulman model has the form

$$\tau^{1/n} = \tau_y^{1/n} + k\left(\dot{\gamma}\mu_p\right)^{1/m}. \tag{1}$$

Let us consider the influence of inaccuracy in determining the parameter τ_y on the flow characteristics and flow rate for different m and n. Let us introduce the designations for the dimensionless shear stress and the flow velocity $\tilde{\tau}_0 = \dfrac{\tau}{\tau_y}$, $\nu = \dfrac{\dot{\gamma}\mu_p}{\tau_y}$.

For $m = n = 2$, equation (1) takes the form $\tilde{\tau}_0 = \left(1+\sqrt{\nu}\right)^2$, (2)

for $m = n = 3$, equation (1) takes the form $\tilde{\tau}_0 = \left(1+\sqrt[3]{\nu}\right)^3$. (3)

The graphs (Figure 1 a, b) show the flow curves for large and small values of the dimensionless flow velocity ν (here $\tilde{\tau}_{02}$ - formula (2), $\tilde{\tau}_{03}$ - formula (3).

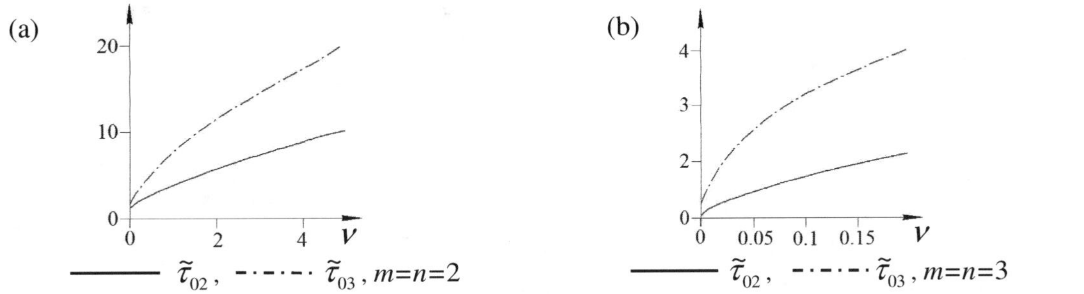

Figure 1. Flow curves based on the Casson model: a), b) – flows at large and small values of dimensionless flow velocity ν, respectively.

To draw up the flow velocity profile, $u = u(r)$ equation (1) at n=m=2 is transformed by integration to the form

$$U(\rho) = \frac{1-\rho^2}{2} + \tau_0(1-\rho) - \frac{4}{3}\sqrt{\tau_0}\cdot(1-\rho^{3/2}), \tag{4}$$

where $\rho = \frac{r}{R}$ is the dimensionless radius; R is the radius of the pipe; $\tau_0 = \frac{\tau_y}{\tau_w}$ is relative stress; τ_w is shear stress on the pipe wall with radius R; τ_y is shear stress at the boundary of the shear region. $U(\rho) = u(\rho)\cdot\frac{\mu_p}{R\cdot\tau_w}$ When the equation is derived, it is assumed that on the walls of the pipe $\rho = 1,\ u = 0$. Graph $\Delta U(\rho)$ for various τ_0 is shown in Figure 2. At m=n=3, equation 1 is converted to the form

$$U(\rho) = \frac{1}{2} - \frac{\rho^2}{2} - \frac{6}{5}\tau^{1/3}(1-\rho^{5/3}) + \frac{9}{4}\tau_0^{2/3}(1-\rho^{4/3}) - \tau_0(1-\rho). \tag{5}$$

The corresponding graph of speed profiles $U(\rho)$ is shown in Figure 3.

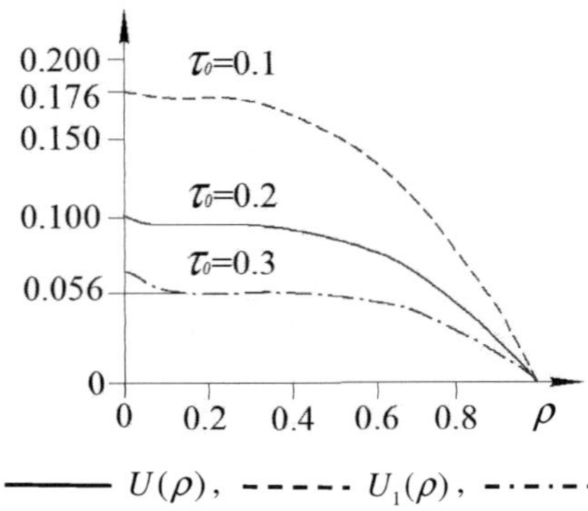

Figure 2. Dimensionless velocity profiles $U(\rho)$ at various speeds τ_0.

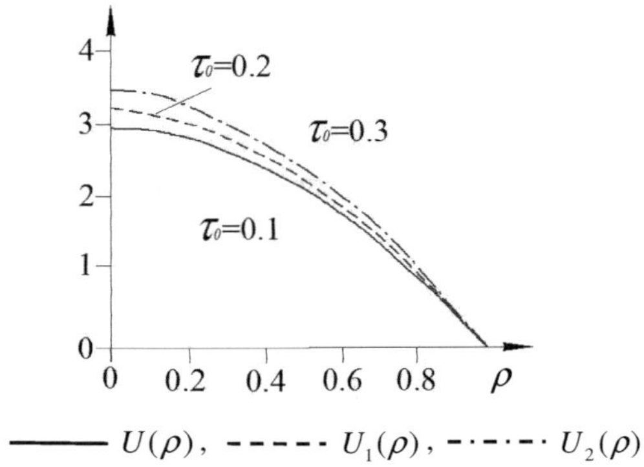

Figure 3. Dimensionless velocity profiles $U(\rho)$ of the Casson model at m=n=3.

In the case of m=3 n=2, the equation of the Casson model is written as

$$\tau^{1/2} = \tau_y^{1/2} + k(\dot{\gamma}\mu_p)^{1/3}. \tag{6}$$

Here k is a coefficient that is introduced to "align" the dimensions of the terms.

After equation (6) is written in dimensionless form and integrating by ρ, the expression is obtained for the dimensionless flow velocity $\tilde{U}(\rho) = u(\rho)\cdot\frac{k\cdot\mu_p}{R\cdot\tau_w^{3/2}}$

$$\Delta U(\rho) = 0.4(1-\rho^{5/2}) + 1.5\tau_0^{1/2}(\rho^2-1) + 2\tau_0(1-\rho^{3/2}) + \tau_0^{3/2}(\rho-1). \tag{7}$$

Graphs of flow velocity profiles $U(\rho)$ for different τ_0 are shown in Figure 4.

When formulas for velocity profiles are available, it is easy to obtain flow rate formulas for various indicators m and n. Now we get the formulas for the flow velocity according to the Caisson model:

a) m=n=2
$$Q=\frac{\pi\cdot R^3\tau_w}{\mu_p}\cdot\left[\frac{1}{2}+\frac{2}{3}\tau_0-\frac{4}{7}\tau_0\right]=\frac{\pi\cdot R^3\tau_w}{\mu_p}\cdot W_1(\tau_0), \quad (8)$$

b) m=n=3
$$Q=\frac{\pi\cdot R^3\tau_w}{\mu_p}\cdot\left[\frac{1}{4}-\frac{6}{11}\tau_0^{1/3}+\frac{9}{10}\tau_0^{2/3}-\frac{1}{3}\tau_0\right]=\frac{\pi\cdot R^3\tau_w}{\mu_p}\cdot W_2(\tau_0), \quad (9)$$

c) m=3, n=2
$$Q=B\cdot\left[\frac{1}{9}-\frac{3}{8}\tau_0^{1/2}+\frac{3}{7}\tau_0-\frac{1}{6}\tau_0^{3/2}\right]=B\cdot W_3(\tau_0), \quad (10)$$

where $B=\frac{2\pi\cdot R^3\tau_w^{3/2}}{k\cdot\mu_p}$.

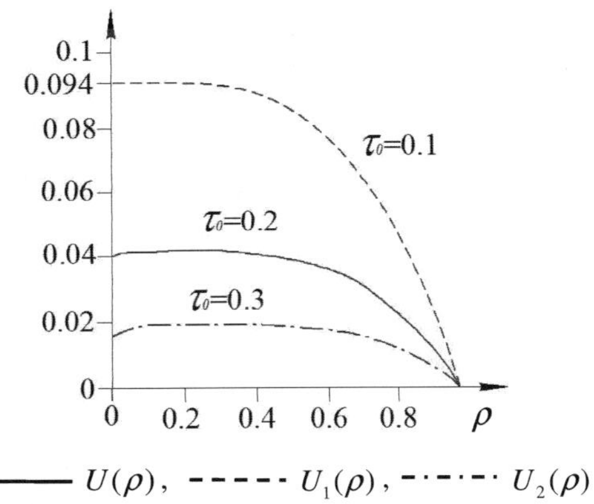

Figure 4. Dimensionless velocity profiles $U(\rho)$ at m=3, n=2.

For the Bingham model of viscous fluid flow, the flow formula has the form [1, 2, 4]

$$Q=\frac{\pi\cdot R^3\tau_w}{\mu_p}\cdot\left[\frac{1}{4}-\frac{1}{3}\tau_0+\frac{\tau_0^4}{12}\right]=\frac{\pi\cdot R^3\tau_w}{\mu_p}\cdot W_\delta(\tau_0). \quad (11)$$

Let us consider the effect of a change of τ_0 on the maximum flow velocity in the Casson model. The minimum possible value of τ_y is taken as the base value: $\tau_{y\min}=\tau_{ym}$. Let the possible change of τ_y be $\Delta\tau_y$, i.e. the current value of the elastic limit will be equal to $\tau_y=\tau_{ym}+\Delta\tau_y$. The parameter $\tau_0=\frac{\tau_y}{\tau_w}$ will be $\tau_0=\tau_{0m}+\Delta\tau_0$. If it is accepted that $\Delta\tau_y=0.1\tau_y$, then $\tau_0=1.1\tau_{0m}$ where $\tau_{0m}=\frac{\tau_{ym}}{\tau_w}$. The expression for the maximum velocity (at $\rho=0$) will have the form:

at m=n=2
$$U'(\rho_1\tau_0)=0.5+\tau_{0m}-1.333\sqrt{\tau_{0m}}, \quad (12)$$

at m=3, n=2
$$U'(\rho_1\tau_0)=0.4-1.5\tau_{0m}^{1/2}+2\tau_{0m}-\tau_{0m}^{3/2}. \quad (13)$$

If instead of τ_{0m} $\tau_0=1.1\tau_{0m}$ is used, the expressions for the maximum velocities will take the form:

at m=n=2
$$U(\rho_1\tau_0)=0.5+1.1\tau_{0m}-1.398\sqrt{\tau_{0m}}, \quad (14)$$

at m=3, n=2 $$U(\rho_1\tau_0) = 0.4 - 1.573\tau_{0m}^{1/2} + 2.2\tau_{0m} - 1.154\tau_{0m}^{3/2}. \quad (15)$$

The relative value of the maximum velocity is determined by the formula

$$\Delta V = \frac{U(\rho_1\tau_0) - U'(\rho_1\tau_0)}{U'(\rho_1\tau_0)}. \quad (16)$$

Velocity curves are shown in (Figure 5 a, b).

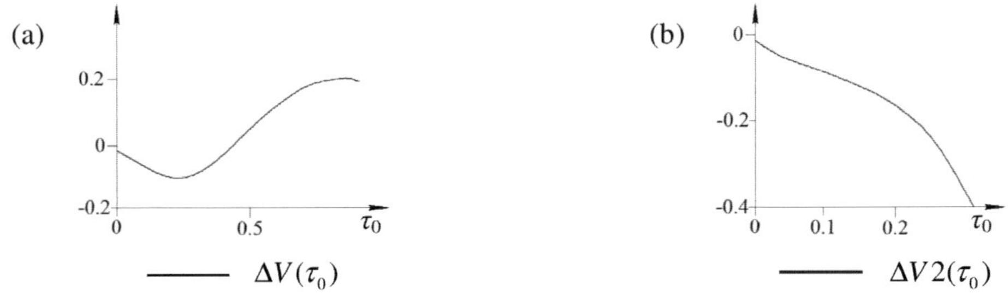

Figure 5. Maximum relative velocities of profiles according to the Casson model ΔV at m=n=2 and m=3, n=2: a) – maximum dimensionless speed ΔV at m=n=2; b) – maximum dimensionless speed ΔV at m=n=3.

In the case of m=3, n=2, the expression for ΔV at $\tau_0 \approx 0.45, \rho = 0$ and at $\tau_0 \approx 0.41$, $\rho = 0.1$ the flow curve is broken. Therefore, the flow process can be described using the Casson model at m=3, n=2 only for small values of τ_0.

Effect of the change of τ_0 on the flow rate. The dependence of the flow rate on the parameter τ_0 is described by the second multipliers in formulas (8 - 11), which are denoted as $\Delta W(\tau_0)$. Let us further consider the corresponding dependencies. The relative change of flow rate is determined by the formula

$$\Delta W = \frac{W(\tau_0) - W'(\tau_0)}{W'(\tau_0)}, \quad (17)$$

here, as in the previous case, $\tau_0 = \tau_{0m} + \Delta\tau_0 = 1.1\tau_{0m}$. The results of estimations are presented in graphical form (Figure 6 a, b).

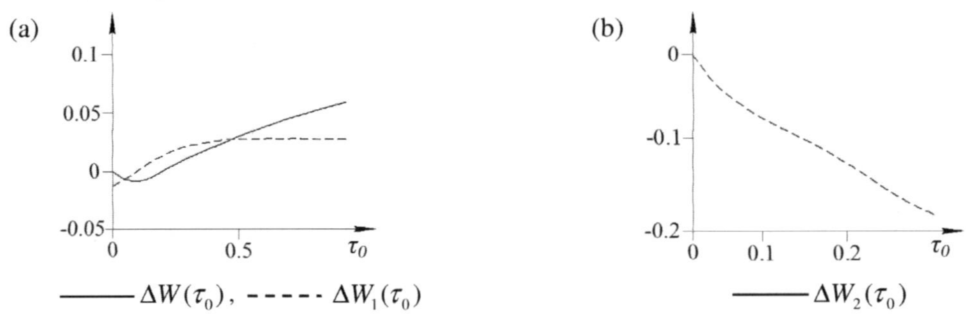

Figure 6. Relative change in flow rate ΔW according to the Casson model: a) – dependence of the flow rate ΔW on the parameter τ_0 at m=n=2; b) – dependence of the flow rate ΔW on the parameter τ_0 at m=n=3.

5. Conclusions

Velocity profiles do not have any clearly defined horizontal sections at different τ_0, which makes it difficult to determine the value of the parameter τ_y.

1. The impact of inaccuracy in determining τ_y on the maximum velocity value is insignificant:
- for the Bingham model, a 10% increase of τ_y leads to a 4.9% decrease in maximum velocity;
- for the Casson model, a 10% increase of τ_y changes the maximum velocity by no more than 5% (*m=3, n=2* is a special case).
2. The effect of a change of τ_y (not more than by 10%) on flow rate is also insignificant – not more than 5%.

References

[1] Wilkinson W L 1964 Nen'yutonovskiye zhidkost *Non-Newton fluids* (Moscow: Mir) p 216
[2] Smolsky B M, Shulman Z P, Gorislavets V M 1970 Rheodynamics and heat transfer of nonlinear viscoplastic materials (Minsk: Science and technology) p 443
[3] Gnoyevoy A V Osnovy teorii techeniy bingamovskikh sred, *Basics of theory of flow of Bingham media.* (Moscow: FIZMATLIT) p 272
[4] Smolskiy B M, Shulman Z P, Gorislavets V M, 1970 Reodinamika i teploobmen nelineyno - vyazkoplastichnykh materialov *Rheodynamics and heat transfer of non-linear viscoplastic materials* (Minsk: Nauka I tekhnika) p 240
[5] Prager W 1963 Vvedeniye v mekhaniku sploshnykh sred, *Introduction into mechanics of continuous media* (Moscow: Inostrannaya literature) p 406
[6] Gorislavets V M, Smolsky V M, Shulman Z P 1968 Convective heat transfer in laminar flow of composite materials in a round tube (Minsk: Science and Technology, in collection of articles "*Heat and mass transfer*") 1968.Vol **3** pp 182-214
[7] Weipert D 1993 Rheologie der Lebensmittel (Hamburg: *Behrs*) p 620
[8] Lipscomb G G, Denn M M 1984 Flow of Bingam fluids ill complex geometries (*Journal of Non-Newtonian Fluid Mechanics)* V **14** p 337-346
[9] Eirich F 1962 Rheology Theory and applications (M Publishing house of Foreign Literature) p 824
[10] Casson N 1959 A Flow Equation for Pigment Oil Suspensions of the *In Mill CC Ed Rheology of Disperse Systems* (Pergamon Press Oxford) pp 84-102
[11] Krieger I M, Dougherty T J 1959 A Mechanism for non-Newtonian flow in suspensions of rigid spheres (Trans Soc Rheol) Vol 3 pp 137-148
[12] Zadorozhnyi A A, Kovrevski A P 2017 Analysis of the flow of Bingham fluids though circular pipelines (*Collection of research papers of* UkrDUZT*)* Issue **168** ISSN(p) 1994-7852 ISSN Online 2413-3795 pp 44-49
[13] Zadorozhnyi A A, Kovrevski A P, Chovnyuk Y V, Remarchuk N P 2018 Features of the flow of liquids of variable viscosity by the pipeline of a various form of tranverse section *Collection of scientific works of the international scientific conference (Technology and transport infrastructure* Kharkiv UkrDUZT*)* pp 24-25
[14] Zadorozhnyi A, Kovrevski A, Chovnyuk Y, Remarchuk N 2018 Flow of a Bingham Fluid Through Circular Pipes with Variable Viscosity Coefficient Along the Pipe Length (*Science Publishing Corporation Publisher of International Academic Journal*) Vol **7** No 4.3 Special Issue 3 pp 100-104
[15] Zadorozhnyi A A 2020 Architectural construction using technologies based on knowledge about the flow of Bingham plastic fluids in various, pipelines *IOP Conf Series (Materials Science and Engineering)* ***907*** *(2020) 012028 IOP Publishing doi:10.1088/1757-899X/907/1/012028* pp 1-6

AUTHOR INDEX

Institute of Physics
Dirac House, Temple Back
Bristol BS1 6BE UK

ISSN: 1757-8981
ISBN 978-1-7138-4551-5
9 781713 845515

CURRAN ASSOCIATES INC.
proceedings.com